Recent Advances in Noncommutative Algebra and Geometry

CONTEMPORARY MATHEMATICS

801

Recent Advances in Noncommutative Algebra and Geometry

Conference in Honor of S. Paul Smith
on the Occasion of His 65th Birthday
Recent Advances and New Directions in the Interplay of
Noncommutative Algebra and Geometry
June 20–24, 2022
University of Washington, Seattle, WA

K. A. Brown
T. J. Hodges
M. Vancliff
J. J. Zhang
Editors

AMS AMERICAN MATHEMATICAL SOCIETY

2020 *Mathematics Subject Classification.* Primary 16-XX, 14A22, 15A66, 17B37, 20G42; Secondary 14H52, 17B38, 20G42, 81R60.

For additional information and updates on this book, visit
www.ams.org/bookpages/conm-801

Library of Congress Cataloging-in-Publication Data

Names: Brown, Kenneth A. (Kenneth Alexander), 1951- editor.

Title: Recent advances in noncommutative algebra and geometry : a conference in honor of S. Paul Smith on the occasion of his 65th birthday, Recent Advances and New Directions in the Interplay of Noncommutative Algebra and Geometry, June 20-24, 2022, University of Washington Seattle, Washington / K.A. Brown, T.J. Hodges, M. Vancliff, J.J. Zhang, editors.

Description: Providence, Rhode Island : American Mathematical Society, [2024] | Series: Contemporary mathematics, 0271-4132 ; volume 801 | Includes bibliographical references.

Identifiers: LCCN 2023047021 | ISBN 9781470472399 (paperback) | 9781470476328 (ebook)

Subjects: LCSH: Noncommutative algebras–Congresses. | Noncommutative differential geometry–Congresses. | AMS: Associative rings and algebras. | Algebraic geometry – Foundations – Noncommutative algebraic geometry. | Linear and multilinear algebra; matrix theory – Basic linear algebra – Clifford algebras, spinors. | Nonassociative rings and algebras – Lie algebras and Lie superalgebras – Quantum groups (quantized enveloping algebras) and related deformations. | Group theory and generalizations – Linear algebraic groups and related topics – Quantum groups (quantized function algebras) and their representations. | Algebraic geometry – Curves – Elliptic curves. | Nonassociative rings and algebras – Lie algebras and Lie superalgebras – Poisson algebras. | Group theory and generalizations – Linear algebraic groups and related topics – Quantum groups (quantized function algebras) and their representations. | Quantum theory – Groups and algebras in quantum theory – Noncommutative geometry.

Classification: LCC QA251.4 .R43 2024 | DDC 512/.46–dc23/eng/20240209

LC record available at https://lccn.loc.gov/2023047021

---

10 9 8 7 6 5 4 3 2 1   29 28 27 26 25 24

Photo courtesy of S. Paul Smith

These proceedings are dedicated to Paul Smith on the occasion of his 65th birthday.

Contents

Preface

This volume forms the proceedings of a conference entitled "Recent Advances and New Directions in the Interplay of Noncommutative Algebra and Geometry", that was held in honor of S. Paul Smith on the occasion of his 65th Birthday, at the University of Washington, Seattle, June 20-24, 2022. This conference was originally scheduled to be held in 2020, but, owing to the Covid-19 pandemic, it was postponed until 2022. The organizing committee consisted of Ken Brown, Tim Hodges, Ron Irving, Michaela Vancliff, and James Zhang.

The first article outlines some of Paul Smith's contributions to mathematics. The subsequent articles in the volume, by some of the leading experts in the areas of noncommutative algebra, noncommutative algebraic geometry, and representation theory, reflect the wide interests of Paul. Specific topics include: Hopf algebras and quantum groups, the elliptic algebras of Feigin and Odesskii, Calabi-Yau algebras, AS-regular algebras and noncommutative algebraic geometry, representation theory and deformation theory, noncommutative ring theory and Lie theory, and more.

The conference was supported by generous grants from the National Science Foundation (award number DMS-1953148), the National Security Agency (award number H98230-20-1-0372), and the Milliman Fund of the Department of Mathematics at the University of Washington. We greatly appreciate the support of all these organizations.

We would like to express our sincere appreciation to our friend Ron Irving, who was a co-organizer of the conference, for his strong support before, during and after the conference. He was instrumental in contributing to the success of the conference with his unwavering energy, time commitment and attention to detail. We also warmly thank staff members and graduate students in the Department of Mathematics at the University of Washington for their help, especially the Department Administrator, Michael Munz, for his help and patience in processing the reimbursements to the funded conference participants. Lastly, we are grateful to the conference attendees for their interesting presentations, stimulating discussions and refreshing enthusiasm.

Ken Brown, Tim Hodges, Michaela Vancliff and James Zhang (editors)

S. Paul Smith's contributions to mathematics

Paul Smith completed his Ph.D. under the supervision of John McConnell at the University of Leeds in 1981. From 1981 until 1983, he held a postdoctoral position at the University of Southern California, followed by a research fellowship at the University of Warwick (funded by the UK SERC), before moving on to hold a position at the University of Washington from 1986 until his retirement in 2021. Over the last 40 years, he has had a profound influence on the field of noncommutative algebra and geometry through his advocacy of certain research areas, his extensive publications and his generous mentoring of graduate students and junior mathematicians.

Paul's research interests changed over the years, but a consistent theme was his belief that noncommutative algebra should be both elegant and relevant to other branches of mathematics. His early work was in enveloping algebras of Lie algebras and differential operators. After a brief involvement with quantum groups, he moved on to what would be the major focus of his research in the second chapter of his career, namely noncommutative projective geometry.

It is hard to overstate Paul's role in the development of noncommutative algebraic geometry; in particular, his contributions to understanding the structure of the Sklyanin algebras, the elliptic algebras of Feigin and Odesskii, and AS-regular algebras in general. Although Paul has retired from the University of Washington, he did not retire from research, so this part of the story is far from over.

We have all learned a great deal from Paul and have benefited from working with him during some part of his career. Below we provide some highlights of Paul's research.

Enveloping algebras and differential operators.

In the late 1970s, the study of enveloping algebras of finite-dimensional Lie algebras was reaching its peak and, at the same time, ring theorists were starting to exploit (Gabriel-Rentschler) Krull dimension as a dimension-theoretic tool. In his Ph.D. thesis, Smith took on the problem of computing the Krull dimension of the enveloping algebra, $U(\mathfrak{g})$, of a finite-dimensional complex semisimple Lie algebra. He was able to show in [**2**] that $\mathrm{Kdim}(U(\mathfrak{sl}(2,\mathbb{C}))) = 2$. In fact, it is not hard to see that, for any semisimple Lie algebra $\mathfrak{g}$ with Borel subalgebra $\mathfrak{b}$, $\dim_{\mathbb{C}}(\mathfrak{b}) \leq \mathrm{Kdim}(U(\mathfrak{g}))$, so Smith's achievement was to show that this lower bound is also an upper bound for $\mathfrak{sl}(2,\mathbb{C})$. The conjecture that this is true in general was an open problem for many years before being confirmed 21 years later by Levasseur in [**97**], using a result from [**83**] on the ring of globally defined differential operators on the basic affine space associated with $\mathfrak{g}$.

The crux of [**2**] is a lemma relating the Gelfand-Kirillov dimension of a finitely generated module over an enveloping algebra to its Krull dimension, an insight which was further exploited in [**4**], where upper bounds for the Krull dimension of factor rings (especially primitive ones) of $U(\mathfrak{g})$, where $\mathfrak{g}$ is semisimple, were obtained. This focus on primitive factor rings reflected important developments in Lie theory which continued throughout the 1980s, and Smith contributed further to this endeavour in [**9**], joint with Hodges, where, in a lovely ring-theoretic application of the Beilinson-Bernstein equivalence of categories [**82**], they obtained an upper bound on the global dimension, $\mathrm{gldim}(U(\mathfrak{g})/P)$, of those factor rings by a minimal primitive ideal P for which the global dimension is finite. Their work, combined with earlier results of Levasseur [**95**] and Joseph and Stafford [**93**], yielded complete knowledge of $\mathrm{gldim}(U(\mathfrak{g})/P)$ for all semisimple Lie algebras $\mathfrak{g}$ and all minimal primitive ideals P.

As already indicated, the research described above required significant input from work on differential operators, and indeed this topic was a second focus of Smith's mathematics in the first decade after his Ph.D. Amongst a number of significant contributions in this area, a highlight was [**15**], joint with Stafford, on the algebra $\mathcal{D}(X)$ of (global) differential operators on an irreducible affine curve X over an algebraically closed field k of characteristic 0. When X is smooth (of arbitrary dimension), $\mathcal{D}(X)$ is a simple affine noetherian domain, generated as a k-algebra by the coordinate ring $\mathcal{O}(X)$ of X and its module of derivations, but when X is singular $k\langle\mathcal{O}(X), \mathrm{Der}_k(\mathcal{O}(X))\rangle \subsetneq \mathcal{D}(X)$ in general. Let X be an irreducible affine curve with normalization $\tilde{X}$, and let $\pi : \tilde{X} \longrightarrow X$ be the projection map. Smith and Stafford proved that $\mathcal{D}(X)$ is an affine noetherian domain with a unique minimal nonzero ideal J, and $\dim_k(\mathcal{D}(X)/J) < \infty$. Moreover $\mathcal{D}(X)$ is simple if and only if π is injective if and only if $\mathcal{D}(X)$ is hereditary if and only if $\mathcal{D}(X)$ and $\mathcal{D}(\tilde{X})$ are Morita equivalent. The proofs exploit the beautiful interplay between $\mathcal{D}(X)$ and $\mathcal{D}(\tilde{X})$ when both are canonically identified with subalgebras of $\mathcal{D}(K)$, where K is the quotient field of $\mathcal{O}(X)$. This paper has stimulated much subsequent work, with 40 MathSciNet citations by May 2023, including further papers by Smith and Stafford themselves. For example, the close relations between rings of differential operators and semisimple Lie algebras revealed by [**82**] were further demonstrated and exploited in a series of important papers [**16**], [**18**], [**96**] by (combinations of) Levasseur, Smith and Stafford. To give one other example of the hinterland of [**15**], the nature of the unique largest finite-dimensional factor ring $\mathcal{D}(X)/J$ discovered in the Smith-Stafford paper was clarified in [**84**] and [**85**].

In [**11**] and [**13**] Smith initiated the study of the algebraic properties of the algebra $\mathcal{D}(X)$ of differential operators of an affine variety X over an algebraically closed field of positive characteristic. In contrast to characteristic zero, these algebras are essentially never noetherian; nevertheless, Smith was able to show that when X is smooth, the global dimension of $\mathcal{D}(X)$ equals the dimension of X, just as in characteristic zero.

Towards the end of the 1980s, Smith became interested in the exciting new field of quantum groups, and he played an important role in drawing noncommutative algebraists into this fascinating new subject. His survey article on quantum groups [**25**] was the first in a series of influential survey texts which have been an important feature of the second part of his career. For himself, the key immediate research impact was in the direction of deformations of enveloping algebras, rather than towards Hopf algebras – thus [**20**] studied algebras $\mathbb{C}\langle E, F, H\rangle$ in which the usual relation $EF - FE = H$ of $\mathfrak{sl}(2, \mathbb{C})$ is replaced by $EF - FE = f(H)$ where $f(H)$ is a polynomial of positive degree. He showed that many, but not all, of the properties of $U(\mathfrak{sl}(2, \mathbb{C}))$ survive in the deformed algebras. The ideas developed in this context were key to many subsequent developments, important for instance in the discovery and study of noncommutative Kleinian singularities, down-up algebras, generalized Weyl algebras and ambiskew polynomial algebras. Smith himself, however, chose to focus much of his later research on graded algebras, in particular on the strand of noncommutative algebraic geometry initiated by Artin and Schelter with [**75**], and to which we now turn.

Artin-Schelter regular algebras.

This section discusses how, and why, Smith became interested in studying algebras via geometric techniques, treating noncommutative graded algebras as being, in essence, homogeneous coordinate rings of (noncommutative) projective varieties.

In the late 1980s, guided by his interest in quantum groups, Smith learned of the paper [**106**], by Sklyanin, concerning an algebra that had appeared in the field of quantum inverse scattering theory, and which seemed to be impervious to the methods used by ring theorists at the time. Smith and Stafford wondered if the new geometric machinery that had been introduced by Artin, Tate and Van den Bergh [**77, 78**] might be useful for studying Sklyanin's algebra.[1]

Sklyanin had conjectured that his algebra had the same Hilbert series as the polynomial ring on four variables, but had only been able to compute the dimension of the degree-i component of the algebra for $i \leq 3$. Using the methods of [**77, 78**], Smith and Stafford proved in [**24**] that Sklyanin's algebra is a noetherian ring without zero divisors, and has the same Hilbert series, and many of the same homological properties, as the polynomial ring on four variables.

A key step in Smith and Stafford's analysis in [**24**] was the classification of the *point modules* over the algebra; such modules are 0-linear modules, where a d-linear module is a cyclic graded module with Hilbert series the same as that of the polynomial ring on $d+1$ variables. They proved that the isomorphism classes of point modules over Sklyanin's algebra are parametrized by the union of an elliptic curve E, embedded as a degree-4 normal curve in $\mathbb{P}^3$, and the four points that are the vertices of the four singular quadrics that contain E. Moreover, they proved that the algebra determines, and is determined by, the geometric data consisting of E, the four vertices, and a translation automorphism σ of E.

Shortly thereafter, in [**27**], Levasseur and Smith classified the *line modules* over Sklyanin's algebra, that is, the 1-linear modules. They proved that the isomorphism classes of line modules over Sklyanin's algebra are naturally parametrized by those lines in $\mathbb{P}^3$ that meet E with multiplicity two (that is, the secant lines to E), which can also be characterized as the lines that lie on the quadrics that contain E. Using the classification of the point modules and line modules in [**24, 27**], Smith and Staniszkis classified, in [**28**], all the finite-dimensional simple modules in the case where σ has infinite order. The main idea in their work was the result that every finite-dimensional simple module over Sklyanin's algebra is a quotient of a line module (regardless of the order of σ). In 1993, Smith circulated the manuscript [**69**], in which he proved that Sklyanin's algebra is a finitely generated module over its center if and only if σ has finite order, and Smith and Tate determined, in [**30**], the

[1] We refer to the algebra in [**106**] as *Sklyanin's algebra*, reserving the term *the Sklyanin algebra* for the generalizations that were defined in [**87, 101**] of the algebra in [**106**].

center of the algebra in the case that σ has finite order. These results and more are summarized in Smith's survey article [**31**]. Many years later, Smith began working with Chirvasitu and Kanda on algebras that are generalizations of Sklyanin's algebra, and that work is discussed below in the section concerning the elliptic algebras of Feigin and Odesskii.

Given the interplay between Sklyanin's algebra and the geometric data (E, σ), Smith became convinced that the new geometric ideas of Artin, Tate and Van den Bergh would provide a means to develop a fruitful relationship between noncommutative algebra and algebraic geometry, reminiscent of the productive marriage that had existed for centuries between commutative algebra and algebraic geometry. Consequently, Smith became interested in algebras that could reasonably be viewed as noncommutative analogs of polynomial rings. Technical conditions to specify such an algebra had been proposed by Artin and Schelter in [**75**], and such an algebra is now called an *AS-regular algebra* or *Artin-Schelter regular algebra*. The work of classifying such algebras having global dimension three that are generated by degree-1 elements was begun in [**75**], and the new geometric methods in [**77, 78**] were introduced for the purpose of completing that task. Smith and Stafford proved in [**24**] that Sklyanin's algebra is AS-regular of global dimension four with Hilbert series the same as that of the polynomial ring on four variables (in spite of the algebra not having a PBW-basis), and so Sklyanin's algebra is often viewed as a noncommutative analog of a polynomial ring, and Smith et al showed in [**24, 27**] that Sklyanin's algebra is amenable to the geometric methods of Artin, Tate and Van den Bergh. It should be emphasized that many of the results proved in [**24, 27, 28**] for Sklyanin's algebra are applicable to more general AS-regular algebras.

Smith explored these ideas further in the privately-circulated draft of a book he began writing on noncommutative algebraic geometry. A major theme in the draft is the motivating viewpoint that a "reasonable" (e.g., noetherian, connected) graded $\Bbbk$-algebra A (where $\Bbbk$ is an algebraically closed field) that has Hilbert series $H(t) = 1/(1-t)^n$ and satisfies "good" homological properties (e.g., AS-regularity) should be viewed as the homogeneous coordinate ring of a noncommutative analog of $\mathbb{P}^{n-1}$. He considers this perspective to be even more justified if A is a member of a flat family of algebras that is parametrized by an irreducible scheme, where the family contains the polynomial ring on n variables. The main idea builds on tools developed in [**74, 80**] by using a certain quotient category of graded modules. Specifically, for an algebra A as above, one considers the category of graded A-modules modulo the full subcategory consisting of the direct limits of finite-dimensional graded A-modules. Viewed in this quotient category, point modules are irreducible objects, and the overriding theme is to treat this quotient category as an analog of the category of quasi-coherent sheaves on $\mathbb{P}^{n-1}$, with the point modules being the skyscraper sheaves at the points of the noncommutative space. Continuing this categorical approach, Smith and Van den Bergh introduced several noncommutative analogs of commutative geometric concepts in their paper [**54**], such as the notion of a noncommutative quadric surface and what it means for such a surface to be smooth.

With the above viewpoint, the homogenization H of the enveloping algebra of the Lie algebra $\mathfrak{sl}(2)$ can be treated as a noncommutative analog of the polynomial ring on four variables. In [**26**], Le Bruyn and Smith proved that the line modules over H are directly related to the Verma modules over $\mathfrak{sl}(2)$ and, hence, to the irreducible representations of $\mathfrak{sl}(2)$. That work inspired Le Bruyn and Van den Bergh, in [**94**], to apply a similar geometric approach to more general Lie algebras. Smith's interest in $\mathfrak{sl}(2)$, quantized analogs thereof and their interplay with algebraic geometry were further explored by him with Chirvasitu and Wong in [**61**], in which they examine the relationship between $U_q(\mathfrak{sl}(2))$ and certain degenerations of Sklyanin's algebra through the lens of (noncommutative) algebraic geometry, proving that the line modules are related to the Verma modules over $U_q(\mathfrak{sl}(2))$, and that the line modules correspond to the lines on the quadrics that contain a certain degree-4 curve (echoing the situation with Sklyanin's algebra).

At about the time that Le Bruyn and Van den Bergh worked on [**94**], they and Smith observed that many known (at the time) quadratic AS-regular algebras of global dimension four mapped onto AS-regular algebras of global dimension three via factoring out the ideal generated by a homogeneous degree 1 normal element. Thus, Le Bruyn, Smith and Van den Bergh classified in [**33**] all the quadratic AS-regular algebras of global dimension four that map onto AS-regular algebras of global dimension three via factoring out the ideal generated by a homogeneous degree 1 central element.

Turning to line modules, Shelton and Vancliff proved in [**104, 105**] that the scheme that parametrizes the line modules over a quadratic AS-regular algebra of global dimension four that has six generic defining relations has dimension one and determines the defining relations of the algebra. Chirvasitu, Smith and Vancliff built on that work by proving in [**67**] that if the scheme that parametrizes the line modules over a quadratic AS-regular algebra of global dimension four has dimension one, then, viewed via the Plücker embedding in $\mathbb{P}^5$, the scheme has degree twenty. Consequently, the scheme that parametrizes the line modules over a generic quadratic AS-regular algebra of global dimension four with Hilbert series $H(t) = 1/(1-t)^4$ has degree twenty when viewed via the Plücker embedding in $\mathbb{P}^5$.

As detailed above, the philosophy that there should be a two-way flow of ideas between geometry and noncommutative algebra has influenced much of Smith's approach to his research since the mid-1990s. This aspect of Smith's activity has inspired many researchers, including new generations of algebraists, to cultivate a thriving, growing relationship between these two fields. This perspective has guided many researchers to study homological questions, Koszulity questions, and related concepts, and these aspects of Smith's research are discussed below.

Homological aspects of noncommutative algebras.

Smith has always been interested in homological properties of various classes of noncommutative algebras. Cumulatively, the work outlined in this subsection indicates how Smith contributed to the gradual realization through the 1990s that the concepts of AS-regularity and Auslander regularity introduced earlier in specific settings by Artin, Schelter, Auslander and Bridger [**75, 81**] had much wider potential. In 1996 Smith published his most cited paper [**32**], in which he established several fundamental results relating AS-regularity and Koszulness for connected graded algebras. Notably, let A be a noetherian connected graded Koszul algebra, then A is AS-regular if and only if its Koszul dual $A^!$ is a Frobenius algebra. This result was later extended to the non-noetherian non-Koszul case by Lu, Palmieri, Wu and Zhang in [**98**]. Smith also proved that the contravariant functor $\operatorname{Ext}^*_A(-,k)$ is a duality between the categories of graded modules having linear resolutions over A

and $A^!$, respectively. His insight to consider graded modules that have linear resolutions has been used widely by mathematicians in noncommutative algebra, representation theory of quivers, and noncommutative algebraic geometry. In [**34**], Smith (together with Zhang) proved that every connected graded self-injective algebra is finite-dimensional and Frobenius. During the same period, Smith (with Ajitabh and Zhang) studied the Auslander-Gorenstein condition and purity (respectively, essential purity) of a minimal injective resolution of an algebra [**36, 38**]. In the PI case, every Auslander-Gorenstein, Cohen-Macaulay ring satisfying a polynomial identity has a pure injective resolution. This latter result is particularly notable since an example of Artin and Stafford [**36**] shows that this assertion fails when the PI condition is removed. Another interesting result is that every noetherian Cohen-Macaulay algebra with finite injective dimension has a quasi-Frobenius quotient ring. Smith also proved a step-duality between left modules and right modules over an Auslander-Gorenstein ring [**38**, Theorem 1.2].

Foundations and scope of noncommutative algebraic geometry.

In the roughly ten-year period beginning in the late nineties, two themes pervaded Smith's research. Firstly, inspired by the general perspective of [**109**], Smith generalized a number of foundational geometric notions from algebraic geometry to the noncommutative setting. Among the notions studied are integrality [**40**], relevant to Artin's conjecture on the birational classification of noncommutative surfaces, maps between noncommutative spaces [**43**], with a view towards defining "the" category of noncommutative spaces, and weakly closed subspaces of a noncommutative space, specializing to the fundamental notion of *closed point* [**41**]. As Smith discovered, in the noncommutative world, there exist closed points that behave like curves with respect to intersection theory, an idea yet to be fully appreciated. As with many of his ideas, this insight arose from Smith's unique perspective coming from his vast experience with classical noncommutative ring theory, Lie theory and his analysis of Sklyanin's algebra. Other highlights of his exploration of foundations are [**42**] and [**37**] (written with Zhang), in which the notions of *fiber* and *curve* are generalized to accommodate noncommutative phenomena. Despite the spectacular results [**76**], [**103**] authoritatively classifying noncommutative curves, [**42**] and [**37**] exhibit examples outside of the "expected" framework of the subject, and point to much more mysterious behavior to be studied.

Another major theme of Smith's research during this period was the discovery and study of extremely novel examples of noncommutative spaces. Each example suggests fruitful vistas that are yet to be fully investigated. Among Smith's many papers describing novel examples, we list only a few highlights here.

In [**55**], Smith constructed a homogeneous coordinate ring for the space of Penrose tilings, inspired by Connes' analysis of this space in terms of noncommutative differential geometry. This suggests there is a fruitful interplay between noncommutative algebraic geometry and differential geometry, a theme he explores in detail in [**53**]. Once again, this is a promising area of study that has not yet received the attention it deserves.

In [**70**], Smith undertook the analysis of an algebra A constructed from a 7-dimensional irreducible representation of the complex semisimple Lie algebra of type G_2, arising in string theory. Among other things, Smith proved that this novel algebra is Koszul and 3-Calabi-Yau, but not noetherian. This work is characteristic of many of Smith's papers – because of his broad perspective, he was able to view A through many lenses, and exploit each.

Smith's paper, [**46**], was directly inspired by noncommutative algebraic geometry. In that paper, he studied, in part, weighted polynomial rings. For such rings, the notion of quotient category (discussed above in the section concerning AS-regular algebras) is equivalent to the category of quasi-coherent sheaves on a certain stack-theoretic quotient of a projective space. This perspective was continued in [**47**], in which he provided a novel viewpoint on the graded first Weyl algebra, where the generators x and d/dx are assigned degrees 1 and -1 respectively. For that situation, he proved that the quotient category of graded modules over the Weyl algebra is equivalent to the category of quasi-coherent sheaves on a certain stack-theoretic quotient of the affine line.

In [**50**], Smith studied another noncommutative ring R arising from string theory, generated by an element x in degree one and y in degree two subject to relations $x^5 = yxy$ and $y^2 = xyx$. To analyze the ring, Smith showed it is a twisted homogenous coordinate ring [**79**] on a degree-six del Pezzo surface X obtained by blowing up $\mathbb{P}^2$ at three non-collinear points. The data defining the twisted homogenous coordinate ring comes naturally from geometric properties of X. From the identification of R as a twisted homogeneous coordinate ring, Smith deduced that R is a noetherian domain, a finite module over its center, and is homologically well behaved. The paper illustrates Smith's remarkable insight, as well as his ability to apply techniques from noncommutative algebraic geometry in unexpected situations.

In [**71**], Smith studied when certain subalgebras of skew Laurent extensions of the field of fractions of a smooth projective surface X over a field k are free subalgebras on at least two variables. Denoting the extension by $k(X)[t^{\pm 1}; \sigma]$, where σ is an automorphism of X, Smith showed that such a free subalgebra exists if and only if the topological entropy of σ is positive. The paper is unusual among results in noncommutative algebraic geometry since it invokes notions from dynamical systems theory.

All of these projects are illustrative of Smith's desire to push the boundaries of the subject and fearlessly explore connections with other areas, inspiring others to take less tentative steps in their investigations.

Twisted Calabi-Yau algebras and twisted superpotential.

The notion of a Calabi-Yau algebra was introduced by Ginzburg in [**89**] in 2006. Since then, this concept and its twisted variants have become one of the central themes in both noncommutative algebra and noncommutative algebraic geometry. Examples of (twisted) Calabi-Yau algebras include connected graded AS-regular algebras and AS-regular Hopf algebras. Calabi-Yau algebras and the associated superpotential have been used extensively in mathematical physics, leading Smith to work on twisted Calabi-Yau algebras and twisted superpotentials in [**57, 60, 63**]. Following a nice observation of Dubois-Violette in [**86**], every N-Koszul AS-regular graded algebra is isomorphic to a derivation quotient algebra determined uniquely by a twisted superpotential. Therefore the corresponding superpotential plays an important role in the study of N-Koszul AS-regular algebras. In [**57**], Mori and Smith studied graded automorphism groups, homological determinants, and graded twists in terms of the superpotential and provided several useful formulae for computation. In [**60**], Mori and Smith classified 3-Calabi-Yau algebras with three generators in terms of superpotentials. This classification provides a platform for testing other open questions in graded ring theory and leads to a complete classification of AS-regular algebras of global dimension three by explicit generators and relations in [**91**,**92**,**99**]. In [**63**], Chirvasitu, Kanda and Smith exploited superpotential methods to construct new classes of graded AS-regular algebras of dimension four from 3-dimensional examples.

The elliptic algebras of Feigin and Odesskii.

In 1989, in [**87**,**101**], Feigin and Odesskii defined a family of graded algebras $Q_{n,k}(E,\eta)$, which we call the *elliptic algebras* of Feigin and Odesskii, that depend on an elliptic curve $(E,+)$ with fixed group structure, coprime integers $n > k \geq 1$, and a point $\eta \in E$ that is not an n-torsion point. These algebras have played an important role in noncommutative algebraic geometry, especially in the study of AS-regular algebras.

Let $\tau \in \mathbb{C}$ be a point of the upper-half complex plane and let E denote the elliptic curve $E = \mathbb{C}/(\mathbb{Z}+\mathbb{Z}\tau)$. The algebra $Q_{n,k}(E,\eta)$ is defined to be the graded $\mathbb{C}$-algebra generated by n variables x_i $(i \in \mathbb{Z}/n\mathbb{Z})$ with n^2 quadratic relations

$$\text{(E1)} \qquad \sum_{r\in\mathbb{Z}/n\mathbb{Z}} \frac{\theta_{j-i+(k-1)r}(0)}{\theta_{j-i-r}(-\eta)\theta_{kr}(\eta)}\, x_{j-r}x_{i+r}, \qquad (i,j \in \mathbb{Z}/n\mathbb{Z})$$

where θ_i $(i \in \mathbb{Z}/n\mathbb{Z})$ are theta functions of order n, quasi-periodic with respect to the lattice $\mathbb{Z}+\mathbb{Z}\tau$ (see [**64**] for the precise definition). Often (conjecturally, always), the span of the degree-2 defining relations of $Q_{n,k}(E,\eta)$ has dimension equal to that of the span of the degree-2 defining relations of the polynomial ring on n variables.

Although the value $\eta = 0$ cannot be substituted directly into (E1) (as the denominator in (E1) becomes zero in this case), there is a way ([**65**]) to extend the definition of $Q_{n,k}(E,\eta)$ such that (i) it holds for all $\eta \in E$, and (ii) $Q_{n,k}(E,0)$ is a polynomial ring on n variables. Thus, $\{Q_{n,k}(E,\eta)\}_{\eta\in E}$ is a family of noncommutative deformations of a polynomial ring on n variables.

If $k = 1$, the algebras $Q_{n,1}(E,\eta)$ are often called (higher-dimensional) Sklyanin algebras, owing to $Q_{4,1}(E,\eta)$ being the algebra in Sklyanin's paper, [**106**]. The results, [**24**], of Smith and Stafford, that were discussed above in the section on AS-regular algebras, were generalized in [**108**] to $Q_{n,1}(E,\eta)$ by Tate and Van den Bergh, who proved that $Q_{n,1}(E,\eta)$ is AS-regular of global dimension n and that it has many other homological properties in common with the polynomial ring on n variables. Smith proved in [**29**] that, for all $n \geq 5$, the scheme that parametrizes the point modules over $Q_{n,1}(E,\eta)$ is E, for all η not belonging to the nm-torsion subgroup of E for any $m \in \{1,\dots,n-2\}$.

In [**108**], Tate and Van den Bergh generalized the results of Levasseur and Smith in [**27**] (described in the section above on AS-regular algebras) by proving that, for any n, every secant plane to E gives a d-linear module over $Q_{n,1}(E,\eta)$ (but other d-linear modules can also exist). Inspired by these results, Staniszkis classified in [**107**] all the d-linear modules over $Q_{n,1}(E,\eta)$ by proving that they are in bijection with those d-planes in $\mathbb{P}^{n-1}$ that are either secant to E or are singular loci of certain quadrics of rank $n-d-1$ that contain E. Her results explain why four of the point modules of $Q_{4,1}(E,\eta)$ correspond to points not on E, yet the point modules of all other $Q_{n,1}(E,\eta)$ correspond only to points that lie on E.

The situation for $k > 1$ is very different from the case $k = 1$. In particular, prior to 2017, most of the work regarding $Q_{n,k}(E, \eta)$, where $k > 1$, was performed by Feigin, Odesskii and their coauthors (cf. [**87, 88, 100, 101**]), but some of their results are not applicable to all cases. Another difference is that there is not, in general, a geometric description of $Q_{n,k}(E, \eta)$ if $k > 1$, unlike the case for $k = 1$. In 2017, convinced of the ground-breaking importance of the work of Feigin and Odesskii, Smith embarked, with Chirvasitu and Kanda, on a research program to thoroughly investigate the algebras $Q_{n,k}(E, \eta)$.

The first paper, [**65**], of the program clarified the definition of the algebra $Q_{n,k}(E, \eta)$, since, as mentioned above, the original definition was not well defined if $\eta \in E$ is an n-torsion point. Thus, in [**64**], several possible definitions are provided for $Q_{n,k}(E, \eta)$ that are well defined for all $\eta \in E$, and some of the definitions are shown to be equivalent. For *all* the definitions to be equivalent, the Hilbert series of $Q_{n,k}(E, \eta)$ needs to be the same as that of the polynomial ring on n variables, and Chirvasitu, Kanda and Smith proved this is the case for very general $\eta \in E$ in their fourth paper, [**66**] (see below); the claim for all $\eta \in E$ is still a conjecture at the time of writing this article. They also proved that, if $\zeta \in E$ is an n-torsion point, then $Q_{n,k}(E, \eta + \zeta)$ is isomorphic to a twist of $Q_{n,k}(E, \eta)$.

The program's second paper, [**72**], investigates a certain variety, now called the *characteristic variety* of $Q_{n,k}(E, \eta)$, that was introduced by Feigin and Odesskii, the definition of which depends on the length g of the negative continued fraction

$$\text{(E2)} \qquad \frac{n}{k} = n_1 - \cfrac{1}{n_2 - \cfrac{1}{\cdots - \cfrac{1}{n_g}}}$$

for n/k, where $2 \leq n_i \in \mathbb{Z}$ for all $i \in \{1, \ldots, g\}$. Chirvasitu, Kanda and Smith proved that each point of the characteristic variety determines a point module over $Q_{n,k}(E, \eta)$.

The third paper, [**64**], in the program studies a canonical map from $Q_{n,k}(E, \eta)$ to the homogeneous coordinate ring B of the characteristic variety of $Q_{n,k}(E, \eta)$. The main result of the paper proves that the characteristic variety is a quotient of the finite product E^g by the action of a certain finite group. Two special cases of the values of $n_1, \ldots, n_g$ in (E2) are considered in [**64**]: one case being where $n_i \geq 3$ for all $i \in \{1, \ldots, g\}$, in which case the characteristic variety is E^g, and the second being where $(n_1, \ldots, n_g)$ is either $(m, 2, \ldots, 2)$ or $(2, \ldots, 2, m)$, for any $m \geq 3$, in which case the characteristic variety is the gth symmetric power $S^g E$ of E. Chirvasitu, Kanda and Smith proved that, in these two cases, the algebra B is generated by degree 1 elements, and that this is equivalent to the canonical map from $Q_{n,k}(E, \eta)$ to B being surjective. Moreover, they prove that, in these cases, the defining relations of B have degree at most three, and that, if $n = 2k + 1$, the characteristic variety is the kth symmetric power $S^k E$ of E. Additionally, if $\eta \in E$ is not a 2-torsion point, then the degree-2 defining relations of $Q_{2k+1,k}(E, \eta)$ are exactly those elements of degree two that vanish on the graph of a certain automorphism, thereby giving $Q_{2k+1,k}(E, \eta)$ a geometric description. These results are beautiful generalizations of the analogous results for $Q_{3,1}(E, \eta)$ in [**77**] and for $Q_{4,1}(E, \eta)$ in [**24**], and are reminiscent of the situation in general in [**77**] for AS-regular algebras of global dimension three.

The next paper, [**66**], in the program studies several fundamental properties of $Q_{n,k}(E, \eta)$, some of which were predicted by Feigin and Odesskii. In particular, it is proved that, if $\eta \in E$ is not a torsion point, then $Q_{n,k}(E,\eta)$ has the same Hilbert series as that of the polynomial ring on n variables and has global dimension n, that it is a Koszul algebra, and that, for all but countably many η (where E, n and k are fixed), $Q_{n,k}(E,\eta)$ is an AS-regular algebra. These results are (partial) generalizations of $Q_{n,1}(E,\eta)$ having these properties, proved by Tate and Van den Bergh in [**108**]. The fact that the defining relations of $Q_{n,k}(E,\eta)$ can be given in terms of a solution to the quantum Yang-Baxter equation was an essential ingredient in proving the main results of [**66**].

The most recent manuscript, [**73**], in the program solves a long-standing open problem regarding determination of the (projective) symplectic leaves for a certain Poission structure defined by Feigin and Odesskii in [**88**], and independently defined by Polishchuk in [**102**] (it was shown by Hua and Polishchuk in [**90**] that the Poisson structures are the same). The main idea is based on $\{Q_{n,1}(E,\eta)\}_\eta$ being a family of flat deformations of $Q_{n,1}(E,0) = \mathbb{C}[x_0,\ldots,x_{n-1}]$, so the semi-classical limit defines a Poisson bracket on $\mathbb{P}^{n-1} = \mathbf{Proj}(\mathbb{C}[x_0,\ldots,x_{n-1}])$. In [**73**], a complete description of the symplectic leaves of the Poisson bracket is given in terms of higher secant varieties to E.

This program by Chirvasitu, Kanda and Smith of thoroughly investigating the elliptic algebras of Feigin and Odesskii is still ongoing, so more papers are expected in the near future.

Supervision of Ph.D. students and postdoctoral fellows.

Paul has supervised nine Ph.D. students: Cody Holdaway, Christopher McMurdie, Izuru Mori, Adam Nyman, Patrick Perkins, Gautam Sisodia, Joanna Staniszkis, Michaela Vancliff and Min Wu; and he has mentored three postdoctoral scholars: Alexandru Chirvăsitu, Ryo Kanda and James Zhang.

Ken Brown, Tim Hodges, Ryo Kanda, Adam Nyman,
Michaela Vancliff and James Zhang

Bibliography

[1] S. P. Smith, *The primitive factor rings of the enveloping algebra of* sl(2, **C**), J. London Math. Soc. (2) **24** (1981), no. 1, 97–108, DOI 10.1112/jlms/s2-24.1.97. MR623674

[2] S. P. Smith, *Krull dimension of the enveloping algebra of* sl(2, **C**), J. Algebra **71** (1981), no. 1, 189–194, DOI 10.1016/0021-8693(81)90114-9. MR627433

[3] S. P. Smith, *An example of a ring Morita equivalent to the Weyl algebra* A_1, J. Algebra **73** (1981), no. 2, 552–555, DOI 10.1016/0021-8693(81)90334-3. MR640048

[4] S. P. Smith, *Krull dimension of factor rings of the enveloping algebra of a semisimple Lie algebra*, Math. Proc. Cambridge Philos. Soc. **93** (1983), no. 3, 459–466, DOI 10.1017/S030500410006076X. MR698349

[5] S. P. Smith, *Central localization and Gel′fand-Kirillov dimension*, Israel J. Math. **46** (1983), no. 1-2, 33–39, DOI 10.1007/BF02760621. MR727021

[6] S. P. Smith, *Gel′fand-Kirillov dimension of rings of formal differential operators on affine varieties*, Proc. Amer. Math. Soc. **90** (1984), no. 1, 1–8, DOI 10.2307/2044657. MR722404

[7] T. J. Hodges and S. P. Smith, *Sheaves of noncommutative algebras and the Beilinson-Bernstein equivalence of categories*, Proc. Amer. Math. Soc. **93** (1985), no. 3, 379–386, DOI 10.2307/2045596. MR773985

[8] K. A. Brown and S. P. Smith, *Bimodules over a solvable algebraic Lie algebra*, Quart. J. Math. Oxford Ser. (2) **36** (1985), no. 142, 129–139, DOI 10.1093/qmath/36.2.129. MR790475

[9] Timothy J. Hodges and S. Paul Smith, *On the global dimension of certain primitive factors of the enveloping algebra of a semisimple Lie algebra*, J. London Math. Soc. (2) **32** (1985), no. 3, 411–418, DOI 10.1112/jlms/s2-32.3.411. MR825916

[10] S. P. Smith, *Differential operators on commutative algebras*, Ring theory (Antwerp, 1985), Lecture Notes in Math., vol. 1197, Springer, Berlin, 1986, pp. 165–177, DOI 10.1007/BFb0076323. MR859394

[11] S. P. Smith, *Differential operators on the affine and projective lines in characteristic* $p > 0$, Séminaire d'algèbre Paul Dubreil et Marie-Paule Malliavin, 37ème année (Paris, 1985), Lecture Notes in Math., vol. 1220, Springer, Berlin, 1986, pp. 157–177, DOI 10.1007/BFb0099511. MR926303

[12] R. Hart and S. P. Smith, *Differential operators on some singular surfaces*, Bull. London Math. Soc. **19** (1987), no. 2, 145–148, DOI 10.1112/blms/19.2.145. MR872128

[13] S. P. Smith, *The global homological dimension of the ring of differential operators on a nonsingular variety over a field of positive characteristic*, J. Algebra **107** (1987), no. 1, 98–105, DOI 10.1016/0021-8693(87)90076-7. MR883872

[14] S. P. Smith, *Curves, differential operators and finite-dimensional algebras*, Séminaire d'algèbre Paul Dubreil et Marie-Paule Malliavin (Paris, 1986), Lecture Notes in Math., vol. 1296, Springer, Berlin, 1987, pp. 158–176, DOI 10.1007/BFb0078525. MR932054

[15] S. P. Smith and J. T. Stafford, *Differential operators on an affine curve*, Proc. London Math. Soc. (3) **56** (1988), no. 2, 229–259, DOI 10.1112/plms/s3-56.2.229. MR922654

[16] T. Levasseur and S. P. Smith, *Primitive ideals and nilpotent orbits in type* G_2, J. Algebra **114** (1988), no. 1, 81–105, DOI 10.1016/0021-8693(88)90214-1. (Corrigendum: 118 (1988), no. 1, 261. MR0961336). MR0931902

[17] S. P. Smith, *The simple* $\mathcal{D}$*-module associated to the intersection homology complex for a class of plane curves*, J. Pure Appl. Algebra **50** (1988), no. 3, 287–294, DOI 10.1016/0022-4049(88)90106-5. MR938620

[18] T. Levasseur, S. P. Smith, and J. T. Stafford, *The minimal nilpotent orbit, the Joseph ideal, and differential operators*, J. Algebra **116** (1988), no. 2, 480–501, DOI 10.1016/0021-8693(88)90231-1. MR953165

[19] S. P. Smith, *Can the Weyl algebra be a fixed ring?*, Proc. Amer. Math. Soc. **107** (1989), no. 3, 587–589, DOI 10.2307/2048153. MR962247

[20] S. P. Smith, *A class of algebras similar to the enveloping algebra of* sl(2), Trans. Amer. Math. Soc. **322** (1990), no. 1, 285–314, DOI 10.2307/2001532. MR972706

[21] S. P. Smith, *Overrings of primitive factor rings of* $U(\mathrm{sl}(2,\mathbf{C}))$, J. Pure Appl. Algebra **63** (1990), no. 2, 207–218, DOI 10.1016/0022-4049(90)90027-F. MR1043751

[22] S. Montgomery and S. Paul Smith, *Skew derivations and* $U_q(\mathrm{sl}(2))$*: Hopf algebras*, Israel J. Math. **72** (1990), no. 1-2, 158–166, DOI 10.1007/BF02764618. MR1098987

[23] S. Paul Smith, *Polynomial solutions to constant coefficient differential equations*, Trans. Amer. Math. Soc. **329** (1992), no. 2, 551–569, DOI 10.2307/2153951. MR1013339

[24] S. P. Smith and J. T. Stafford, *Regularity of the four-dimensional Sklyanin algebra*, Compositio Math. **83** (1992), no. 3, 259–289. MR1175941

[25] S. P. Smith, *Quantum groups: an introduction and survey for ring theorists*, Noncommutative rings (Berkeley, CA, 1989), Math. Sci. Res. Inst. Publ., vol. 24, Springer, New York, 1992, pp. 131–178, DOI 10.1007/978-1-4613-9736-6_6. MR1230220

[26] Lieven Le Bruyn and S. P. Smith, *Homogenized* $\mathfrak{sl}(2)$, Proc. Amer. Math. Soc. **118** (1993), no. 3, 725–730, DOI 10.2307/2160112. MR1136235

[27] Thierry Levasseur and S. Paul Smith, *Modules over the* 4*-dimensional Sklyanin algebra* (English, with English and French summaries), Bull. Soc. Math. France **121** (1993), no. 1, 35–90. MR1207244

[28] S. Paul Smith and J. M. Staniszkis, *Irreducible representations of the* 4*-dimensional Sklyanin algebra at points of infinite order*, J. Algebra **160** (1993), no. 1, 57–86, DOI 10.1006/jabr.1993.1178. MR1237078

[29] S. P. Smith, *Point modules over Sklyanin algebras*, Math. Z. **215** (1994), no. 2, 169–177, DOI 10.1007/BF02571706. MR1259453

[30] S. P. Smith and J. Tate, *The center of the* 3*-dimensional and* 4*-dimensional Sklyanin algebras*, Proceedings of Conference on Algebraic Geometry and Ring Theory in honor of Michael Artin, Part I (Antwerp, 1992), K-Theory **8** (1994), no. 1, 19–63, DOI 10.1007/BF00962089. MR1273835

[31] S. P. Smith, *The four-dimensional Sklyanin algebras*, Proceedings of Conference on Algebraic Geometry and Ring Theory in honor of Michael Artin, Part I (Antwerp, 1992), K-Theory **8** (1994), no. 1, 65–80, DOI 10.1007/BF00962090. MR1273836

[32] S. Paul Smith, *Some finite-dimensional algebras related to elliptic curves*, Representation theory of algebras and related topics (Mexico City, 1994), CMS Conf. Proc., vol. 19, Amer. Math. Soc., Providence, RI, 1996, pp. 315–348. MR1388568

[33] Lieven Le Bruyn, S. P. Smith, and Michel Van den Bergh, *Central extensions of three-dimensional Artin-Schelter regular algebras*, Math. Z. **222** (1996), no. 2, 171–212, DOI 10.1007/PL00004532. MR1429334

[34] S. Paul Smith and James J. Zhang, *Self-injective connected algebras*, Comm. Algebra **25** (1997), no. 7, 2243–2248, DOI 10.1080/00927879708825986. MR1451692

[35] S. Paul Smith and James J. Zhang, *A remark on Gelfand-Kirillov dimension*, Proc. Amer. Math. Soc. **126** (1998), no. 2, 349–352, DOI 10.1090/S0002-9939-98-04074-X. MR1415339

[36] K. Ajitabh, S. P. Smith, and J. J. Zhang, *Auslander-Gorenstein rings*, Comm. Algebra **26** (1998), no. 7, 2159–2180, DOI 10.1080/00927879808826267. MR1626582

[37] S. Paul Smith and James J. Zhang, *Curves on quasi-schemes*, Algebr. Represent. Theory **1** (1998), no. 4, 311–351, DOI 10.1023/A:1009984608942. MR1683617

[38] K. Ajitabh, S. P. Smith, and J. J. Zhang, *Injective resolutions of some regular rings*, J. Pure Appl. Algebra **140** (1999), no. 1, 1–21, DOI 10.1016/S0022-4049(99)00049-3. MR1700566

[39] I. Mori and S. Paul Smith, *Bézout's theorem for non-commutative projective spaces*, J. Pure Appl. Algebra **157** (2001), no. 2-3, 279–299, DOI 10.1016/S0022-4049(00)00012-8. MR1812056

[40] S. Paul Smith, *Integral non-commutative spaces*, J. Algebra **246** (2001), no. 2, 793–810, DOI 10.1006/jabr.2001.8957. MR1872125

[41] S. Paul Smith, *Subspaces of non-commutative spaces*, Trans. Amer. Math. Soc. **354** (2002), no. 6, 2131–2171, DOI 10.1090/S0002-9947-02-02963-X. MR1885647

[42] S. Paul Smith and James J. Zhang, *Fibers in Ore extensions*, Algebr. Represent. Theory **5** (2002), no. 4, 411–431, DOI 10.1023/A:1020183419932. MR1930971

[43] S. Paul Smith, *Maps between non-commutative spaces*, Trans. Amer. Math. Soc. **356** (2004), no. 7, 2927–2944, DOI 10.1090/S0002-9947-03-03411-1. (Corrigendum: 368 (2016), no. 11, 8295–8302. MR354681). MR2052602

[44] Iain Gordon and S. Paul Smith, *Representations of symplectic reflection algebras and resolutions of deformations of symplectic quotient singularities*, Math. Ann. **330** (2004), no. 1, 185–200, DOI 10.1007/s00208-004-0545-y. MR2091684

[45] Izuru Mori and S. Paul Smith, *The Grothendieck group of a quantum projective space bundle*, K-Theory **37** (2006), no. 3, 263–289, DOI 10.1007/s10977-006-0020-5. MR2273460

[46] S. Paul Smith, *Computation of the Grothendieck and Picard groups of a toric DM stack $\mathcal{X}$ by using a homogeneous coordinate ring for $\mathcal{X}$*, Glasg. Math. J. **53** (2011), no. 1, 97–113, DOI 10.1017/S001708951000056X. MR2747137

[47] S. Paul Smith, *A quotient stack related to the Weyl algebra*, J. Algebra **345** (2011), 1–48, DOI 10.1016/j.jalgebra.2011.08.014. MR2842052

[48] M. Blunk, S. J. Sierra, and S. Paul Smith, *A derived equivalence for a degree 6 del Pezzo surface over an arbitrary field*, J. K-Theory **8** (2011), no. 3, 481–492, DOI 10.1017/is010011013jkt134. MR2863422

[49] Cody Holdaway and S. Paul Smith, *An equivalence of categories for graded modules over monomial algebras and path algebras of quivers*, J. Algebra **353** (2012), 249–260, DOI 10.1016/j.jalgebra.2011.11.033. (Corrigendum: 357 (2012), 319-321. MR2905258). MR2872446

[50] S. Paul Smith, *A non-commutative homogeneous coordinate ring for the degree six del Pezzo surface*, J. Algebra **354** (2012), 95–109, DOI 10.1016/j.jalgebra.2011.11.030. MR2879225

[51] S. Paul Smith, *"Degenerate" 3-dimensional Sklyanin algebras are monomial algebras*, J. Algebra **358** (2012), 74–86, DOI 10.1016/j.jalgebra.2012.01.039. MR2905019

[52] S. Paul Smith, *Category equivalences involving graded modules over path algebras of quivers*, Adv. Math. **230** (2012), no. 4-6, 1780–1810, DOI 10.1016/j.aim.2012.03.031. MR2927354

[53] Edwin Beggs and S. Paul Smith, *Non-commutative complex differential geometry*, J. Geom. Phys. **72** (2013), 7–33, DOI 10.1016/j.geomphys.2013.03.018. MR3073899

[54] S. Paul Smith and Michel Van den Bergh, *Noncommutative quadric surfaces*, J. Noncommut. Geom. **7** (2013), no. 3, 817–856, DOI 10.4171/JNCG/136. MR3108697

[55] S. Paul Smith, *The space of Penrose tilings and the noncommutative curve with homogeneous coordinate ring $k\langle x,y\rangle/(y^2)$*, J. Noncommut. Geom. **8** (2014), no. 2, 541–586, DOI 10.4171/JNCG/164. MR3275041

[56] Gautam Sisodia and S. Paul Smith, *The Grothendieck group of non-commutative non-noetherian analogues of $\mathbb{P}^1$ and regular algebras of global dimension two*, J. Algebra **426** (2015), 188–210, DOI 10.1016/j.jalgebra.2014.11.024. MR3301905

[57] Izuru Mori and S. Paul Smith, *m-Koszul Artin-Schelter regular algebras*, J. Algebra **446** (2016), 373–399, DOI 10.1016/j.jalgebra.2015.09.016. (Corrigendum: 493 (2018), 500-501. MR3715223). MR3421098

[58] A. Nyman and S. P. Smith, *A generalization of Watts's theorem: right exact functors on module categories*, Comm. Algebra **44** (2016), no. 7, 3160–3170, DOI 10.1080/00927872.2015.1065873. MR3507176

[59] Alex Chirvasitu and S. Paul Smith, *Exotic elliptic algebras of dimension 4*, Adv. Math. **309** (2017), 558–623, DOI 10.1016/j.aim.2017.01.010. With an appendix by Derek Tomlin. MR3607286

[60] Izuru Mori and S. Paul Smith, *The classification of 3-Calabi-Yau algebras with 3 generators and 3 quadratic relations*, Math. Z. **287** (2017), no. 1-2, 215–241, DOI 10.1007/s00209-016-1824-5. MR3694675

[61] Alex Chirvasitu, S. Paul Smith, and Liang Ze Wong, *Noncommutative geometry of homogenized quantum $\mathfrak{sl}(2,\mathbb{C})$*, Pacific J. Math. **292** (2018), no. 2, 305–354, DOI 10.2140/pjm.2018.292.305. MR3733976

[62] Alex Chirvasitu and S. Paul Smith, *Exotic elliptic algebras*, Trans. Amer. Math. Soc. **371** (2019), no. 1, 279–333, DOI 10.1090/tran/7341. MR3885145

[63] Alex Chirvasitu, Ryo Kanda, and S. Paul Smith, *New Artin-Schelter regular and Calabi-Yau algebras via normal extensions*, Trans. Amer. Math. Soc. **372** (2019), no. 6, 3947–3983, DOI 10.1090/tran/7672. MR4009424

[64] Alex Chirvasitu, Ryo Kanda, and S. Paul Smith, *Maps from Feigin and Odesskii's elliptic algebras to twisted homogeneous coordinate rings*, Forum Math. Sigma **9** (2021), Paper No. e4, 50, DOI 10.1017/fms.2020.60. MR4202489

[65] Alex Chirvasitu, Ryo Kanda, and S. Paul Smith, *Feigin and Odesskii's elliptic algebras*, J. Algebra **581** (2021), 173–225, DOI 10.1016/j.jalgebra.2021.04.009. MR4249759

[66] Alex Chirvasitu, Ryo Kanda, and S. Paul Smith, *Elliptic R-matrices and Feigin and Odesskii's elliptic algebras*, Selecta Math. (N.S.) **29** (2023), no. 2, Paper No. 31, 81, DOI 10.1007/s00029-023-00827-0. MR4563694

[67] Alex Chirvasitu, S. Paul Smith, and Michaela Vancliff, *A geometric invariant of 6-dimensional subspaces of* 4×4 *matrices*, Proc. Amer. Math. Soc. **148** (2020), no. 3, 915–928, DOI 10.1090/proc/14294. MR4055923

[68] Alexandru Chirvasitu and S. Paul Smith, *Some algebras having relations like those for the 4-dimensional Sklyanin algebras*, J. Korean Math. Soc. **60** (2023), no. 4, 745–777, DOI 10.4134/JKMS.j220200. MR4609549

[69] S. P. Smith, *Simple Modules over the 4-Dimensional Sklyanin Algebras at Points of Finite Order.* preprint (1993) `arXiv:1802.06023`.

[70] S. P. Smith, *A 3-Calabi-Yau algebra with* G_2 *symmetry constructed from the octonions.* preprint (2011) `arXiv:1104.3824`.

[71] S. P. Smith, *Free algebras arising from positive-entropy automorphisms of surfaces.* preprint (2013) `arXiv:1308.3902`.

[72] A. Chirvasitu, R. Kanda, S. P. Smith, *The characteristic variety for Feigin and Odesskii's elliptic algebras,* `arXiv:1903.11798`

[73] A. Chirvasitu, R. Kanda, S. P. Smith, *The symplectic leaves for the elliptic Poisson bracket on projective space defined by Feigin-Odesskii and Polishchuk,* `arXiv:2210.13042`.

[74] Michael Artin, *Geometry of quantum planes*, Azumaya algebras, actions, and modules (Bloomington, IN, 1990), Contemp. Math., vol. 124, Amer. Math. Soc., Providence, RI, 1992, pp. 1–15, DOI 10.1090/conm/124/1144023. MR1144023

[75] Michael Artin and William F. Schelter, *Graded algebras of global dimension* 3, Adv. in Math. **66** (1987), no. 2, 171–216, DOI 10.1016/0001-8708(87)90034-X. MR917738

[76] M. Artin and J. T. Stafford, *Noncommutative graded domains with quadratic growth*, Invent. Math. **122** (1995), no. 2, 231–276, DOI 10.1007/BF01231444. MR1358976

[77] M. Artin, J. Tate, and M. Van den Bergh, *Some algebras associated to automorphisms of elliptic curves*, The Grothendieck Festschrift, Vol. I, Progr. Math., vol. 86, Birkhäuser Boston, Boston, MA, 1990, pp. 33–85. MR1086882

[78] M. Artin, J. Tate, and M. Van den Bergh, *Modules over regular algebras of dimension* 3, Invent. Math. **106** (1991), no. 2, 335–388, DOI 10.1007/BF01243916. MR1128218

[79] M. Artin and M. Van den Bergh, *Twisted homogeneous coordinate rings*, J. Algebra **133** (1990), no. 2, 249–271, DOI 10.1016/0021-8693(90)90269-T. MR1067406

[80] M. Artin and J. J. Zhang, *Noncommutative projective schemes*, Adv. Math. **109** (1994), no. 2, 228–287, DOI 10.1006/aima.1994.1087. MR1304753

[81] Auslander, M.; Bridger, M. *Stable module theory.* Mem. Amer. Math. Soc., **94**, American Mathematical Society, Providence, RI, 1969.

[82] Alexandre Beĭlinson and Joseph Bernstein, *Localisation de g-modules* (French, with English summary), C. R. Acad. Sci. Paris Sér. I Math. **292** (1981), no. 1, 15–18. MR610137

[83] Roman Bezrukavnikov, Alexander Braverman, and Leonid Positselskii, *Gluing of abelian categories and differential operators on the basic affine space*, J. Inst. Math. Jussieu **1** (2002), no. 4, 543–557, DOI 10.1017/S1474748002000154. MR1954436

[84] Kenneth A. Brown, *The Artin algebras associated with differential operators on singular affine curves*, Math. Z. **206** (1991), no. 3, 423–442, DOI 10.1007/BF02571353. MR1095764

[85] R. C. Cannings and M. P. Holland, *Differential operators, n-branch curve singularities and the n-subspace problem*, Trans. Amer. Math. Soc. **347** (1995), no. 4, 1439–1451, DOI 10.2307/2154824. MR1273480

[86] Michel Dubois-Violette, *Multilinear forms and graded algebras*, J. Algebra **317** (2007), no. 1, 198–225, DOI 10.1016/j.jalgebra.2007.02.007. MR2360146

[87] B. L. Feigin, A. V. Odesskii, *Sklyanin algebras associated with an elliptic curve.* Preprint deposited with Institute of Theoretical Physics of the Academy of Sciences of the Ukrainian SSR (1989), 33 pages.

[88] B. L. Feĭgin and A. V. Odesskiĭ, *Vector bundles on an elliptic curve and Sklyanin algebras*, Topics in quantum groups and finite-type invariants, Amer. Math. Soc. Transl. Ser. 2, vol. 185, Amer. Math. Soc., Providence, RI, 1998, pp. 65–84, DOI 10.1090/trans2/185/04. MR1736164

[89] V. Ginzburg, Calabi-Yau algebras, preprint, arXiv:math/0612139v3.

[90] Zheng Hua and Alexander Polishchuk, *Shifted Poisson structures and moduli spaces of complexes*, Adv. Math. **338** (2018), 991–1037, DOI 10.1016/j.aim.2018.09.018. MR3861721

[91] Ayako Itaba and Masaki Matsuno, *Defining relations of 3-dimensional quadratic AS-regular algebras*, Math. J. Okayama Univ. **63** (2021), 61–86. MR4174859

[92] Ayako Itaba and Masaki Matsuno, *AS-regularity of geometric algebras of plane cubic curves*, J. Aust. Math. Soc. **112** (2022), no. 2, 193–217, DOI 10.1017/S1446788721000070. MR4389808

[93] A. Joseph and J. T. Stafford, *Modules of $\mathfrak{k}$-finite vectors over semisimple Lie algebras*, Proc. London Math. Soc. (3) **49** (1984), no. 2, 361–384, DOI 10.1112/plms/s3-49.2.361. MR748996

[94] Lieven Le Bruyn and Michel Van den Bergh, *On quantum spaces of Lie algebras*, Proc. Amer. Math. Soc. **119** (1993), no. 2, 407–414, DOI 10.2307/2159921. MR1149975

[95] Thierry Levasseur, *Dimension injective des quotients primitifs minimaux de l'algèbre enveloppante d'une algèbre de Lie semi-simple* (French, with English summary), C. R. Acad. Sci. Paris Sér. I Math. **292** (1981), no. 7, 385–387. MR609758

[96] T. Levasseur and J. T. Stafford, *Rings of differential operators on classical rings of invariants*, Mem. Amer. Math. Soc. **81** (1989), no. 412, vi+117, DOI 10.1090/memo/0412. MR988083

[97] Thierry Levasseur, *Krull dimension of the enveloping algebra of a semisimple Lie algebra*, Proc. Amer. Math. Soc. **130** (2002), no. 12, 3519–3523, DOI 10.1090/S0002-9939-02-06507-3. MR1918828

[98] Di Ming Lu, John H. Palmieri, Quan Shui Wu, and James J. Zhang, *Koszul equivalences in A_∞-algebras*, New York J. Math. **14** (2008), 325–378. MR2430869

[99] Masaki Matsuno, *A complete classification of 3-dimensional quadratic AS-regular algebras of type EC*, Canad. Math. Bull. **64** (2021), no. 1, 123–141, DOI 10.4153/S0008439520000302. MR4242997

[100] A. V. Odesskiĭ, *Elliptic algebras* (Russian, with Russian summary), Uspekhi Mat. Nauk **57** (2002), no. 6(348), 87–122, DOI 10.1070/RM2002v057n06ABEH000573; English transl., Russian Math. Surveys **57** (2002), no. 6, 1127–1162. MR1991863

[101] A. V. Odesskiĭ and B. L. Feĭgin, *Sklyanin's elliptic algebras* (Russian), Funktsional. Anal. i Prilozhen. **23** (1989), no. 3, 45–54, 96, DOI 10.1007/BF01079526; English transl., Funct. Anal. Appl. **23** (1989), no. 3, 207–214 (1990). MR1026987

[102] Polishchuk, A. *Poisson structures and birational morphisms associated with bundles on elliptic curves.* Internat. Math. Res. Notices (1998), no. **13**, 683–703.

[103] I. Reiten and M. Van den Bergh, *Noetherian hereditary abelian categories satisfying Serre duality*, J. Amer. Math. Soc. **15** (2002), no. 2, 295–366, DOI 10.1090/S0894-0347-02-00387-9. MR1887637
Math. Soc. **65** (2002) No. 3, 575–590.

[104] Brad Shelton and Michaela Vancliff, *Schemes of line modules. I*, J. London Math. Soc. (2) **65** (2002), no. 3, 575–590, DOI 10.1112/S0024610702003186. MR1895734

[105] Brad Shelton and Michaela Vancliff, *Schemes of line modules. II*, Comm. Algebra **30** (2002), no. 5, 2535–2552, DOI 10.1081/AGB-120003569. MR1904652

[106] E. K. Sklyanin, *Some algebraic structures connected with the Yang-Baxter equation* (Russian), Funktsional. Anal. i Prilozhen. **16** (1982), no. 4, 27–34, 96. MR684124

[107] Joanna M. Staniszkis, *Linear modules over Sklyanin algebras*, J. London Math. Soc. (2) **53** (1996), no. 3, 464–478, DOI 10.1112/jlms/53.3.464. MR1396711

[108] John Tate and Michel van den Bergh, *Homological properties of Sklyanin algebras*, Invent. Math. **124** (1996), no. 1-3, 619–647, DOI 10.1007/s002220050065. MR1369430

[109] Michel Van den Bergh, *Blowing up of non-commutative smooth surfaces*, Mem. Amer. Math. Soc. **154** (2001), no. 734, x+140, DOI 10.1090/memo/0734. MR1846352

Contemporary Mathematics
Volume **801**, 2024
https://doi.org/10.1090/conm/801/16035

Ore extensions of commutative rings and the Dixmier-Moeglin equivalence

Jason P. Bell, Léon Burkhardt, and Nicholas Priebe

Abstract. We consider Ore extensions of the form $T := R[x; \sigma, \delta]$ with R a commutative integral domain that is finitely generated over a field k. We show that if T has Gelfand-Kirillov dimension less than four then a prime ideal $P \in \mathrm{Spec}(T)$ is primitive if and only if $\{P\}$ is locally closed in $\mathrm{Spec}(T)$, if and only if the Goldie ring of quotients of T/P has centre that is an algebraic extension of k. We also show that there are examples for which these equivalences do not all hold for T of integer Gelfand-Kirillov dimension greater than or equal to 4.

1. Introduction

In the study of rings, it is of special importance to understand the irreducible representations of a ring; that is, to understand its simple left or right modules. This approach is especially fruitful in the study of finite groups where it allows one to decompose the group algebra of a finite group over an algebraically closed field as a finite product of matrix rings, assuming that the characteristic of the field does not divide the order of the group. In general, however, it is a difficult problem to classify the irreducible representations of a ring and so one often settles for a coarser understanding of the representation theory of an algebra by instead classifying its primitive ideals; that is, the ideals that annihilate a simple left module of the ring.

In general, the set of annihilators of simple left modules and the set of annihilators of simple right modules do not coincide. Bergman [**Ber64**], while still an undergraduate, gave an example of a ring in which (0) is a right primitive but not a left primitive ideal; in practice, however, the two sets often agree and so we will not concern ourselves with such pathological examples.

The primitive ideals of a ring are necessarily prime and thus $\mathrm{Prim}(R)$, the set of primitive ideals of R, forms a distinguished subset of $\mathrm{Spec}(R)$. Our objective then becomes to identify the prime ideals in $\mathrm{Spec}(R)$ that are primitive. A fundamental result in this direction is the pioneering work of Dixmier and Moeglin [**Dix77**, **Moe80**], who together showed that if L is a finite-dimensional complex Lie algebra then the primitive ideals of the enveloping algebra $U(L)$ are just the prime ideals of $\mathrm{Spec}(U(L))$ that are locally closed in the Zariski topology; in addition, they showed that the primitive prime ideals are exactly those that are *rational*; that is, primes

2020 *Mathematics Subject Classification.* Primary 16D60, 16A20, 16A32.
Key words and phrases. primitive ideals, Dixmier-Moeglin equivalence, prime spectrum.

P such that the Goldie ring of quotients of $U(L)/P$ has centre exactly the base field $\mathbb{C}$. In general, for a Noetherian algebra R over a field k, a prime ideal P of R is rational if $\mathrm{Frac}(R/P)$ has centre that is an algebraic extension of the base field k, where for a prime Goldie ring S we let $\mathrm{Frac}(S)$ denote its Goldie ring of quotients, which is a simple Artinian ring that plays the role that the field of fractions plays for commutative domains.

Thus Dixmier and Moeglin's fundamental result says that for $P \in \mathrm{Spec}(U(L))$ we have the following equivalences:

$$P \text{ **locally closed**} \iff P \text{ **primitive**} \iff P \text{ **rational**}.$$

These equivalences have since been shown to hold for other classes of rings, including (and this list is by no means exhaustive) affine algebras satisfying a polynomial identity (PI algebras, for short) [**Von96**, 2.6], group algebras of nilpotent-by-finite groups [**Zal71**], many quantum algebras [**GL00**], [**BG01**, II.8.5], affine cocommutative Hopf algebras of finite Gelfand-Kirillov dimension in characteristic zero [**BL14**], Hopf Ore extensions of affine commutative Hopf algebras [**BSM18**], Noetherian twisted homogeneous coordinate rings of complex curves and surfaces [**BRS10**], Hopf algebras of Gelfand-Kirillov dimension two (under mild homological assumptions) [**GZ10**], finitely generated complex connected graded domains of Gelfand-Kirillov dimension two that are generated in degree one (cf. Artin and Stafford [**AS95**] and [**BRS10**]).

An important remark, which we make use of throughout the paper, is that for algebras that satisfy the Nullstellensatz, to prove that the above equivalences hold, it suffices to prove that rational prime ideals are necessarily locally closed (see, for example, [**Bel18**, Theorem 2.3]). For background on the Nullstellensatz, we refer the reader to [**BG01**, II.7]. In particular, the algebras we consider in this paper all satisfy the Nullstellensatz by a result of Irving [**Irv79**, Theorem 2] and so we will make use of the following facts.

FACT 1.1. Let R be a commutative Noetherian Jacobson ring. Then an Ore extension $R[x;\sigma,\delta]$ satisfies the Nullstellensatz.

FACT 1.2. Let T be a Noetherian algebra that satisfies the Nullstellensatz. Then if

$$P \text{ rational} \implies P \text{ locally closed} \tag{1.1}$$

for all $P \in \mathrm{Spec}(T)$ then T satisfies the Dixmier-Moeglin equivalence.

In honour of the work of Dixmier and Moeglin, we now say that a ring R satisfies the *Dixmier-Moeglin equivalence* if a prime ideal P is primitive if and only if it is rational, if and only if $\{P\}$ is locally closed in $\mathrm{Spec}(R)$.

Lorenz [**Lor77**] shortly after the work of Dixmier and Moeglin gave an example of a polycyclic group that is not nilpotent-by finite whose group algebra does not satisfy the Dixmier-Moeglin equivalence. Concretely, his example is an algebra of the form $R[z^{\pm 1};\sigma]$ where R is the ring of Laurent polynomials over $\mathbb{C}$ in two variables. In his example, (0) is a primitive ideal of the group algebra that is not locally closed. Later in [**BLLM17**], it was shown that for each integer $d \geq 4$ there is a finitely generated commutative integral domain R over the complex numbers with a $\mathbb{C}$-linear derivation such that $R[x;\delta]$ is a domain of Gelfand-Kirillov dimension d, which does not satisfy the Dixmier-Moeglin equivalence. In light of these examples, it is of special significance to ask when an Ore extension of a commutative ring satisfies the Dixmier-Moeglin equivalence. We show that in some sense the counterexamples of [**BLLM17**] are optimal, by showing that for Gelfand-Kirillov dimension < 4 the Dixmier-Moeglin equivalence holds. For background related to growth and Gelfand-Kirillov dimension, we refer the interested reader to the book of Krause and Lenagan [**KL00**].

THEOREM 1.3. *Let k be a field of characteristic zero, let R be a finitely generated commutative k-algebra that is a domain, let σ be a k-algebra automorphism of R, and let δ be a k-linear σ-derivation of R. Then if $T = R[x;\sigma,\delta]$ has Gelfand-Kirillov dimension strictly less than four then T satisfies the Dixmier-Moeglin equivalence.*

In the case that we have a skew extension of derivation type $T = R[x;\delta]$ (that is, an Ore extension in which the automorphism σ is the identity), we can show a stronger result: that if R is a finitely generated prime Noetherian algebra of Gelfand-Kirillov dimension < 3 that satisfies both the Nullstellensatz and the Dixmier-Moeglin equivalence, then T does too (see Theorem 3.6).

The outline of this paper is as follows. In §2, we consider Ore extensions of Gelfand-Kirillov dimension two; in §3 we look at Ore extensions of derivation type and prove Theorem 3.6; finally, in §4 we prove Theorem 1.3.

We make the remark that throughout this paper, we will study rings of the form $R[x;\sigma,\delta]$ and we will often consider related rings of the form $S[x;\sigma',\delta']$ where S is either a homomorphic image of, a subring of, or a localization of R, and the maps σ' and δ' are induced from σ and δ in a canonical way. In this case, as an abuse of notation, we will write $S[x;\sigma,\delta]$ with the understanding that, strictly speaking, the maps σ and δ are not technically maps from S to itself, but can be obtained from the original maps σ and δ on R in a natural way.

2. Ore extensions of commutative algebras of Gelfand-Kirillov dimension one

Given a commutative Noetherian ring R and an Ore extension $T := R[x;\sigma,\delta]$ of R, Goodearl [**Goo92**, Theorem 3.1] showed that for $P \in \mathrm{Spec}(T)$ with $I := P \cap R$, we have that the ideal P falls into one of the following three cases:

(a) T/P is commutative;
(b) T/P is not commutative and I is a (σ,δ)-prime ideal that is σ-prime;
(c) T/P is not commutative and I is a (σ,δ)-prime ideal that is δ-prime but not σ-prime.

Throughout this paper we will refer to the primes of $\mathrm{Spec}(T)$ as being of types (a)–(c), making reference to the above trichotomy of Goodearl. We note that in case (c), Goodearl shows in addition that R/I has a unique associated prime, which contains $(1-\sigma)(R)$.

DEFINITION 2.1. Let k be a field and let R be a finitely generated k-algebra. We say that a subspace V of R is a *frame* of R if $1 \in V$, V is finite-dimensional, and V generates R as a k-algebra.

We begin by stating an important result of Zhang.

PROPOSITION 2.2 ([**Zha97**, Lemma 4.1]). *Let R be a finitely generated k-algebra and suppose that σ is a k-algebra automorphism of R that preserves a frame V and that δ is a σ-derivation. Then if $W = V + kx \subseteq T := R[x;\sigma,\delta]$ then W is a* *frame of T and there is a positive integer $L > 0$ such that* $\dim(W^n) \le n \cdot \dim(V^{Ln})$ *for all $n \ge 1$. In particular,* $\mathrm{GKdim}(R[x;\sigma,\delta]) = \mathrm{GKdim}(R) + 1$.

PROOF. Zhang's proof shows that if we choose L so that $\delta(V) \subseteq V^L$, then the above inequality holds. □

LEMMA 2.3. *Let k be an algebraically closed field and let R be a finitely generated commutative k-algebra of Gelfand-Kirillov dimension one that is a domain. If σ is a k-algebra automorphism of R then σ preserves a frame.*

PROOF. Let $\overline{R}$ denote the integral closure of R in $\mathrm{Frac}(R)$. Then σ lifts to an automorphism of $\overline{R}$. Let $X = \mathrm{Spec}(\overline{R})$, which is a smooth irreducible affine curve over k, and let τ denote the automorphism of X corresponding to σ. Then X can be embedded as an open subset of a projective variety Y, and replacing Y by its normalization if necessary (doing this does not affect the open normal subset X), we may assume that Y is a smooth projective curve and τ induces a birational map of Y, which is in fact an automorphism since Y is a smooth curve.

If Y has genus ≥ 2 then τ has finite order on Y and so σ is a finite-order automorphism of R and so σ preserves a frame. If Y has genus 1 then Y is an elliptic curve and since $Y \setminus X$ is finite and preserved by τ, some iterate of τ has a fixed point. Then after a change of coordinates, we may assume this fixed point is the identity of Y and so τ is a group automorphism of Y and thus again has finite order and so once again σ preserves a frame.

Hence we may assume that Y is of genus zero and thus $Y \cong \mathbb{P}^1$. Then if

$$\#(Y \setminus X) \ge 3$$

then some iterate of τ has at least three fixed points and thus is the identity, so once again σ has finite order and we are done. Thus we may assume that $\#(Y \setminus X) \in \{0,1,2\}$ and that $Y = \mathbb{P}^1$, and since X is affine and Y is projective, $Y \setminus X$ is non-empty. Then after a change of coordinates we may assume that $Y \setminus X$ is either $\{\infty\}$ or $\{0,\infty\}$ and so $X = \mathbb{A}^1$ or $X = \mathbb{A}^1 \setminus \{0\}$ and so σ is either linear or affine linear and so the induced map on $\overline{R}$ preserves a frame. Then if V is a frame of R then V is contained is some σ-invariant frame W of $\overline{R}$ and so $W \cap R$ is a frame of R that is invariant under σ. The result follows. □

LEMMA 2.4. *Let k be an algebraically closed field and let R be a finitely generated reduced commutative k-algebra of Gelfand-Kirillov dimension at most one and let σ be a k-algebra automorphism of R. Then σ preserves a frame of R. In particular every one-step Ore extension of R has Gelfand-Kirillov dimension two.*

PROOF. If R is of Gelfand-Kirillov dimension zero then R is finite-dimensional and so R is a frame. Thus we may assume R has Gelfand-Kirillov dimension one. Let $P_1, \ldots, P_s$ denote the minimal primes of R. Then σ permutes these primes and hence there is some $m \ge 1$ such that $\sigma^m(P_i) = P_i$ for $i = 1, \ldots, s$. If σ^m preserves a frame W then σ preserves the frame $\sum_{i=0}^{m-1} \sigma^i(W)$ and thus it suffices to show that σ^m preserves a frame. Now R/P_i is a domain of Gelfand-Kirillov dimension one and so σ^m preserves a frame W_i of R/P_i for $i = 1, \ldots, s$ by Lemma 2.3. Since the intersection of the P_i is (0), R embeds in $\prod_{i=1}^s R/P_i$ via the map π which sends r to $(r + P_1, \ldots, r + P_s)$. We let V be a frame for R. Then there is some N such that $\pi_i(V) \subseteq W_i^N$ for $i = 1, \ldots, s$, where $\pi_i : R \to R/P_i$ is the canonical surjection $r \mapsto r + P_i$. We let $W = \{r \in R \colon \pi(r) \in W_1^N \times \cdots \times W_s^N\}$. Then since π is injective, W is finite-dimensional and by construction W contains V and so W is a frame for R. Note that if $r \in W$ then $\pi(\sigma^m(r)) = (\sigma^m(r) + P_1, \ldots, \sigma^m(r) + P_s)$. Since $W_1, \ldots, W_s$ are σ^m-invariant frames of $R/P_1, \ldots, R/P_s$ respectively, we see that if $r \in W$ then $\sigma^m(r) \in W$, and so W is a σ^m-invariant frame of R. The "in particular" clause now follows from Proposition 2.2. □

REMARK 2.5. In general, a k-algebra automorphism of a finitely generated commutative k-algebra of Gelfand-Kirillov dimension one need not preserve a frame. For example if $R = k[x, y^{\pm 1}]/(x^2)$ and σ is the k-algebra automorphism of R given by $\sigma(x) = xy$ and $\sigma(y) = y$ then σ does not preserve a frame of R since $\{\sigma^i(x) : i \geq 0\} = \{xy^i : i \geq 0\}$ which spans an infinite-dimensional subspace of R, and so x cannot be in a σ-invariant frame of R.

PROPOSITION 2.6. *Let k be an algebraically closed field, let R be a finitely generated commutative k-algebra, let σ be a k-algebra automorphism, and let δ be a k-linear σ-derivation of R. If $T := R[x; \sigma, \delta]$ has Gelfand-Kirillov dimension < 3 and R is either reduced or σ preserves a frame then T satisfies the Dixmier-Moeglin equivalence.*

PROOF. Since $3 > \mathrm{GKdim}(T) \geq \mathrm{GKdim}(R) + 1$, we see R is either of Gelfand-Kirillov dimension zero or one. Hence by Lemma 2.4, σ preserves a frame if R is reduced. Thus we may assume that σ preserves a frame. By Fact 1.2, it suffices to show that rational primes of T are locally closed.

Let P be a prime ideal of T. Then if T/P is commutative then P is locally closed if and only if P is rational and thus it suffices to consider prime ideals P in $\mathrm{Spec}(T)$ of types (b) and (c). In these cases we can let $I = R \cap P$, which is (σ, δ)-prime, and replace R by R/I and T by $(R/I)[x; \sigma, \delta]$. Then T/P is a prime homomorphic image of $R[x; \sigma, \delta]$, which is now a prime ring [**Goo92**, Prop. 3.3]. If after these modifications R is of Gelfand-Kirillov dimension zero then R is finite-dimensional and so T is of Gelfand-Kirillov dimension 1 and thus satisfies a polynomial identity by the Small-Stafford-Warfield theorem [**SSW85**]. It follows that T satisfies the Dixmier-Moeglin equivalence in this case. Thus we may assume that $\mathrm{GKdim}(R) = 1$ and $\mathrm{GKdim}(T) = 2$. Then since σ preserves a frame of R and since R has linear growth we also have that T is of quadratic growth by Proposition 2.2.

Let P be a rational prime ideal of T. We must show that P is locally closed. As above, we may assume that $P \cap R = (0)$ and that P is of type (b) or (c) and R has Gelfand-Kirillov dimension one. Then if T has Gelfand-Kirillov dimension ≤ 1 or if (0) is not rational then T is PI [**BS10**] and hence satisfies the Dixmier-Moeglin equivalence. Thus we may assume that T is of quadratic growth and that (0) is a rational prime ideal. Since T/P is of Gelfand-Kirillov dimension ≤ 1 for P nonzero, it satisfies a polynomial identity and hence satisfies the Dixmier-Moeglin equivalence. So it suffices to show that (0) is locally closed to obtain the Dixmier-Moeglin equivalence in this case.

We pick a nonzero commutator $\theta \in T$. Then T has finitely many primes $Q_1, \ldots, Q_s$ of co-Gelfand-Kirillov dimension one [**Bel10**]. We claim that every nonzero prime ideal Q of T either contains some Q_i or it contains θ. To see this, note that if Q is of type (a) then it contains θ and so we may assume that Q is of type (b) or (c). Since T has quadratic growth and is prime, T/Q is either of Gelfand-Kirillov dimension zero or one; if the Gelfand-Kirillov dimension is one, then Q

must in fact be some Q_i by construction. Otherwise, T/Q is finite-dimensional and then $J := R \cap Q$ has finite codimension in R. It follows that we may assume R/J is finite-dimensional and that Q is either of type (b) or (c) in $\mathrm{Spec}(T)$ (and so T/Q is not commutative).

Let $S = T/JT$, which is a prime algebra of Gelfand-Kirillov dimension one. Moreover T/Q is a homomorphic image of T/JT, and so $Q \supseteq JT$. But since JT is a prime ideal of T of co-Gelfand-Kirillov dimension 1 we have $JT = Q_i$ for some i and so the result follows. Thus we see that the intersection of the nonzero prime ideals of T is nonzero and so (0) is locally closed. The result follows. □

3. Ore extensions of derivation type

In this section we consider Ore extensions of the form $T := R[x; \delta]$ where R is a finitely generated Noetherian algebra of Gelfand-Kirillov dimension < 3 that satisfies the Nullstellensatz. We show under these conditions that the Ore extension T also satisfies the Dixmier-Moeglin equivalence as long as R satisfies the Dixmier-Moeglin equivalence. The assumption that R be of Gelfand-Kirillov dimension < 3 is necessary as there are examples of affine commutative algebras R of Gelfand-Kirillov dimension three such that $R[x; \delta]$ does not satisfy the Dixmier-Moeglin equivalence [**BLLM17**]. On the other hand, if δ preserves a frame V of R and, in addition, R is a $\mathbb{C}$-algebra then $R[x; \delta]$ satisfies the Dixmier-Moeglin equivalence [**BWW17**, Theorem 1.6]. We first show there is a connection between having bounded matrix images and satisfying the Dixmier-Moeglin equivalence for Noetherian algebras of Gelfand-Kirillov dimension two. We recall that if k is an algebraically closed field and A is a k-algebra then A has *bounded matrix images* if the set of n for which there exists a surjective k-algebra homomorphism from A to $M_n(k)$ is finite. More generally, if k is not algebraically closed, we say that A has bounded matrix images if $A \otimes_k \bar{k}$ has bounded matrix images. Small has asked whether a prime Noetherian algebra of quadratic growth necessarily has bounded matrix images (cf. [**Bel03**, Question 3.1]), and there are currently no known examples of such an algebra with unbounded matrix images.

PROPOSITION 3.1. *Let k be a field and let R be a prime Noetherian k-algebra of Gelfand-Kirillov dimension less than three that satisfies the Nullstellensatz. Then R satisfies the Dixmier-Moeglin equivalence if and only if R has bounded matrix images.*

PROOF. By Fact 1.2, to show that R satisfies the Dixmier-Moeglin equivalence, it suffices to show that rational primes of R are locally closed.

Suppose that R has unbounded matrix images. Then let $\mathcal{S}$ denote the set of prime ideals of finite codimension. Then

$$I := \bigcap_{Q \in \mathcal{S}} Q$$

has the property that R/I has unbounded matrix images and thus R/I must have Gelfand-Kirillov dimension at least two, since otherwise it would have Gelfand-Kirillov dimension ≤ 1 and hence would be PI [**SSW85**], which contradicts the fact that it has unbounded matrix images. It follows that $I = (0)$ since $\mathrm{GKdim}(R/J) \leq \mathrm{GKdim}(R) - 1$ for a nonzero ideal J of a prime Noetherian algebra. It follows that (0) is not locally closed in $\mathrm{Spec}(R)$. If R satisfies the Dixmier-Moeglin equivalence then (0) cannot be rational and so R is PI [**BS10**]. Thus we see that if R has unbounded matrix images then R does not satisfy the Dixmier-Moeglin equivalence.

Now suppose that R has bounded matrix images. If R is PI then it satisfies the Dixmier-Moeglin equivalence [**Von96**] and so we may assume that R is not PI and so (0) is a rational prime ideal [**BS10**]. We now claim that (0) is locally closed in $\mathrm{Spec}(R)$. To see this, notice that if P is a nonzero prime ideal of R then R/P is of Gelfand-Kirillov dimension ≤ 1 and hence is PI. Since R has bounded matrix images, there is some m such that R/P satisfies the m-th standard identity. In particular, since R does not satisfy this identity, there is some nonzero element of R that is in every nonzero prime ideal of R and so (0) is locally closed. Next notice that if P is a nonzero rational ideal then R/P has Gelfand-Kirillov dimension ≤ 1 and thus is PI. Since P is rational it follows that P is maximal and thus locally closed. It follows that R satisfies the Dixmier-Moeglin equivalence if R has bounded matrix images. The result follows. $\square$

We now consider rings of the form $R[x;\delta]$ when R is PI. We first give a preliminary result.

LEMMA 3.2. *Let k be a field and let R be a finitely generated commutative k-algebra of Gelfand-Kirillov dimension ≤ 2 that is an integral domain. Then if δ is a k-linear derivation of R such that* (0) *is a rational prime of $R[x;\delta]$ then the intersection of the nonzero δ-invariant prime ideals of R is nonzero.*

PROOF. If k is of characteristic $p > 0$, then $\delta(a^p) = 0$ for every $a \in R$ and so $R[x;\delta]$ is a finite module over its centre, and hence $R[x;\delta]$ is PI and so it satisfies the Dixmier-Moeglin equivalence. Thus we may assume without loss of generality that k has characteristic zero. By Fact 1.2 it suffices to show that rational prime ideals are locally closed.

Every nonzero prime ideal of R is either height one or maximal (or both if R has Gelfand-Kirillov dimension 1). Since (0) is rational, a result of Jouanolou [**Jou78**] (see also [**BLLM17**, Theorem 6.1] for showing the connection) gives that there are only finitely many height one δ-invariant prime ideals of R. Hence it suffices to show that the intersection of the δ-invariant maximal ideals of R is nonzero. Since (0) is a rational prime ideal of $R[x;\delta]$, there is some $y \in R$ such that $\delta(y) \neq 0$. Let P be a δ-invariant maximal ideal of R. Then R/P is a finite extension of k and hence there is an irreducible polynomial $p(t) \in k[t]$ such that $p(y) \in P$. Then since P is δ-invariant, we see $\delta(p(y)) = p'(y)\delta(y) \in P$. Then $p'(t)$ and $p(t)$ have gcd one and so $\delta(y) \in P$. Thus the intersection of the δ-invariant maximal ideals of R contains $\delta(y)$ and so the result follows. $\square$

The following result are well-known, but we supply proofs for the reader's convenience.

LEMMA 3.3. *Let k be a field and let R be a prime Noetherian k-algebra with a k-linear derivation δ. Then we have the following:*

(i) *if* (0) *is a rational prime of $R[x;\delta]$ then $\mathrm{Frac}(R)[x;\delta]$ is a simple ring and so every nonzero ideal of $R[x;\sigma,\delta]$ intersects R non-trivially.*
(ii) *if P is a prime ideal of $R[x;\delta]$ then $R \cap P$ is a δ-invariant prime ideal of R.*

PROOF. For (i), observe that $\mathrm{Frac}(R)$ is an simple Artinian ring. Let I be a nonzero ideal of $\mathrm{Frac}(R)[x;\delta]$. Then we may pick a nonzero element $q = a_d x^d + \cdots + a_1 x + a_0 \in I$ with $a_d, \ldots, a_1, a_0 \in \mathrm{Frac}(R)$ and $a_d \neq 0$ and d minimal.

Since $\mathrm{Frac}(R)$ is simple, there exist $u_1, \ldots, u_s, v_1, \ldots, v_s \in \mathrm{Frac}(R)$ such that $\sum u_i a_d v_i = 1$. In particular, $\sum u_i q v_i$ is a monic polynomial of degree d in I, and so we may assume without loss of generality that q is monic. But now it is easily checked that $[q, x] \in I$ has strictly smaller degree than d and hence must be zero. Similarly, since q is monic, $[q, a] \in I$ has degree less than d for all $a \in R$ and so we see that q is central in $\mathrm{Frac}(R)]x;\delta]$. But by assumption (0) is a rational prime of $\mathrm{Frac}(R)[x;\delta]$, so the centre is necessarily algebraic over k; in particular, q must be of degree 0, and so $1 \in I$. This shows (i).

For (ii), let P be a prime ideal of $R[x;\delta]$ and let $I = P \cap R$. Then I is δ-invariant and since R is Noetherian, there is a finite set of primes $Q_1, \ldots, Q_t$ that are minimal above I. Then $Q_1, \ldots, Q_t$ are δ-invariant by [**MR01**, 14.2.3]. But the image of $Q_1 \cap \cdots \cap Q_t$ in R/I is the nil radical, which is nilpotent since R/I is Noetherian. If follows that if $J_i = Q_i R[x;\delta]$, then some is some n such that $(J_1 \cdots J_t)^n \subseteq IR[x;\delta] \subseteq P$. But since P is prime, we must have $J_i \subseteq P$ for some i and so $J_i \cap R = Q_i \subseteq I$. Since $I \subseteq Q_i$, we have equality and so I is prime. □

LEMMA 3.4. *Let R be a finitely generated prime PI algebra of Gelfand-Kirillov dimension < 3. If (0) is a rational prime ideal of $T := R[x;\delta]$ then (0) is locally closed.*

PROOF. A finitely generated prime PI algebra has integer Gelfand-Kirillov dimension, so in fact R has Gelfand-Kirillov dimension at most two. Since (0) is rational, we have $\mathrm{Frac}(R)[x;\delta]$ is simple by Lemma 3.3(i), and so every nonzero prime ideal P of T intersects R non-trivially and by Lemma 3.3(ii), $I := R \cap P$ is a δ-invariant prime ideal of R. Thus it suffices to show that the intersection of the nonzero δ-invariant prime ideals of R is nonzero.

Let $Z := Z(R)$ denote the centre of R. Then Z is a domain and there exists a nonzero $z \in Z$ such that $R[1/z]$ is a free $Z[1/z]$-module of finite rank [**ASZ99**, Prop. 4.4(3)].

Since $R[1/z]$ is a finitely generated algebra, $Z[1/z]$ is also finitely generated by the Artin-Tate lemma (see [**MS81**, Proposition 2]). Then δ induces a derivation of $R[1/z]$, which then restricts to a derivation of $Z[1/z]$. Then since (0) is a rational prime ideal of $R[1/z][x;\delta]$, (0) is also a rational prime ideal of $Z[1/z][x;\delta]$ [**Let89**, Corollary 1.2].

Now by Lemma 3.2 we have that the intersection of the nonzero δ-invariant prime ideals of $Z[1/z]$ is nonzero and hence contains a nonzero element u. We now claim that u is in every nonzero δ-invariant prime ideal of $R[1/z]$. To see this, let P be a nonzero δ-invariant prime ideal of $R[1/z]$. Then $I := P \cap Z[1/z]$ is nonzero by Posner's theorem and is evidently δ-invariant. Then let $Q_1, \ldots, Q_s$ be the minimal prime ideals above I. Then by [**MR01**, 14.2.3], $Q_1, \ldots, Q_s$ are all δ-invariant and hence $u \in \sqrt{I} = Q_1 \cap \cdots \cap Q_s$. It follows that $u^n \in I$ for some n. But now u is central in R and since I is prime we then have $u \in I$ and hence $u \in P$. Thus we obtain the claim.

Finally, note that every nonzero δ-invariant prime ideal of R either contains z, or it survives in the localization $R[1/z]$, and hence contains u, and so the intersection of the nonzero δ-invariant prime ideals of R is nonzero, since R is itself a prime ring. □

PROPOSITION 3.5. *Let k be a field, let R be a prime Noetherian k-algebra of Gelfand-Kirillov dimension less than three, and let δ be a k-linear derivation of R. If R satisfies both the Nullstellensatz and the Dixmier-Moeglin equivalence and if R is not PI then $T := R[x;\delta]$ satisfies the Dixmier-Moeglin equivalence whenever T satisfies the Nullstellensatz.*

PROOF. We first consider the case when (0) is a rational prime ideal of T. In this case $\mathrm{Frac}(R)[x;\delta]$ is simple by Lemma 3.3(i), and hence every nonzero ideal of T intersects R non-trivially. In particular, if P is a nonzero prime ideal of T then by Lemma 3.3(ii), $P \cap R$ is a nonzero δ-invariant prime ideal of R and so the intersection of the nonzero prime ideals of T contains the intersection of the nonzero prime ideals of R.

Since R is not PI, (0) is a rational prime ideal of R [**BS10**] and so the intersection of the nonzero prime ideals of R is nonzero, since R satisfies the Dixmier-Moeglin equivalence. Thus (0) is locally closed in $\mathrm{Spec}(T)$. So we have shown that if (0) is rational then (0) is locally closed. Now suppose that P is a nonzero rational prime ideal of T and let $I = P \cap R$. Then $I \neq (0)$ and by Lemma 3.3(ii), I is a δ-invariant prime ideal of R. Then R/I is a prime ring of Gelfand-Kirillov dimension ≤ 1 and T/P is a homomorphic image of $(R/I)[x;\delta]$. By the Small-Warfield theorem [**SW84**], R/I is a finite module over its centre, Z, and Z is integral domain invariant under δ. Moreover, Z is a finitely generated algebra by the Artin-Tate lemma (see [**MS81**, Proposition 2]). Now let $P' = P \cap Z[x;\delta]$. Since $Z[x;\delta]$ is an integral domain of Gelfand-Kirillov dimension two, we see that if $P' \neq (0)$ then $Z[x;\delta]/P'$ has Gelfand-Kirillov dimension at most one, and since T/P is a finite module over $Z[x;\delta]/P'$, T/P is PI and so if P is rational then it is locally closed. Thus we may assume that $P' = (0)$ and since $(R/I)[x;\delta]$ is prime, we then have $P = (0)$. Then since (0) is rational in T/P, we must have (0) is rational as a prime ideal of $Z[x;\delta]$.

Then by Lemma 3.4 we see that (0) is locally closed in $Z[x;\delta]$ (and hence also primitive). We claim that this gives (0) is locally closed in $(R/I)[x;\delta]$. To see this, observe that since R/I is a finitely generated Z-module, we can invert some nonzero element f of Z so that $(R/I)[1/f]$ is a free $Z[1/f]$-module. Moreover, δ extends to $(R/I)[1/f]$. Then by [**Let89**, Theorem 2.4], (0) is a locally closed ideal of $(R/I)[1/f][x;\delta]$ as it is a finite free module over $Z[1/f][x;\delta]$, and so by Posner's theorem there is a nonzero element $a \in Z$ such that nonzero prime ideal of $(R/I)[1/f][x;\delta]$ contains a. But since f is central, a prime ideal contains a power of f if and only if it contains f. Thus we see that the intersection of the nonzero prime ideals of $(R/I)[x;\delta]$ must contain af. Hence we have shown that rational prime ideals of T are locally closed when (0) is rational prime ideal of T.

Next we consider the case when (0) is not a rational prime ideal of T and let P be a rational prime ideal of T. If $I := P \cap R \neq (0)$, then as in the preceding case we have that T/P is a homomorphic image of $(R/I)[x;\delta]$ and since R/I has Gelfand-Kirillov dimension ≤ 1, we see that $(R/I)[x;\delta]$ satisfies the Dixmier-Moeglin equivalence as before by reducing to Proposition 2.6. Thus we may assume that $P \cap R = (0)$. Then P survives in the localization $\mathrm{Frac}(R)[x;\delta]$ and generates a nonzero prime ideal $\widetilde{P}$. Then since $\mathrm{Frac}(R)$ is simple, $\widetilde{P}$ contains a nonzero monic polynomial in

x and so $\mathrm{Frac}(R)[x;\delta]/\widetilde{P}$ is a finite $\mathrm{Frac}(R)$-module and thus is Artinian; since it is prime, it is simple. It follows that every prime ideal Q that strictly contains P must intersect R non-trivially and by Lemma 3.3(ii), $Q\cap R$ is then a nonzero prime ideal of R. As before, since R is not PI, (0) is a rational prime ideal of R and since R satisfies the Dixmier-Moeglin equivalence, we see that the intersection of the nonzero ideals of R is nonzero and so the intersection of the prime ideals of T that strictly contain P is nonzero. The result follows. □

THEOREM 3.6. *Let k be a field, let R be a finitely generated prime Noetherian k-algebra of Gelfand-Kirillov dimension less than three that satisfies the Nullstellensatz, and let δ be a k-linear derivation of R. If R satisfies the Dixmier-Moeglin equivalence and if $T = R[x, \delta]$ satisfies the Nullstellensatz then T also satisfies the Dixmier-Moeglin equivalence.*

PROOF. If R is not PI this follows from Proposition 3.5. Thus we may assume that R is PI and it suffices to show that a rational prime ideal of T is locally closed. If (0) is a rational prime ideal of T then (0) is locally closed by Lemma 3.4. Thus it suffices to show that every nonzero rational prime ideal P is locally closed. Let $I := P\cap R$. If $I = (0)$, then P survives in the localization $\mathrm{Frac}(R)[x;\delta]$, and since P is nonzero, the corresponding ideal, $\tilde{P}$, in $\mathrm{Frac}(R)[x;\delta]$ contains a monic polynomial in x and so $\mathrm{Frac}(R)[x;\delta]/\tilde{P}$ is a finite module over $\mathrm{Frac}(R)$ and hence is PI, which contradicts the fact that P is rational unless R is simple and hence of Gelfand-Kirillov dimension zero. In this case T is of Gelfand-Kirillov dimension one and hence PI and so T satisfies the Dixmier-Moeglin equivalence in this case. On the other hand, if I is nonzero then T/P is a homomorphic image of $S := (R/I)[x;\delta]$, and R/I is a finitely generated prime algebra of Gelfand-Kirillov dimension ≤ 1 and is PI. If (0) is a rational prime ideal of S then (0) is locally closed by Lemma 3.4. If (0) is not rational then since S is of Gelfand-Kirillov dimension two, it is PI [**BS10**]. But P corresponds to a rational prime ideal P' of S with $P'\cap(R/I) = (0)$, and so this is impossible since S is PI. The result follows. □

4. Proof of Theorem 1.3

In this section we give the proof of Theorem 1.3. We first need two preliminary lemmas.

LEMMA 4.1. *Let k be an algebraically closed field, let R be a commutative k-algebra with a frame-preserving k-algebra automorphism σ and σ-derivation δ and let $T = R[x;\sigma,\delta]$. If σ is not the identity map and δ is not inner and R is σ-prime then $R_0 := \{r \in R\colon \sigma(r) = r\}$ is central in T.*

PROOF. Suppose that σ is not the identity map. Then since σ preserves a frame, either there is some nonzero $f \in R$ with $\sigma(f) = \lambda f$ and $\lambda \neq 1$ or there is some $f \in R$ with $\sigma(f)-f = 1$. In the latter case, δ is inner, so we may assume there is some nonzero f with $\sigma(f) = \lambda f$ with $\lambda \neq 1$. Now by [**BSM18**, Lemma 4.14] we have f is regular. Now let $r \in R_0$. Then expanding both sides of $\delta(rf) = \delta(fr)$ and since $\sigma(r) = r$ and $\sigma(f) = \lambda f$ we then see that $(\sigma(f) - f)\delta(r) = 0$ and so $\delta(r) = 0$, since $\lambda \neq 1$ and f is regular. Now this means every element of R_0 commutes with x and so R_0 is central in T, as claimed. □

LEMMA 4.2. *Let k be an algebraically closed field, let R be a finitely generated commutative k-algebra, let σ be a frame-preserving k-algebra automorphism of R,* *and let δ be a σ-derivation of R. Suppose that R is σ-prime, σ is not the identity map, δ is not inner, (0) is a rational prime of T, and suppose that there exists a prime ideal of T of type (c). Then there is a nonzero δ-prime ideal I of R such that $P \cap R = I$ for every prime ideal P of type (c).*

PROOF. Let P be a rational prime ideal of T of type (c). Goodearl shows there is a unique associated prime ideal $Q = Q(P)$ of R/I and $\sigma(r) - r \in Q$ for all $r \in R$. Let W be a σ-invariant frame of R. Then since (0) is rational and R is σ-prime, by Lemma 4.1 we have that $R_0 = k$, where $R_0 = \{r\colon \sigma(r) = r\}$. It follows that $(\sigma - \mathrm{id})|_W$ has one-dimensional kernel and so the image of $(\sigma - \mathrm{id})|_W$ has codimension one in W. Thus $W/Q \cap W$ is one-dimensional. Since this is true for every σ-invariant frame and since $R = \bigcup W^n$, we then see that R/Q is a 1-dimensional k-vector space. Furthermore this argument shows that

$$Q = \mathrm{Span}_k\{r - \sigma(r)\colon r \in R\}.$$

Hence we have shown that if P and P' are two primes of type (c) then

$$Q(P) = Q(P') = \mathrm{Span}_k\{r - \sigma(r)\colon r \in R\}.$$

Now suppose that we have two distinct primes P and P' of type (c). Then let $Q = Q(P) = Q(P')$ and let $I = R \cap P$ and $I' = R \cap P'$. Goodearl [**Goo92**, Proposition 3.11] shows that I and I' are the largest δ-prime ideals contained in Q and so $I = I'$. The result follows. □

PROPOSITION 4.3. *Let k be an algebraically closed field, let R be a finitely generated reduced commutative k-algebra of Gelfand-Kirillov dimension two, let σ be a k-algebra automorphism of R that preserves a frame, and let δ be a k-linear σ-derivation of R. Then $T := R[x;\sigma,\delta]$ satisfies the Dixmier-Moeglin equivalence.*

PROOF. If σ is the identity then the result follows by Theorem 3.6, since every prime ideal of T intersects R in a prime ideal, and so we may assume that R is prime in this case; if δ is inner, then $R[x;\sigma,\delta] \cong R[x;\sigma]$ and the result follows by [**BWW17**, Theorem 1.6].[1] Thus for the remainder of the proof, we assume that σ is not the identity and δ is not inner.

We now let P be a rational prime ideal of T. We consider each case in Goodearl's trichotomy, and in each case show that P is locally closed.

Case I. P is a rational prime of type (a).

In this case T/P is commutative and hence P must be maximal since P is rational.

Case II. P is a rational prime of type (b).

In this case, we let $I = R \cap P$, and we may assume that T/P is not commutative by the preceding case. Since the radical of I is σ-invariant and I is σ-prime, we see that I is radical and so R/I is reduced. First observe that if R/I is of Gelfand-Kirillov dimension ≤ 1, then since T/P is a homomorphic image of $S := T/IT \cong (R/I)[x;\sigma,\delta]$ and since σ preserves a frame, we have S is of Gelfand-Kirillov dimension at most two and thus satisfies the Dixmier-Moeglin equivalence by Proposition 2.6. It follows that P is locally closed in $\mathrm{Spec}(T)$ when R/I has

[1] Although the theorem in [**BWW17**] has a hypothesis that the base field be uncountable, this is only used to ensure that $R[x;\sigma]$ satisfies the Nullstellensatz. But since R is commutative, we can use Fact 1.1 in this case and then follow the original proof.

Gelfand-Kirillov dimension at most one, and so we assume for the remainder of this case that R/I has Gelfand-Kirillov dimension two.

It now suffices to show that the intersection of the prime ideals that properly contain P of each type (a)–(c) properly contains P. The intersection of prime ideals of type (c) that contain P again properly contains P by Lemma 4.2; the intersection of primes that contain P of type (a) again properly contains P since it contains all commutators and T/P is not commutative by assumption. Finally, we show that the intersection of primes of type (b) that properly contain P properly contains P. By Lemma 4.1, $R_0 = \{r \in R \colon \sigma(r) = r\}$ is equal to k and since σ preserves a frame and δ is not inner, there is some nonzero f in R such that $\sigma(f) = \lambda f$ with $\lambda \neq 1$. Then since $\sigma(f) - f$ is invertible in $R[1/f]$ we have δ is inner in $R[1/f][x;\sigma,\delta]$ and so $R[1/f][x;\sigma,\delta] \cong R[1/f][x;\sigma]$. By [**BSM18**, Lemma 4.14] we have f is regular mod $I := P \cap R$ unless $f \in I$, and so the intersection of primes P of type (b) for which f is not regular mod $P \cap R$ contains the element f. Hence it suffices to show that the intersection of nonzero prime ideals of type (b) for which f is regular mod $I := P \cap R$ is nonzero. But since σ preserves a frame, $R[1/f][x;\sigma]$ satisfies the Dixmier-Moeglin equivalence [**BWW17**] and so the intersection of prime ideals P of type (b) for which f is regular mod $P \cap R$ is nonzero. The result follows.

Case III. P is of type (c).

As in Case II, we may assume that $I := P \cap R$ has the property that R/I has Gelfand-Kirillov dimension two. Since R is reduced and Noetherian, it has a finite set of minimal prime ideals $\{Q_1, \ldots, Q_t\}$, and since R is reduced, it has radical ideal (0) and so the intersection of the Q_i's is zero. Moreover since R/I has the same Gelfand-Kirillov dimension as R, after relabelling we may assume that I is contained in Q_1. Then by Goodearl's trichotomy, I has a unique associated prime and since R is commutative and Noetherian, the minimal primes above I are necessarily associated. Thus Q_1 is radical over I. Now let $J = \{r \in R \colon Ir = (0)\}$. Now if $t = 1$, then since R is reduced, we see it is a domain and $I = Q_1 = (0)$ and so I is σ-prime and δ-prime, so we are done by Case II.

Notice next if $t > 1$ then since $I \subseteq Q_1$, J is nonempty since it contains the intersection $Q_2 \cap \cdots \cap Q_t \neq (0)$. It is then straightforward to verify that the fact that I is a (σ,δ)-ideal forces J to be a (σ,δ)-ideal. Similarly, we can let I' denote the set of elements $r \in R$ with $rJ = (0)$, which again is easily seen to be a (σ,δ)-ideal. Then $I' \supseteq I$ and since Q_1 is radical over I, we then see $(Q_1 J)^m = (0)$ for some m. But R is reduced and so this gives $Q_1 J = (0)$ and so $I' \supseteq Q_1$. But since $J \supseteq Q_2 \cap \cdots \cap Q_t$, the annihilator of J is contained in the annihilator of $Q_2 \cap \cdots \cap Q_t$, which is Q_1. Thus $I' = Q_1$ and since I' is a (σ,δ)-ideal that is radical over I and since I is δ-prime, we see that $I' = I$ and so $I = Q_1$, and so I is again σ-prime and the result follows from Case II. □

PROOF OF THEOREM 1.3. Suppose first that R is an integral domain of Gelfand-Kirillov dimension ≥ 2. Since T has Gelfand-Kirillov dimension less than 4, a result of Zhang [**Zha97**] shows that σ preserves a frame of R and that R has Gelfand-Kirillov dimension two. Since T satisfies the Nullstellensatz, to prove that T satisfies the Dixmier-Moeglin equivalence, it suffices to show that rational prime ideals of T are locally closed in $\mathrm{Spec}(T)$. Since R is reduced and k is of characteristic zero (and hence perfect), we have $R_K := R \otimes_k K$ is reduced by [**Bou59**, Chapitre 5, §15.5, Théorème 3(d)], and it is a finitely generated K-algebra whenever K is a field extension of k, and $T_K := T \otimes_k K \cong R_K[x;\sigma,\delta]$ is left Noetherian, Jacobson, and satisfies the Nullstellensatz. Hence to show that T satisfies the Dixmier-Moeglin equivalence, using Irving-Small reduction (see [**Row88**, Theorem 4.2.27]), it suffices to show that T_K satisfies the Dixmier-Moeglin equivalence for every algebraically closed extension of k. Since σ preserves a K-frame of R_K and R_K is reduced, T_K satisfies the Dixmier-Moeglin equivalence by Proposition 4.3, and so the result now follows in this case.

Thus we may assume that R has Gelfand-Kirillov dimension zero or one. If R has Gelfand-Kirillov dimension zero, then T is of Gelfand-Kirillov dimension one and hence is PI and so the result holds [**Von96**]. Thus we may assume that R has Gelfand-Kirillov dimension one. Then $R \otimes_k K$ is a reduced K algebra whenever K is an algebraically closed extension of k and so the map induced by σ preserves a frame of $R \otimes_k K$ by Lemma 2.4 and so using Irving-Small reduction and Proposition 4.3 again gives the result. □

REMARK 4.4. We note that the characteristic zero hypothesis is used when invoking the Irving-Small reduction techniques.

5. Errata to [BWW17]

We finish this paper by taking the opportunity to list some corrections to the paper [**BWW17**], which in many ways this paper builds upon. The first-named author is indebted to Zhipeng Chen, who discovered the following issues.

(1) In the Assumptions on page 7, we should also have the assumption: all prime ideals of R are completely prime and that (0) is a rational prime ideal of $R[x;T]$. In particular, Lemmas 3.2 and 3.3 should have these hypotheses (which are implicitly used in the proofs).
(2) In Lemma 3.4, there should be the assumption that R satisfies the Dixmier-Moeglin equivalence.

Acknowledgment

We thank the anonymous referee for many helpful comments and suggestions.

References

[ASZ99] M. Artin, L. W. Small, and J. J. Zhang, *Generic flatness for strongly Noetherian algebras*, J. Algebra **221** (1999), no. 2, 579–610, DOI 10.1006/jabr.1999.7997. MR1728399

[AS95] M. Artin and J. T. Stafford, *Noncommutative graded domains with quadratic growth*, Invent. Math. **122** (1995), no. 2, 231–276, DOI 10.1007/BF01231444. MR1358976

[Bel03] Jason P. Bell, *Examples in finite Gel′fand-Kirillov dimension*, J. Algebra **263** (2003), no. 1, 159–175, DOI 10.1016/S0021-8693(03)00021-8. MR1974084

[Bel10] Jason P. Bell, *A dichotomy result for prime algebras of Gelfand-Kirillov dimension two*, J. Algebra **324** (2010), no. 4, 831–840, DOI 10.1016/j.jalgebra.2010.03.028. MR2651571

[Bel18] Jason Bell, *On the importance of being primitive* (English, with English and Spanish summaries), Rev. Colombiana Mat. **53** (2019), no. suppl., 87–112. MR4053367

[BLLM17] Jason Bell, Stéphane Launois, Omar León Sánchez, and Rahim Moosa, *Poisson algebras via model theory and differential-algebraic geometry*, J. Eur. Math. Soc. (JEMS) **19** (2017), no. 7, 2019–2049, DOI 10.4171/JEMS/712. MR3656478

[BSM18] Jason Bell, Omar León Sánchez, and Rahim Moosa, *D-groups and the Dixmier-Moeglin equivalence*, Algebra Number Theory **12** (2018), no. 2, 343–378, DOI 10.2140/ant.2018.12.343. MR3803706

[BL14] Jason P. Bell and Wing Hong Leung, *The Dixmier-Moeglin equivalence for cocommutative Hopf algebras of finite Gelfand-Kirillov dimension*, Algebr. Represent. Theory **17** (2014), no. 6, 1843–1852, DOI 10.1007/s10468-014-9474-y. MR3284332

[BRS10] J. Bell, D. Rogalski, and S. J. Sierra, *The Dixmier-Moeglin equivalence for twisted homogeneous coordinate rings*, Israel J. Math. **180** (2010), 461–507, DOI 10.1007/s11856-010-0111-0. MR2735073

[BS10] Jason P. Bell and Agata Smoktunowicz, *Extended centres of finitely generated prime algebras*, Comm. Algebra **38** (2010), no. 1, 332–345, DOI 10.1080/00927870802254850. MR2597499

[BWW17] Jason P. Bell, Kaiyu Wu, and Shelley Wu, *The Dixmier-Moeglin equivalence for extensions of scalars and Ore extensions*, Groups, rings, group rings, and Hopf algebras, Contemp. Math., vol. 688, Amer. Math. Soc., Providence, RI, 2017, pp. 1–14. MR3649164

[Ber64] G. M. Bergman, *A ring primitive on the right but not on the left*, Proc. Amer. Math. Soc. **15** (1964), 473–475, DOI 10.2307/2034527. MR167497

[Bou59] N. Bourbaki, *Éléments de mathématique. I: Les structures fondamentales de l'analyse. Fascicule XI. Livre II: Algèbre. Chapitre 4: Polynomes et fractions rationnelles. Chapitre 5: Corps commutatifs* (French), Actualités Scientifiques et Industrielles [Current Scientific and Industrial Topics], vol. No. 1102, Hermann, Paris, 1959. Deuxième édition. MR0174550

[BG01] Ken A. Brown and Ken R. Goodearl, *Lectures on algebraic quantum groups*, Advanced Courses in Mathematics. CRM Barcelona, Birkhäuser Verlag, Basel, 2002, DOI 10.1007/978-3-0348-8205-7. MR1898492

[Dix77] J. Dixmier, *Idéaux primitifs dans les algèbres enveloppantes* (French), J. Algebra **48** (1977), no. 1, 96–112, DOI 10.1016/0021-8693(77)90296-4. MR447360

[Goo92] K. R. Goodearl, *Prime ideals in skew polynomial rings and quantized Weyl algebras*, J. Algebra **150** (1992), no. 2, 324–377, DOI 10.1016/S0021-8693(05)80036-5. MR1176901

[GL00] K. R. Goodearl and E. S. Letzter, *The Dixmier-Moeglin equivalence in quantum coordinate rings and quantized Weyl algebras*, Trans. Amer. Math. Soc. **352** (2000), no. 3, 1381–1403, DOI 10.1090/S0002-9947-99-02345-4. MR1615971

[GZ10] K. R. Goodearl and J. J. Zhang, *Noetherian Hopf algebra domains of Gelfand-Kirillov dimension two*, J. Algebra **324** (2010), no. 11, 3131–3168, DOI 10.1016/j.jalgebra.2009.11.001. MR2732991

[Irv79] Ronald S. Irving, *Generic flatness and the Nullstellensatz for Ore extensions*, Comm. Algebra **7** (1979), no. 3, 259–277, DOI 10.1080/00927877908822347. MR519702

[IS80] Ronald S. Irving and Lance W. Small, *On the characterization of primitive ideals in enveloping algebras*, Math. Z. **173** (1980), no. 3, 217–221, DOI 10.1007/BF01159659. MR592369

[Jou78] J. P. Jouanolou, *Hypersurfaces solutions d'une équation de Pfaff analytique* (French), Math. Ann. **232** (1978), no. 3, 239–245, DOI 10.1007/BF01351428. MR481129

[KL00] Günter R. Krause and Thomas H. Lenagan, *Growth of algebras and Gelfand-Kirillov dimension*, Revised edition, Graduate Studies in Mathematics, vol. 22, American Mathematical Society, Providence, RI, 2000, DOI 10.1090/gsm/022. MR1721834

[Let89] Edward Letzter, *Primitive ideals in finite extensions of Noetherian rings*, J. London Math. Soc. (2) **39** (1989), no. 3, 427–435, DOI 10.1112/jlms/s2-39.3.427. MR1002455

[Lor77] Martin Lorenz, *Primitive ideals of group algebras of supersoluble groups*, Math. Ann. **225** (1977), no. 2, 115–122, DOI 10.1007/BF01351715. MR424862

[MR01] J. C. McConnell and J. C. Robson, *Noncommutative Noetherian rings*, Revised edition, Graduate Studies in Mathematics, vol. 30, American Mathematical Society, Providence, RI, 2001. With the cooperation of L. W. Small, DOI 10.1090/gsm/030. MR1811901

[Moe80] C. Moeglin, *Idéaux bilatères des algèbres enveloppantes* (French, with English summary), Bull. Soc. Math. France **108** (1980), no. 2, 143–186. MR606087

[MS81] S. Montgomery and L. W. Small, *Fixed rings of Noetherian rings*, Bull. London Math. Soc. **13** (1981), no. 1, 33–38, DOI 10.1112/blms/13.1.33. MR599637

[Row88] Louis H. Rowen, *Ring theory. Vol. II*, Pure and Applied Mathematics, vol. 128, Academic Press, Inc., Boston, MA, 1988. MR945718

[SW84] L. W. Small and R. B. Warfield Jr., *Prime affine algebras of Gel'fand-Kirillov dimension one*, J. Algebra **91** (1984), no. 2, 386–389, DOI 10.1016/0021-8693(84)90110-8. MR769581

[SSW85] L. W. Small, J. T. Stafford, and R. B. Warfield Jr., *Affine algebras of Gel'fand-Kirillov dimension one are PI*, Math. Proc. Cambridge Philos. Soc. **97** (1985), no. 3, 407–414, DOI 10.1017/S0305004100062976. MR778674

[Von96] Nikolaus Vonessen, *Actions of algebraic groups on the spectrum of rational ideals*, J. Algebra **182** (1996), no. 2, 383–400, DOI 10.1006/jabr.1996.0176. MR1391589

[Zal71] A. E. Zalesskiĭ, *The irreducible representations of finitely generated nilpotent groups without torsion* (Russian), Mat. Zametki **9** (1971), 199–210. MR302769

[Zha97] James J. Zhang, *A note on GK dimension of skew polynomial extensions*, Proc. Amer. Math. Soc. **125** (1997), no. 2, 363–373, DOI 10.1090/S0002-9939-97-03602-2. MR1350966

Department of Pure Mathematics, University of Waterloo, Waterloo, Ontario, Canada, N2L 3G1

Email address: jpbell@uwaterloo.ca

Departement of Mathematics, ETH Zurich, Rämistrasse 101, 8092 Zurich, Switzerland

Email address: lburkhardt@student.ethz.ch

Department of Pure Mathematics, University of Waterloo, Waterloo, Ontario, Canada, N2L 3G1

Email address: nicholas.priebe@uwaterloo.ca

Contemporary Mathematics
Volume **801**, 2024
https://doi.org/10.1090/conm/801/16082

Deformations of gentle A_∞-algebras

Raf Bocklandt and Jasper van de Kreeke

ABSTRACT. In this paper we calculate the Hochschild cohomology of gentle A_∞-algebras of arc collections on marked surfaces. When the underlying arc collection has no loops or two-cycles, we show that the dgla structure of the Hochschild complex is formal and give an explicit realization of all deformations up to gauge equivalence.

CONTENTS

1. Introduction

To a collection of oriented arcs on a closed surface with marked points, one can associate an A_∞-algebra called the gentle A_∞-algebra. These were introduced in [**5**, **9**] to study wrapped Fukaya categories of punctured surfaces [**1**], where the punctured surface is obtained by removing the marked points. These algebras are also variants of well-known algebras that are studied in representation theory [**2**, **12**, **16**].

In this paper we study the deformation theory of these A_∞-algebras in detail. The intuition from mirror symmetry tells us that deforming these algebras as curved A_∞-algebras should correspond to filling in these punctures with normal points or orbifold points [**6**, **18**].

We will work out this idea as follows. Starting from an arc collection on a closed surface with marked points, we introduce the combinatorial notion of an orbigon and use it to define a family of higher products that count these orbigons. We show that these structures satisfy the curved A_∞-axioms and indeed deform the gentle A_∞-algebra coming from the arc collection.

To show that these deformations solve the deformation problem we look at the Hochschild cohomology. We show that each element in the Hochschild cohomology

2020 *Mathematics Subject Classification.* Primary 16E40, 16S80.
Key words and phrases. Deformation theory, noncommutative algebra, representation theory.
The authors were supported in part by the NWO-grant 'Grant Algebraic methods and structures in the theory of Frobenius manifolds and their applications' (TOP1.17.012).

is determined by its zeroth and first component; in other words its nullary and unary product. In the case that the arc collection has no loops or two-cycles we obtain an explicit basis for the Hochschild cohomology. This enables us to give a description of the Gerstenhaber algebra structure on the Hochschild cohomology.

These computations match work done by Wong [**20**] on the Borel-Moore cohomology for matrix factorizations of dimer models (which is the B-side analogon of our setting, from the perspective of mirror symmetry [**5**]), as well as the theme that Hochschild cohomology of wrapped Fukaya categories agrees with symplectic homology [**7, 17**].

Finally we show that the bracket from the Gerstenhaber structure is formal and conclude that each solution of the Maurer-Cartan equation for the Hochschild cohomology is gauge equivalent to one of the curved A_∞-structures we introduced.

2. Curved gentle algebras

In this section we will introduce curved gentle algebras. These are curved deformations of the gentle A_∞-algebras that were defined in [**5**] and [**9**].

2.1. Arc collections and gentle algebras.

DEFINITION 2.1. A **marked surface** (S, M) is a pair consisting of a compact oriented surface without boundary S and a finite subset of marked points $M \subset S$. We will denote the genus of the surface by g. Furthermore we will assume that $|M| \geq 3$ if S is a sphere and $|M| \geq 1$ otherwise.

An **arc collection** $\mathcal{A}$ is a set of oriented curves $a : [0,1] \to S$ that meet M only at the end points ($a^{-1}(M) = \{0,1\}$) and do not (self)-intersect internally. We say that an arc collection **splits the surface** if the complement of the arcs is the disjoint union of discs. These disks are called the **faces** and we denote the set of faces by F. Each face can be seen as a polygon bounded by arcs.

An arc collection that splits the surface satisfies

- the **no monogons or digons condition [NMD]**, if no single arc or pair of arcs bounds a face. In this case the surface is split in n-gons with $n \geq 3$.
- the **no loops or two cycles condition [NL2]** if every arc has two different end points and no two arcs share more than one end point.
- the **dimer condition**, if the arcs around each disk form an oriented cycle. In case of a dimer, we will call a face positive or negative depending on whether the cycle around it is anticlockwise or clockwise. We can partition F accordingly: $F = F^+ \cup F^-$.

From now on we will assume that [NMD] holds, but in later sections we will sometimes have to assume the stronger [NL2] condition.

Recall that a **quiver** is a four-tuple $Q = (Q_0, Q_1, h, t)$ representing an oriented graph with vertices Q_0, arrows Q_1 and maps $h, t : Q_1 \to Q_0$ that assign to each arrow its head and tail. A quiver is graded by a group G if it comes with a map $|\cdot| : Q_1 \to G$.

The **path algebra** $\mathbb{C}Q$ of a quiver Q is the complex vector space spanned by the paths with as product concatenation of paths if possible and zero otherwise. We write the arrows as going from right to left so $\alpha\beta$ is a genuine path if $t(\alpha) = h(\beta)$. Every vertex of the quiver $v \in Q_0$ gives rise to a path of length zero, which is called the vertex idempotent $\mathbb{1}_v$. These span a semisimple subalgebra which we will denote by $\Bbbk \cong \mathbb{C}^{Q_0}$. If Q is G-graded then $\mathbb{C}Q$ is a graded $\Bbbk$-algebra by assigning to each arrow its G-degree and to each vertex idempotent degree 0.

It is tempting to consider the arc collection itself as a quiver, but we will not do this. Instead we will consider a different quiver, which is obtained by putting a vertex in the middle of each arc.

DEFINITION 2.2. Given an arc collection $\mathcal{A}$ that splits (S, M), we define a $\mathbb{Z}_2$-graded quiver $Q_\mathcal{A}$ as follows.

- The vertices of the quiver are the arcs: $(Q_\mathcal{A})_0 = \mathcal{A}$.
- For each angle of a face we define an arrow that corresponds to the internal anticlockwise angle between consecutive arcs that meet in that corner.
- An arrow has degree zero if both arcs have the same direction at the marked point (both outwards or both inwards), and degree one otherwise. Note that $\mathcal{A}$ is a dimer if and only if all arrows have $\mathbb{Z}_2$-degree 1.

$|\alpha| = 0$ $|\alpha| = 0$ $|\alpha| = 1$ $|\alpha| = 1$

We will denote these 'angle arrows' by Greek letters and use $h(\alpha)$ and $t(\alpha)$ to denote the arcs (vertices) that correspond to the head and tail of the arrows. Each angle arrow also turns around a unique marked point and is contained in a unique face. We will denote these by $m(\alpha)$ and $f(\alpha)$. In this quiver two consecutive arrows either are angles that turn around a common marked point or they correspond to consecutive angles in a face.

$$t(\alpha) = h(\beta) \implies m(\alpha) = m(\beta) \text{ or } f(\alpha) = f(\beta)$$

DEFINITION 2.3. The **gentle algebra** $\mathrm{Gtl}_\mathcal{A}$ of an arc collection $\mathcal{A}$ is the path algebra of $Q_\mathcal{A}$ modulo the ideal of relations spanned by the products of arrows that are consecutive angles in a face.

$$\mathrm{Gtl}_\mathcal{A} = \mathbb{C}Q_\mathcal{A}/\langle \alpha\beta | \alpha \text{ and } \beta \text{ are consecutive angles in a face}\rangle$$

The product rule in this algebra can be illustrated pictorially as follows.

$\alpha\beta \neq 0$ $\alpha\beta = 0$

REMARK 2.4. The number of arrows is $2|\mathcal{A}|$ because in each vertex precisely two arrows arrive and two arrows leave. More precisely for every angle arrow α there are precisely two angle arrows β_1 and β_2 with $h(\alpha) = t(\beta_i)$. One will satisfy $\beta_1\alpha = 0$ and for the other one we have $\beta_2\alpha \neq 0$. Likewise there are two γ_i with $h(\gamma_i) = t(\alpha)$, one with $\alpha\gamma_1 = 0$ and one with $\alpha\gamma_2 \neq 0$. This notion of a gentle algebra refers to this property and derives from representation theory [**2**]. The usual definition of a gentle algebra also entails that the algebra is finite-dimensional but this is not the case for our algebras. It is possible to obtain finite dimesnional algebras by looking at surfaces with marked points on the boundary [**12, 16**].

EXAMPLE 2.5. In the picture below on the left you see an arc collection on a torus with two marked points. It has 4 arcs and two square faces. The quiver $Q_{\mathcal{A}}$ has 4 vertices and 8 angle arrows. The angle arrows all have degree 1. The relations are generated by all paths $\alpha_i\beta_j$ and $\beta_i\alpha_j$, whenever the arrows are composable. This implies that every nonzero path in the gentle algebra is either a sequence of only $\alpha's$ or only $\beta's$. The former turn around the first marked point, while the latter turn around the second marked point.

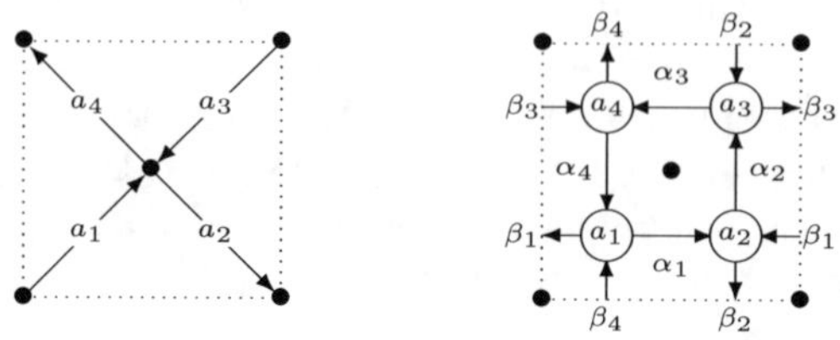

In general, the nonzero paths of a gentle algebra correspond to positive angles between arcs that share a marked point. Every nonzero angle paths is uniquely determined by it sequence of angle arrows and products of angles paths turning around different marked points are zero.

For each $m \in M$ we define

$$\ell_m = \alpha_1 \dots \alpha_k + \dots + \alpha_k \dots \alpha_1$$

where $\alpha_1, \dots, \alpha_k$ are the angle arrows around m ordered in anticlockwise direction. These elements represent single loops around the marked points and they generate the center of the algebra.

LEMMA 2.6.

$$Z(\mathrm{Gtl}_{\mathcal{A}}) = \mathbb{C}[\ell_m | m \in M]/(\ell_i\ell_j | i \neq j).$$

PROOF. One can easily check that the $\alpha\ell_m = \ell_m\alpha$ if $m(\alpha) = m$ and $\alpha\ell_m = \ell_m\alpha = 0$ if $m(\alpha) \neq m$. This also implies that $\ell_u\ell_v = 0$ if $u \neq v$.

Suppose $z = c\beta + \dots$ is a nonzero central element containing the angle path β and let α be the angle arrow that follows β in the cycle around m. Then we have that $\alpha\beta \neq 0$, so if $\alpha z = z\alpha$ then $\alpha\beta$ must end in α and hence β must be a cycle that winds a number of full turns around m. Therefore z will contain ℓ_m^r and the ℓ_m^r form a basis for the part of the center with path length > 0. The length 0 part of the center is $\mathbb{C}$ because the quiver is connected. □

LEMMA 2.7. As a $Z(\mathrm{Gtl}_{\mathcal{A}})$-module the outer ($\mathbb{Z}_2$-graded) $\mathbb{k}$-derivations are generated by the Euler derivations

$$E_\alpha := \alpha\partial_\alpha.$$

PROOF. Let α be an angle arrow. If d is a $\mathbb{k}$-derivation then $h(d\alpha) = h(\alpha)$ and $t(d\alpha) = t(\alpha)$. If $d\alpha$ contains a term γ that does not turn around the same marked point as α, let β be the angle such that $\gamma\beta \neq 0$ is cyclic. We then have that $\alpha\beta = 0$ because they turn around different marked points. Therefore

$$0 = d(\alpha\beta) = \gamma\beta + \dots \pm \alpha d\beta,$$

but this is impossible because these two components turn around different marked points and hence cannot cancel each other.

So suppose $d\alpha$ turns around the same marked point m as α. In that case there are the following possibilities.

(1) If $a = h(\alpha) = t(\alpha)$ then either a forms a monogon (which is excluded by condition [NMD]) or $\alpha = \ell_m$ is a full turn around a marked point m with one arc a arriving.

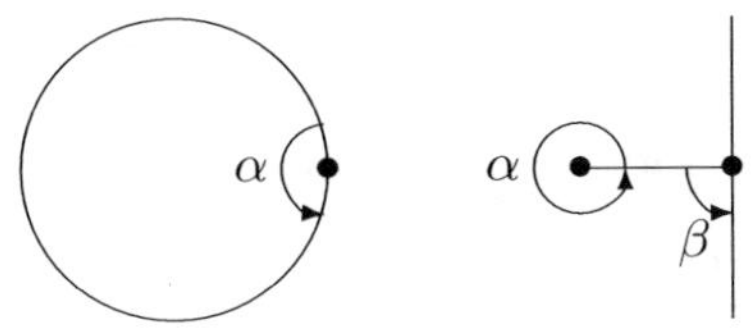

In the latter case $d\alpha = g(\alpha)\mathbb{1}_a$ for some polynomial g. This polynomial cannot have a constant term: let β be the arrow that follows α in the face $f(\alpha)$. Then α and β turn around different marked points so

$$0 = d(\beta\alpha) = (d\beta)\alpha \pm \beta g(\alpha) = \pm g_0\beta.$$

So $d\alpha = (g(\alpha)\alpha^{-1})\alpha = zE_\alpha(\alpha)$ for $z = g(\ell_m)\ell_m^{-1}$.

(2) If $a = h(\alpha) \neq t(\alpha)$ and a is a loop then there is an angle path $\kappa : a \to a$ following α that does not form a full turn around m.

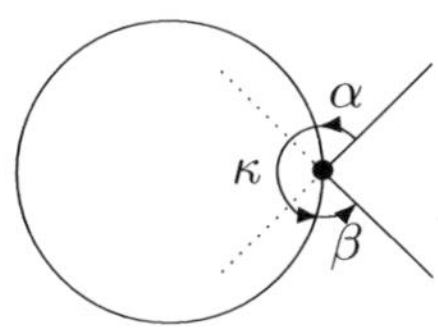

In that case $d\alpha$ can contain a term of the form $\kappa\alpha$. Let β be the angle arrow that directly follows κ. Note that $\beta \neq \alpha$ otherwise the arc a forms a monogon. Because β and α are on different ends of the arc a, we have that

$$d(\beta\alpha) = \beta(c\kappa\alpha + \dots) \pm (d\beta)\alpha = 0.$$

Therefore $d(\beta)$ contains $c\beta\kappa$. The commutator $[c\kappa, -]$ is only nonzero for the angle arrows α and β. By substracting this from d we can make these terms disappear.

(3) If $a = t(\alpha) \neq h(\alpha)$ and a is a loop we can do a similar reasoning as above.

(4) If $a = t(\alpha) \neq h(\alpha) = b$ and a, b are both loops then $d\alpha$ can contain a term of the form $\kappa_1\alpha\kappa_2$. If κ_1 is not a full turn then let β be the angle arrow such that $\beta\kappa_1\alpha\kappa_2 \neq 0$. Because $\beta\alpha \neq 0$ we get

$$d(\beta\alpha) = \beta(c\kappa_1\alpha\kappa_2 + \dots) \pm (d\beta)\alpha = 0.$$

So κ_2 must end in α but then κ_2 is a full turn, so $\kappa_1\alpha\kappa_2 = \ell_m^r\kappa_1\alpha$, and we can make this term disappear with $[\ell_m^r\kappa_1, -]$. If κ_1 is a full turn we can do the same.

(5) If neither $h(\alpha)$ or $t(\alpha)$ are loops then every path in $d(\alpha)$ is of the form $\ell_m^r\alpha$.

From the discussion above we see that we can remove all terms that are not of the form $\ell_m^r\alpha$ by subtracting commutators. □

REMARK 2.8. Lemma 2.6 holds more generally for marked surfaces with boundary, but lemma 2.7 does not hold in this generality (think of the cylinder with one marked point on each boundary circle, this has an arc collection whose gentle algebra is the Kronecker quiver $\circ \implies \circ$). However if we impose the [NL2] condition the lemma still holds.

We end this section with a nice interpretation of Koszul duality in this setting. If $\mathcal{A}$ is an arc collection on (S, M), choose a set of points $F \subset S$ in the centers of the faces. For each arc a draw a perpendicular arc $a^\perp$ that connects the centers of the faces adjacent to a and points inside the left face. The **dual arc collection** $\mathcal{A}^\perp = \{a^\perp \mid a \in \mathcal{A}\}$ forms an arc collection for (S, F).

THEOREM 2.9. The Koszul dual of the gentle algebra is a gentle algebra of the dual arc collection:

$$(\mathrm{Gtl}_{\mathcal{A}})^! \cong \mathrm{Gtl}_{\mathcal{A}^\perp}.$$

PROOF. From Bardzell [**3**] and [**5**] we know that $A = \mathrm{Gtl}_{\mathcal{A}}$ has a bimodule resolution $B^\bullet$ spanned by $A \otimes_{\Bbbk} b \otimes_{\Bbbk} A$ where $b = \beta_l \beta_{l-1} \cdots \beta_1 \in \mathbb{C}Q_{\mathcal{A}}$ is a path of angle arrows such that all products $\beta_i \beta_{i-1}$ are zero. In other words paths that turn around faces. The maps between the terms have the following form

$$1 \otimes b_k \dots b_1 \otimes 1 \mapsto b_k \otimes b_{k-1} \dots b_1 \otimes 1 - (-1)^k 1 \otimes b_k \dots b_2 \otimes b_1$$

Therefore the Koszul dual $\mathrm{Ext}^\bullet_A(\Bbbk, \Bbbk) = \mathrm{H\,Hom}(B^\bullet, {}_L\Bbbk \otimes \Bbbk_R)$ is spanned by the dual basis $b^\vee$. The element $b^\vee$ corresponds to the (right) module extension

$$0 \leftarrow \mathbb{C}\mathbb{1}_{t(\beta_l)} \xleftarrow{\beta_l} \mathbb{C}\mathbb{1}_{h(\beta_l)} \leftarrow \dots \leftarrow \mathbb{C}\mathbb{1}_{t(\beta_1)} \xleftarrow{\beta_1} \mathbb{C}\mathbb{1}_{h(\beta_1)} \leftarrow 0.$$

Note we use right modules because in that way α and $\alpha^\vee$ run in opposite directions. Stitching together two of these module extensions shows that the ext product matches the (reverse) concatenation of paths if the paths turn around the same face. All other products are zero. In particular the ext algebra is spanned by dual angle arrows and $\alpha^\vee \beta^\vee = 0$ if $m(\alpha) = m(\beta)$. Finally if α is an angle arrow then the dual $\alpha^\vee$ in the Koszul dual has degree $1 - |\alpha|$. All this can be realised geometrically by considering the dual angles as angles between perpendicular arcs.

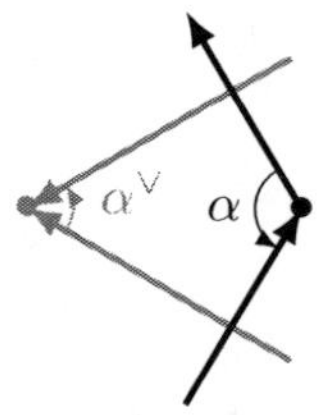

□

REMARK 2.10. In the theorem above we are referring to the classical Koszul dual [**4**] where the gentle algebra is considered as a graded algebra graded by path length. A similar statement holds if we work with Koszul duality for A_∞-algebras but then we need to work with the gentle algebras completed by path length.

2.2. Orbigons. Now let R be a complete local commutative ring over $\mathbb{C}$ with maximal ideal R^+ and residue field $R/R^+ = \mathbb{C}$. For each element $r \in Z(\mathrm{Gtl}_{\mathcal{A}}) \widehat{\otimes} R^+$ we will define a curved R-linear A_∞-structure over $\mathrm{Gtl}_{\mathcal{A}} \widehat{\otimes} R$. When tensored with R/R^+ the result will be the gentle A_∞-algebra as defined in [**5, 9**], so our construction gives a family of deformations of the latter.

To describe these deformations, we need the notion of an orbigon. Morally this is a branched cover from a disk to the surface such that the boundary is mapped to arcs and the branch points are mapped to marked points. We can construct such branched covers by joining faces together. We will define this concept combinatorially and inductively in two steps.

DEFINITION 2.11. **Tree-gons** are certain sequences of angles up to cyclic permutation. The basic tree-gon come from faces: $(\alpha_k, \dots, \alpha_1)$ is a tree-gon if they form the consecutive angles of a face such that $h(\alpha_i) = t(\alpha_{i+1})$.

If $(\alpha_k, \dots, \alpha_1)$ and $(\beta_l, \dots, \beta_1)$ are tree-gons such that $\alpha_1\beta_l \neq 0$ and $\alpha_k\beta_1 \neq 0$ in $\mathrm{Gtl}_\mathcal{A}$ then we define a new tree-gon

$$(\beta_1\alpha_k, \alpha_{k-1}, \dots, \alpha_2, \alpha_1\beta_l, \dots, \beta_2)$$

Geometrically this operation stitches the two tree-gons together over the common arc $h(\beta_l) = t(\alpha_1)$.

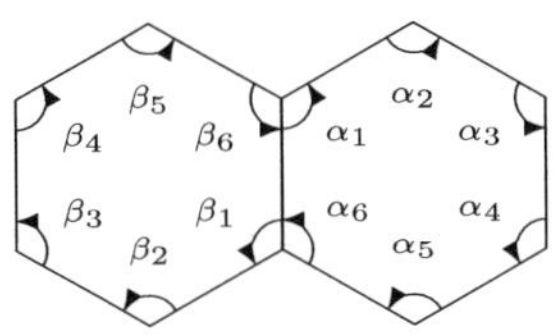

REMARK 2.12. From this definition one can deduce that tree-gons are sequences of internal angles of faces stitched together in a tree-like way. This tree, whose nodes are the faces and whose edges are the stitched arcs, can be reconstructed solely from the angle sequence.

Let $(\gamma_u, \dots, \gamma_1)$ be the sequence of all the indecomposable angle arrows in the tree-gon. Define a permutation σ on $\{1, \dots, u\}$ as follows: for every i there will be a unique shortest nontrivial path $\gamma_j \dots \gamma_{i+1}$ with $i + u \leq j < i$ that lifts to a contractible loop in the universal cover of $S \setminus M$. Set $\sigma(i) = j \mod u$. One can easily show that for a tree-gon we have that $\sigma^2 = \mathrm{Id}$ and $\sigma(i) = i$ if and only if $h(\gamma_i)$ is an arc on the boundary of the tree-gon. This means that the internal arcs are in 1–1 correspondence with the 2-cycles in σ. The nodes of the tree on the other hand correspond to the orbits of the permutation $\sigma \circ (u \dots 1)$. Indeed this permutation follows the angles and crosses over if the arc is internal, and therefore it cycles around the faces. An edge and a node are incident if their orbits intersect.

Now we will allow to fold together two consecutive arcs on the boundary of a tree-gon that are identical. The result is a disk-like shape with internal marked points, so some of the angles sit in the interior of the disk. We will use square brackets to denote what is interior. In general we get sequences of angles separated by square brackets and commas such as $(\alpha, \beta[\gamma[\delta]\epsilon]\zeta[\eta])$. The **reduced sequence** cuts out anything that is between square brackets: e.g. $(\alpha, \beta\zeta)$. Again we work inductively.

DEFINITION 2.13. An **orbigon** is a bracketed cyclic sequence of angles. The basic orbigons are tree-gons without any brackets and if $(\dots, U, \dots)$ is an orbigon for which the reduced sequence of U is an angle that turns r full turns around a marked point m then $(\dots [U] \dots)$ is also an orbigon. The type of the orbigon is a multiset containing all pairs $p = (m, r)$ that were needed to introduce the brackets. Such a pair is also called an **orbifold point**.

Different tree-gons can be folded together to form the same disk, therefore we also need to impose an equivalence relation on the orbigons. Suppose $(\alpha_k, \dots, \alpha_1)$ and $(\beta_l, \dots, \beta_1)$ are tree-gons and

$$(\beta_1\alpha_k|\dots|\alpha_j[\underbrace{\alpha_{j-1}|\dots|\alpha_1}_{U_1}\underbrace{\beta_l|\dots|\beta_{i+1}}_{U_2}]\beta_i|\dots|\beta_2)$$

is an orbigon, where the $|$-separators can be commas or brackets. Then the shift rule shifts the position of a piece between brackets and reverses the order of the two parts U_1, U_2 that are in thee different tree-gons as follows

$$(\beta_1[\underbrace{\beta_l|\dots|\beta_{i+1}}_{U_2}\underbrace{\alpha_{j-1}|\dots|\alpha_1}_{U_1}]\alpha_k|\dots|\alpha_j\beta_i|\dots|\beta_2)$$

where brackets in U_1 that are paired up with a bracket U_2 and vice versa are flipped, for the formula to make sense.

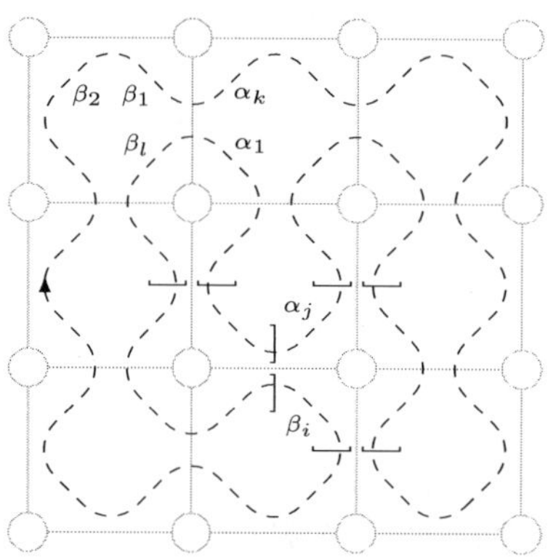

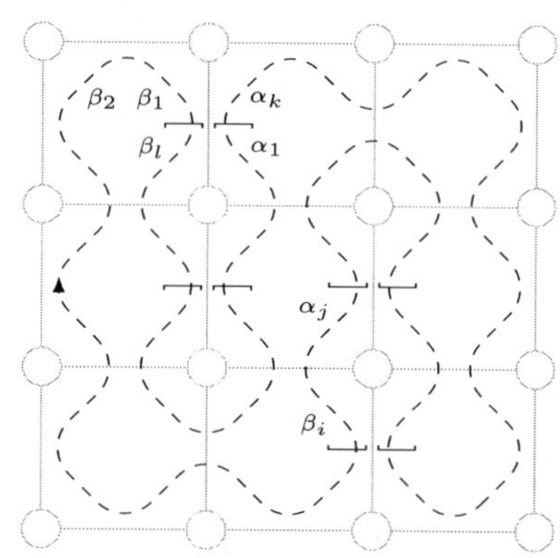

The **shift rule** generates an equivalence relation on the orbigons that preserves the type and the reduced sequence.

REMARK 2.14. Every tree-gon corresponds to a tree with nodes labeled by faces and edges labeled by arcs. Each folding operation adds another edge to the graph corresponding to the arc $a = h(U)$. In this way we end up with a graph of which this tree is a spanning tree. The graph, which we will sometimes refer to as the **face graph**, is planar. It divides the plane into regions which can be labeled by a pair (m, r) from the type multiset. Clearly the equivalence relation changes the underlying spanning tree while keeping the face graph the same. For the example above the face graph is a grid with 3×3 nodes and the spanning trees are indicated in bold.

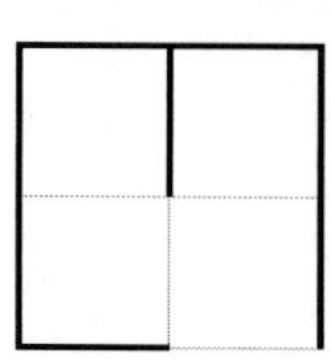

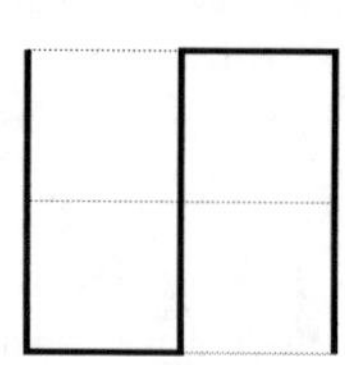

The spanning tree of the face graph completely determines the unreduced sequences of the orbigon because it is the sequence of angles that runs around the tree. It also determines the brackets: each time you cross an edge of the face graph that is not in the spanning tree you open or close a bracket depending on whether you crossed it the first or the second time.

LEMMA 2.15. Two orbigons are equivalent if and only if their face graphs are equivalent as labeled planar graphs.

PROOF. By construction the shift rule does not change the underlying face graph. Now suppose that the two orbigons have isomorphic face graphs, then we have to show that we can move from one spanning tree T_1 to another spanning tree T_2 via the shift rules. If a is an edge (or dually an arc) in T_1 not contained in T_2 then if we remove a, T_1 will split in two parts (or dually tree-gons) denote the angles in the first tree-gon by $\alpha_k, \dots, \alpha_1$ and those in the second tree-gon by $\beta_l, \dots, \beta_1$ such that $a = t(\beta_1) = h(\alpha_k)$.

Because T_2 is connected it must contain an arc b that connects the two parts of $T_1 \setminus \{a\}$. Now find the indices i, j such that $b = h(\beta_i) = t(\alpha_j)$ and perform the corresponding shift rule. The result will be an orbigon with a spanning tree equal to $T_1 \setminus \{a\} \cup \{b\}$, which is one edge closer to T_2. Keep repeating this procedure until the spanning tree is T_2. □

REMARK 2.16. Note that we can also stitch orbigons together over a common arc in their reduced sequences by stitching the underlying tree-gons and transferring the brackets to the stiched tree-gon. The face graph of the new orbigon consists of the two graphs of the smaller orbigons joined together by one edge labeled by the common arc. Moreover because the common arc must be contained in all spanning trees, this implies that the small orbigons are uniquely determined by the big one.

LEMMA 2.17. If $(\alpha_k, \dots, \xi\eta, \dots \alpha_1)$ is the reduced sequence of an orbigon and ξ, η are nontrivial angle paths then there are two possibilities:

A the orbigon is stitched together of two smaller orbigons with reduced sequences $(\beta_r, \dots, \xi)$ and $(\gamma_r, \dots, \eta)$ such that $\gamma_r\beta_r = \alpha_r$ for some r.

B the orbigon is folded together from an orbigon with reduced sequence $(\alpha_k, \dots, \xi, \ell^r_m h(\eta), \eta, \dots, \alpha_1)$

in both cases these orbigons are unique.

PROOF. Look at the face graph of the orbigon. The arc $t(\xi) = h(\eta)$ will correspond to an edge in this graph. If the graph remains connected after removing this edge then this new graph will be the graph of an orbigon that can be folded to the old orbigon and we are in situation B. Otherwise the old orbigon is stitched together from two smaller orbigons and we are in situation A. □

LEMMA 2.18. For a given type and reduced sequence there are at most a finite number of orbigons (up to equivalence).

PROOF. Each bracketing changes two commas in brackets and introduces an extra pair in the multiset of the type. Therefore the length of the unreduced sequence (the number of commas and brackets) is 2 times the size of the multiset longer than the unreduced sequence (the number of commas). □

LEMMA 2.19. If the arc collection satisfies

- the no monogon or digon condition [NMD] then all tree-gons have length at least 3.

- the no loops or two-cycles condition [NL2] then all orbigons have reduced sequences of length at least 3.

PROOF. If [NMD] holds then all faces are at least 3-gons. Gluing an n-gon to an m-gon results in an $n+m-2$-gon, and $n+m-2>2$ if $n,m>2$. Furthermore, an orbigon with a reduced sequence of length one or two would give rise to a loop or two-cycle of arcs. □

2.3. A_∞-structures on the gentle algebra. Fix a commutative ring R. Recall that if A is a $\mathbb{Z}$ or $\mathbb{Z}/2\mathbb{Z}$-graded projective R-module then a curved A_∞-structure is a collection of R-linear maps

$$\mu^k : A^{\otimes_R i} \to A$$

of degree $2-k$, satisfying the curved A_∞-axioms:

$$\sum_{k\geq l\geq m\geq 0} (-1)^{\|x_m\|+\dots+\|x_1\|}\mu(x_k,\dots,\mu(x_l,\dots,x_{m-1}),x_m,\dots,x_1)=0.$$

Here $\|x\|$ is shorthand for the **shifted degree**: $\|x\| = |x|-1$. If $l=m$ then we interpret the middle $\mu()$ as the element $\mu^0 := \mu^0(1) \in A$. This is called the curvature and if it is zero the structure is called uncurved.

If A is an algebra, we say that μ is an **extension of the product** if $a\cdot b = (-1)^{\deg b}\mu^2(a,b)$. If $A = \mathbb{C}Q/I\otimes R$ comes from a path algebra of a quiver we will take tensor products over $\Bbbk\otimes R$ instead of over R and we ask that the vertex idempotents $\mathbb{1}_v$ are **strict**: all products $\mu^{\neq 2}$ for which one of the entries is such an idempotent are zero.

Now let R be a local nilpotent or complete ring over $\mathbb{C}$ with maximal ideal R^+. Fix an element $r\in Z(\mathrm{Gtl}_\mathcal{A})\widehat{\otimes}R^+$, where $\widehat{\otimes}$ denotes the R^+-adic completed tensor product. We will write this element as

$$r_0 + \sum_m r_m(\ell_m)$$

where $r_0\in R^+$, the $r_m(t)\in R^+[\![t]\!]$ are R^+-valued power series without constant term such that the reductions $r_m \in (R^+/(R^+)^k)[\![t]\!]$ are actually polynomials for every $k\geq 1$. We will write the jth coefficient of $r_m(t)$ as r_p where $p=(m,j)$ is viewed as an orbifold point. We will also write ℓ_p as shorthand for ℓ_m^j. We are now ready to define a family of extensions of the gentle algebra.

DEFINITION 2.20. For any $r\in Z(\mathrm{Gtl}_\mathcal{A})\widehat{\otimes}R^+$ we define ${}^r\mu_\bullet$ on $\mathrm{Gtl}_\mathcal{A}\widehat{\otimes}R$ as follows

- For each orbifold point $p=(m,j)$ we define a nullary product
$${}^r\mu_p^0 : \Bbbk \to \mathrm{Gtl}_\mathcal{A}\widehat{\otimes}R^+ : 1 \mapsto r_p\ell_m^j$$
- For each orbigon ψ with reduced sequence $(\alpha_k,\dots,\alpha_1)$ of length k we define a k-ary product ${}^r\mu_\psi$. If β and γ are angles such that $\beta\alpha_{k+i}\neq 0$ and $\alpha_i\gamma\neq 0$ then we set
$${}^r\mu_\psi(\beta\alpha_{k+i},\dots,\alpha_i) = r_\psi\beta \text{ and } {}^r\mu_\psi(\alpha_{k+i},\dots,\alpha_i\gamma) = (-1)^{|\gamma|}r_\psi\gamma.$$
where r_ψ is the product of all r_p with p running over the orbifold points in the type of ψ. If the type is empty (i.e. ψ is a tree-gon) then we put $r_\psi = 1$. All other ${}^r\mu_\psi(\beta_1,\dots,\beta_n)$ where the β_i are angles are zero.

The total product is the sum of all these products together with an overall curvature coming from r_0:

$$\begin{aligned}
{}^r\mu^0 &= r_0\mu_{\mathbb{1}}^0 + \sum_p {}^r\mu_p \\
{}^r\mu^1 &= \sum_{|\psi|=1} {}^r\mu_\psi \\
{}^r\mu^2 &= {}^r\mu^2_{\text{ord}} + \sum_{|\psi|=2} {}^r\mu_\psi \\
{}^r\mu^k &= \sum_{|\psi|=k} {}^r\mu_\psi.
\end{aligned}$$

where $\mu_{\mathbb{1}}^0$ is the nullary product with $\mu_{\mathbb{1}}^0(1) = 1$ and ${}^r\mu^2_{\text{ord}}$ the R-linear extension of the ordinary product in $\text{Gtl}_\mathcal{A}$. This is well defined because by lemma 2.18 there are only a finite number of k-orbigons for each type and R is nilpotent or complete.

REMARK 2.21. The condition [NMD] implies that there are no tree-gons with length 1 or 2. This means that ${}^r\mu^1 = 0 \mod R^+$ and ${}^r\mu^2 = \mu^2_{\text{Gtl}_\mathcal{A}} \mod R^+$. The condition [NL2] implies that there are no orbigons with length 1 or 2. This means that ${}^r\mu^1 = 0$ and ${}^r\mu^2 = {}^r\mu^2_{\text{ord}}$.

REMARK 2.22. If $r = 0$, the only ${}^r\mu_\psi$ that contribute are those coming from tree-gons and each contributes with a factor $r_\psi = 1$. These are precisely those that correspond to immersed disks that do not cover marked points with their interior. Therefore the definition is equivalent to the definition in [**9**]. In [**5**] the definition of the higher product is given inductively, with an induction step that is analogous to the induction step we used to define a tree-gon. Hence the definition is also equivalent to the definition in [**5**]. From both papers we can conclude that ${}^r\mu$ is an uncurved A_∞-structure if $r = 0$. If $r \neq 0$ then ${}^r\mu$ will be a curved deformation of this uncurved A_∞-structure.

REMARK 2.23. Each higher product removes a sequence of angles that forms an orbigon. If the angle that remains comes from the first entry, we will call that the **front of the product**, if it comes from the last we call it the **back of the product**.

PROPOSITION 2.24. If [NL2] holds then ${}^r\mu$ is a well defined curved A_∞-structure on $\text{Gtl}_\mathcal{A}\widehat{\otimes}R$, which is strict over $\Bbbk = \mathbb{C}^\mathcal{A}$.

PROOF. We need to check the curved A_∞-axioms. The first two axioms

$${}^r\mu^1({}^r\mu^0(1)) = 0,\ {}^r\mu^1({}^r\mu^1(\alpha) + {}^r\mu^2({}^r\mu^0(1),\alpha) - {}^r\mu^2(\alpha,{}^r\mu^0(1)) = 0$$

follow easily from the facts that ${}^r\mu^0(1)$ is a central element and ${}^r\mu^1 = 0$. The third axiom holds because ${}^r\mu_{\text{ord}}$ is associative and there are no ${}^r\mu_3(\dots,{}^r\mu_0(1),\dots)$ terms as this would imply an orbigon with reduced sequence $(\dots[\ell^k_m]\dots)$ of length 1. This contradicts lemma 2.19.

To show the higher axioms, first note that by construction all the ${}^r\mu$ are strict over $\Bbbk$, so we only need to check the axioms when all entries are nontrivial paths. Fix an angle sequence $(\gamma_r,\dots,\gamma_1)$ of length ≥ 4 and assume that there is a nonzero double product

$${}^r\mu_u(\gamma_{s+t-1},\dots,\gamma_{i+t},{}^r\mu_v(\gamma_{i+t-1},\dots,\gamma_i),\gamma_{i-1},\dots,\gamma_1)$$

where ${}^r\mu_u$ can be a ${}^r\mu_{\mathrm{ord}}$ (if $s=2$) or a ${}^r\mu_\psi$ if $s>2$, and ${}^r\mu_v$ can be a ${}^r\mu_p$ (if $t=0$), a ${}^r\mu_{\mathrm{ord}}$ (if $t=2$) or a ${}^r\mu_\psi$ (if $t>2$). Such a product is uniquely characterized by the triple (u,v,i).

Each nonzero double product is a path multiplied with a factor $\pm r_u r_v$ (where we use the convention that $r_{\mathrm{ord}}=1$). We will show there is a unique other triple with a nonzero double product cancel that cancels it.

To do this, we will go through all possible triples (u,v,i) systematically, making a distinction between whether i is 1, s or somewhere in the middle. For each case we draw a diagram of the possible situations and every situation will occur exactly twice (see figure 2.1). Note that most diagrams have two versions depending on whether the (only/outer) higher product has a front or a back. We will only consider the latter cases. The diagrams are named after the operations of type A, B from lemma 2.17 that can be performed on the orbigons to stitch or fold them together.

(1) (ψ,p,i)
- If $i=1$ then ℓ_p contains the back of ${}^r\mu_\psi$ and we are in situation $O1$. We can compensate this term with a term of a similar type but where ℓ_p contains the front:
$${}^r\mu_\psi(\gamma_s,\dots,\gamma_1,{}^r\mu_p) \text{ cancels with } {}^r\mu_\psi({}^r\mu_p,\gamma_s,\dots,\gamma_1)$$
- If $1<i<s$ then ψ has ℓ_p as one of its angles and we are in situation $B1$. We can fold the orbigon together to form a new orbigon ψ' with $r_{\psi'}=r_\psi r_p$. The other term that cancels it is of the form $(\psi',\mathrm{ord},i-1)$.
- If $i=s$ and ℓ_p comes before γ_s then either γ_s is shorter than the back of ${}^r\mu_\psi$ or longer. If it is shorter, we are in situation $B2$. If it is longer than the back, the arc $t(\gamma_1)$ will lie inside ψ. In that case we distinguish type BA if $t(\gamma_1)$ cuts ψ in two pieces, or BB if it opens a second fold at a marked point q. The former is compensated by a terms of type (ψ_1,ψ_2,i), while the latter is compensated by a term (ψ_1,q,i). In both cases ψ_1 includes the orbifold point p and has a front)

(2) (ψ,ord,i)
- If $1<i\le s$ then the inner product ${}^r\mu^2(\gamma_{i+1},\gamma_i)$ is an angle from the orbigon ψ then $h(\gamma_i)$ either opens a fold of ψ ($B1$) or it cuts the orbigon in two ($A1L, A1R, A1M$). The first is compensated by a term of type $(\psi',p,i+1)$, while the latter three by a term of type (ψ_1,ψ_2,j).
- If $i=1$ and γ_1 is part of the back of ${}^r\mu_\psi$ then $h(\gamma_1)$ will cut the back in two. This is situation $O2$ and it is compensated by a term of the form $(\mathrm{ord},\psi,2)$.
If the back of ${}^r\mu_\psi$ is part of γ_1 then $h(\gamma_1)$ either opens a fold of ψ or it cuts the orbigon in two. Just like when $i\neq 1$ above, we are in cases ($B1, A1L, A1R, A1M$) but now with γ_1 equal to the top of the hexagon instead of the bottom. They are also compensated by terms of type $(\psi',p,i+1)$ and (ψ_1,ψ_2,j).

(3) (ord,ψ,i)
- If $i=1$ and ${}^r\mu^2(\gamma_{t+1},{}^r\mu_\psi(\dots,\gamma_1))$ puts something in front of the back of ${}^r\mu_\psi$, three things can happen. If γ_{t+1} is shorter than the first

angle of ψ then $h(\gamma_{t+1})$ either cuts ψ in two ($A2$) or opens a fold of ψ ($B2$). These cases are compensated by terms of the form (ψ_1, ψ_2, j) and (ψ', p, t).
If γ_{t+1} is longer than the first angle we are in situation $O3$ and can compensate it with a term of the same type but with a head

$$^r\mu^2(^r\mu_\psi(\gamma_{t+1}, \dots, \gamma_2), \gamma_1).$$

- If $i = 2$ then $^r\mu^2(^r\mu_\psi(\dots, \gamma_2), \gamma_1)$ adds something to the back of $^r\mu_\psi$ and we are in situation $O2$.

(4) (ψ_1, ψ_2, i)
- If $i = 1$ and $^r\mu_{\psi_2}$ has a back then we are in $A1M$. If $^r\mu_{\psi_2}$ has a front we are in situation $O4$. Then we can commute the order of ψ_1 and ψ_2, this gives two terms that cancel:

$$^r\mu_{\psi_1}(\gamma_{s+t-1}, \dots, {^r\mu_{\psi_2}}(\dots, \gamma_1)) \text{ and } ^r\mu_{\psi_2}(^r\mu_{\psi_1}(\gamma_{s+t-1}, \dots), \dots, \gamma_1).$$

- If $1 < i < s$ and $^r\mu_{\psi_2}$ has a back equal to an angle of ψ_1 we are in situation $A1L$, while if it has a front equal to an angle, we are in situation $A1R$.
- If $i = s$ and $^r\mu_{\psi_2}$ has a front we are in $A1R$. If it has a back then it depends on the back of $^r\mu_{\psi_1}$. If it is at least as long as the last angle of ψ_2 we are in situation $A2$. Otherwise $t(\gamma_1)$ will cut ψ_2 in two pieces (situation AA) or it will open a fold (AB). The former is compensated by a term of type (ψ_1', ψ_2', j) while the latter is compensated by a term (ψ_1', p, j). In both cases the outer product now has a front. Note that type AB and BA only seem to occur once in the list but this is because they are canceled by terms with a front and these are not in the list.

□

REMARK 2.25. The proof is analogous to the one in [**9**], but because of the internal orbifold points extra cases needed to be considered. These are all the cases with a B in their label. The proof can also be extended to the case where [NL2] does not hold but this will include a lot more extra cases.

REMARK 2.26. It is also possible to weight the faces in the orbigons using elements in R^+. Choose an element $s_{(f,i)} \in R^+$ for each face f and $i \in \mathbb{N}$. The weight of a tree-gon is then defined inductively by giving $(\alpha_{ik}, \dots, \alpha_1)$ weight $s_{(f,i)}$ if the angle sequence turns i times around f (note that this definition allows for orbifold faces as well). If you stitch two tree-gons together their weight is defined as the product. The folding operation adds an additional factor $r_{(m,j)}$ in the same way as before.

This gives a family of products $^{r,s}\mu^\bullet$ on $\mathrm{Gtl}_\mathcal{A} \widehat{\otimes} R$ that depends on two parameters sets $r : M \times \mathbb{N} \to R^+$ and $s : M \times \mathbb{N} \to R^+$. Note that the curvature only depends on r not on s.

This same parameter set can be used for the Koszul dual $\mathrm{Gtl}_{\mathcal{A}^\perp}$ but now the role of r and s are reversed because for the Koszul dual the role of the marked points and the faces are swapped. In this case s will give rise to curvature, while r only contributes to the higher products. Note however that if $s_{(f,j)}$ is nonzero for an infinite number of (f, j) we should go to the completed version $\widehat{\mathrm{Gtl}}_{\mathcal{A}^\perp}$ for the curvature to make sense.

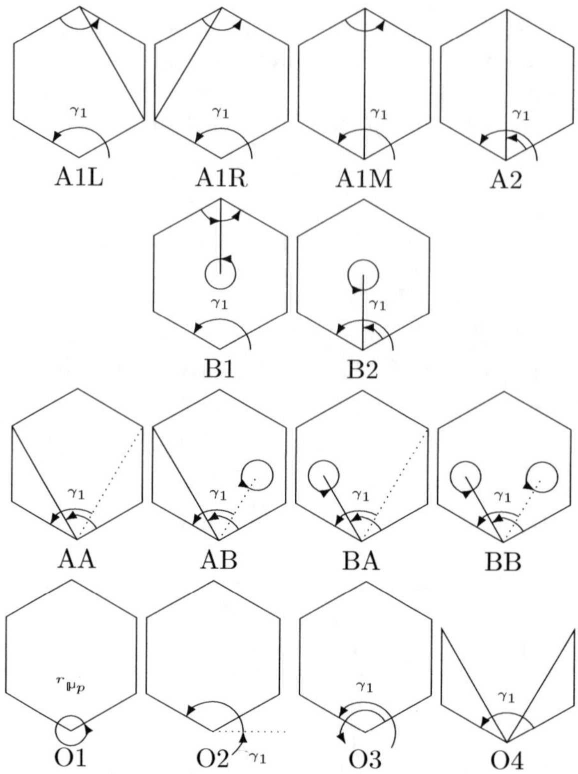

FIGURE 2.1. The possible diagrams that can occur in as double products

3. Hochschild cohomology of gentle A_∞-algebras

3.1. Definitions.

DEFINITION 3.1. If (A, μ) is a $\mathbb{Z}$- or $\mathbb{Z}_2$-graded A_∞-algebra over a semisimple algebra $\Bbbk$, we define the A_∞-Hochschild complex as

$$\mathrm{CC}^\bullet(A) = \mathrm{Hom}_\Bbbk(\bigoplus_{i\geq 0} A[1]^{\otimes_\Bbbk i}, A[1])$$

Every element $\nu \in \mathrm{CC}^j(A)$ can be seen as a collection of n-ary products ν^n of degree $1 + j - n$, one for each $n \in \mathbb{Z}_{\geq 0}$. We will call ν^n the nth component of ν.

REMARK 3.2. Note that the grading on this space is not the classical grading on Hochschild cohomology coming from the number of entries in the products, but the one coming from the degrees of the maps $A[1]^{\otimes_\Bbbk i} \to A[1]$. For this grading the product μ can be seen as an element of $\mathrm{CC}^1(A)$. To avoid confusion we will call this degree the **∞-degree** $\| \bullet \|$. If we work $\mathbb{Z}_2$-graded we will also refer to this grading as the **parity** and denote the homogeneous parts as $\mathrm{CC}^{\text{even}}(A)$ and $\mathrm{CC}^{\text{odd}}(A)$.

- On $\mathrm{CC}^\bullet(A)$ we have a bracket of degree 0:

$$\begin{aligned}[\kappa,\nu](a_r,\dots,a_1) &:= \sum_{0\le i\le j\le r}(-1)^{(\|a_1\|+\dots+\|a_i\|)\|\nu\|}\kappa(a_r,\dots,a_{j+1},\nu(a_j,\dots,a_{i+1}),a_i,\dots,a_1)\\ &- (-1)^{\|\nu\|\|\kappa\|+(\|a_1\|+\dots+\|a_i\|)\|\kappa\|}\sum_{0\le i\le j\le r}\nu(a_r,\dots,a_{j+1},\kappa(a_j,\dots,a_{i+1}),a_i,\dots,a_1).\end{aligned}$$

  The A_∞-axioms for the product μ, can be rephrased as $[\mu,\mu]=0$, which also means that $d=[\mu,-]$ is a differential of degree 1 and the triplet

$$(\mathrm{CC}^\bullet(A), d, [,])$$

  is a differential graded Lie algebra (DGLA). The solutions to the Maurer-Cartan equation

$$d\nu + \frac{1}{2}[\nu,\nu] = 0$$

  describe the deformations of μ as a curved A_∞-structure.
- There is also a second product on $\mathrm{CC}^\bullet(A)$: the cup product.

$$\begin{aligned}(\kappa\smile\nu)(a_r,\dots a_1) &:= \sum_{0\le i\le j\le u\le v\le r}(-1)^{\maltese}\mu(a_r,\dots,\kappa(a_v,\dots,a_{u+1}),\dots,\nu(a_j,\dots,a_{i+1}),\dots,a_1)\end{aligned}$$

  with $\maltese = (\|a_1\|+\dots+\|a_u\|)\|\kappa\| + (\|a_1\|+\dots+\|a_i\|)\|\nu\| + \|\nu\| + 1$. This product has degree 1 and together with d it satisfies the graded Leibniz rule. Therefore

$$(\mathrm{CC}^\bullet(A)[-1], d, \smile)$$

  is a differential graded algebra DGA and if we go to homology the triplet

$$(\mathrm{HH}^\bullet(A)[-1], \smile, [,])$$

  is a Gerstenhaber algebra.

REMARK 3.3. These two classical structures on Hochschild cohomology of ordinary ungraded algebras were analyzed first by Gerstenhaber [**8**]. Both definitions have been extended to A_∞-algebras and are well-known in the literature, see e.g. [**15**].

Mescher [**15**] uses a different sign convention for A_∞-categories. This means his definition of the cup product also has different signs. We have adapted the signs in order to suit the signs in our definition of A_∞-categories. In particular, the cup product as defined is graded symmetric, with respect to a degree shift of 1 on the Hochschild cohomology:

$$\nu\smile\eta = (-1)^{(\|\nu\|-1)(\|\eta\|-1)}\eta\smile\nu. \tag{3.1}$$

The unexpected shift by 1 in this sign rule is desired. In fact, this "shifted graded symmetry" renders the cup product truly graded symmetric with respect to the "traditional" grading of the Hochschild complex, which differs from the A_∞-grading precisely by one. In other words, if one regards an ordinary ungraded algebra and starts grading the Hochschild cohomology at zero (instead of minus one), the cup product becomes graded symmetric.

We have tried to arrange the signs such that the cup product together with the Gerstenhaber bracket turns $\mathrm{HH}(\mathrm{Gtl}_{\mathcal{A}})$ into a Gerstenhaber algebra with correct signs. In order to be a Gerstenhaber algebra, sign conventions of cup product and bracket need to be tuned to each other. The relevant compatiblity condition is the signed Leibniz rule which reads

$$[\nu, \eta \smile \omega] = [\nu, \eta] \smile \omega + (-1)^{\|\nu\|(\|\eta\|-1)} \eta \smile [\nu, \omega].$$

We have tried to arrange the signs with the Leibniz rule in mind, but we have not conducted the tedious checks. We have however chosen the signs in analogy with [**15**]. In particular, the signs are such that the cup product descends to Hochschild cohomology and is graded symmetric, the latter we did check by hand.

REMARK 3.4. If (A, μ) is a strictly unital A_∞-algebra over $\Bbbk$, i.e. $\mu(x_1, \ldots, x_n) = 0$ if $n \neq 2$ and at least one of the $x_i \in \Bbbk$, one can construct the **normalized Hochschild cochain complex** $\underline{\mathrm{CC}}^\bullet(A)$. This is the subspace of all cochains that evaluate to zero if at least one of the entries is in $\Bbbk$. This subspace is closed under all the operations above the embedding is a quasi-isomorphism [**10**]. The Maurer-Cartan equation for $\underline{\mathrm{CC}}^\bullet(A)$ classifies deformations of μ for which $\Bbbk$ remains strict.

We will now specialize to the case where $A = \mathrm{Gtl}_{\mathcal{A}}$ is the gentle algebra of an arc collection $\mathcal{A}$ equipped with $\mu = {}^r\upmu$ for $r = 0$. We have seen that the $\mathrm{Gtl}_{\mathcal{A}}$ has a natural $\mathbb{Z}_2$-grading, so the ∞-degree is a $\mathbb{Z}_2$-degree, which we refer to as the parity. It is possible to lift this $\mathbb{Z}_2$-grading to a $\mathbb{Z}$-grading. To do this we assign to each angle α a degree $\deg \alpha \in \mathbb{Z}$ with the same parity as $|\alpha|$. In order to make this degree compatible with μ we have to ensure that μ has ∞-degree 1. This happens only when the total degree of the angles in a k-gon is $k-2$. As every angle occurs only in one k-gon this can always be done but not canonically.

The gentle A_∞-algebra also has an extra grading coming from the relative first homology

$$G = H_1(S \setminus M, \mathcal{A}, \mathbb{Z}).$$

This group can be described in terms of the angles and faces

$$G = \frac{\bigoplus_{\alpha \in (Q_{\mathcal{A}})_1} \mathbb{Z}\alpha}{\langle \alpha_1 + \cdots + \alpha_k \mid (\alpha_1, \ldots, \alpha_k) \text{ is a face} \rangle}.$$

The group G comes with two natural maps

- $\pi : G \to \mathbb{Z}^{\mathcal{A}} : \alpha \mapsto h(\alpha) - t(\alpha)$
- $\iota : \mathbb{Z}^M \to G : m \mapsto \sum_{\alpha \in \ell_m} \alpha$

We have $\pi \circ \iota = 0$. The image of the π has corank 1, while the kernel of ι is $(1, \ldots, 1)$. Therefore if $\#M \geq 2$ the map ι is nonzero.

If we grade $\mathrm{Gtl}_{\mathcal{A}}$ by giving each angle its corresponding degree in G, it is clear that all products μ have degree $0 \in G$. We can transfer the G-grading to the Hochschild complex and it is easy to see that it contains only nonzero elements for degrees in $\operatorname{Ker} \pi$. Furthermore the natural operations $[,]$, $\smile$ and d all have G-degree 0.

REMARK 3.5. It is also clear from this construction that if deg and deg′ are two different $\mathbb{Z}$-lifts of the $\mathbb{Z}_2$-grading then their difference factors through G because deg and deg′ assign the same degree to a face. More precisely, the possible $\mathbb{Z}$-gradings form a $\mathrm{Hom}(G, 2\mathbb{Z})$-torsor.

LEMMA 3.6. If deg is a $\mathbb{Z}$-lift of the $\mathbb{Z}_2$-degree on $\mathrm{Gtl}_{\mathcal{A}}$ then

$$\sum_{m\in M} \deg \ell_m = 4 - 4g - 2\#M = 2\chi(S,M) < 0.$$

Here g is the genus of the marked surface (S, M) and $\chi(S, M)$ its Euler characteristic (which is negative by assumption).

PROOF.

$$\sum_{m\in M} \deg \ell_m = \sum_{\alpha\in(Q_{\mathcal{A}})_1} \deg \alpha = \sum_{f\in F} \deg f$$
$$= \sum_{f\in F}(2 - \#\{\alpha \in f\}) = 2\#F - 2\#\mathcal{A} = 2(2-2g) - 2\#M.$$

□

3.2. Reduction to the zeroth and first component. In this section we will show that the Hochschild cohomology class of a cocycle can be read off from the 0th and 1st component. In other words, if ν^0 and ν^1 are both zero then ν is a coboundary. Remark 3.4 allows us to restrict to cochains which belong to the normalized Hochschild cohomology. For such classes we will construct an $\epsilon \in \underline{\mathrm{CC}}^\bullet(A)$ such that $\nu - d\epsilon$ evaluates zero on all sequences $(\beta_k, \dots, \beta_1)$ where the β_i are nontrivial angle paths and $t(\beta_i) = h(\beta_{i+1})$. We will often refer to the procedure of adding a coboundary as **gauging**.

DEFINITION 3.7. Let $\beta = (\beta_k, \dots, \beta_1)$ be a sequence of nontrivial angles with $h(\beta_i) = t(\beta_{i+1})$ for $i < k$ (we do not impose that it cycles so $h(\beta_i)$, $t(\beta_1$ can be different).

- We call β **elementary** if all β_i are indecomposable angles (angle arrows).
- An index i is called a **contractible index** if $\beta_{i+1}\beta_i \neq 0$.
- Denote the set of contractible indices by $\mathrm{Contr}(\beta)$ and if $S \subset \mathrm{Contr}(\beta)$ then the **contracted sequence** β^S will be the sequence where consecutive angles separated by a contractible index are multiplied together.
- Every angle sequence is of the form β^S where β is elementary and S a set of contractible indices. Moreover β and S are uniquely determined. Two sequences are of the **same type** if they are contractions of the same elementary sequence.
- The **total length** of a sequence will be the length of its underlying indecomposable sequence.
- For an elementary sequence we have that $\mathrm{Contr}(\beta) = \{\}$ if and only if it consists of consecutive angles of a polygonal face. Therefore we will call such sequences **polygon sequences**.

LEMMA 3.8. Let $\nu \in \mathrm{Ker}(d) \subset \underline{\mathrm{CC}}^\bullet(\mathrm{Gtl}_{\mathcal{A}})$ with (A) $\nu^0 = \nu^1 = 0$. Then ν can be gauged to satisfy additionally that (B) $\nu^2(\alpha, \beta) = 0$ for all pairs of angle paths with $\alpha\beta \neq 0$. During the gauging procedure, the value of ν does not change on polygon sequences.

PROOF. We will construct an $\epsilon = \epsilon^1$ such that $\nu^2(\alpha, \beta) = (d\epsilon)^2(\alpha, \beta)$ if $\alpha\beta \neq 0$. Set $\epsilon^1(\alpha) = 0$ for every indecomposable angle α. Now define inductively ϵ^1 for decomposable angles (angle paths) by the rule

$$\epsilon^1(\alpha\beta) = \epsilon^1(\alpha)\beta + \alpha\epsilon^1(\beta) - (-1)^{|\beta|}\nu^2(\alpha, \beta). \tag{3.2}$$

Let us check that extending ϵ^1 according to this rule is well-defined, that is, $\epsilon^1((\alpha\beta)\gamma) = \epsilon^1(\alpha(\beta\gamma))$. Indeed,

$$\begin{aligned}
&\big(\epsilon^1(\alpha\beta)\gamma+\alpha\beta\epsilon^1(\gamma)-(-1)^{|\gamma|}\nu^2(\alpha\beta,\gamma)\big)-\big(\epsilon^1(\alpha)\beta\gamma+\alpha\epsilon^1(\beta\gamma)-(-1)^{|\beta\gamma|}\nu^2(\alpha,\beta\gamma)\big)\\
&= \epsilon^1(\alpha)\beta\gamma + \alpha\epsilon^1(\beta)\gamma - (-1)^{|\beta|}\nu^2(\alpha,\beta)\gamma + \alpha\beta\epsilon^1(\gamma) - (-1)^{|\gamma|}\nu^2(\alpha\beta,\gamma)\\
&\quad - \epsilon^1(\alpha)\beta\gamma - \alpha\epsilon^1(\beta)\gamma - \alpha\beta\epsilon^1(\gamma) + (-1)^{|\gamma|}\alpha\nu^2(\beta,\gamma) + (-1)^{|\beta\gamma|}\nu^2(\alpha,\beta\gamma)\\
&= -(-1)^{|\beta|}\nu^2(\alpha,\beta)\gamma-(-1)^{|\gamma|}\nu^2(\alpha\beta,\gamma)+(-1)^{|\gamma|}\alpha\nu^2(\beta,\gamma)+(-1)^{|\beta\gamma|}\nu^2(\alpha,\beta\gamma)\\
&= (-1)^{|\beta|}(d\nu)(\alpha,\beta,\gamma) = 0.
\end{aligned}$$

In the final row we have used that $\nu^0 = \nu^1 = 0$. We conclude that for angles α, β with $\alpha\beta \neq 0$ we have

$$(d\epsilon)(\alpha,\beta) = (-1)^{|\beta|}\epsilon^1(\alpha)\beta + (-1)^{|\beta|}\alpha\epsilon^1(\beta) - (-1)^{|\beta|}\epsilon^1(\alpha\beta) = \nu^2(\alpha,\beta)$$

as desired. Furthermore $(d\epsilon)^0 = (d\epsilon)^1 = 0$ and if $(\alpha_k,\dots,\alpha_1)$ is a polygon sequence, then

$$(d\epsilon)(\alpha_k,\dots,\alpha_1) = \epsilon(\mu(\alpha_k,\dots,\alpha_1)) + \mu(\dots,\epsilon(\alpha_i),\dots) = 0.$$

because all α_i are indecomposable so $\epsilon(\alpha_i) = 0$. □

LEMMA 3.9. *Let $\nu \in \mathrm{Ker}(d) \subset \underline{\mathrm{CC}}^\bullet(\mathrm{Gtl}_{\mathcal{A}})$ with (A) $\nu^0 = \nu^1 = 0$ and (B) $\nu^2(\alpha,\beta) = 0$ if $\alpha\beta \neq 0$. Then ν can be gauged to (C) evaluate zero on polygonal sequences without affecting the conditions (A), (B).*

PROOF. We start off with a little remark that is important to follow the argument. Note that when going around a polygon $(\alpha_N,\dots,\alpha_1)$ the boundary arcs can be oriented clockwise or anticlockwise. If two consecutive arcs $a_i = t(\alpha_i)$ and $a_{i+1} = h(\alpha_i)$ have the same orientation then $|\alpha_i| = 1$ and $\|\alpha_i\| = 0$.

$\|\alpha\| = 1$ $\|\alpha\| = 1$ $\|\alpha\| = 0$ $\|\alpha\| = 0$

Therefore if we have a polygonal sequence $(\alpha_{i+k},\dots,\alpha_{i+1})$ then the orientation of $h(\alpha_{i+k})$ and $t(\alpha_{i+1})$ will be the same (different) if the total shifted degree $\|\alpha_{i+k}\| + \dots + \|\alpha_{i+1}\|$ is zero (one).

Working with the shifted degree is useful because the shifted degree of $\|\nu(\alpha_{i+k},\dots,\alpha_{i+1})\|$ is the total shifted degree $\|\alpha_{i+k}\| + \dots + \|\alpha_{i+1}\|$ plus the parity of ν. With this in mind we will distinguish two cases depending on the parity of ν.

- The parity of ν is odd. We proceed by induction. Regard an elementary polygon $(\alpha_N,\dots,\alpha_1)$ of length N and assume ν already vanishes on its polygon sequences of length $\leq k-1$. We will simultaneously gauge away ν on the polygon sequences of length k of this polygon.

  First note that an arc can occur at most twice on the boundary of a face and if it does it must be oriented once clockwise and once anticlockwise around the face because the surface is orientable. This implies that if $\nu(\alpha_{i+k},\dots,\alpha_{i+1})$ contains an identity, then k must be a multiple of N. Indeed, if $k < N$ and there is an identity $\mathbb{1}_a$ in the result then $t(\alpha_{i+1}) = h(\alpha_{i+k}) = a$. Because the two orientations of a around the face are different the shifted degree of $(\alpha_{i+k}, \ldots, \alpha_{i+1})$ is odd. As ν is odd, $\nu(\alpha_{i+k}, \ldots, \alpha_{i+1})$ will be even and hence it cannot contain $\mathbb{1}_a$ (whose shifted degree is odd).

Let us now assume k is not a multiple of N. Then we can write $\nu(\alpha_{i+k}, \ldots, \alpha_{i+1}) = \alpha_{i+k}\delta^{(i+k)} + \gamma^{(i+1)}\alpha_{i+1}$. Note that

$$\begin{aligned}
0 =& (d\nu)(\alpha_{i+k+1}, \ldots, \alpha_{i+1})\\
=& (-1)^{\|\alpha_{i+k}\|+\cdots+\|\alpha_{i+1}\|}\alpha_{i+k+1}(\alpha_{i+k}\delta^{(i+k)} + \gamma^{(i+1)}\alpha_{i+1})\\
&+ (-1)^{\|\alpha_{i+1}\|+|\alpha_{i+1}|}(\alpha_{i+k+1}\delta^{(i+k+1)} - \gamma^{(i+2)}\alpha_{i+2})\alpha_{i+1}\\
=& \alpha_{i+k+1}((-1)^{\|\alpha_{i+k}\|+\cdots+\|\alpha_{i+1}\|}\gamma^{(i+1)} - \delta^{(i+k+1)})\alpha_{i+1}.
\end{aligned}$$

In evaluating the first row we have used the induction hypothesis. Independent of arc directions, the angles α_{i+k+1} and $\gamma^{(i+1)}$ are composable and $\delta^{(i+k+1)}$ and α_{i+1} are composable. We deduce

$$\delta^{(i+k+1)} = (-1)^{\|\alpha_{i+k}\|+\cdots+\|\alpha_{i+1}\|}\gamma^{(i+1)}$$

for all i. Set $\epsilon(\alpha_{i+k-1}, \ldots, \alpha_{i+1}) = (-1)^{\|\alpha_{i+k-1}\|+\cdots+\|\alpha_{i+1}\|+1}\delta^{(i+k)}$ for all i. Then

$$\begin{aligned}
(d\epsilon)(\alpha_{i+k}, \ldots, \alpha_{i+1}) &= \alpha_{i+k}\delta^{(i+k)} + (-1)^{\|\alpha_{i+k}\|+\cdots+\|\alpha_{i+2}\|+1+|\alpha_{i+1}|}\delta^{(i+k+1)}\alpha_{i+1}\\
&= \alpha_{i+k}\delta^{(i+k)} + \gamma^{(i+1)}\alpha_{i+1} = \nu(\alpha_{i+k}, \ldots, \alpha_{i+1}).
\end{aligned}$$

We have $(d\epsilon)^0 = (d\epsilon)^1 = 0$. Let us now check that $d\epsilon$ does not affect any other elementary polygon sequences or the sequences in the same polygon with length $l \leq k-1$. Regarding shorter sequences in the same polygon, we have

$$(d\epsilon)(\alpha_{i+l}, \ldots, \alpha_{i+1}) = \mu(\ldots, \epsilon, \ldots) + \epsilon(\ldots, \mu(\alpha_{i+s+tN}, \ldots, \alpha_{i+s+1}), \ldots)$$

where N is the length of the polygon. The first summand vanishes, since $l \leq k-1$. In the second summand, the inner μ may only yield an identity and ϵ vanishes. For sequences in other polygons, similar arguments apply. We have safeguarded that $\nu^{0,1} = 0$ is preserved when gauging by ϵ. For $k > 2$, also $(d\epsilon)^2$ vanishes. In case $k = 2$ the $(d\epsilon)^2(\alpha, \beta)$ may be nonzero in rare cases, but we fix this by applying Lemma 3.8, without changing ν on polygon sequences.

Finally, let us treat the case where k is a multiple of the length N of the polygon. Apart from the identities, we can gauge everything in $\nu(\alpha_{i+k}, \ldots, \alpha_{i+1})$ away as above. It remains to check out the identities. Write $\nu(\alpha_{i+k}, \ldots, \alpha_{i+1}) = c_i \mathbb{1}_{a_i}$. Note we have

$$\begin{aligned}
0 &= (d\nu)(\alpha_{i+k+1}, \ldots, \alpha_{i+1})\\
&= \alpha_{i+k+1}\nu(\alpha_{i+k}, \ldots, \alpha_{i+1}) - \nu(\alpha_{i+k+1}, \ldots, \alpha_{i+2})\alpha_{i+1}\\
&= c_i\alpha_{i+k+1} - c_{i+1}\alpha_{i+1} = (c_i - c_{i+1})\alpha_{i+1}.
\end{aligned}$$

Here we have used that ν already vanishes on sequences of the same polygon of length $\leq k-1$. We obtain that all c_i around the polygon are equal. Denote this value by c. Choose an angle α_1 in the polygon and define

$\epsilon(\alpha_{k-N+1},\dots,\alpha_1) := c\alpha_1$. As desired,

$$\begin{aligned}(d\epsilon)(\alpha_{i+k},\dots,\alpha_{i+1}) &= \mu(\dots,\epsilon(\alpha_{k+1},\dots,\alpha_1),\dots)\\ &= c\mu(\alpha_{i+N},\dots,\alpha_{i+1}) = c\mathbb{1}_{a_i}.\end{aligned}$$

In the first row we have used that the sequence $\alpha_1,\dots,\alpha_{k-N+1}$ of length $k-N+1$ appears precisely once in the sequence $\alpha_{i+k},\dots,\alpha_{i+1}$ and hence an inner ϵ can be applied precisely once. We see that $(d\epsilon)^0 = (d\epsilon)^1 = 0$. Let us check that $d\epsilon$ vanishes on any polygon sequence $\beta_1,\dots,\beta_l$ of length $l \leq k-1$ in the same polygon. Indeed,

$$(d\epsilon)(\beta_l,\dots,\beta_1) = \epsilon(\dots,\mu(\dots),\dots) + \mu^2(\dots,\epsilon(\dots),\dots) + \mu^{\geq 3}(\dots,\epsilon(\dots),\dots).$$

The first summand vanishes, because the inner μ only gives identities. The second summand vanishes, because the result of the inner ϵ is a multiple of the first, equivalently last input of its input sequence, hence not composable with the input angle of the outer μ^2. The third summand vanishes, because ϵ consumes $k-N+1$ inputs and the outer $\mu^{\geq 3}$ needs $N-1$ more inputs, while the sequence $\beta_1,\dots,\beta_l$ has length only $l \leq k-1$. Similarly, one checks that $d\epsilon$ vanishes entirely on polygon sequences in other polygons. For $k > N$, we have $(d\epsilon)^2 = 0$. For $k = N$, it may happen that $(d\epsilon)^2 \neq 0$, but we fix this by applying Lemma 3.8.

In total, we have gauged ν infinitely many times during this proof. However, the ϵ gauges have higher and higher input length. Moreover we only invoke Lemma 3.8 finitely many times, this means that the total sum of the gauges is defined in $\mathrm{Hom}_{\mathbb{k}}(\bigoplus_{i\geq 0} A[1]^{\otimes_{\mathbb{k}} i}, A[1])$ and we conclude it is a Hochschild cochain.

- The parity of ν is even. As in the odd case, we proceed again by induction over the length k of the polygon sequence.

  Let us first check for possible identities in $\nu(\alpha_{i+k},\dots,\alpha_{i+1})$. In case k is a multiple of N, the source arc of α_{i+1} is equal to the target arc of α_{i+k}, in particular $\|\nu(\alpha_{i+k},\dots,\alpha_{i+1})\|$ is even and does not contain identities. In case k is not a multiple of N, we observe

$$(d\nu)(\alpha_{i+k+1},\alpha_{i+k},\dots,\alpha_{i+1}) = \alpha_{i+k+1}\nu(\alpha_{i+k},\dots,\alpha_{i+1}) + \nu(\alpha_{i+k+1},\dots,\alpha_{i+2})\alpha_{i+1}.$$

  Since k is not a multiple of N, we have $\alpha_{i+k+1} \neq \alpha_{i+1}$ and conclude that $\nu(\alpha_{i+k},\dots,\alpha_{i+1})$ contains no identities.

  We now proceed with gauging $\nu(\alpha_{i+k},\dots,\alpha_{i+1})$ to zero. By abuse of wording, let us say "head" and "tail" of a polygon's arc a to mean the vertex at the clockwise end and at the counterclockwise end of a. For every i we write

$$\nu(\alpha_{i+k},\dots,\alpha_{i+1}) = \kappa_i + \lambda_i,$$

  where κ_i contains the angle paths that starts at the "head" of a_{i+1} and λ_i starts at the "tail" of $a_{i+1} = t(\alpha_{i+1})$.

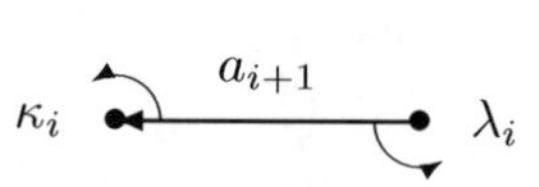

We show that λ_i necessarily vanishes. Regard the source arc a_{i+1} of α_{i+1} and target arc a_{i+k+1} of α_{i+k}. We have

$$0 = (d\nu)(\alpha_{i+k+1}, \ldots, \alpha_{i+1}) = \alpha_{i+k+1}(\kappa_i + \lambda_i) + (\kappa_{i+1} + \lambda_{i+1})\alpha_{i+1}.$$

We know κ_i is odd or even, depending on whether a_{i+1} and a_{i+k+1} are equally oriented with respect to the polygon or not. Since κ_i always starts at the "head" of a_{i+1}, we deduce from its degree that it always ends at the "tail" of a_{i+k+1}. In particular $\alpha_{i+k+1}\kappa_i = 0$. Similarly, λ_i starts at the "tail" of a_{i+1}, hence ends at the "head" of a_{i+k+1}, whether a_{i+1} and a_{i+k+1} are oriented equally or not. In particular α_{i+k+1} and λ_i are composable. But α_{i+1} starts at the "head" of a_{i+1}, while λ_i starts at the "tail" of a_{i+1}, hence the summands $\alpha_{i+k+1}\lambda_i$ and $(\kappa_{i+1} + \lambda_{i+1})\alpha_{i+1}$ consist of disjoint sets of angles. Since the whole sum is supposed to vanish, we conclude $\lambda_i = 0$.

Since κ_i starts at the side of α_{i+1} and $\nu(\alpha_{i+k}, \ldots, \alpha_{i+1})$ contains no identities, we can write $\kappa_i = \gamma_i \alpha_{i+1}$. We aim at gauging $\nu(\alpha_{i+k}, \ldots, \alpha_{i+1})$ to zero. Let us first gauge away all terms except the possible scalar multiple of α_{i+1}, which comes from a possible identity in γ_i. Put

$$\epsilon(\alpha_{i+k}, \ldots, \alpha_{i+2}) := \gamma_i.$$

Then

$$(d\epsilon)(\alpha_{i+k}, \ldots, \alpha_{i+1}) = \gamma_i \alpha_{i+1} + \alpha_{i+k}\gamma_{i-1}.$$

Apart from identities in γ_{i-1}, the angles α_{i+k} and γ_{i-1} are not composable. To see this, consider two cases. If a_i and a_{i+k} are oriented opposite, then $\nu(\alpha_{i+k-1}, \ldots, \alpha_i) = \kappa_{i-1}$ is even. Hence γ_{i-1} enters a_{i+k} at the opposite side of where α_{i+k} leaves. If a_i and a_{i+k} are oriented equally, then κ_{i-1} is odd and γ_{i-1} enters a_{i+k} still at the opposite side of where α_{i+k} leaves. Either way, we have $\alpha_{i+k}\gamma_{i-1} = 0$ and $\gamma_i\alpha_{i+1} = \kappa_i$ remains as desired.

Let us check that $d\epsilon$ vanishes on polygon sequences in other polygons and on polygon sequences shorter than k in the same polygon. Indeed,

$$(d\epsilon)(\beta_l, \ldots, \beta_1) = \epsilon(\ldots, \mu(\ldots), \ldots) + \mu(\ldots, \epsilon(\ldots), \ldots).$$

In the first summand, the inner μ can only give identities, on which ϵ vanishes. In the second summand, the input sequence of the inner ϵ must be from the same polygon as $\alpha_1, \ldots, \alpha_N$ and of length precisely $k-1$. Since the outer μ is $\mu^{\geq 2}$, this means that $\beta_1, \ldots, \beta_l$ has length at least k, which was not to be assumed.

Finally, let us gauge away the remaining scalar multiples of α_{i+1} in $\nu(\alpha_{i+k}, \ldots, \alpha_{i+1})$. For $i \in \mathbb{Z}/N\mathbb{Z}$, write $\nu(\alpha_{i+k}, \ldots, \alpha_{i+1}) = c_i\alpha_{i+1}$. If $c_i \neq 0$, we deduce that the target arc of α_{i+k} is equal to the target arc of α_{i+1}, that is, $a_{i+k+1} = a_{i+2}$. Moreover, ν is even and hence a_{i+k+1} and a_{i+2} are oriented equally. In other words, a_{i+k+1} and a_{i+2} are equal as arcs and are oriented equally in the polygon. As we remarked in the odd case, this is only possible if $i+k+1 = i+2$ in $\mathbb{Z}/N\mathbb{Z}$. We conclude that if some c_i does not vanish, then $k = lN + 1$ for some $l \geq 1$.

Assuming $k = lN + 1$, let us show that the different c_i along the polygon are related. Indeed,

$$0 = (d\nu)(\alpha_{(l+1)N}, \ldots, \alpha_1) = \sum_{i=1}^{N} c_i \mathbb{1}_{a_1} + \mu^2(\alpha_{(l+1)N}, \nu(\ldots)) + \mu^2(\nu(\ldots), \alpha_1).$$

We have used that an inner ν can be placed in precisely N ways, replicating the corresponding angle α_i and forming a disk together with the remaining angles. Moreover, an inner μ is impossible, since it only yields identities. An outer μ^2 is possible, but gives non-empty angles only which can be distinguished from the identities. We deduce that the sum of all c_i vanishes. Put

$$\epsilon(\alpha_{i+lN}, \ldots, \alpha_{i+1}) := -\left(\sum_{j=1}^{i-1} c_j\right) \mathbb{1}_{a_{i+1}}.$$

Then

$$(d\epsilon)(\alpha_{i+lN+1}, \ldots, \alpha_{i+1}) = -\sum_{j=1}^{i-1} c_j \alpha_{i+lN+1} + \sum_{j=1}^{i} c_j \alpha_{i+1}.$$

This definition makes sense for $i \in \mathbb{Z}/N\mathbb{Z}$, since the sum over c_i vanishes. The sums cancel each other because $\alpha_{i+lN+1} = \alpha_{i+1}$, and $c_i\alpha_{i+1}$ remains as desired. It is standard to check that $d\epsilon$ does not have values on shorter polygon sequences or other polygons.

□

LEMMA 3.10. Let $\nu \in \operatorname{Ker}(d) \subset \underline{\mathrm{CC}}^\bullet(\mathrm{Gtl}_{\mathcal{A}}$ with (A) $\nu^0 = \nu^1 = 0$, (B) $\nu(\alpha, \beta) = 0$ if $\alpha\beta \neq 0$ and suppose (C) ν vanishes on polygonal sequences. Then ν can be gauged to zero on all angle sequences.

PROOF. We prove the statement by gauging ν to zero inductively on the total length of the angle sequences. After each step of gauging, we prove that all additional assumptions (A-C) still hold.

We already know $\nu^1 = 0$ and if (α, β) is a sequence of length two then either $\alpha\beta \neq 0$ of (α, β) is a polygon sequence, so by (B) and (C) we are done for total length at most 2.

Now let $(\alpha_k, \ldots, \alpha_1)$ be any sequence of $k \geq 3$ angles $\alpha_i : a_i \to a_{i+1}$. By induction, we assume that ν already vanishes on all shorter sequences. Since ν already vanishes on polygon sequences, we can assume $(\alpha_k, \ldots, \alpha_1)$ is not a polygon sequence.

We will gauge ν on $(\alpha_k, \ldots, \alpha_1)$ simultaneously with all other sequences of the same type and therefore we will assume $(\alpha_k, \ldots, \alpha_1)$ is elementary. Now $(\alpha_k, \ldots, \alpha_1)$ consists of indecomposable angles of which at least one pair is composable (otherwise it would be a polygon sequence). Orientations of arcs will play no role in this proof, apart from determining the degrees of the angles. For simplicity, let us assume all α_i have odd degree. (This happens in case of a dimer.) Similarly, ν is either of odd or even parity, and the signs we write are for the odd case.

Let $\mathrm{Contr}(\beta)$ be the set of all contractible indices and $s = \max(\mathrm{Contr}(\beta))$. We will now gauge all β^S for $S \subset \mathrm{Contr}(\beta)$ simultaneously. Put

$$\epsilon(\beta^{\{s\}\cup T}) = (-1)^{|T|}\nu(\beta^T), \quad T \subset \mathrm{Contr}(\beta) \setminus \{s\}.$$

We show that $(d\epsilon)(\beta^S) = \nu(\beta^S)$ for all $S \subset \mathrm{Contr}(\beta)$.

- In case $s \notin S$, we have
$$(d\epsilon)(\beta^S) = \epsilon(\beta^{\{s\}\cup S}) = (-1)^{|S|}\nu(\beta^S).$$
by definition. Note that any $\epsilon(\dots, \mu^{\geq 3}, \dots)$ and $\mu(\dots, \epsilon, \dots)$ vanish since their input sequence of ϵ is shorter than $(\alpha_k, \dots, \alpha_1)$.
- In case $s \in S$, first observe that for $K \subset \mathrm{Contr}(\beta)$ we have
$$\sum_{t \notin K} (-1)^{1+|K_{<t}|}\nu(\beta^{K\cup\{t\}}) = (d\nu)(\beta^K) \mp \nu(\dots, \mu^{\geq 3}, \dots) \mp \mu(\dots, \nu, \dots) = 0.$$
Here $|K_{<t}|$ denotes the number of indices in K smaller than t. Indeed, the second and third summands vanish because the input sequence of ν is shorter than β. Now using this observation for $K = S \setminus \{s\}$ we get
$$\begin{aligned}(d\epsilon)(\beta^S) &= -\sum_{t \notin S}(-1)^{1+|S_{<t}|}\epsilon(\beta^{S\cup\{t\}}) \\ &= \sum_{t\notin S}(-1)^{|S\setminus\{s\}\cup\{t\}|+|S_{<t}|}\nu(\beta^{S\setminus\{s\}\cup\{t\}}) \\ &= \sum_{t \notin S\setminus\{s\}} (-1)^{|S|+1}(-1)^{|(S\setminus\{s\})_{<t}|+1}\nu(\beta^{S\setminus\{s\}\cup\{t\}}) \\ &\quad - (-1)^{|S|+1}(-1)^{|S_{<s}|+1}\nu(\beta^{S\setminus\{s\}\cup\{s\}}) \\ &= 0 + \nu(\beta^S)\end{aligned}$$
as desired. In the first row, we have used that all $\epsilon(\dots, \mu^{\geq 3}, \dots)$ and $\mu(\dots, \epsilon, \dots)$ vanish since their input sequence of ϵ is shorter than $(\alpha_k, \dots, \alpha_1)$.

Finally, $(d\epsilon)^0 = (d\epsilon)^1 = 0$ and $d\epsilon$ vanishes on polygon sequences $\gamma_l, \dots, \gamma_1$. Indeed,
$$(d\epsilon)(\gamma_l, \dots, \gamma_1) = \epsilon(\dots, \mu, \dots) + \mu(\dots, \epsilon, \dots). \tag{3.3}$$
The first summand vanishes because the inner μ yields only vertex idempotents. The second summand vanishes, since the input sequence of ϵ consists only of indecomposable angles, while a sequence $\beta^{\{s\}\cup T}$ on which ϵ is defined has at least one decomposable angle.

Next, note that $d\epsilon$ vanishes if an input is a vertex idempotent. Now assume $\gamma_l, \dots, \gamma_1$ is any sequence of non-empty angles of total length less than or equal to that of $\alpha_k, \dots, \alpha_1$. Consider the sum (3.3) again. The first summand vanishes in case of $\mu^{\geq 3}$ because then the input sequence of ϵ has less total length than $\alpha_1, \dots, \alpha_k$. If the first summand does not vanish for some μ^2, we conclude that the indecomposable constituents of the sequence are equal to $\alpha_1, \dots, \alpha_k$. In other words, it is one of those sequences we have just gauged correctly already. The second summand vanishes, since the input sequence of ϵ is shorter than $\alpha_k, \dots, \alpha_1$. □

THEOREM 3.11. Let $\nu \in \mathrm{Ker}(d)$ with $\nu^{0,1} = 0$ then ν is zero in $\mathrm{HH}^\bullet(\mathrm{Gtl}_A)$.

PROOF. This is an immediate consequence of the previous lemmas. □

This theorem implies that the cohomology class of the cocycles can be read off from its zeroth and first components.

LEMMA 3.12. If ν is a cocycle then

- ν^0 is a central element,
- if $\nu^0 = 0$ then ν^1 is a derivation.

PROOF. We have that $\nu^0(1)$ is a central element because

$$0 = (d\nu)^1(\alpha) = \mu^2(\alpha, \nu^0(1)) + (-1)^{\|\alpha\|}\mu^2(\nu^0(1), \alpha).$$

If $\nu^0 = 0$ then ν^1 is a derivation because

$$0 = (-1)^{|\beta|}(d\nu)(\alpha, \beta) = \nu^1(\alpha)\beta + \alpha\nu^1(\beta) - \nu(\alpha\beta),$$
$$0 = \nu^1(\mathbb{1}_a)\mathbb{1}_a + \mathbb{1}_a\nu^1(\mathbb{1}_a) - \nu(\mathbb{1}_a).$$

□

Depending on the parity of the cocycle we can be even more precise: odd cocycles are determined by ν^0, which is a central element, while even cocycles are determined by ν^1, viewed as an outer derivation of $\mathrm{Gtl}_{\mathcal{A}}$.

THEOREM 3.13. Let $\mathcal{A}$ be an arc collection.

(1) The map

$$\mathrm{HH}^{\mathrm{even}}(\mathrm{Gtl}_{\mathcal{A}}) \to \mathrm{OutDer}(\mathrm{Gtl}_{\mathcal{A}}) : \nu \mapsto \nu^1$$

is a well-defined embedding.

(2) The map

$$\mathrm{HH}^{\mathrm{odd}}(\mathrm{Gtl}_{\mathcal{A}}) \to Z(\mathrm{Gtl}_{\mathcal{A}}) : \nu \mapsto \nu^0(1)$$

is a well-defined embedding.

PROOF. For the first statement note that the center only has cycles of even degree, which represent odd maps $A[1]^{\otimes 0} \to A[1]$. Therefore $\nu^0 = 0$ if the parity of ν is even and hence by the previous lemma ν^1 is indeed a derivation. The map is also well defined because

$$(d\kappa)^1(\alpha) = \mu^1(\kappa^1(\alpha)) + \mu^2(\kappa^0, \alpha) - \mu^2(\alpha, \kappa^0) = [\kappa^0, \alpha].$$

Furthermore if ν^1 is inner and $\nu^0 = 0$ then we can find a κ such that $(\nu - d\kappa)^1 = 0$ zero. Because $(\nu - d\kappa)^0$ is zero for degree reasons we have by theorem 3.11 that $\nu = 0$ in $\mathrm{HH}^{\mathrm{even}}(\mathrm{Gtl}_{\mathcal{A}})$. Therefore the map is an embedding.

For the second statement, first note that this map is well-defined because $(d\kappa)^0 = \mu^1 \circ \kappa^0 = 0$. Now suppose that $\nu^0(1) = 0$, then we will construct an even ϵ such that $(d\epsilon)^1(\alpha) = \nu^1(\alpha)$ for all indecomposable angles. Because $\nu^0(1) = (d\epsilon)^0 = 0$ both $(d\epsilon)^1$ and ν^1 are derivations and therefore $(d\epsilon)^1 = \nu^1$. This means that $(\nu - d\epsilon)^{\leq 1} = 0$ and hence $\nu = 0$ in $\mathrm{HH}^{\mathrm{odd}}(\mathrm{Gtl}_{\mathcal{A}})$. Therefore the map is an embedding.

To construct ϵ, let us first make sure that $\nu^1(\alpha)$ does not include the identity. Indeed let $\alpha : a \to a$ be an indecomposable angle. Since a is not contractible by assumption, α is not the only angle in the polygon. In particular, α does not go from head to tail of a or from tail to head of a. We conclude α is even and $\nu^1(\alpha)$ is odd, hence does not include an identity.

Regard the polygon that α sits in. By abuse of wording, let us say "head" and "tail" of a polygon's arc a to mean the vertex at the clockwise end and at the counterclockwise end of a. Independent of arrow directions, $\nu^1(\alpha)$ can be decomposed into a part running from the tail of a to the tail of b and a part running from the

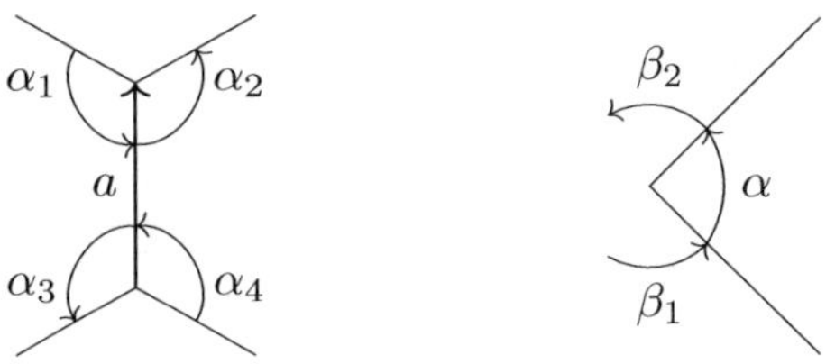

(a) Surrounding angles (b) Surrounding angles

FIGURE 3.1.

head of a to the head of b. Write this decomposition as $\nu^1(\alpha) = \alpha\delta^{(\alpha)} + \gamma^{(\alpha)}\alpha$. Whether α is even or odd, $\delta^{(\alpha)}$ and $\gamma^{(\alpha)}$ are always odd since ν^1 is.

Now let a be an arc. Denote by α_1, α_2, α_3, α_4 the angles incident at a as in 3.1a. We have

$$\begin{aligned} 0 = (d\nu)(\alpha_2, \alpha_4) &= (-1)^{\|\alpha_4\|}\mu^2(\nu^1(\alpha_2), \alpha_4) + \mu^2(\alpha_2, \nu^1(\alpha_4)) + \nu^1(\mu^2(\alpha_2, \alpha_4)) \\ &= -\gamma^{(\alpha_2)}\alpha_2\alpha_4 - \alpha_2\delta^{(\alpha_2)}\alpha_4 + (-1)^{\|\alpha_4\|}\alpha_2\gamma^{(\alpha_4)}\alpha_4 + (-1)^{\|\alpha_4\|}\alpha_2\alpha_4\delta^{(\alpha_4)} + 0 \\ &= \alpha_2((-1)^{\|\alpha_4\|}\gamma^{(\alpha_4)} - \delta^{(\alpha_2)})\alpha_4. \end{aligned}$$

We conclude $\delta^{(\alpha_2)} = (-1)^{\|\alpha_4\|}\gamma^{(\alpha_4)}$. Note that we have used that $\alpha_2\alpha_4 = 0$ and that $\gamma^{(\alpha_4)}$ and $\delta^{(\alpha_2)}$ both run from the tail of a to the head of a, and that α_4 ends at the tail of a and α_2 starts at the head of a. Similarly,

$$\begin{aligned} 0 = (d\nu)(\alpha_3, \alpha_1) &= (-1)^{\|\alpha_1\|}\mu^2(\nu^1(\alpha_3), \alpha_1) + \mu^2(\alpha_3, \nu^1(\alpha_1)) + \nu^1(\mu^2(\alpha_3, \alpha_1)) \\ &= -\alpha_3\delta^{(\alpha_3)}\alpha_1 - \gamma^{(\alpha_3)}\alpha_3\alpha_1 + (-1)^{\|\alpha_1\|}\alpha_3\alpha_1\delta^{(\alpha_1)} + (-1)^{\|\alpha_1\|}\alpha_3\gamma^{(\alpha_1)}\alpha_1 + 0 \\ &= \alpha_3(-\delta^{(\alpha_3)} + (-1)^{\|\alpha_1\|}\gamma^{(\alpha_1)})\alpha_1. \end{aligned}$$

We conclude $\delta^{(\alpha_3)} = (-1)^{\|\alpha_1\|}\gamma^{(\alpha_1)}$. Let us now put

$$\epsilon_a^0 := (-1)^{\|\alpha_4\|+1}\gamma^{(\alpha_4)} - \delta^{(\alpha_3)} = (-1)^{\|\alpha_1\|+1}\gamma^{(\alpha_1)} - \delta^{(\alpha_2)}.$$

The expression can be read independent on the arrow direction of a: It stays invariant under rotating the labels α_1, α_2, α_3 and α_4 by 180 degrees. In other words, ϵ_a^0 can be seen either as the (signed) difference of γ and δ of its incident angles at its head or at its tail. This makes it easy to check $(d\epsilon)(\alpha) = \nu^1(\alpha)$ for any indecomposable angle $\alpha : a \to b$. Denote by β_1 and β_2 its predecessor and successor indecomposable angles around the same puncture as in Figure 3.1b. Then

$$\begin{aligned} (d\epsilon)(\alpha) &= (-1)^{|\alpha|}\epsilon_b^0\alpha - \alpha\epsilon_a^0 \\ &= (-1)^{|\alpha|}((-1)^{\|\alpha\|+1}\gamma^{(\alpha)} - \delta^{(\beta_2)})\alpha - \alpha((-1)^{\|\beta_1\|+1}\gamma^{(\beta_1)} - \delta^{(\alpha)}) \\ &= \gamma^{(\alpha)}\alpha + (-1)^{\|\alpha\|}\delta^{(\beta_2)}\alpha + (-1)^{\|\beta_1\|}\alpha\gamma^{(\beta_1)} + \alpha\delta^{(\alpha)} \\ &= \gamma^{(\alpha)}\alpha + \alpha\delta^{(\alpha)} = \nu^1(\alpha). \end{aligned}$$

In the penultimate equality, we have used that $\delta^{(\beta_2)}$ ends where α starts and is odd, therefore not composable with α. Similarly $\alpha\gamma^{(\beta_1)} = 0$. We conclude that $(d\epsilon)^1 = \nu^1$ on indecomposable angles. □

REMARK 3.14. If ν is an even cocycle then ν^0 is zero, but if ν has odd parity then ν^1 need not to be zero or not even a derivation. An example of this occurs when m is a marked point surrounded by a loop arc. This loop gives rise to an orbigon of length 1, so the corresponding product A_∞-deformation has a nontrivial ${}^r\mathbb{\mu}^1$ if $r_{(m,1)} \neq 0$ and therefore $\nu_{(m,1),o}$, as defined in definition 3.15, will be an odd cocycle with nontrivial ν^1.

3.3. A set of generators. In this section we will describe a special set of Hochschild classes that span the Hochschild cohomology.

The first class we need is the unit class $\nu_{\mathbb{1}}$, which corresponds to a single nullary product: $\nu_{\mathbb{1}}^0 : \mathbb{k} \to A[1] : \mathbb{1}_a \mapsto \mathbb{1}_a$. Clearly this is a Hochschild cycle because the $\mathbb{1}_a$ are strict idempotents. Its parity is odd.

DEFINITION 3.15. For each orbifold point $p = (m, r) \in M \times \mathbb{N}$ we define its orbifold point homology class as

$$\nu_{p,o} = \frac{1}{\hbar}({}^r\mathbb{\mu} - \mu) \text{ with } r = \ell_m^r \hbar \in Z(A) \otimes \mathbb{C}[\hbar]/(\hbar^2).$$

LEMMA 3.16. $\nu_{p,o}$ is a cocycle of odd parity.

PROOF. These are clearly cocycles because ${}^r\mathbb{\mu}$ satisfies the Maurer-Cartan equation, which reduces to $d({}^r\mathbb{\mu}) = 0$ in the first order. □

A second set of cocycles are characterized as follows. Fix a map $\lambda : (Q_{\mathcal{A}})_1 \to \mathbb{C}$. Define the first component of ν_λ to be the derivation such that $\nu_\lambda(\alpha) = \lambda_\alpha \alpha$ and set $\nu_\lambda^{\neq 1} = 0$. The derivation condition implies that for a path $\beta = \beta_k \dots \beta_1$ we have $\nu_\lambda(\beta) = \sum_i \lambda_{\beta_i} \beta$, so it makes sense to define $\lambda_\beta := \sum_i \lambda_{\beta_i}$.

LEMMA 3.17. ν_λ is a cocycle if and only if for every polygon $(\alpha_N, \dots, \alpha_1)$ we have $\sum_i \lambda_{\alpha_i} = 0$.

PROOF. The condition that $\sum_i \lambda_{\alpha_i} = 0$ for every polygon implies that $\sum_{\alpha_i} \lambda_{\alpha_i} = 0$ for every tree-gon $(\alpha_k, \dots, \alpha_1)$. Therefore

$$(d\nu_\lambda)(\alpha_1, \dots, \alpha_k\beta) = (\sum \lambda_{\alpha_i} + \lambda_\beta)\mu(\alpha_k, \dots, \alpha_1\beta) - \nu_\lambda(\beta) = 0.$$

On the other hand if $\nu^{\neq 1} = 0$ then we know from lemma 3.12 that ν must be a derivation, while for every polygon $d(\nu_\lambda)(\alpha_N, \dots, \alpha_1)$=0 implies that $\sum_i \lambda_{\alpha_i}$=0. □

The set of all ν_λ with these properties form a vector space S. If two elements in S differ by a commutator with an element in $\mathbb{k}$ they will represent the same homology class.

LEMMA 3.18. $\dim S/[\mathbb{k}, -] = \dim \mathrm{H}^1(S, M, \mathbb{C}) = 2g - 1 + |M|$.

PROOF. The dimension of S is $2\#\mathcal{A} - \#\{\text{faces}\}$ because there are 2 angles arriving in each arc and every face gives one linear condition. All these conditions are linearly independent because every angle occurs only in one face. The kernel of the map $\mathbb{k} \to [\mathbb{k}, -]$ is $\mathbb{C}$, so the image of $[\mathbb{k}, -]$ in S is of dimension $\#\mathcal{A} - 1$ and therefore

$$\dim S/[\mathbb{k}, -] = 2\#\mathcal{A} - \#\{\text{faces}\} - \#\mathcal{A} + 1 = 2g - 2 + |M| + 1.$$

□

It is easy to find a basis for $S/[\Bbbk, -]$. Look again at Figure 3.1a. For each arc a there are 4 angles: two α_2, α_3 leaving a and two α_1, α_2 arriving in a. The angles α_2 and α_4 sit in the face on the right of a and α_1,α_3 in the face on the left.

We define $\nu_{a,L}$ to be the derivation with

$$\lambda(\alpha_1) = 1, \lambda(\alpha_3) = -1$$

while all other entries of λ are zero. Similarly we define $\nu_{a,R}$ with

$$\lambda(\alpha_2) = 1, \lambda(\alpha_4) = -1.$$

Note that because $[\mathbb{1}_a, -] = \nu_{a,L} - \nu_{a,R}$ we have that up to homology $\nu_{a,L} = \nu_{a,R}$, so we will drop the subscript R, L and set $\nu_a := \nu_{a,L}$.

LEMMA 3.19. If $\mathcal{T}$ is a set of arcs that forms a spanning tree in the face graph then

$$\{\nu_a \mid a \in \mathcal{A} \setminus \mathcal{T}\}$$

is a basis for $S/[\Bbbk, -]$.

PROOF. The sum of the ν_a where a runs over the arcs going around a given face is zero. Because $\mathcal{T}$ is a spanning tree, the $\{\nu_a \mid a \in \mathcal{A} \setminus T\}$ are independent. Moreover because a spanning tree has $\#\{\text{faces}\} - 1$ arcs, the cardinality of this set is the dimension of $S/[\Bbbk, -]$. □

REMARK 3.20. We will call the ν_a and more general the ν_λ, which are linear combinations of the ν_a, **arc classes**.

Let $p = (m, j)$ be an orbifold point and a be any arc arriving (or leaving) the marked point. We now define

$$\nu_{p,e} := \nu_{p,o} \smile \nu_a \text{ or } -\nu_a \text{ if } a \text{ leaves } m.$$

Note that if a is not a loop we have that $\nu_{p,e}$ is zero for every angle arrow except for α_1 (or α_3 in the case a leaves m).

LEMMA 3.21. If the arc collection $\mathcal{A}$ satisfies [NL2] then the homology class of $\nu_{p,e}$ does not depend on the choice of a.

PROOF. Let a and b be neighboring arcs incident at the same puncture, such that b comes after a in clockwise order. Let $\epsilon = \epsilon^0$ be the odd cochain given by $\epsilon^0 = \ell_m^j \mathbb{1}_a$. For any angle α winding around puncture $\mathrm{m}(\alpha)$ we have

$$d(\epsilon)(\alpha) = \begin{cases} +\ell_m^j \alpha & t(\alpha) = a \text{ and } \mathrm{m}(\alpha) = m \\ -\ell_m^j \alpha & h(\alpha) = a \text{ and } \mathrm{m}(\alpha) = m \\ 0 & \text{otherwise} \end{cases}$$

Because a is not a loop, the two nonzero cases each happen for just one α_1 and α_3 angle. We conclude

$$[d(\epsilon)]^1 = [\nu_{p,e} \smile \nu_a - \nu_{p,e} \smile \nu_b]^1.$$

In other words the difference between $\nu_{p,e} \smile \nu_a$ and $\nu_{p,e} \smile \nu_b$ is homotopic to a cocycle κ with $\kappa^0, \kappa^1 = 0$. In combination with theorem 3.11 this means that two consecutive arcs around m define the same $\nu_{p,e}$-class, and hence all arcs around m do. □

THEOREM 3.22. Let $\mathcal{A}$ be an arc collection and $\mathcal{T}$ a spanning tree in the face graph.

(1) The cocycles $\nu_{\mathbb{1}}$, $\nu_{p,o}$ where p runs over all orbifold points form a basis for $\mathrm{HH}^{\mathrm{odd}}(\mathrm{Gtl}_{\mathcal{A}})$.
(2) If $\mathcal{A}$ satisfies [NL2] then the cocycles $\nu_{p,e}$, ν_a where p runs over all orbifold points and $a \in \mathcal{A} \setminus \mathcal{T}$ form a basis for $\mathrm{HH}^{\mathrm{even}}(\mathrm{Gtl}_{\mathcal{A}})$.

PROOF. The zeroth components of $\nu_{\mathbb{1}}$, $\nu_{p,o}$ form a basis for $Z(\mathrm{Gtl}_{\mathcal{A}})$, so the embedding $\mathrm{HH}^{\mathrm{odd}}(\mathrm{Gtl}_{\mathcal{A}}) \to Z(\mathrm{Gtl}_{\mathcal{A}})$ in lemma 3.13 is a bijection.

To show the second part, note that the fact that there are no loops or two-cycles implies that all nonzero angle paths with the same head and tail as α turn around the same marked point as α and are of the form $\ell_m^r\alpha$.

If ν is G-homogeneous this means that there are two possibilities

- If the degree is 0 then ν must be a linear combination of the ν_a.
- $\nu(\alpha) = \lambda_\alpha \ell_m^j \alpha$ with $j > 0$ then the only other β for which $\nu(\beta) \neq 0$ must turn around m. Up to an inner derivation this ν is determined by the sum $\sum \lambda_\alpha$ where α runs over the indecomposables turning around m.

□

3.4. Gerstenhaber structure product on Hochschild cohomology. In this section, we compute the Gerstenhaber algebra structure on $\mathrm{HH}(\mathrm{Gtl}_{\mathcal{A}})$. In other words, we determine the bracket and the cup product on Hochschild cohomology. The bracket agrees with the computations by Wong for the Borel-Moore cohomology of matrix factorizations for dimer models [**20**], which is conjecturally equivalent to $\mathrm{HH}(\mathrm{Gtl}_{\mathcal{A}})$.

To describe the bracket and cup product, we use the basis for odd and even Hochschild cohomology constructed earlier: the odd classes $\nu_{(m,j),o}$ for a puncture $m \in M$ and $j \geq 0$, the even classes $\nu_{(m,j),e}$, and the arc classes ν_λ with $\lambda: (Q_{\mathcal{A}})_1 \to \mathbb{C}$.

We start with the cup-product

PROPOSITION 3.23. Let $m, n \in M$ be two marked points, let $i, j \geq 1$ be two indices, and ν_λ, ν_κ two arc classes. Then the cup product in cohomology reads as follows:

$$\begin{aligned}
\nu_{(m,i),o} \smile \nu_{(n,j),o} &= \delta_{mn}\nu_{(m,i+j),o},\\
\nu_{(m,i),o} \smile \nu_{(n,j),e} &= \delta_{mn}\nu_{(m,i+j),e},\\
\nu_{(m,i),o} \smile \nu_\lambda &= \lambda_{\ell_m}\nu_{(m,i),e},\\
\nu_\kappa \smile \nu_\lambda &= 0,\\
\nu_{(p,i),e} \smile \nu_\lambda &= 0.
\end{aligned}$$

PROOF. We compute the cup products of the given Hochschild cocycles first on chain level. Then we compute their projection to cohomology. In fact, for the odd products it suffices to compute the curvature component $(\nu \smile \eta)^0$ and for the even products it suffices to compute the first component $(\nu \smile \eta)^1$. Also recall that the parity of $\smile$ itself is odd, so the product of two classes with the same (different) parity is odd (even). We are now ready to start the calculations. For the first identity, regard

$$(\nu_{(m,i),o} \smile \nu_{(n,j),o})^0 = \mu^2(\nu^0_{(m,i),o}(1), \nu^0_{(n,j),o}(1)) = \delta_{mn}\mu^2(\ell^i_m, \ell^j_n) = \ell^{i+j}_m.$$

This is precisely the curvature of the Hochschild cohomology class $\nu_{(m,i+j),o}$ and hence projects to it. For the second identity and third identity one can do a similar calculation but now one has to look a the first component. Alternatively, we can use the definition of $\nu_{(p,i),e}$ in combination with the associativity of $\smile$ on the Hochschild cohomology.

For the last two identities we again have to look at the zeroth component and these are trivially zero because both factors have trivial curvature. □

PROPOSITION 3.24. Let $m, n \in M$ be two marked points, let $i, j \geq 1$ be two indices, and ν_λ, ν_κ two arc classes. Then the Gerstenhaber bracket in cohomology reads as follows:

$$\begin{aligned}
[\nu_{(m,i),o}, \nu_{(n,j),o}] &= 0, \\
[\nu_{(m,i),e}, \nu_{(n,j),o}] &= \delta_{mn} \cdot j \cdot \nu_{(m,i+j),o}, \\
[\nu_{(m,i),e}, \nu_{(n,j),e}] &= \delta_{mn} \cdot (j - i) \cdot \nu_{(m,i+j),e}, \\
[\nu_\lambda, \nu_{(m,i),o}] &= i\lambda_{\ell_m} \cdot \nu_{(m,i),o}, \\
[\nu_\lambda, \nu_{(m,i),e}] &= i\lambda_{\ell_m} \cdot \nu_{(m,i),e}, \\
[\nu_\kappa, \nu_\lambda] &= 0.
\end{aligned}$$

PROOF. The strategy for this calculation is the same as for the cup product. We calculate the zeroth or first component of the bracket.

$$\begin{aligned}
[\nu, \eta]^0 &= \nu^1(\eta^0) - (-1)^{\|\nu\|\|\eta\|}\eta^1(\nu^0), \\
[\nu, \eta]^1(\alpha) &= \nu^1(\eta^1(\alpha)) + (-1)^{\|\eta\|\|\alpha\|}\nu^2(\eta^0, \alpha) + \nu^2(\alpha, \eta^0) \\
&\quad - (-1)^{\|\nu\|\|\eta\|}\big(\eta^1(\nu^1(\alpha)) + (-1)^{\|\nu\|\|\alpha\|}\eta^2(\nu^0, \alpha) + \eta^2(\alpha, \nu^0)\big).
\end{aligned}$$

The main difference is now that the bracket has even parity, so we need to check the zeroth component whenever the two entries have different parity. In that case the result is the outer derivation of the even class applied to the central element of the odd class. This takes care of the second and fourth identity.

The first component of the bracket of two even classes is the commutator of their outer derivations because they have no ν^0. This takes care of the third and the last two identities.

Finally the bracket of two odd classes is zero because they have no ν^1, ν^2 if [NL2] holds. If [NL2] fails and α is an indecomposable angle then $\nu^1(\alpha)$ can only be nonzero if $a = h(\alpha)$ is a loop around a puncture but then the arc collection does not split the surface. Secondly for our deformed products a term of the form ${}^r\mu^2(\ell_p, \alpha)$ is always cancelled by ${}^r\mu^2(\alpha, \ell_p)$ and therefore the same holds for the odd classes. This implies that $[\nu, \eta]^1(\alpha)$ is zero for all indecomposable α and because $[\nu, \eta]^1$ is a derivation it is identically zero. □

The Lie-bracket on $\mathrm{HH}^\bullet(\mathrm{Gtl}_\mathcal{A})$ is part of an L_∞-structure. This is a graded vector space L together with brackets $[, \ldots,]^k : L^{\otimes k} \to L$ of degree $2 - k$ which satisfy the L_∞-axioms [**19**].

If $(K^\bullet, d, [])$ is a DGLA then we can construct an L_∞-structure $[,]^\bullet$ on the homology such that $(HK^\bullet, 0, [,]^\bullet)$ is L_∞-quasi-isomorphic to $(K^\bullet, d, [])$.

These products can be constructed using the homotopy transfer lemma (see e.g. [**14**]). The construction can be summarized as follows:

- Split $K^\bullet$ as a direct sum of graded vector spaces

$$K^\bullet = H \oplus I \oplus R$$

such that $\mathrm{Im} d = I$ and $\mathrm{Ker}\, d = H \oplus I$. In this way $H \cong HK^\bullet$ and d restricts to an isomorphism $d_{IR} : R \to I$.

- Let h be the map
$$h : H \oplus I \oplus R \to H \oplus I \oplus R : (u, v, w) \mapsto (0, 0, d^{-1}_{R \to I}(v)).$$
In this way the projection onto H becomes $\pi = \mathbb{1} - dh - hd$.
- for $x_1, \dots, x_k \in H$ we define
$$[x_1, \dots, x_k]^k = \sum_t c_t \pi[\dots, h[x_i, x_j], \dots,]$$
as a linear combination of all possible ways of fully bracketed expressions where each internal bracket is composed with an h. (The precise coefficients c_t will not matter in our discussion).

In our case we will take $K = \mathrm{CC}(\mathrm{Gtl}_{\mathcal{A}})$, $H \cong \mathrm{HH}(\mathrm{Gtl}_{\mathcal{A}})$ to be the span of the classes we constructed, and we take I and R to be compatible with the G-grading.

THEOREM 3.25. If $\mathcal{A}$ satisfies [NL2] then we can choose a split such that all higher brackets are zero.

PROOF. First note that $[\nu_{\mathbb{1}}, -] = 0$ in $\underline{\mathrm{CC}}(\mathrm{Gtl}_{\mathcal{A}})$, so $[\nu_{\mathbb{1}}, -, \dots, -]^k$ will also be zero.

To show that the other brackets are zero, pick any $\mathbb{Z}$-grading of the gentle algebra. Then we can assume that that the bracket $[,]^k$ has degree $2 - k$ for this grading and degree 0 for the $G = \mathrm{H}(S \setminus M, \mathcal{A}, \mathbb{Z})$-grading. As we assumed that $\mathcal{A}$ has no loops, there are at least two punctures so the G-degree of ℓ_m is nontrivial for each marked point.

If all the entries of the bracket have G-degree 0 (which means that they are all arc classes, which have $\mathbb{Z}$-degree 0), then the homotopy transfer lemma tells us that the product must be zero because all pairs $[\nu_\lambda, \nu_\kappa]$ are already zero in $\mathrm{CC}(\mathrm{Gtl}_{\mathcal{A}})$.

Suppose that there is at least one $\nu_{p,e}$ in one entry. We make the following distinctions:

- All the entries have G-degree a multiple of $\deg_G \ell_m$. In that case the G-degree of the result will be $r \deg_G \ell_m$ for some $r \in \mathbb{N}$ and therefore the $\mathbb{Z}$-degree must be either $r \deg_{\mathbb{Z}} \ell_m$ if the result is even or $r \deg_{\mathbb{Z}} \ell_m - 1$ if the result is odd. The total $\mathbb{Z}$-degree of the entries is at most $r \deg_{\mathbb{Z}} \ell_m$ and this only happens if all entries are all even. Because the product has degree $2 - k$ it can only be nonzero for degree reasons if $2 - k \geq -1$, or in other words if $k = 2, 3$.
- Some entries have degrees that turn around different punctures. If the result has G-degree $\deg_G \ell^j_m$ then the G-degree of the entries must also be $\deg_G \ell^j_m$ but if there are degrees for more than one marked point all marked points must appear because the only relation between the degrees of the ℓ_m is that $\sum_{m \in M} \deg_G \ell_m = 0$. On the other hand by lemma 3.6 we have that $\sum_m \deg_{\mathbb{Z}} \ell_m \leq -1$, so the $\mathbb{Z}$-degree of the result must be $\leq \deg_{\mathbb{Z}} \ell^j_m + 2 - k - 1$. This is impossible if $k \geq 3$.

From the discussions above we see that because of degree reasons the only nontrivial higher brackets are of the form
$$[\nu_{(m,i_1),e}, \nu_{(m,i_2),e}, \nu_{(m,i_3),e}] = \lambda \nu_{(m,i_1+i_2+i_3),o}$$

or cases where one or more $\nu_{(m,i_j),e}$ are ν_λs. To show that these products are all zero, we have to look at the homotopy transfer construction in detail. The triple product can be written down in terms of expressions of the form

$$\pi[h[\nu_{(m,i_1),e},\nu_{(m,i_2),e}],\nu_{(m,i_3),e}]$$

where π is the projection onto homology. We will now argue why such terms are zero.

From our computation in 3.24 we know that

$$[\nu_{(m,i_1),e},\nu_{(m,i_2),e}]^0 = 0$$
$$[\nu_{(m,i_1),e},\nu_{(m,i_2),e}]^1 = ((i_2-i_1)\nu_{(m,i_1+i_1),e})^1$$

So $[\nu_{(m,i_1),e},\nu_{(m,i_2),e}] = (i_2-i_1)\nu_{(m,i_1+i_2),e} + d\kappa$ with $(d\kappa)^{\leq 1} = 0$. From the construction in section 3.2, we can choose κ such that $\kappa^0 = 0$. For what follows, choose $R^{\mathrm{odd}} = R_1 \oplus R_2$ (in a G-graded way) such that $R_1^{\neq 0} = R_2^0 = 0$ and $R_1 \cap Z(\mathrm{Gtl}_{\mathcal{A}}) = 0$. With respect to the direct sum $\mathrm{CC}^{\mathrm{odd}} = H^{\mathrm{odd}} \oplus I^{\mathrm{odd}} \oplus R_1 \oplus R_2$ write $\kappa = h + i + r_1 + r_2$. Then $0 = \kappa^0 = h^0 + r_1^0$, hence $r_1 = r_1^0 = -h^0 \in R_1 \cap Z(\mathrm{Gtl}_{\mathcal{A}}) = 0$ and we conclude $\kappa = i + r_2$. Simply subtracting i from κ keeps $\kappa^0 = 0$ and brings κ into $R_2 \subset R$. Finally

$$[h[\nu_{(m,i_1),e},\nu_{(m,i_2),e}],\nu_{(m,i_3),e}]^0 = [\kappa,\nu_{(m,i_3),e}]^0 = \nu_{(m,i_3),e}(\kappa^0) \pm \kappa^1(\nu^0_{(m,i_3),e}) = 0$$

Therefore all contributions are zero. □

COROLLARY 3.26. If $\mathcal{A}$ has no loops or two-cycles then $\mathrm{CC}(\mathrm{Gtl}_{\mathcal{A}}), d, [,]$ is formal. In other words there is an L_∞-quasi-isomorphism between $\mathrm{CC}(\mathrm{Gtl}_{\mathcal{A}}), d, [,]$ and $\mathrm{HH}(\mathrm{Gtl}_{\mathcal{A}}), 0, [,]$.

3.5. Classifying curved deformations. As an application of our computations we will show that every curved deformation of $\mathrm{Gtl}_{\mathcal{A}}$ is equivalent to one of the ${}^r\mu$. Remember that if (A,μ) is an A_∞-algebra over $\Bbbk$ and R is a complete local Noetherian unital commutative $\mathbb{C}$-algebra with maximal ideal R^+ and residue field $R/R^+ = \mathbb{C}$ then a curved deformation is an odd element $\nu \in \mathrm{Hom}(\bigoplus_i A^{\otimes_\Bbbk^i}, A)\widehat{\otimes} R^+$ such that $\mu+\nu$ satisfies the curved A_∞-axioms. As we indicated before this equation is equivalent to the Maurer-Cartan equation for the Hochschild cochain complex together with the Gerstenhaber bracket [**11**].

THEOREM 3.27. If [NL2] holds, then every curved A_∞-deformation of $\mathrm{Gtl}_{\mathcal{A}}$ over R is equivalent to one of the ${}^r\mu$.

PROOF. We give a proof in terms of deformation functors, in the language of [**13**]. In short, we interpret our explicit class of deformations ${}^r\mu$ as a functor of Artin rings. Gauging by even elements acts on the values of this functor and lands in Maurer-Cartan elements. The first part of the proof deals with the case of R being Artinian. In the second part of the proof, we pass to the non-Artinian case.

Let Art denote the category of Artinian local Noetherian unital commutative rings over $\mathbb{C}$ with residue field $\mathbb{C}$, with morphisms being local ($\phi(R^+) \subset S^+$) and unital ($\phi(1_R) = 1_S$). We build three functors $G, F, \mathrm{MC} : \mathrm{Art} \to \mathrm{Set}$. The functor G is the gauge group functor, F is the functor of our deformation parameters r, and MC is the standard Maurer-Cartan functor. More precisely, define

$$G(R) := \exp(\mathrm{CC}^{\mathrm{even}}(\mathrm{Gtl}_{\mathcal{A}})\widehat{\otimes}R^+),$$
$$F(R) := Z(\mathrm{Gtl}_{\mathcal{A}})\widehat{\otimes}R^+,$$
$$\mathrm{MC}(R) := \mathrm{MC}(\mathrm{CC}(\mathrm{Gtl}_{\mathcal{A}}), R).$$

All three assignments come with natural restriction maps $G(R) \to G(S)$, $F(R) \to F(S)$ and $\mathrm{MC}(R) \to \mathrm{MC}(S)$ for every morphism $\phi : R \to S$ in Art. It is standard to check that all three are deformation functors in the sense of [**13**, Definition 2.5]. In fact, G and F as well as their product functor $G \times F$ are smooth (unobstructed) in the sense of [**13**, Definition 2.8]. Regard now the morphism of functors given by

$$\Phi(R) : G(R) \times F(R) \to \mathrm{MC}(R), (g, r) \mapsto g. \,^r\mu.$$

Let us define obstruction theories $O_{G\times F}$ and O_{MC} for $G \times F$ and MC by defining both $O_{G\times F} := 0$ and $O_{\mathrm{MC}} := 0$ to be trivial. We will now show that Φ is smooth by applying [**13**, Proposition 2.17]. In the terminology of that paper, we have to check three items: (1) $\Phi(\mathbb{C}[\epsilon])$ is surjective where $\mathbb{C}[\epsilon] = \mathbb{C}[X]/(X^2)$ is the ring of dual numbers, (2) the obstruction theory $O_{G\times F} = 0$ is complete, (3) the morphism between obstruction theories $O_{G\times F} \to O_{\mathrm{MC}}$ is injective and compatible with Φ.

We check the three conditions. (1) Let $\epsilon c \in \mathrm{MC}(\mathbb{C}[\epsilon])$. Then c is a Hochschild cocycle and can be gauged by some $g \in G(\mathbb{C}[\epsilon])$ to be equal to an $^r\mu$ for some $r \in \epsilon Z(\mathrm{Gtl}_{\mathcal{A}})$. In other words, we have $\Phi(\mathbb{C}[\epsilon])(g, r) = \epsilon c$. This proves $\Phi(\mathbb{C}[\epsilon])$ surjective. Item (2) and item (3) are trivial because $G \times F$ is smooth.

The standard smoothness criterion [**13**, Proposition 2.17] implies that Φ is smooth. Putting $S = \mathbb{C}$ in the definition of smoothness implies that $\Phi(R) : G(R) \times F(R) \to \mathrm{MC}(R)$ is surjective for every $R \in$ Art. In other words, every deformation of $\mathrm{Gtl}_{\mathcal{A}}$ over $R \in$ Art is equivalent to an $^r\mu$ by gauge equivalence.

In the second part of the proof, we generalize to the non-Artinian case. Let R be a complete local Noetherian unital commutative $\mathbb{C}$-algebra with residue field $\mathbb{C}$. Let $\mu \in \mathrm{MC}(\mathrm{CC}(\mathrm{Gtl}_{\mathcal{A}}), R)$ be a deformation over R, meaning a Maurer-Cartan element in the completed tensor product $\mu \in \mathrm{CC}(\mathrm{Gtl}_{\mathcal{A}})\widehat{\otimes}R^+$. Our aim is to show that μ is gauge equivalent to some $^r\mu$ with $r \in Z(\mathrm{Gtl}_{\mathcal{A}})\widehat{\otimes}R^+$.

Our strategy is to truncate μ to $R/(R^+)^i$ for every i and use the first part of the proof to construct an element r_i and a gauge g_i. We use smoothness of Φ to force both sequences (r_i) and (g_i) to converge.

Put $\mu_i := \Pi_i(\mu) \in \mathrm{MC}(R/(R^+)^i)$. We shall construct sequences $r_i \in F(R/(R^+)^i)$ and $g_i \in G(R/(R^+)^i)$ such that (1) $\Phi(R/(R^+)^i)(g_i, r_i) = \mu_i$, (2) $\Pi_i(r_{i+1}) = r_i$ and (3) $\Pi_i(g_{i+1}) = g_i$ for all $i \in \mathbb{N}$. For the induction base $i = 1$, let $g_1 := 1 \in G(\mathbb{C})$ and $r_1 := 0 \in F(\mathbb{C})$. Since $\Phi(\mathbb{C})(g_1, r_1) = 0 = \mu_1 \in \mathrm{MC}(\mathbb{C})$, the three conditions are satisfied at $i = 1$.

For induction hypothesis, assume the sequences have already been constructed until index i. Since Φ is smooth, we have a surjection

$$G(R/(R^+)^{i+1} \times F(R/(R^+)^{i+1})$$
$$\twoheadrightarrow (G(R/(R^+)^i) \times F(R/(R^+)^i)) \times_{\mathrm{MC}(R/(R^+)^i)} \mathrm{MC}(R/(R^+)^{i+1}).$$

Pick (g_i, r_i, μ_{i+1}) on the right hand side. Indeed, $\Phi(R/(R^+)^i)(g_i, r_i) = \mu_i = \Pi_i(\mu_{i+1})$ by assumption and construction. By surjectivity there is a lift (g_{i+1}, r_{i+1}) such that (1) $\Phi(R/(R^+)^{i+1})(g_{i+1}, r_{i+1}) = \mu^{i+1}$ and (2) $\Pi_i(r_{i+1}) = r_i$ and (3) $\Pi_i(g_{i+1}) = g_i$. This finishes the induction step.

Finally, we have constructed the desired sequences (r_i) and (g_i). Since $\Pi_i(r_{i+1}) = r_i$, the sequence r_i converges to some $r \in Z(\mathrm{Gtl}_\mathcal{A})\widehat{\otimes}R^+$ and g_i converges to some $g \in \exp(\mathrm{CC}^{\mathrm{even}}(\mathrm{Gtl}_\mathcal{A})\widehat{\otimes}R^+)$. Within $\mathrm{MC}(R/(R^+)^i)$ we have

$$\Pi_i(g.{}^r\mu) = \Phi(R/(R^+)^i)(g_i, r_i) = \mu_i = \Pi_i(\mu), \quad \forall i \in \mathbb{N}.$$

Passing to the limit gives that $g.{}^r\mu = \mu$ within $\mathrm{MC}(R)$. In other words, μ is gauge equivalent to ${}^r\mu$. □

REMARK 3.28. In remark 2.26 we extended the notion of orbigons to allow weights on the faces and the marked points. This allows us to construct curved deformations ${}^{r,s}\mu$ of the (completed) gentle algebra without its A_∞-structure coming from the Wrapped Fukaya category. It is also possible to show that every deformation of the completed gentle algebra is equivalent to one of these forms. This nicely fits into the framework of Koszul duality because Koszul dual A_∞-algebras have the same deformation theory.

References

[1] Mohammed Abouzaid, Denis Auroux, Alexander I. Efimov, Ludmil Katzarkov, and Dmitri Orlov, *Homological mirror symmetry for punctured spheres*, J. Amer. Math. Soc. **26** (2013), no. 4, 1051–1083, DOI 10.1090/S0894-0347-2013-00770-5. MR3073884

[2] Ibrahim Assem, Thomas Brüstle, Gabrielle Charbonneau-Jodoin, and Pierre-Guy Plamondon, *Gentle algebras arising from surface triangulations*, Algebra Number Theory **4** (2010), no. 2, 201–229, DOI 10.2140/ant.2010.4.201. MR2592019

[3] Michael J. Bardzell, *The alternating syzygy behavior of monomial algebras*, J. Algebra **188** (1997), no. 1, 69–89, DOI 10.1006/jabr.1996.6813. MR1432347

[4] Alexander Beilinson, Victor Ginzburg, and Wolfgang Soergel, *Koszul duality patterns in representation theory*, J. Amer. Math. Soc. **9** (1996), no. 2, 473–527, DOI 10.1090/S0894-0347-96-00192-0. MR1322847

[5] Raf Bocklandt, *Noncommutative mirror symmetry for punctured surfaces*, Trans. Amer. Math. Soc. **368** (2016), no. 1, 429–469, DOI 10.1090/tran/6375. With an appendix by Mohammed Abouzaid. MR3413869

[6] Alexander I. Efimov, *Homological mirror symmetry for curves of higher genus*, Adv. Math. **230** (2012), no. 2, 493–530, DOI 10.1016/j.aim.2012.02.022. MR2914956

[7] Sheel Ganatra, *Symplectic Cohomology and Duality for the Wrapped Fukaya Category*, ProQuest LLC, Ann Arbor, MI, 2012. Thesis (Ph.D.)–Massachusetts Institute of Technology. MR3121862

[8] Murray Gerstenhaber, *The cohomology structure of an associative ring*, Ann. of Math. (2) **78** (1963), 267–288, DOI 10.2307/1970343. MR161898

[9] F. Haiden, L. Katzarkov, and M. Kontsevich, *Flat surfaces and stability structures*, Publ. Math. Inst. Hautes Études Sci. **126** (2017), 247–318, DOI 10.1007/s10240-017-0095-y. MR3735868

[10] Bernhard Keller, *A-infinity algebras, modules and functor categories*, Trends in representation theory of algebras and related topics, Contemp. Math., vol. 406, Amer. Math. Soc., Providence, RI, 2006, pp. 67–93, DOI 10.1090/conm/406/07654. MR2258042

[11] Maxim Kontsevich and Yan Soibelman, *Deformation theory*, Livre en préparation (2002).

[12] Yankı Lekili and Alexander Polishchuk, *Derived equivalences of gentle algebras via Fukaya categories*, Math. Ann. **376** (2020), no. 1-2, 187–225, DOI 10.1007/s00208-019-01894-5. MR4055159

[13] Marco Manetti, *Deformation theory via differential graded Lie algebras*, Algebraic Geometry Seminars, 1998–1999 (Italian) (Pisa), Scuola Norm. Sup., Pisa, 1999, pp. 21–48. MR1754793

[14] Martin Markl, *Transferring A_∞ (strongly homotopy associative) structures*, Rend. Circ. Mat. Palermo (2) Suppl. **79** (2006), 139–151. MR2287133

[15] Stephan Mescher, *A primer on a-infinity-algebras and their hochschild homology*, preprint arXiv:1601.03963 (2016).

[16] Sebastian Opper, Pierre-Guy Plamondon, and Sibylle Schroll, *A geometric model for the derived category of gentle algebras*, preprint `arXiv:1801.09659` (2018).
[17] Alexander F. Ritter and Ivan Smith, *The monotone wrapped fukaya category and the open-closed string map*, `arXiv:1201.5880` (2012).
[18] Paul Seidel, *Homological mirror symmetry for the genus two curve*, J. Algebraic Geom. **20** (2011), no. 4, 727–769, DOI 10.1090/S1056-3911-10-00550-3. MR2819674
[19] Jim Stasheff, *L-infinity and A-infinity structures*, High. Struct. **3** (2019), no. 1, 292–326. MR3939050
[20] Michael Wong, *Dimer models and Hochschild cohomology*, J. Algebra **585** (2021), 207–279, DOI 10.1016/j.jalgebra.2021.06.003. MR4273810

KdV Institute, University of Amsterdam, Amsterdam, The Netherlands
Email address: `r.r.j.bocklandt@uva.nl`

KdV Institute, University of Amsterdam, Amsterdam, The Netherlands
Email address: `j.d.c.vandekreeke3@uva.nl`

Contemporary Mathematics
Volume **801**, 2024
https://doi.org/10.1090/conm/801/16036

The prime spectrum of the Drinfeld double of the Jordan plane

K. A. Brown and J. T. Stafford

Abstract. The Hopf algebra $\mathcal{D}$ which is the subject of this paper can be viewed as a Drinfeld double of the bosonisation of the Jordan plane. Its prime and primitive spectra are completely determined. As a corollary of this analysis it is shown that $\mathcal{D}$ satisfies the Dixmier-Moeglin Equivalence, leading to the formulation of a conjecture on the validity of this equivalence for pointed Noetherian Hopf algebras.

1. Introduction

1.1. Throughout, k will denote an algebraically closed field of characteristic 0. The Hopf k-algebra $\mathcal{D}$ of the title was defined and some initial properties were derived in [**2**], with further results proved in [**1,3**]. Our focus here is on the prime and primitive spectra of $\mathcal{D}$, which we completely determine. The Hopf algebra $\mathcal{D}$ is a pointed affine noetherian domain of Gelfand-Kirillov dimension 6 whose definition is recalled in §2.1. It is a beautiful algebra with a number of striking properties which make it worthy of study from several perspectives, three of which we briefly outline in §§1.3-1.5. First we summarise our results and explain where they are located.

1.2. Results. It was proved in [**1**, Theorem 4.10], and explained in detail here in Lemma 2.1 and Theorem 2.2(iv), that the centre $Z(\mathcal{D})$ of $\mathcal{D}$ is generated by elements z, ω and θ with $z\theta = \omega^2$. Thus Maxspec($Z(\mathcal{D})$) has one singular point, namely $\mathfrak{m}_0 := \langle z, \omega, \theta\rangle$. There is one other distinctive maximal ideal of the centre, namely

$$\mathfrak{m}_+ \;:=\; Z(\mathcal{D}) \cap \mathcal{D}^+ \;=\; \langle z - 16, \omega + 16, \theta - 16\rangle,$$

where $\mathcal{D}^+$ is the augmentation ideal of $\mathcal{D}$. Finally, let $\mathcal{K}$ denote the kernel of the Hopf algebra surjection $\pi : \mathcal{D} \longrightarrow U(\mathfrak{sl}(2, k))$, mentioned in §1.4, below, so $\mathcal{K}$ is a Hopf ideal which is described in Theorem 2.2(ii).

The main results of this paper are given by the following theorems and give a complete description of the prime and primitive ideals of $\mathcal{D}$.

2020 *Mathematics Subject Classification.* Primary 16T05, 16D25; Secondary 16T20, 16S40, 17B37.

Both authors were partially supported by Leverhulme Emeritus Fellowships, respectively EM-2017-081 and EM-2019-015. The authors thank Nicolás Andruskiewitsch, Ivan Angiono and Hector Pena Pollastri for helpful comments and for sharing early versions of their work.

THEOREM 1.1. *Keep the above notation. Then the set of primitive prime ideals,* Privspec$(\mathcal{D})$*, of* $\mathcal{D}$ *consists of:*

(I) *the maximal ideals* $\mathfrak{m}\mathcal{D}$ *for* $\mathfrak{m} \in \mathrm{Maxspec}(Z(\mathcal{D}))$*, with* $\mathfrak{m} \neq \mathfrak{m}_0, \mathfrak{m}_+$*;*
(II) *the primitive ideals containing* $\mathfrak{m}_+\mathcal{D}$*, namely* $\mathfrak{m}_+\mathcal{D}$ *itself together with*

$$\mathcal{P} := \pi^{-1}(\mathrm{Privspec}(U(\mathfrak{sl}(2,k))));$$

(III) *the unique prime ideal* P_0 *minimal over* $\mathfrak{m}_0\mathcal{D}$*.*

The prime ideal P_0 in Theorem 1.1 is defined in (4.11) and satisfies the extra property that $\mathfrak{m}_0\mathcal{D} = (P_0)^2 \subsetneq P_0$.

THEOREM 1.2. *In the above notation, the non-primitive prime ideals of* $\mathcal{D}$ *are:*

(A) $\{0\}$, $\mathcal{K}$*;*
(B) *the principal prime ideals* $\mathfrak{p}\mathcal{D}$ *for every height one prime* $\mathfrak{p}$ *of* $Z(\mathcal{D})$ *except* $\mathfrak{p}_1 = \langle z, \omega\rangle$ *and* $\mathfrak{p}_2 = \langle \theta, \omega\rangle$*;*
(C) *height one primes* P_1, P_2*, with* $\mathfrak{p}_i\mathcal{D} = P_i^2 \subsetneq P_i$ *for* $i = 1, 2$*, with each* P_i *generated by a normal (but not central) element. Moreover,* $P_1 + P_2 = P_0$*.*

THEOREM 1.3. *Retain the above notation.*

(a) *Every non-primitive prime is completely prime. Every primitive ideal is completely prime, except the co-Artinian maximal ideals (other than the counit), which form a subset of* $\mathcal{P}$*.*
(b) *Every prime ideal* P*, apart from the co-Artinian maximal ideals and (possibly)* P_0*, has* $\mathcal{D}/P$ *birationally equivalent to a Weyl algebra* $A_n(\mathbb{K})$*, where* $1 \leq n \leq 2$ *and* $\mathbb{K}$ *is a field of transcendence degree at most 2 over* k*.*

Theorem 1.1 is proved in Section 4, see in particular Subsection 4.5, while Theorem 1.2 is proved in Theorem 5.3. Finally, Theorem 1.3 is proved in Subsection 5.2. Some questions and a conjecture (Conjecture 5.6) are scattered through the paper.

1.3. Hopf algebras in duality. The full Drinfeld double $\mathcal{D}(H) = H \bowtie H^\circ$ of an infinite dimensional Hopf algebra H may often be unwieldy due to H having "too many" finite dimensional representations and thus leading to an unmanageably large finite dual H°. This has generated significant recent interest in constructing doubles $\mathcal{D}(H) := H \bowtie H'$ where H' is some suitable Hopf subalgebra of H°; see for example [**9**, **20**] and the papers listed in §1.1. Much is at present unclear: for example, what is an appropriate definition of a "suitable" Hopf subalgebra H'; and does a suitable algebra H' always exist? The double $\mathcal{D}$ of the Jordan plane is a test case for these and other questions. In particular, some of the desirable properties exhibited by $\mathcal{D}$ may form a paradigm for what one might aim for in defining doubles in more general settings.

1.4. Representation theory. A striking feature of the representation theory of $\mathcal{D}$ is the fact, proved in [**1**, Theorem 3.11] and recalled here in Theorem 2.2(iii), that $U(\mathfrak{sl}(2,k))$ is a quotient Hopf algebra of $\mathcal{D}$ and the finite dimensional irreducible $\mathcal{D}$-modules are precisely the finite dimensional irreducible $U(\mathfrak{sl}(2,k))$-modules. This immediately suggests a plethora of questions, some of which are addressed in [**3**], where Verma $\mathcal{D}$-modules are introduced. But many others remain untouched: for instance, what can be said about the category of locally finite dimensional $\mathcal{D}$-modules, and for each primitive ideal P of $\mathcal{D}$ can one find a canonical irreducible module (hopefully a factor of a Verma module) whose annihilator equals P? The first step in these questions is to classify the primitive ideals of $\mathcal{D}$, as we do here.

1.5. Dixmier-Moeglin equivalence. Let R be a noetherian k-algebra. Then *R satisfies the Dixmier-Moeglin Equivalence* if the following three statements are equivalent for every prime ideal P of R:

(i) P is a primitive ideal;
(ii) P is *rational* in the sense that the centre of the Goldie ring of quotients $Q(R/P)$ equals k;
(iii) P is *locally closed* in the sense that $P \subsetneqq R \cap \bigcap\{Q : P \subsetneqq Q,\ Q \text{ prime}\}$.

As we prove, this condition is indeed satisfied by $\mathcal{D}$.

THEOREM 1.4 (See Corollary 5.5). *Keep the notation from §1.2. Then $\mathcal{D}$ satisfies the Dixmier-Moeglin Equivalence.*

The validity of the Dixmier-Moeglin Equivalence for an algebra R yields simultaneous representation-theoretic, algebraic and topological characterisations of the primitive ideals amongst the prime ideals of R. This equivalence is a feature of some but not all Hopf k-algebras; see, for example, [**5**, **6**] for discussions of when it holds for a noetherian (Hopf) algebra. As we saw above $\mathcal{D}$ satisfies the equivalence. See §5.2 for more details, where we also give a rather ambitious Conjecture 5.6, proposing a general result encompassing all affine Noetherian pointed Hopf $\mathbb{C}$-algebras.

1.6. Notation. Throughout, all vector spaces and all unadorned tensor products are understood to be over the base field k. We denote the comultiplication of a Hopf algebra H by Δ and its augmentation ideal by H^+. The Gelfand-Kirillov, or GK dimension of an object X is denoted by $\mathrm{GKdim}(X)$, while the global (homological) dimension, respectively injective dimension of a ring R is denoted by $\mathrm{gldim}(R)$, respectively $\mathrm{injdim}(R)$. For precision, we specify that in the *Ore extension* $T = S[v; \sigma, \partial]$, multiplication is defined by

$$vs = s^{\sigma} v + \partial(s) \qquad \text{for } s \in S. \tag{1.1}$$

It follows that ∂ is a σ-derivation in the sense that $\partial(ab) = a^{\sigma}\partial(b) + \partial(a)b$. This follows the conventions of, for example, [**17**, p.34].

2. Preliminaries

2.1. Definitions and notation. The following definitions and notation from [**1**] will remain in play throughout the paper. First, the *Jordan plane* is

$$J \;:=\; k\langle x, y : [y, x] = -\tfrac{1}{2}x^2\rangle,$$

with *bosonization*

$$\mathcal{H} \;:=\; J\#C_\infty = J\#\langle g^{\pm 1}\rangle, \text{ where } gxg^{-1} = x,\ gyg^{-1} = y + x.$$

Then the *Drinfeld double* of J is defined to be $\mathcal{D} := \mathcal{H}\langle u, v, \zeta\rangle$, with additional relations as follows:

$$\begin{aligned}
&[u, v] = \tfrac{1}{2}u^2;\ \ [\zeta, v] = -v;\ \ [\zeta, u] = -u;\ [u, y] = 1 - g;\\
&[v, x] = 1 - g + xu;\ \ [v, y] = yu - g\zeta;\ [v, g] = gu;\ \ [\zeta, y] = y;\ \ [\zeta, x] = x;\\
&[x, u] = [x, g] = [u, g] = [\zeta, g] = 0.
\end{aligned}$$

The coalgebra structure, which will mostly not concern us here, is determined for $\mathcal{H}$ by specifying that g is grouplike and x and y are $(g, 1)$−primitive; and then extended to $\mathcal{D}$ by setting u and ζ to be primitive and $\Delta(v) = v\otimes 1 + 1\otimes v + \zeta\otimes u$.

Observe that $K := k\langle u, v, \zeta\rangle$ is a Hopf subalgebra of $\mathcal{D}$ and in fact, as one can see from the PBW theorem for $\mathcal{D}$ as described in [**2**, Proposition 2.3(ii)], $\mathcal{D} = \mathcal{H} \otimes_k K$ as vector spaces. As noted in [**2**, Lemma 2.2] there is a non-degenerate skew pairing between $\mathcal{H}$ and K which yields the multiplication relations between these subalgebras as in [**14**].

2.2. Initial results. We gather together in Theorem 2.2 some of the main results of [**2**] and [**1**]. We must first define some elements of $\mathcal{D}$, as follows. Set

$$q := ux + 2(1+g), \quad \text{and} \quad s := xv + uy + (-\frac{1}{2}ux + g - 1)\zeta - 2(g+1). \tag{2.1}$$

The following lemma is partly explicit, partly implicit, in [**1**, §4]. Given a k-algebra automorphism σ of a k-algebra H, we say that the element h of H is *σ-normal* if $ha = \sigma(a)h$ for all $a \in H$.

LEMMA 2.1. *Keep the above notation.*

(i) *q and s are both σ-normal, where σ is the automorphism of $\mathcal{D}$ defined by*

$$\sigma(y) = y + \tfrac{1}{2}x, \quad \sigma(v) = v - \tfrac{1}{2}u,$$

with σ acting as the identity on the other generators of $\mathcal{D}$.

(ii) *σ^2 equals conjugation by g on $\mathcal{D}$; that is, $\sigma^2(h) = ghg^{-1}$ for all $h \in \mathcal{D}$.*

(iii) *The elements $z := q^2g^{-1}$, $\theta := s^2g^{-1}$ and $\omega := qsg^{-1}$ are in the centre $Z(\mathcal{D})$ of $\mathcal{D}$.*

PROOF. (i) and (ii) are easy checks, and (iii) is immediate from (i) and (ii). □

THEOREM 2.2. *Retain the notation introduced above.*

(i) [**2**, Proposition 2.7(i)] *$\mathcal{O}(G) := k\langle x, u, g^{\pm 1}\rangle$ is a normal commutative Hopf subalgebra of $\mathcal{D}$, with $G = ((k,+) \times (k,+)) \rtimes (k^*, \times)$.*

(ii) [**2**, Proposition 2.7(ii)] *$\mathcal{K} = \mathcal{D}\mathcal{O}(G)^+$ is a Hopf ideal of $\mathcal{D}$, with an isomorphism of Hopf algebras*

$$\mathcal{D}/\mathcal{D}\mathcal{O}(G)^+ \cong U(\mathfrak{sl}_2(k)). \tag{2.2}$$

(iii) [**1**, Theorem 3.11] *The finite dimensional irreducible $\mathcal{D}$-modules are the finite dimensional irreducible $U(\mathfrak{sl}_2(k))$-modules given by the epimorphism (2.2).*

(iv) [**1**, Theorem 4.10] *With the notation from Lemma 2.1, the centre of $\mathcal{D}$ is*

$$Z(\mathcal{D}) = k\langle z, \omega, \theta : z\theta = \omega^2\rangle, \tag{2.3}$$

(v) [**1**, Remark 2.2] *$\mathcal{D}$ is pointed.* □

We'll need the following labelling of the maximal ideals of $Z(\mathcal{D})$. Note that here there are two maximal ideals of $Z(\mathcal{D})$ which require particular attention.

NOTATION 2.3. (i) By Theorem 2.2((iv), Maxspec($Z(\mathcal{D})$) consists of $\{\mathfrak{m}_{(\alpha,\gamma)} := \langle z - \alpha^2\gamma^{-1}, \omega - \alpha, \theta - \gamma\rangle : \alpha \in k,\ \gamma \in k^*\} \,\dot\cup\, \{\mathfrak{m}_\beta := \langle z - \beta, \omega, \theta\rangle : \beta \in k\}$. Note that $\mathfrak{m}_{(\alpha,\gamma)}$ can be simplified to $\mathfrak{m}_{(\alpha,\gamma)} = \langle \omega - \alpha, \theta - \gamma\rangle$, while $\mathfrak{m}_\beta = \langle z - \beta, \omega\rangle$ when $\beta \neq 0$.

(ii) It is easy to calculate using the definition of the counit that

$$\mathcal{D}^+ \cap Z(\mathcal{D}) = \mathcal{O}(G)^+\mathcal{D} \cap Z(\mathcal{D}) = \mathfrak{m}_{(-16,16)}.$$

We thus denote $\mathfrak{m}_{(-16,16)}$ by $\mathfrak{m}_+$.

(iii) It is clear that the singular locus of $Z(\mathcal{D})$ is $\{\mathfrak{m}_0\}$.

3. Ring-theoretic preparations

In this section we assemble some properties needed in the analysis of the primitive spectrum of $\mathcal{D}$. The proofs are most easily approached by viewing $\mathcal{D}$ as an iterated Hopf Ore extension starting not from the base field k but from the commutative normal Hopf subalgebra $\mathcal{O}(G) = k\langle x, u, g^{\pm 1}\rangle$ of Theorem 2.2(i). More precisely:

PROPOSITION 3.1. *$\mathcal{D}$ is an iterated Ore extension*

$$\mathcal{D} = \mathcal{O}(G)[y; \delta_1][\zeta; \delta_2][v; \tau, \delta_3], \tag{3.1}$$

where the derivations δ_1 and δ_2, the automorphism τ and the τ-derivation δ_3 can be read off from the defining relations of $\mathcal{D}$ given in Subsection 2.1.

In particular, $\mathcal{D}$ is a noetherian domain.

PROOF. Use the proof of [**2**, Proposition 1.6] to show that $\mathcal{D}$ has basis

$$\{g^a x^b u^c y^d \zeta^e v^f : a \in \mathbb{Z}, b, \dots, f \in \mathbb{N}\}.$$

Then the form of the relations (2.1) combined with [**12**, Theorem 1, p.438] show that it is indeed an Ore extension. □

Although it will be not needed in this paper, the description (3.1) even describes $\mathcal{D}$ as an Iterated Hopf Ore Extension (IHOE), in the sense that each extension in that formula is itself a Hopf algebra. It also shows that, by setting

$$\deg x = \deg u = \deg g = \deg g^{-1} = 0;$$
$$\deg y = \deg \zeta = \deg v = 1,$$

one obtains a filtration $\mathcal{F}$ on $\mathcal{D}$ with associated graded algebra

$$\operatorname{gr}_{\mathcal{F}} \mathcal{D} = \mathcal{O}(G)[\overline{y}, \overline{\zeta}, \overline{v}]. \tag{3.2}$$

So $\operatorname{gr}_{\mathcal{F}} \mathcal{D}$ is a commutative polynomial algebra in 6 variables with one variable inverted.

3.1. Homological properties. In this subsection we note that $\mathcal{D}$ has certain useful homological properties, and we begin with the relevant definitions. A ring A is called *Auslander Gorenstein* if it has finite injective dimension and satisfies the *Gorenstein condition*: if $p < q$ are non-negative integers and M is a finitely generated A-module, then $\operatorname{Ext}^p_A(N, A) = 0$ for every submodule N of $\operatorname{Ext}^q_A(M, A)$. The ring A is *Auslander regular* if it is Auslander Gorenstein of finite global dimension. Set $j_A(M) = \min\{r : \operatorname{Ext}^r_A(M, A) \neq 0\}$ for the *homological grade* of M. Then an Auslander Gorenstein ring A of finite GK dimension is called *GK-Cohen-Macaulay* (or just CM), provided that $j_A(M) + \operatorname{GKdim}(M) = \operatorname{GKdim}(A)$ holds for each such M. Obviously affine commutative regular rings are both Auslander regular and CM.

PROPOSITION 3.2. (i) *$\mathcal{D}$ is Auslander regular and CM.*

(ii) *$\mathcal{D}$ is AS regular in the sense of, say,* [**19**].

(iii) $\operatorname{GKdim}(\mathcal{D}) = 6 = \operatorname{gldim}(\mathcal{D})$.

(iv) *GK dimension is an exact function on finitely generated $\mathcal{D}$-modules.*

PROOF. (i) By [**7**, Remark, p.157] the filtration $\mathcal{F}$ is Zariskian and so the result follows from [**7**, Theorems 3.8, 3.9 and Remark, p. 165].

(ii) This is immediate from (i) and [**11**, Lemma 6.1].

(iii) By [**22**, Corollary 1.4], $\mathrm{GKdim}(\mathcal{D}) = \mathrm{GKdim}(\mathrm{gr}_{\mathcal{F}}(\mathcal{D})) = 6$.

By Proposition 3.1 and [**21**, Theorem 7.5.3(i)], we have $\mathrm{gldim}(\mathcal{D}) \le 6$. By [**1**, Theorem 3.11] $\mathcal{D}$ has a finite dimensional module, say M and the CM condition implies that M has homological dimension ≥ 6. Hence $\mathrm{gldim}(\mathcal{D}) = 6$.

(iv) Since $j_{\mathcal{D}}$ is exact on finitely generated $\mathcal{D}$-modules by [**19**, Theorem 2.3], this follows from the CM condition. □

□

3.2. Key lemma. The following lemma will be crucial in our analysis of the primitive spectrum of $\mathcal{D}$. In its proof, given an ideal B of a noetherian ring S, we denote by $\sqrt{B}$ the ideal of S such that $\sqrt{B}/B$ is the nilradical of S/B.

LEMMA 3.3. *Let M be a finitely generated (right or left) $\mathcal{D}$-module such that either $\mathrm{Ann}_{k[x]}(M) \neq 0$ or $\mathrm{Ann}_{k[u]}(M) \neq 0$. Then*

(i) *there exists $r \ge 1$ such that*

$$(\mathfrak{m}_+\mathcal{D})^r \subseteq (\mathcal{O}(G)^+\mathcal{D})^r \subseteq \mathrm{Ann}_{\mathcal{D}}(M);$$

(ii) $\mathrm{GKdim}(M) \le 3$.

PROOF. (i) Let $I := \mathrm{Ann}_{\mathcal{D}}(M)$, an ideal of $\mathcal{D}$. Assume that $I \cap k[x] \neq 0$, the proof in the other case being exactly similar, but with $k\langle u, v\rangle$ replacing J. One easily confirms that every non-zero prime ideal of the Jordan plane $J = k\langle x, y\rangle$ contains x. Therefore, since $I \cap k[x] \neq 0$, there exists $N \ge 1$ such that $x^N \in I \cap k[x] \subseteq I \cap \mathcal{O}(G)$. Since $\mathcal{O}(G)$ is commutative,

$$x \in \sqrt{(I \cap \mathcal{O}(G))}. \tag{3.3}$$

Since I is an ideal of $\mathcal{D}$, $[v, I] \subseteq I$; moreover, from the defining relations of $\mathcal{D}$ and the fact that $\mathcal{O}(G) = k\langle x, u, g^{\pm 1}\rangle$, $[v, \mathcal{O}(G)] \subseteq \mathcal{O}(G)$. Therefore

$$[v, I \cap \mathcal{O}(G)] \subseteq I \cap \mathcal{O}(G). \tag{3.4}$$

Since k has characteristic 0 it follows from (3.4) and [**17**, Lemma 3.20] that

$$[v, \sqrt{(I \cap \mathcal{O}(G))}] \subseteq \sqrt{(I \cap \mathcal{O}(G))}. \tag{3.5}$$

By (3.3) and (3.5)

$$[v, x] = 1 - g + xu \in \sqrt{(I \cap \mathcal{O}(G))},$$

so that $(1 - g) \in \sqrt{(I \cap \mathcal{O}(G))}$. Then

$$[v, g - 1] = [v, g] = gu \in \sqrt{(I \cap \mathcal{O}(G))},$$

so that $u \in \sqrt{(I \cap \mathcal{O}(G))}$. Since $\mathcal{O}(G)^+$ is generated by x, u and $g - 1$ we deduce that $\mathcal{O}(G)^+\mathcal{D} \subseteq \sqrt{I}$, proving (i).

(ii) By (i) M is a finitely generated $\mathcal{D}/(\mathcal{O}(G)^+\mathcal{D})^r$-module for some $r \ge 1$. Since $\mathcal{D}/\mathcal{O}(G)^+\mathcal{D} \cong U(\mathfrak{sl}(2, k)$ by Theorem 2.2(ii), and so has GK dimension 3, (ii) follows from this and Proposition 3.2(iv). □

3.3. Ore localisations of $\mathcal{D}$. To help in the analysis of its primitive spectrum we need four Ore localisations of $\mathcal{D}$. The first of these is described in [**1**, Theorem 4.8], and the others are similar. These sets are described as follows:

DEFINITION 3.4. Label the following four subsets of $\mathcal{D}$:

$$A := \{q^i : i \ge 0\} \dot\cup \{x^j : j \ge 0\}, \quad B := \{s^i : i \ge 0\} \dot\cup \{x^j : j \ge 0\},$$
$$C := \{q^i : i \ge 0\} \dot\cup \{u^j : j \ge 0\}, \quad D := \{s^i : i \ge 0\} \dot\cup \{u^j : j \ge 0\}.$$

LEMMA 3.5. (1) *The elements x and u act ad-locally-nilpotently on $\mathcal{D}$. Consequently, $\{x^i : i \geq 0\}$ and $\{u^i : i \geq 0\}$ are Ore sets in $\mathcal{D}$.*

(2) *For each $\Omega \in \{A, B, C, D\}$ the set Ω is an Ore set of regular elements of $\mathcal{D}$, and we write the corresponding localisation as $\mathcal{D}_{(\Omega)}$.*

PROOF. (1) For x this is proved in [**1**, Lemma 4.3(i)]. The claim for u is a similar easy consequence of the defining relations of $\mathcal{D}$.

(2) Localising at the powers of q is the same as localising at the powers of q^2 or even at the powers of $z = q^2g^{-1}$ (use the fact that $q^{-1} = q \cdot q^{-2}$ and that g is a unit). Thus, for $\Omega = A$ or $\Omega = C$ and appealing to Lemma 2.1(iii), we can replace q by the central element z. Similarly in the other two cases we can replace s by the central element θ. Thus in each case we wish to localise at one central and one locally ad-nilpotent element in the domain $\mathcal{D}$. Thus it is indeed an Ore set of regular elements. $\square$

Thus each of the four rings $\mathcal{D}_{(\Omega)}$ is a subalgebra of the quotient division algebra $Q(\mathcal{D})$ of $\mathcal{D}$ that contains $\mathcal{D}$. As we next show, each of these rings is a localisation of the second Weyl algebra over a commutative ring.

NOTATION 3.6. (i) In $\mathcal{D}_{(A)}$, set $p_A := -2q^{-1}x^{-1}y$, $q_A := q$, $t_A := qx^{-1}$, and $\eta_A := -xq^{-1}\zeta$.
(ii) In $\mathcal{D}_{(B)}$, set $p_B := -2s^{-1}x^{-1}y$, $q_B := s$, $t_B := sx^{-1}$. $\eta_B := -xs^{-1}\zeta$.
(iii) In $\mathcal{D}_{(C)}$, set $p_C := 2q^{-1}u^{-1}v$, $q_C := q$, $t_C := q^{-1}u^{-1}$, $\eta_C := -uq\zeta$.
(iv) In $\mathcal{D}_{(D)}$, set $p_D := 2s^{-1}u^{-1}v$, $q_D := s$, $t_D := s^{-1}u^{-1}$, $\eta_D := us\zeta$.
We further set $z_\Omega := z$ for $\Omega = A$ and $\Omega = C$ but $z_\Omega := \theta$ when $\Omega = B, D$.

The motivation behind the above definitions becomes clear from the following lemma. For $\Omega = A$, this was obtained in the proof of [**1**, Theorem 4.8]. The claims regarding the other elements can be checked by a similar direct calculation.

LEMMA 3.7. *Let $\Omega \in \{A, B, C, D\}$. Then we have the following relations in $Q(\mathcal{D})$:*

$$[p_\Omega, q_\Omega] = 1 = [\eta_\Omega, t_\Omega], \quad \textit{with all other brackets being zero;} \quad \square$$

When $\Omega = A$ the following result is given in [**1**, Theorem 4.8], although we give a proof that works for all 4 cases simultaneously.

THEOREM 3.8. *For each $\Omega \in \{A, B, C, D\}$, the localisation $\mathcal{D}_{(\Omega)}$ is a localised Weyl algebra over its centre. More precisely:*

$$\mathcal{D}_{(\Omega)} = A_2^{(\Omega)}(k) \otimes S_{(\Omega)},$$

where $A_2^{(\Omega)}(k)$ denotes the localisation of the second Weyl algebra over k with generators $p_\Omega, q_\Omega^{\pm 1}, \eta_\Omega, t_\Omega^{\pm 1}$, while $S_{(\Omega)}$ is the commutative ring $S_{(\Omega)} = k[z_\Omega^{\pm 1}, \omega]$.

PROOF. The generators z, ω and θ of $Z(\mathcal{D})$ are given in Lemma 2.1(iii), from which it follows that the subalgebra $S_{(\Omega)}$ of $Q(\mathcal{D})$ is contained in the centre $Z(\mathcal{D}_{(\Omega)})$. Therefore we can consider the subalgebra

$$\mathcal{E}_{(\Omega)} := S_{(\Omega)}\langle p_\Omega, q_\Omega, t_\Omega, \eta_\Omega\rangle \subseteq \mathcal{D}_{(\Omega)}. \tag{3.6}$$

We claim that the inclusion (3.6) is an equality. In order to prove this, check that given generators of $\mathcal{D}_\Omega$ are contained in $\mathcal{E}_{(\Omega)}$. Thus, for example, when $\Omega = A$, one shows that $\{q^{-1}, x^{\pm 1}, y, \zeta, g^{\pm 1}, u, v\} \subset \mathcal{E}_{(A)}$, and similarly in the other cases. Thus,

$\mathcal{E}_{(\Omega)} = \mathcal{D}_{(\Omega)}$, as claimed. As noted in the proof of Lemma 3.5 the localisation of $\mathcal{D}$ at Ω involves inverting one central and one ad-nilpotent element of $\mathcal{D}$. Thus, by Proposition 3.2(iii) and [**18**, Lemma 4.7], $\mathrm{GKdim}(D_{(\Omega)}) = \mathrm{GKdim}(\mathcal{D}) = 6$. We conclude that $\mathrm{GKdim}(\mathcal{E}_{(\Omega)}) = \mathrm{GKdim}(\mathcal{D}) = 6$.

On the other hand, by Lemma 3.7 $\mathcal{E}_{(\Omega)}$ is a factor of the ring

$$V_{(\Omega)} := S_{(\Omega)} \otimes_k A_2^{(\Omega)}(k),$$

which is also a domain of GK-dimension 6. So if $\mathcal{E}_{(\Omega)}$ were a proper factor of $V_{(\Omega)}$, then [**21**, Corollary 8.3.6] would imply that $\mathrm{GKdim}(\mathcal{E}_{(\Omega)}) < 6$, giving a contradiction.

So the only possibility is that $\mathcal{E}_{(\Omega)} \cong V_{(\Omega)} = S_{(\Omega)} \otimes_k A_2^{(\Omega)}(k)$, as required. □

4. The primitive spectrum of $\mathcal{D}$

In this section we describe the primitive spectrum of $\mathcal{D}$. This splits naturally into several cases:

- the primitive ideals not containing $\mathfrak{m}_+$ or $\mathfrak{m}_0$; these are the generic ones;
- the ideal $\mathfrak{m}_+\mathcal{D}$, which is also primitive;
- the ideal $\mathfrak{m}_0\mathcal{D}$, for which $\sqrt{\mathfrak{m}_0\mathcal{D}}$ is a unique prime ideal P_0;
- finally, P_0 is also maximal.

The details are given in the next four subsections with the results being combined in Subsection 4.5.

In this section and in Section 5.1 we will without further reference use of the yoga for prime ideals of Noetherian rings under Ore localisation as described in, for example, [**17**, Theorems 10.18 and 10.20]. We use Notation 2.3 to describe the maximal ideals of $Z(\mathcal{D})$ and Definition 3.4 to define Ore sets in $\mathcal{D}$.

4.1. The generic minimal primitives.

We begin by looking at the generic case.

THEOREM 4.1. *Let $\mathfrak{m}$ be a maximal ideal of $Z(\mathcal{D})$ with $\mathfrak{m} \neq \mathfrak{m}_+$ and $\mathfrak{m} \neq \mathfrak{m}_0$. Then the following are true.*

(i) *$\mathfrak{m}\mathcal{D}$ is a completely prime maximal ideal of $\mathcal{D}$.*
(ii) *The localisation of $\mathcal{D}/\mathfrak{m}\mathcal{D}$ at the powers of (the image of) either x or u is isomorphic to a localised Weyl algebra $A_2^{(\Omega)}(k)$, where $\Omega \in \{A, B, C, D\}$.*
(iii) $\mathrm{GKdim}(\mathcal{D}/\mathfrak{m}\mathcal{D}) = 4$.
(iv) *$\mathfrak{m}\mathcal{D}$ is generated by a central regular sequence of length 2.*
(v) *$\mathcal{D}/\mathfrak{m}\mathcal{D}$ is CM and is Auslander Gorenstein with* $\mathrm{injdim}(\mathcal{D}/\mathfrak{m}\mathcal{D}) < 4$.

PROOF. (i), (ii) By Notation 2.3, $\mathfrak{m} = \langle z-\alpha, \omega-\beta, \theta-\gamma\rangle$ with $\alpha,\beta,\gamma \in k$ and $\alpha\gamma = \beta^2$. Moreover, thanks to the hypothesis on $\mathfrak{m}$, either (a) $\alpha \neq 0$ or (b) $\gamma \neq 0$.

Assume (a). We prove (ii) for the localisation at the powers of x. (The arguments for powers of u are exactly similar, but using the Ore sets C and D rather than A and B.) Using the notation of §3.3 and applying Theorem 3.8, we see that $\mathfrak{m}\mathcal{D}_{(A)}$ is a maximal ideal of $\mathcal{D}_{(A)}$. Observe that, since $A := \{q^i, x^j : i, j \geq 0\}$ and

$$z = q^2 g^{-1} \equiv \alpha \neq 0 \bmod(\mathfrak{m}\mathcal{D}),$$

$A_2^{(A)}(k)$ is isomorphic to the localisation of $\mathcal{D}/\mathfrak{m}\mathcal{D}$ at the powers of x. This suffices to prove (ii) in case (a).

Define

$$P_{\mathfrak{m}} := \mathfrak{m}\mathcal{D}_{(A)} \cap \mathcal{D},$$

so that $P_{\mathfrak{m}}$ is a completely prime ideal of $\mathcal{D}$ with $\mathfrak{m}\mathcal{D} \subseteq P_{\mathfrak{m}}$. By definition of $P_{\mathfrak{m}}$,

$$\mathfrak{m}\mathcal{D}_{(A)} = P_{\mathfrak{m}}\mathcal{D}_{(A)}. \tag{4.1}$$

We claim that in fact

$$P_{\mathfrak{m}} = \mathfrak{m}\mathcal{D}. \tag{4.2}$$

Since $\mathcal{D}$ is (left) noetherian there exist $e_1, \ldots, e_t \in P_{\mathfrak{m}}$ such that $P_{\mathfrak{m}} = \mathfrak{m}\mathcal{D} + \sum_{i=1}^{t} \mathcal{D}e_i$. By (4.1), for each $i = 1, \ldots, t$ there exist $f_i \in \mathfrak{m}\mathcal{D}$ and $s_i \in \mathbb{Z}_{\geq 0}$ such that

$$e_i = f_i x^{-s_i}. \tag{4.3}$$

Define $s := \max\{s_i : 1 \leq i \leq t\} \in \mathbb{Z}_{\geq 0}$, and

$$I := \{\tau \in \mathcal{D} : P_{\mathfrak{m}}\tau \subseteq \mathfrak{m}\mathcal{D}\}.$$

Thus I is an ideal of $\mathcal{D}$ containing $\mathfrak{m}\mathcal{D}$ and, by (4.3), $x^s \in I$. If $s = 0$ then $I = \mathcal{D}$; otherwise we see from Lemma 3.3 that $(\mathfrak{m}_+)^r \subset I$ for some $r \geq 1$. Since also $\mathfrak{m} \subseteq I$ and $\mathfrak{m} \neq \mathfrak{m}_+$ by hypothesis, it follows that $I = \mathcal{D}$, and (4.2) is proved.

In case (a) it remains only to prove that $P_{\mathfrak{m}}$ is a maximal ideal of $\mathcal{D}$. Suppose then that J is an ideal of $\mathcal{D}$ with $P_{\mathfrak{m}} \subsetneq J$. Then $J\mathcal{D}_{(A)} = D_{(A)}$ by the maximality of the ideal $P_{\mathfrak{m}}\mathcal{D}_{(A)}$ of $\mathcal{D}_{(A)}$. Again using the fact that $q + \mathfrak{m}\mathcal{D}$ is a unit of $\mathcal{D}/\mathfrak{m}\mathcal{D}$ we see that $x^s \in J$ for some $s \geq 1$. Then, as before, Lemma 3.3 implies that $J = \mathcal{D}$.

Suppose that (b) holds rather than (a). Then the element s is a unit mod$(\mathfrak{m}\mathcal{D})$, so we use the same argument as for (a), but working with $\mathcal{D}_{(C)}$ rather than $\mathcal{D}_{(A)}$.

(iii) By (ii) and [**18**, Example 3.7 and Theorem 4.9] the localisation of $\mathcal{D}/\mathfrak{m}\mathcal{D}$ at the powers of x has GK dimension 4. Since $\mathrm{ad}(x)$ acts nilpotently on $\mathcal{D}/\mathfrak{m}\mathcal{D}$ by Lemma 3.5, it follows from [**18**, Theorem 4.9] that $\mathrm{GKdim}(\mathcal{D}/\mathfrak{m}\mathcal{D}) = 4$.

(iv) Again we assume (a) that $z - \alpha \in \mathfrak{m}$ for $\alpha \in k \setminus \{0\}$, the proof in case (b) being similar. We can begin a regular central sequence in $\mathfrak{m}\mathcal{D}$ with $z - \alpha$. Since $\mathcal{D}$ is CM of GK-dimension 6 by Proposition 3.2(i, iii), it follows from [**16**, Theorem 7.2(b)] that $\mathcal{D}/(z-\alpha)\mathcal{D}$ is CM of GK-dimension 5. Moreover, by [**19**, Remark 2.4] the CM property ensures that $\mathcal{D}/(z-\alpha)\mathcal{D}$ is GK-homogeneous; that is, it contains no non-zero ideal with GK-dimension strictly less than 5. Since $Z(\mathcal{D})/(z-\alpha)Z(\mathcal{D})$ is a polynomial algebra we can choose $y \in \mathfrak{m}$ such that $\mathfrak{m} = \langle z-\alpha, y\rangle$. If $y + (z-\alpha)\mathcal{D}$ is a zero divisor in $\mathcal{D}/(z-\alpha)\mathcal{D}$ we obtain a non-zero ideal of $\mathcal{D}/(z-\alpha)\mathcal{D}$ killed by $\mathfrak{m}\mathcal{D}$, contradicting the GK-homogeneity of $\mathcal{D}/(z-\alpha)\mathcal{D}$ in view of (iii). Thus $\{z-\alpha, y\}$ is a regular central sequence in $\mathfrak{m}\mathcal{D}$.

(v) Since $\mathcal{D}$ is CM by Proposition 3.2(i), $R = \mathcal{D}/\mathfrak{m}\mathcal{D}$ is CM with $\mathrm{GKdim}(R) = 4$ by (iv) and two applications of [**16**, Theorem 7.2(b)]. The Auslander Gorenstein property is given by (iv) and [**19**, §3.4, Remark (3)]. As R is simple it cannot have a finite dimensional module. Hence $\mathrm{injdim}(R) < 4$ follows from the next lemma. □

The following observation is well-known.

LEMMA 4.2. *Let R be a noetherian, Auslander Gorenstein, CM ring and write* $\mathrm{GKdim}(R) = m$. *Then* $\mathrm{injdim}(R) \leq m$. *Moreover* $\mathrm{injdim}(R) = m \iff R$ *has a finite dimensional representation.*

PROOF. Let $n = \text{injdim}(R)$ and pick a finitely generated R-module M such that $\text{Ext}_R^n(M,R) \neq 0$. By the Auslander condition and the spectral sequence [**19**, Theorem 2.2] $j(E^{nn}(M)) = n$ for $E^{nn} = \text{Ext}^n(\text{Ext}^n(M,R),R)$. By the CM property $\text{GKdim}(E^{nn}(M)) = m - n$ and the result follows easily. □

4.2. Non-generic minimal primitives (I) - $\mathfrak{m}_+$. The next case to consider is $\mathfrak{m}\mathcal{D}$ for $\mathfrak{m} = \mathfrak{m}_+$, as we do here. Recall from Notation 2.3(ii) that $\mathfrak{m}^+ = \mathcal{D}^+ \cap Z(\mathcal{D}) = \langle \omega + 16, \theta - 16 \rangle$.

We start with a subsidiary result, which works for any field k of characteristic zero.

THEOREM 4.3. *$\mathcal{D}$ is a Jacobson ring that satisfies the Nullstellensatz, in other words:*

(i) *every prime ideal of $\mathcal{D}$ is an intersection of primitive ideals;*
(ii) *for every simple $\mathcal{D}$-module M, $\text{End}_{\mathcal{D}}(M)$ is algebraic over k. In particular, every primitive ideal of $\mathcal{D}$ contains a maximal ideal of $Z(\mathcal{D})$.*

PROOF. By (3.2), $\mathcal{D}$ has a filtration $\mathcal{F}$ such that the associated graded ring $\text{gr}_{\mathcal{F}}(\mathcal{D})$ is a commutative affine ring. Hence by [**22**, Corollary 1.7] there is a second filtration $\mathcal{G}$ by finite dimensional k-subspaces of $\mathcal{D}$ such that $\text{gr}_{\mathcal{F}}(\mathcal{D})$ is also a commutative and affine ring. The result now follows from [**4**, Theorem 0.4]. □

THEOREM 4.4. (i) *$\mathfrak{m}_+\mathcal{D}$ is a completely prime, primitive ideal of $\mathcal{D}$.*

(ii) *The localisation of $\mathcal{D}/\mathfrak{m}_+\mathcal{D}$ at the powers of x or the powers of u is a localisation of the Weyl algebra $A_2(k)$ at powers of a generator.*
(iii) *$\mathfrak{m}_+\mathcal{D}$ is generated by a regular central sequence of length 2.*
(iv) *$\mathcal{D}/\mathfrak{m}_+\mathcal{D}$ is Auslander Gorenstein and CM with*

$$\text{GKdim}(\mathcal{D}/\mathfrak{m}_+\mathcal{D}) = 4 = \text{injdim}(\mathcal{D}/\mathfrak{m}_+\mathcal{D}).$$

(v) *Every prime ideal P of $\mathcal{D}$ which strictly contains $\mathfrak{m}_+\mathcal{D}$ satisfies*

$$\mathcal{O}(G)^+\mathcal{D} \subseteq P,$$

so the space of such primes P is homeomorphic to $\text{Spec}(U(\mathfrak{sl}(2,k))$.

PROOF. Recall that $q_A = q$. Since $q^2 \equiv 16g \not\equiv 0 \bmod(\mathfrak{m}_+\mathcal{D})$, Theorem 3.8 implies that $\mathfrak{m}_+\mathcal{D}_{(A)}$ is a maximal ideal of $\mathcal{D}_{(A)}$, with $\mathcal{D}/\mathfrak{m}_+\mathcal{D}_{(A)} \cong A_2^{(A)}(k)$. Therefore, defining $P_+ := \mathfrak{m}_+\mathcal{D}_{(A)} \cap \mathcal{D}$, we deduce that P_+ is a completely prime ideal of $\mathcal{D}$ with

$$\mathfrak{m}_+\mathcal{D} \subseteq P_+. \tag{4.4}$$

We will eventually show that (4.4) is an equality.

As in the proofs of Theorem 4.1(i),(ii), let I be the right annihilator in $\mathcal{D}$ of $P_+/\mathfrak{m}_+\mathcal{D}$. Then I contains $\mathfrak{m}_+\mathcal{D}$ and a power of x, and hence, by Lemma 3.3,

$$(\mathcal{O}(G)^+\mathcal{D})^r \subseteq I \qquad \text{for some } r \in \mathbb{Z}_{\geq 1}, \tag{4.5}$$

In particular, $\text{GKdim}(\mathcal{D}/I) \leq 3$ by Lemma 3.3(ii). Therefore, by [**18**, Proposition 5.1(d)]

$$\text{GKdim}(P_+/\mathfrak{m}_+\mathcal{D}) \leq 3. \tag{4.6}$$

Recall that $\text{GKdim}(A_2^{(A)}(k)) = 4$ by [**18**, Example 7.3 and Theorem 4.9], so that

$$\text{GKdim}(\mathcal{D}/P_+) = 4 \tag{4.7}$$

by [**18**, Theorem 4.9]. Thus, from (4.6), (4.7) and Proposition 3.2(iv) it follows that

$$\text{GKdim}(\mathcal{D}/\mathfrak{m}_+\mathcal{D}) \;=\; 4. \tag{4.8}$$

By Proposition 3.2, $\mathcal{D}$ is CM and Auslander regular, with $\text{gldim}(\mathcal{D}) = 6 = \text{GKdim}(\mathcal{D})$. It therefore follows from the CM property of $\mathcal{D}$ together with (4.8) that

$$j_{\mathcal{D}}(\mathcal{D}/\mathfrak{m}_+\mathcal{D}) \;=\; 6 - 4 \;=\; 2. \tag{4.9}$$

From (4.9) and [**8**, Proposition 3.6] we deduce that the maximum length of a regular sequence of elements of $\mathfrak{m}_+$ on $\mathcal{D}$ is precisely 2; in particular any choice of a generating pair of elements of $\mathfrak{m}_+$, for example, $\{z - 16, \omega + 16\}$, is a regular sequence on $\mathcal{D}$. Therefore, by two applications of [**16**, Theorem 7.2(b)],

$$\mathcal{D}/\mathfrak{m}_+\mathcal{D} \text{ is CM of GK-dimension 4.} \tag{4.10}$$

Similarly, two applications of [**19**, §3.4, Remark (3)] show that $\mathcal{D}/\mathfrak{m}_+\mathcal{D}$ is Auslander Gorenstein. By Lemma 3.3(i) and Theorem 2.2(ii), $U(\mathfrak{sl}(2,k)) \cong \mathcal{D}/\mathcal{D}\mathcal{O}(G)^+$ is a factor of $\mathcal{D}/\mathfrak{m}_+\mathcal{D}$ and so $\mathcal{D}/\mathfrak{m}_+\mathcal{D}$ has a non-zero finite dimensional module, M. Thus, by Lemma 4.2, $\text{injdim}(\mathcal{D}/\mathfrak{m}_+\mathcal{D}) = 4$.

By [**19**, Remark 2.4], again, the CM property for $\mathcal{D}/\mathfrak{m}_+\mathcal{D}$ implies that $\mathcal{D}/\mathfrak{m}_+\mathcal{D}$ is GK-homogeneous. Therefore we may conclude from (4.6) that P_+ does indeed equal $\mathfrak{m}_+\mathcal{D}$. This proves (i) - (iv), with the exception of showing that $\mathfrak{m}_+\mathcal{D}$ is primitive.

(v) Let Q be a prime ideal of $\mathcal{D}$ with $\mathfrak{m}_+\mathcal{D} \subsetneq Q$. As already noted, q is congruent to a unit mod$(\mathfrak{m}_+\mathcal{D})$. Then $Q\mathcal{D}_{(A)} = \mathcal{D}_{(A)}$ by *(ii)*, so Q must contain a power of x. Hence, by Lemma 3.3, $\mathcal{O}(G)^+\mathcal{D} \subseteq Q$, as required.

Finally, to see that $\mathfrak{m}_+\mathcal{D}$ is primitive note that (v) shows that it is locally closed. Hence it is primitive by Theorem 4.3(i). □

4.3. Non-generic minimal primitives (II) - $\mathfrak{m}_0$. In this subsection we begin our study of the ideal $\mathfrak{m}_0\mathcal{D}$. Recall the definition of q, s from (2.1) and, from Notation 2.3(iii), that $\mathfrak{m}_0 := \langle q^2g^{-1},\, qsg^{-1},\, s^2g^{-1}\rangle$ is the unique singular point of $\text{Maxspec}(Z(\mathcal{D}))$. Clearly the right ideal

$$P_0 \;:=\; q\mathcal{D} + s\mathcal{D} \tag{4.11}$$

is a two-sided ideal of $\mathcal{D}$ since q and s are normal in $\mathcal{D}$ by Lemma 2.1. Moreover,

$$\mathfrak{m}_0\mathcal{D} = P_0^2 \subset P_0 \subseteq \sqrt{\mathfrak{m}_0\mathcal{D}}.$$

As part of the next proposition we see that P_0 is completely prime, so the second inclusion above is an equality. In fact P_0 is a maximal ideal, but this is more difficult to prove, and is delayed until §4.4.

PROPOSITION 4.5. *Retain the above notation, and set $T := \mathcal{D}/P_0$.*

(i) *T is a localisation of a 4-step iterated Ore extension of k, namely*

$$T \;=\; \big((k[u,x]\langle(ux+2)^{-1}\rangle)[y;\partial_1]\big)[v;\sigma,\partial_2],$$

where u and x commute,

$$\partial_1(u) = -\tfrac{1}{2}ux - 2, \quad \partial_1(x) = -\tfrac{1}{2}x^2,$$
$$\partial_2(u) = -\tfrac{1}{2}u^2, \quad \partial_2(x) = \tfrac{3}{2}ux + 2, \quad \partial_2(y) = \tfrac{3}{2}uy - 2,$$

and $\sigma(y) = y + \frac{1}{2}x$, with $\sigma(x) = x$ and $\sigma(u) = u$.

(ii) $\{q, s\}$ *forms a regular normal sequence of generators of* P_0.
(iii) $\operatorname{gldim}(T) \leq 4 = \operatorname{GKdim}(T)$.
(iv) T *is CM and is an Auslander regular domain.*

PROOF. Throughout the proof we abuse notation by simply denoting the image in T of an element ω of $\mathcal{D}$ by ω when no confusion seems likely.

(i),(ii) Since $q := ux + 2(1+g)$ and $q \equiv 0 \bmod(P_0)$, we can write

$$g \equiv -\tfrac{1}{2}(ux+2) \bmod(P_0), \tag{4.12}$$

so that

$$ux + 2 \text{ is a unit in } T. \tag{4.13}$$

Using (4.12) we find that, $\bmod(P_0)$,

$$s \ := \ xv + uy + (-\tfrac{1}{2}ux + g - 1)\zeta - 2(g+1) \ \equiv \ xv + uy + 2g\zeta - 2g - 2,$$

so that, since $s \in P_0$,

$$\zeta \ \equiv \ -\tfrac{1}{2}g^{-1}(ux + xv + uy) \quad \bmod(P_0) \tag{4.14}$$

It follows from (4.12), (4.13) and (4.14) that

$$T \ = \ k\langle u, x, (ux+2)^{-1}, y, v\rangle. \tag{4.15}$$

The relations for $\mathcal{D}$ given in §2.1 immediately imply the following relations for the generators for T listed in (4.15)

$$[u,x] = 0, \qquad [y,x] = -\tfrac{1}{2}x^2, \qquad [v,x] = \tfrac{3}{2}ux + 2,$$
$$[y,u] = -\tfrac{1}{2}ux - 2, \qquad [v,u] = -\tfrac{1}{2}u^2, \qquad [v,y] = \tfrac{3}{2}uy + \tfrac{1}{2}xv - 2.$$

Clearly the iterated Ore extension of $k[u,x]\langle(ux+2)^{-1}\rangle$ defined in (i), which we temporarily label $\widehat{T}$, satisfies precisely these relations, so there is an algebra epimorphism Φ from $\widehat{T}$ onto T.

We next show that Φ is an isomorphism, which we do by computing $\operatorname{GKdim}(T)$. First note that $\operatorname{GKdim}(\widehat{T}) = 4$ by [**18**, Theorem 12.3.1], since it is a PBW extension in 2 variables of $k[u,x]\langle(ux+2)^{-1}\rangle$. Thus, certainly $\operatorname{GKdim}(T) \leq 4$. On the other hand $\mathcal{D}$ is CM of GK-dimension 6 by Proposition 3.2(i, iii). Hence, because q is a regular normal element of $\mathcal{D}$ by Lemma 2.1, $\mathcal{D}/q\mathcal{D}$ is CM of GK-dimension 5 by [**16**, Theorem 7.2(b) and its proof]. Moreover $\mathcal{D}/q\mathcal{D}$ is GK-homogeneous by [**19**, Remark 2.4]. Since $\operatorname{GKdim}(T) \leq 4$, this ensures that

$$s \text{ cannot be a zero-divisor } \bmod(q\mathcal{D}). \tag{4.16}$$

Since $P_0 \ := \ q\mathcal{D} + s\mathcal{D}$, a second application of [**16**, Theorem 7.2(b) and its proof] yields $\operatorname{GKdim}(T) = 4$ and also shows that

$$T \text{ is CM.} \tag{4.17}$$

Since $\widehat{T}$ is a domain, the equality $\operatorname{GKdim}(\widehat{T}) = 4 = \operatorname{GKdim}(T)$, combined with [**18**, Proposition 3.15], shows that $\widehat{T} = T$. Thus (i) is proved, with (ii) also following thanks to (4.16).

(iii) By (i), T is a 2-step iterated Ore extension of $k[u.x]\langle(ux+2)^{-1}\rangle$, and so two applications of [**21**, Theorem 7.5.3(i)] gives $\operatorname{gldim}(T) \leq 4$.

(iv) That T is a domain is clear from (i), while the CM property was proved in (4.17). The Auslander Gorenstein property holds for $\mathcal{D}$ by Proposition 3.2(i).

Thus, by (ii) and two applications of [**16**, Theorem 7.2(a)], T is also Auslander Gorenstein and it is then Auslander regular by (iii). $\square$

We remark that, by Lemma 4.2 and Theorem 2.2(iii) it follows that $\mathrm{gldim}(T) < 4$. We do not know the exact value of $\mathrm{gldim}(T)$.

4.4. Maximality of P_0. Let $T := \mathcal{D}/P_0$ as in Proposition 5.3. Define also the following subalgebras of T:

$$R := k\langle u, x, (ux+2)^{-1}\rangle, \text{ and } S := R[y; \partial_1],$$

so that $T = S[v; \sigma, \partial_2]$. It is important to note that, by the formulæ in Proposition 4.5, R is preserved by the σ-derivation ∂_2. Moreover, since $\sigma_{|R}$ is the identity, ∂_2 actually restricts to a derivation on R.

It is much easier to determine when an Ore extension is simple if the ring is a differential operator ring, in the sense that the defining automorphism is actually the identity. Thus we will reduce to that case. The idea follows from Lemma 2.1 which shows that σ^2 is given by the inner automorphism τ_g in the sense that $\sigma(s) = \tau_g(s) = gsg^{-1}$ for suitable $g \in S$. We will therefore extend R, S and T by $\sqrt{g}$ and show that σ is then inner, and so can be removed. The details are given in the next few results, culminating in Proposition 4.9.

NOTATION 4.6. In the algebraic closure of R, set $h = (ux+2)^{-\frac{1}{2}}$. Write $\widetilde{R} = R\langle h\rangle = k\langle u, x, h, h^{-1}\rangle$. We extend the ∂_i to derivations on $\widetilde{R}$ by the usual rules for fractional powers:

$$\partial(h) = (-\frac{1}{2})(ux+2)^{-1}h\partial(ux+2),$$

for $\partial = \partial_1, \partial_2$. Set $\widetilde{S} = \widetilde{R}[y; \partial_1]$. Finally, we can extend σ to $\widetilde{R}$ and $\widetilde{S}$ by setting $\sigma(h) = h$. Then both σ and ∂_2 are naturally defined on $\widetilde{S}$ as an automorphism, respectively σ-derivation and so $\widetilde{T} = \widetilde{S}[v; \sigma, \partial_2]$ is a well-defined Ore extension of $\widetilde{S}$.

The following observation will prove useful.

LEMMA 4.7. *$\widetilde{S}$ is a free left and right S module on basis $\{1, h\}$. Similarly, $\widetilde{T}$ is a free left and right T module on basis $\{1, h\}$.*

PROOF. As $h^2 = (ux+2)^{-1} \in R$, the construction of $\widetilde{R}$ ensures that $\widetilde{R}$ is a free left and right R-module on basis $\{1, h\}$. We can then write

$$\widetilde{S} = \bigoplus_{i=0}^{\infty} \widetilde{R}y^i = \bigoplus Ry^i \oplus \bigoplus Rhy^i = \bigoplus Ry^i \oplus \bigoplus Ry^ih.$$

Collecting terms shows that $\widetilde{S} = S \oplus Sh$. As S is a domain this is necessarily a direct sum of free modules. The same argument works for $\widetilde{T}$. $\square$

LEMMA 4.8. *On $\widetilde{S}$, σ is the inner automorphism $\tau_{h^{-1}}$; thus $\sigma(f) = h^{-1}fh$ for $f \in \widetilde{S}$.*

PROOF. Since $\widetilde{R}$ is a commutative ring on which σ is the identity, the lemma holds trivially on $\widetilde{R}$. It therefore just remains to check that the automorphisms agree on y. To prove this, we rewrite $h^{-1}yh$ as follows.

$$\begin{aligned} h^{-1}yh &= (ux+2)^{\frac{1}{2}}y(ux+2)^{-\frac{1}{2}} \\ &= (ux+2)^{\frac{1}{2}}(ux+2)^{-\frac{1}{2}}y \;+\; (ux+2)^{\frac{1}{2}}\cdot\partial_1\big((ux+2)^{-\frac{1}{2}}\big) \\ &= y \;+\; (ux+2)^{\frac{1}{2}}(-\tfrac{1}{2})(ux+2)^{-\frac{3}{2}}\cdot\partial_1((ux+2)) \\ &= y \;-\; \tfrac{1}{2}(ux+2)^{-1}\Big((-\tfrac{1}{2}ux-2)x-u(\tfrac{1}{2}x^2)\Big) \\ &= y \;-\; \tfrac{1}{2}(ux+2)^{-1}\big(-(ux+2)x\big) \\ &= y\;+\tfrac{1}{2}x \;=\; \sigma(y); \end{aligned}$$

as required. □

PROPOSITION 4.9. *Set $\alpha = hv$. Then $\widetilde{T}$ is the Ore extension $\widetilde{T} = \widetilde{S}[\alpha;\widetilde{\partial}_2]$ where $\widetilde{\partial}_2$ is the derivation of $\widetilde{S}$ defined by $\widetilde{\partial}_2(s) = h\partial_2(s)$ for $s \in \widetilde{S}$; thus*

$$\widetilde{\partial}_2(u) = -\tfrac{1}{2}hu^2, \quad \widetilde{\partial}_2(x) = h(\tfrac{3}{2}ux+2) \quad \textit{and} \quad \widetilde{\partial}_2(y) = h((\tfrac{3}{2}uy-2).$$

As such, $\widetilde{T}$ is a noetherian domain.

PROOF. This is a formal computation. Indeed, for $s \in \widetilde{S}$, Lemma 4.8 implies that $\sigma(s) = h^{-1}sh$. Equivalently,

$$\begin{aligned} \alpha s &= hvs = h\sigma(s)v + h\partial_2(s) \\ &= hh^{-1}shv + h\partial_2(s) = s\alpha + h\partial_2(s). \end{aligned} \tag{4.18}$$

Therefore, since $\widetilde{T} = \widetilde{S}[v;\sigma,\partial_2] = \bigoplus \widetilde{S}v^i$, we see that $\widetilde{T} = \bigoplus \widetilde{S}\alpha^i$. Since $\widetilde{T}$ is a domain, combining this with (4.18) and [**12**, Theorem 1, p.438] gives the desired conclusion. □

Our next aim will be to show that the ring $\widetilde{T}$ is a simple domain, after which it is easy to prove the same conclusion for T. We start with some preparatory results.

LEMMA 4.10. *If there exists a non-zero $(\partial_1,\widetilde{\partial}_2)$-invariant ideal I in $\widetilde{R}$, then there exists a non-zero $(\partial_1,\widetilde{\partial}_2)$-invariant prime ideal P in $\widetilde{R}$.*

PROOF. Using [**17**, Lemma 3.18(b)] twice, clearly $I\widetilde{S}$ is a proper non-zero ideal of $\widetilde{S}$ and then $I\widetilde{T}$ is a proper nonzero ideal of $\widetilde{T}$. Pick a prime ideal $Q \supseteq I\widetilde{T}$. Then, by [**17**, Lemmata 3.18 and 3.21], twice, $Q_1 = Q \cap \widetilde{S}$ is a $\widetilde{\partial}_2$-invariant prime ideal of $\widetilde{S}$ and hence $Q_2 = Q_1 \cap \widetilde{R}$ is a ∂_1-invariant prime ideal of $\widetilde{R}$. However, since $\widetilde{R}$ and Q_1 are both $\widetilde{\partial}_2$-invariant, so is Q_2. Thus, $P = Q_2$ is the desired prime ideal. □

PROPOSITION 4.11. *There is no proper, non-zero $(\partial_1,\widetilde{\partial}_2)$-invariant ideal I in $\widetilde{R}$.*

PROOF. Suppose that there exists such an ideal I. By Lemma 4.10 we can and will assume that I is a prime ideal. Suppose, first, that $(xu+\lambda) \in I$, for some $\lambda \in k$. Then

$$I \ni \partial_1(xu+\lambda) \;=\; (-\tfrac{1}{2}ux-2)x - (\tfrac{1}{2}x^2)u \;=\; -(ux+2)x$$

and

$$I \ni \widetilde{\partial}_2(xu+\lambda) \;=\; h\big(-\tfrac{1}{2}u^2x + (\tfrac{3}{2}ux+2)u\big) \;=\; h\big(ux^2+2u\big) \;=\; h(ux+2)u.$$

As $(xu+2)^{-1} = h^2 \in \widetilde{R}$, clearly $\lambda \neq 2$ and so the two equations imply that $x \in I$, respectively $u \in I$. Thus, $I = x\widetilde{R} + u\widetilde{R}$. But, now $I \ni \partial_1(u) = -\frac{1}{2}ux - 2$ and so $I = \widetilde{R}$, a contradiction. We conclude that

(4.19) $$I \cap \mathcal{C} = \emptyset \quad \text{for } \mathcal{C} = \{(xu+\lambda) : \lambda \in k^*\}$$

Since I is a prime ideal it follows that $\mathcal{C} \subseteq \mathcal{C}(I)$ and hence that $I_{\mathcal{C}}$ is a proper prime ideal of the localisation $\widetilde{R}_{\mathcal{C}}$.

Next, if $I_{\mathcal{C}} \ni f = f(u)$ for some $f(u) \in k[u]$, then $I_{\mathcal{C}} \ni \partial_1(f) = -\frac{1}{2}(ux+4)\frac{df}{du}$. Hence $\frac{df}{du} \in I_{\mathcal{C}}$. By induction on $\deg f$, this implies that $I_{\mathcal{C}} = \widetilde{R}_{\mathcal{C}}$, a contradiction. Thus $I_{\mathcal{C}} \cap k[u] = 0$ and so we can further localise at $\mathcal{S} = k[u]^* := k[u] \smallsetminus \{0\}$, and conclude that $I_{\mathcal{CS}}$ is a proper prime ideal of $\widetilde{R}_{\mathcal{CS}}$. Now consider $\widetilde{R}_{\mathcal{CS}}$. We have $\widetilde{R} = k\langle u, x, h, h^{-1}\rangle$ and $h^{-2} = (ux+2)$ whence $x = u^{-1}(h^{-2}-2)$. Thus $\widetilde{R}_{\mathcal{CS}} = \widetilde{R}_{\mathcal{SC}}$ is a localisation of $k(u)[h, h^{-1}]$.

The advantage of working in $\widetilde{R}_{\mathcal{CS}}$ is that we can simplify our derivation $\widetilde{\partial}_2$. On $\widetilde{R}$ and $\widetilde{R}_{\mathcal{CS}}$ write $\partial_u = \frac{\partial}{\partial u}$ and $\partial_x = \frac{\partial}{\partial x}$. Then, as derivations on either ring,

$$\partial_1 = -(\tfrac{1}{2}xu+2)\partial_u \;-\; \tfrac{1}{2}x^2\partial_x \qquad \text{while} \qquad \widetilde{\partial}_2 \;=\; -\tfrac{1}{2}hu^2\partial_u \;+\; h(\tfrac{3}{2}ux+2)\partial_x.$$

We now set $\mu := -hu^2(ux+4)^{-1}$ and take

$$\widetilde{\partial}_2' \;:=\; \widetilde{\partial}_2 + \mu\partial_1 = \Big(-\tfrac{1}{2}hu^2 + \mu(-\tfrac{1}{2}ux-2)\Big)\partial_u \;+\; \Big(h(\tfrac{3}{2}ux+2) - \tfrac{1}{2}x^2\mu\Big)\partial_x.$$

This element μ has been chosen so that the coefficient of ∂_u in $\widetilde{\partial}_2'$ is

$$-\tfrac{1}{2}(ux+4)^{-1}\Big(hu^2(ux+4) - (ux+4)hu^2\Big) \;=\; 0.$$

Therefore,

$$\begin{aligned}
\widetilde{\partial}_2' &= \Big(h(\tfrac{3}{2}ux+2) \;+\; \tfrac{1}{2}hx^2u^2(ux+4)^{-1}\Big)\partial_x \\
&= (ux+4)^{-1}h\Big((\tfrac{3}{2}ux+2)(ux+4) + \tfrac{1}{2}x^2u^2\Big)\partial_x \\
&= (ux+4)^{-1}h\Big(2u^2x^2 + 8ux + 8\Big)\partial_x \\
&= \beta\partial_x \qquad \text{for} \quad \beta \;:=\; 2(xu+4)^{-1}(ux+2)^2h.
\end{aligned}$$

Since $I_{\mathcal{CS}}$ is invariant under both ∂_1 and $\widetilde{\partial}_2$, it is also invariant under $\widetilde{\partial}_2'$. Since β is a unit in $\widetilde{R}_{\mathcal{CS}}$, it follows that

(4.20) $I_{\mathcal{CS}}$ is also invariant under $\beta^{-1}\widetilde{\partial}_2' = \partial_x$.

Thus, by (4.20) and the expression given above for ∂_1, $I_{\mathcal{CS}}$ is invariant under $(\frac{1}{2}ux+2)\partial_u$, and therefore under ∂_u since $\frac{1}{2}ux+2$ is a unit. So $I_{\mathcal{CS}}$ is invariant under ∂_u and ∂_x. Since $\widetilde{R}_{\mathcal{CS}}$ is a localisation of $k[u,x]$ this forces $I_{\mathcal{CS}} = \widetilde{R}_{\mathcal{CS}}$, giving the required contradiction. □

In order to pass between T and $\widetilde{T}$ we need:

LEMMA 4.12. *If $\widetilde{T}$ is a simple ring then so is T.*

PROOF. Suppose that T has a proper ideal J. Then $X = \widetilde{T}/J\widetilde{T}$ is a $(T, \widetilde{T})$-bimodule. Moreover, by Lemma 4.7 $\widetilde{T}$ is a finitely generated left T-module and so X is a finitely generated left T-module; say $X = \sum_{i=1}^{r} Tx_i$. Then, as $\widetilde{T}$ is an Ore domain, $\operatorname{ann}_{\widetilde{T}}(X) = \bigcap_i \operatorname{ann}_{\widetilde{T}}(x_i) \neq 0$. Since $\widetilde{T}$ is a simple ring this implies that $\operatorname{ann}_{\widetilde{T}}(X) = \widetilde{T}$ and hence that $X = 0$. In other words, $J\widetilde{T} = \widetilde{T}$.

On the other hand, by Lemma 4.7, $\widetilde{T} = T + Th$ is a free left T-module and so $J\widetilde{T} = J \oplus Jh \neq \widetilde{T}$. This contradiction proves the lemma. □

We now put everything together and prove the main result of this subsection.

THEOREM 4.13. *T is a simple ring.*

PROOF. By Lemma 4.12 it suffices to prove that $\widetilde{T}$ is simple. By [**21**, Theorem 1.8.4] applied to $\widetilde{T} = \widetilde{S}[\alpha; \widetilde{\partial}_2]$, we need to prove

(a) $\widetilde{\partial}_2$ is not an inner derivation on $\widetilde{S}$, and

(b) $\widetilde{S}$ has no proper $\widetilde{\partial}_2$-invariant ideals.

Now, as $\partial_1(x) = -\frac{1}{2}x^2$, the right ideal $x\widetilde{S}$ is a proper two-sided ideal of $\widetilde{S}$. As such, it is preserved by any inner derivation of $\widetilde{S}$. But $\widetilde{\partial}_2(x) = h(\frac{3}{2}ux + 2) \notin x\widetilde{S}$, this means $\widetilde{\partial}_2$ cannot be an inner derivation of $\widetilde{S}$ and so (a) holds.

Suppose that $\widetilde{S}$ has a proper $\widetilde{\partial}_2$-invariant ideal I. Then, by [**17**, Lemma 3.18], $K = I \cap \widetilde{R}$ is a ∂_1-invariant ideal of $\widetilde{R}$, while by [**17**, Lemma 3.19], $K \neq 0$. Since both I and $\widetilde{R}$ are both $\widetilde{\partial}_2$-invariant, so is K. In other words, K is a proper $(\partial_1, \widetilde{\partial}_2)$-invariant ideal of $\widetilde{R}$. This contradicts Proposition 4.11. Thus (b) holds and so [**21**, Theorem 1.8.4] implies that $\widetilde{T}$ is simple. □

REMARK 4.14. We end the subsection by noting that $\widetilde{T}$ is obviously birational to the Weyl algebra A_2. We do not know if the same is true for T itself.

4.5. The shape of the primitive spectrum of $\mathcal{D}$. In this subsection we combine the earlier results of this section to prove Theorem 1.1. By Theorem 4.3, every primitive ideal P of $\mathcal{D}$ contains a maximal ideal of $Z(\mathcal{D})$. Thus $\operatorname{Privspec}(\mathcal{D})$ is the disjoint union

$$\operatorname{Privspec}(\mathcal{D}) = \dot{\bigcup}_{\mathfrak{m} \in \operatorname{Maxspec}(Z(\mathcal{D}))} \mathcal{V}(\mathfrak{m}) \tag{4.21}$$

where $\mathcal{V}(\mathfrak{m}) = \{P \in \operatorname{Privspec}(\mathcal{D}) : \mathfrak{m} \subseteq P\}$. There are thus 3 cases, corresponding to §§4.1, 4.2 and 4.3.

(I) $\mathcal{V}(\mathfrak{m})$, where $\mathfrak{m} \in \operatorname{Maxspec}(Z(\mathcal{D}))$ with $\mathfrak{m} \neq \mathfrak{m}_+$ and $\mathfrak{m} \neq \mathfrak{m}_0$. By Theorem 4.1, $\mathcal{V}(\mathfrak{m}) = \{\mathfrak{m}\mathcal{D}\}$ is a single generic maximal ideal of $\mathcal{D}$. Moreover $\mathcal{D}/\mathfrak{m}\mathcal{D}$ is birationally equivalent to the second Weyl algebra, with other properties as listed in that theorem.

(II) $\mathcal{V}(\mathfrak{m}_+)$. By Theorem 4.4, this consists of $\mathfrak{m}_+\mathcal{D}$, together with

$$\mathcal{V}(\mathcal{O}(G)^+\mathcal{D}) := \{P \in \operatorname{Privspec}(\mathcal{D}) : \mathcal{O}(G)^+\mathcal{D} \subset P\},$$

which is homeomorphic to $\operatorname{Privspec}(U(\mathfrak{sl}(2,k)))$ by Theorem 2.2(ii).

Recall that $\operatorname{Privspec}(U(\mathfrak{sl}(2,k)))$ is composed of the co-Artinian maximal ideals $\{M_n : n \in \mathbb{Z}_{\geq 1}\}$, where $M_n = \operatorname{Ann}(V_n)$, V_n being the n-dimensional irreducible $U(\mathfrak{sl}_2(k))$−module, together with the *minimal primitives* of $U(\mathfrak{sl}(2,k))$; that is, the ideals $(\Omega - \lambda)U(\mathfrak{sl}(2,k)) : \lambda \in k\}$, where Ω is the Casimir element. Each M_n

contains one such minimal primitive and each minimal primitive is contained in at most one M_n; the remaining minimal primitives are also maximal. Note that $\mathcal{O}(G)^+\mathcal{D}$ is prime but not primitive since $\mathcal{D}/\mathcal{O}(G)^+\mathcal{D} \cong U(\mathfrak{sl}(2,k))$ and this domain satisfies the Nullstellensatz and has non-trivial centre $k[\Omega]$.

(III) $\mathcal{V}(\mathfrak{m}_0)$. This is the singleton $\{P_0 = q\mathcal{D} + s\mathcal{D} = \sqrt{\mathfrak{m}_0}\}$, by Proposition 4.5 and Theorem 4.13.

5. Prime ideals and the Dixmier-Moeglin equivalence

In this section we prove Theorem 1.2 from the introduction, which describes the prime ideals of $\mathcal{D}$, and we discuss the Dixmier-Moeglin equivalence for $\mathcal{D}$.

5.1. The prime spectrum of $\mathcal{D}$. We need the following lemmas for the proof of the main result, Theorem 5.3.

LEMMA 5.1. *Let P be a nonzero prime ideal of $\mathcal{D}$. Then $P \cap Z(\mathcal{D}) \neq \{0\}$.*

PROOF. If $x^i \in P$ for some $i \geq 0$ then $\mathcal{O}(G)^+\mathcal{D} \subseteq P$ by Lemma 3.3 applied with $M = \mathcal{D}/P$, and therefore $\mathfrak{m}^+ = \mathcal{O}(G)^+\mathcal{D} \cap Z(\mathcal{D}) \subseteq P$, proving the lemma for P. So we may assume that $\{x^i : i \geq 0\} \cap P = \emptyset$. Similarly, we may assume that $\{q^j : j \geq 0\} \cap P = \emptyset$, since otherwise $0 \neq q^n g^{-2n} \in P \cap Z(\mathcal{D})$ for some $n \geq 0$ and again the result follows for P.

Hence, using Notation 3.6 and Theorem 3.8, $P\mathcal{D}_{(A)}$ survives as a non-zero proper ideal of $\mathcal{D}_{(A)} = \mathcal{D}\langle q^{-1}, x^{-1}\rangle \;=\; A_2^{(A)}(k) \otimes_k S_{(A)}$, where $A_2^{(A)}(k)$ is a localised Weyl algebra and $S_{(A)} = k[z^{\pm 1}, \omega]$. In particular,

$$P\mathcal{D}_{(A)} \;=\; (P\mathcal{D}_{(A)} \cap S_{(A)})\mathcal{D}_{(A)}. \tag{5.1}$$

By [**17**, Theorem 10.20] and the discussion in the first paragraph of this proof, $P = P\mathcal{D}_{(A)} \cap \mathcal{D}$, and therefore

$$P \cap Z(\mathcal{D}) \;=\; P\mathcal{D}_{(A)} \cap Z(\mathcal{D}) \;=\; (P\mathcal{D}_{(A)} \cap S_{(A)}) \cap Z(\mathcal{D}). \tag{5.2}$$

Since the $Z(\mathcal{D})$-module $S_{(A)}/Z(\mathcal{D})$ is $\{z^i\}$-torsion, that is $\{q^i g^{-2i}\}$-torsion, it follows from (5.1) and (5.2) that $P \cap Z(\mathcal{D}) \neq \{0\}$ as required. □

Note that, since k is algebraically closed of characteristic 0, the defining relation $z\theta = \omega^2$ of $Z(\mathcal{D})$ can be rewritten using a linear change of variables as the quadratic form $X^2 + Y^2 = Z^2$. Thus a proof of the next result can be found at [**15**, p.51 and Proposition 11.4].

LEMMA 5.2. *All height one primes of $Z(\mathcal{D})$ are principal except $\mathfrak{p}_1 := \langle z, \omega\rangle$ and $\mathfrak{p}_2 := \langle \theta, \omega\rangle$.* □

Here is the main result of this section, where the notation of Lemma 5.2 is used in Part (ii). This proves Theorem 1.2 from the introduction.

THEOREM 5.3. *Let P be a prime but not primitive ideal of $\mathcal{D}$.*

(i) *There are the following three possibilities for P.*
   (a) $P = \{0\}$.
   (b) $P = \mathcal{O}(G)^+\mathcal{D}$.
   (c) *P has height one and is minimal over $(P \cap Z(\mathcal{D}))\mathcal{D}$ for a height one prime ideal $P \cap Z(D)$ of $Z(\mathcal{D})$.*

(ii) *In case (c), if* $P \cap Z(\mathcal{D}) = \mathfrak{p}_i$ *for* $i = 1$, *resp.* $i = 2$, *then* $P = q\mathcal{D}$, *resp.* $P = s\mathcal{D}$. *The remaining primes in case (c) are precisely the set*

$$\{P : P = f\mathcal{D}\},$$

as f *ranges through the equivalence classes of irreducible elements of* $Z(\mathcal{D})$ *other than the associates of* z, ω, θ.

REMARK 5.4. By Theorem 2.2(ii), $\mathcal{O}(G)^+\mathcal{D} = \mathcal{K}$ in the notation from the introduction.

PROOF. Note first that $\{0\}$ is completely prime by Proposition 3.1, and is not primitive, because $\mathcal{D}$ satisfies the Nullstellensatz by Theorem 4.3 and $Z(\mathcal{D}) \neq k$. This covers case (a).

Let P be a non-zero prime but not primitive ideal of $\mathcal{D}$. By Lemma 5.1,

$$\{0\} \neq \mathfrak{p} := P \cap Z(\mathcal{D}).$$

If $\mathfrak{p} = \mathfrak{m}_+$ then Theorem 4.4 together with the discussion at §4.5(**II**) shows that the only possibility is $P = \mathcal{O}(G)^+\mathcal{D}$, which is completely prime but is again not primitive thanks to the Nullstellensatz, since $Z(U(\mathfrak{sl}(2,k))) \neq k$. This is case (b).

If $\mathfrak{p} = \mathfrak{m}_0$ then $P = P_0$, which is maximal by Theorem 4.13, so this case can't happen. Similarly, $\mathfrak{p}$ is any maximal ideal of $Z(\mathcal{D})$ apart from $\mathfrak{m}_+$ or $\mathfrak{m}_0$, then $P = \mathfrak{p}\mathcal{D}$ is a maximal ideal of $\mathcal{D}$ by Theorem 4.1(i), which again gives a contradiction.

So we are left with the case when $\mathfrak{p}$ has height one. Assume first that $\mathfrak{p} = fZ(\mathcal{D})$ is principal. Then, by Lemma 5.2, $z = q^2g^{-1} \notin P$, and $\{x^i : i \geq 0\} \cap P = \emptyset$ by Lemma 3.2 Therefore, using Notation 3.6 and Theorem 3.8

$$\mathfrak{p}\mathcal{D}_{(A)} = (P \cap S_{(A)})\mathcal{D}_{(A)} = P\mathcal{D}_{(A)}.$$

We claim that $P = \mathfrak{p}\mathcal{D}$. To see this, note that $\mathfrak{p}\mathcal{D} = f\mathcal{D}$ is principal, so that $\mathcal{D}/\mathfrak{p}\mathcal{D}$ is CM of GK-dimension 5, by [**16**, Theorem 7.2(b) and its proof], and GK-homogeneous by [**19**, §3.4, Remark (3)]. Now $P/\mathfrak{p}\mathcal{D}$ it is killed by $\mathfrak{p}\mathcal{D}$ and by a power of q or a power of x, and so has GK-dimension less than 5, respectively by Theorem 4.1(iii) and 4.4(iv) or by Lemma 3.2. This forces $P/\mathfrak{p}\mathcal{D} = \{0\}$ and so $P = \mathfrak{p}\mathcal{D}$, as claimed.

Suppose finally that $\mathfrak{p} = \mathfrak{p}_1$ or $\mathfrak{p} = \mathfrak{p}_2$. In the first case, since q is a normal element of $\mathcal{D}$ by Lemma 2.1, $q \in \sqrt{\mathfrak{p}\mathcal{D}}$. Thus

$$q\mathcal{D} \subseteq P. \tag{5.3}$$

We claim that (5.3) is an equality. To see this, note that $s \notin P$, since otherwise $P \cap Z(\mathcal{D}) = \mathfrak{m}_0$, which is ruled out by hypothesis. Moreover $\{x^j : j \geq 0\} \cap P = \emptyset$ by Lemma 3.2. So we can localise at the Ore set $B = \{s^i x^j : i, j \geq 0\}$ of Definition 3.4 and pass to the localised Weyl algebra $D_{(B)} = A_2^{(B)}(k) \otimes S_{(B)}$ of Theorem 3.8. However, $P\mathcal{D}_{(B)}$ and $q\mathcal{D}_{(B)}$ have the same intersection with the centre $S_{(B)}$, namely $\omega\theta^{-1}S_{(B)} = \mathfrak{p}_1 S_{(B)}$. Therefore $P\mathcal{D}_{(B)} = q\mathcal{D}_{(B)}$ since the ideals of $D_{(B)}$ are centrally generated. Therefore $P/q\mathcal{D}$ is B-torsion, so, if it is not zero, it contains a nonzero element which is either killed by q and by s, or by q and x. As in the previous paragraph $\mathcal{D}/q\mathcal{D}$ is GK-homogeneous of GK-dimension 5, and so has no such non-zero torsion submodule, proving that (5.3) is an equality.

If $\mathfrak{p} = \mathfrak{p}_2$ then the argument to show that $P = s\mathcal{D}$ is similar, but using the Ore set A; it is left to the reader. □

5.2. The Dixmier-Moeglin equivalence. The following gives evidence in favour of [**6**, Conjecture 1.3], which proposes that an affine noetherian Hopf $\mathbb{C}$-algebra of finite GK dimension should satisfy the Dixmier-Moeglin equivalence. See [**5**, **10**] for a further discussion of this property. The next result proves Theorem 1.4.

COROLLARY 5.5. *$\mathcal{D}$ satisfies the Dixmier-Moeglin equivalence.*

PROOF. We check first using the description of the primitive spectrum in §4.5 that every primitive ideal is locally closed. For classes **(I)** and **(III)** this is clear since all these primitive ideals are maximal. The primitive ideals in **(II)** are homeomorphic to the primitive spectrum of $U(\mathfrak{sl}(2,k))$, and the latter algebra satisfies the equivalence by [**23**]. Thus, by [**10**, Lemma II.7.15], it only remains to show that every rational prime ideal P is primitive.

The non-primitive prime ideals P are listed in Theorem 5.3 and it is easy to check that none of them is rational. In more detail, recall from Theorem 2.2(iv) that $Z(\mathcal{D})$ is a ring of Krull dimension 2. Thus in cases (a) and (c) of Theorem 5.3, some non-zero central element of $\mathcal{D}$ remains non-zero in $\mathcal{D}/P$ and hence in the quotient ring $Q(\mathcal{D}/P)$. This leaves the case when $P = \mathcal{O}(G)^+\mathcal{D}$. But here $\mathcal{D}/\mathcal{O}(G)^+\mathcal{D} \cong U(\mathfrak{sl}_2(k))$ by Theorem 2.2(ii). Thus the Casimir element of $U(\mathfrak{sl}_2(k))$ provides the nonzero central element of $Q(\mathcal{D}/P)$, which completes the proof. □

In order to complete the proof of the results claimed in the introduction, it remains to prove Theorem 1.3, but this is now easy.

Proof of Theorem 1.3. With one exception this is proved in the results of the last two sections that describe the prime ideals of $\mathcal{D}$. The exception is the claim that all the completely prime factors of $\mathcal{D}$ (with the possible exception of $\mathcal{D}/P_0$, as noted in Remark 4.14) are birationally equivalent to Weyl algebras. For the primitive ideals P strictly containing $\mathcal{O}(G)^+\mathcal{D}$ this follows from [**13**, Remarque 7.1]. For the other prime ideals, this is clear from the description of the prime ideals in the last two sections. □

Based on little more than the known results and counterexamples for group algebras and enveloping algebras, the theorem [**6**] for the cocommutative case, the recent work of Sierra and Walton on the noetherian property for enveloping algebras [**24**], together with the above result and other isolated examples, we are tempted to propose the following conjecture as a strengthening in the pointed setting of [**6**, Conjecture 1.3], as much in the hope of stimulating the discovery of counterexamples as in expectation of a positive result.

CONJECTURE 5.6. *Let H be an affine noetherian pointed Hopf $\mathbb{C}$-algebra. Then the following are equivalent:*

(1) $\mathrm{GKdim}(H)$ *is finite.*
(2) *H satisfies the Dixmier-Moeglin Equivalence.*
(3) *The group $G(H)$ of grouplikes of H is nilpotent-by-finite.*

References

[1] Nicolás Andruskiewitsch, François Dumas, and Héctor Martín Peña Pollastri, *On the double of the Jordan plane*, Ark. Mat. **60** (2022), no. 2, 213–229, DOI 10.4310/arkiv.2022.v60.n2.a1. MR4500364

[2] Nicolás Andruskiewitsch and Héctor Peña Pollastri, *On the restricted Jordan plane in odd characteristic*, J. Algebra Appl. **20** (2021), no. 1, Paper No. 2140012, 26, DOI 10.1142/S0219498821400120. MR4209966

[3] N. Andruskiewitsch and H. M. Pena Pollastri, *On the finite-dimensional representations of the double of the Jordan plane*, Preprint, arXiv:2211.01581, 2022.

[4] M. Artin, L. W. Small, and J. J. Zhang, *Generic flatness for strongly Noetherian algebras*, J. Algebra **221** (1999), no. 2, 579–610, DOI 10.1006/jabr.1999.7997. MR1728399

[5] Jason Bell, *On the importance of being primitive* (English, with English and Spanish summaries), Rev. Colombiana Mat. **53** (2019), no. suppl., 87–112. MR4053367

[6] Jason P. Bell and Wing Hong Leung, *The Dixmier-Moeglin equivalence for cocommutative Hopf algebras of finite Gelfand-Kirillov dimension*, Algebr. Represent. Theory **17** (2014), no. 6, 1843–1852, DOI 10.1007/s10468-014-9474-y. MR3284332

[7] Jan-Erik Björk, *The Auslander condition on Noetherian rings*, Séminaire d'Algèbre Paul Dubreil et Marie-Paul Malliavin, 39ème Année (Paris, 1987/1988), Lecture Notes in Math., vol. 1404, Springer, Berlin, 1989, pp. 137–173, DOI 10.1007/BFb0084075. MR1035224

[8] Kenneth A. Brown, *Unruffled extensions and flatness over central subalgebras*, J. Algebra **284** (2005), no. 2, 771–800, DOI 10.1016/j.jalgebra.2004.10.008. MR2114579

[9] K. A. Brown, M. Couto, and A. Jahn, *The finite dual of commutative-by-finite Hopf algebras*, Glasg. Math. J. **65** (2023), no. 1, 62–89, DOI 10.1017/S0017089522000052. MR4518762

[10] Ken A. Brown and Ken R. Goodearl, *Lectures on algebraic quantum groups*, Advanced Courses in Mathematics. CRM Barcelona, Birkhäuser Verlag, Basel, 2002, DOI 10.1007/978-3-0348-8205-7. MR1898492

[11] K. A. Brown and J. J. Zhang, *Dualising complexes and twisted Hochschild (co)homology for Noetherian Hopf algebras*, J. Algebra **320** (2008), no. 5, 1814–1850, DOI 10.1016/j.jalgebra.2007.03.050. MR2437632

[12] P. M. Cohn, *Algebra, Vol. II*, Wiley, 1977.

[13] Jacques Dixmier, *Quotients simples de l'algèbre enveloppante de* $\mathfrak{sl}_2$ (French), J. Algebra **24** (1973), 551–564, DOI 10.1016/0021-8693(73)90127-0. MR310031

[14] Yukio Doi and Mitsuhiro Takeuchi, *Multiplication alteration by two-cocycles—the quantum version*, Comm. Algebra **22** (1994), no. 14, 5715–5732, DOI 10.1080/00927879408825158. MR1298746

[15] Robert M. Fossum, *The divisor class group of a Krull domain*, Ergebnisse der Mathematik und ihrer Grenzgebiete [Results in Mathematics and Related Areas], Band 74, Springer-Verlag, New York-Heidelberg, 1973. MR382254

[16] K. R. Goodearl and T. H. Lenagan, *Primitive ideals in quantum* SL_3 *and* GL_3, New trends in noncommutative algebra, Contemp. Math., vol. 562, Amer. Math. Soc., Providence, RI, 2012, pp. 115–140, DOI 10.1090/conm/562/11134. MR2905557

[17] K. R. Goodearl and R. B. Warfield Jr., *An introduction to noncommutative Noetherian rings*, 2nd ed., London Mathematical Society Student Texts, vol. 61, Cambridge University Press, Cambridge, 2004, DOI 10.1017/CBO9780511841699. MR2080008

[18] Günter R. Krause and Thomas H. Lenagan, *Growth of algebras and Gelfand-Kirillov dimension*, Revised edition, Graduate Studies in Mathematics, vol. 22, American Mathematical Society, Providence, RI, 2000, DOI 10.1090/gsm/022. MR1721834

[19] Thierry Levasseur, *Some properties of noncommutative regular graded rings*, Glasgow Math. J. **34** (1992), no. 3, 277–300, DOI 10.1017/S0017089500008843. MR1181768

[20] Kangqiao Li and Gongxiang Liu, *Finite duals of affine prime regular Hopf algebras of GK-dimension one*, AIMS Math. **8** (2023), no. 3, 6829–6879. MR4550624

[21] J. C. McConnell and J. C. Robson, *Noncommutative Noetherian rings*, Revised edition, Graduate Studies in Mathematics, vol. 30, American Mathematical Society, Providence, RI, 2001. With the cooperation of L. W. Small, DOI 10.1090/gsm/030. MR1811901

[22] J. C. McConnell and J. T. Stafford, *Gel′fand-Kirillov dimension and associated graded modules*, J. Algebra **125** (1989), no. 1, 197–214, DOI 10.1016/0021-8693(89)90301-3. MR1012671

[23] C. Moeglin, *Idéaux primitifs des algèbres enveloppantes* (French), J. Math. Pures Appl. (9) **59** (1980), no. 3, 265–336. MR604473

[24] Susan J. Sierra and Chelsea Walton, *The universal enveloping algebra of the Witt algebra is not noetherian*, Adv. Math. **262** (2014), 239–260, DOI 10.1016/j.aim.2014.05.007. MR3228429

School of Mathematics and Statistics, University of Glasgow, Glasgow G12 8QQ, Scotland

Email address: ken.brown@glasgow.ac.uk

School of Mathematics, The University of Manchester, Manchester M13 9PL, England

Email address: Toby.Stafford@manchester.ac.uk

Contemporary Mathematics
Volume **801**, 2024
https://doi.org/10.1090/conm/801/16037

Modular properties of elliptic algebras

Alex Chirvasitu, Ryo Kanda, and S. Paul Smith

ABSTRACT. Fix a pair of relatively prime integers $n > k \geq 1$, and a point $(\eta \,|\, \tau) \in \mathbb{C} \times \mathbb{H}$, where $\mathbb{H}$ denotes the upper-half complex plane, and let $\binom{a\ b}{c\ d} \in \mathrm{SL}(2, \mathbb{Z})$. We show that Feigin and Odesskii's elliptic algebras $Q_{n,k}(\eta \,|\, \tau)$ have the property $Q_{n,k}\big(\frac{\eta}{c\tau+d} \,\big|\, \frac{a\tau+b}{c\tau+d}\big) \cong Q_{n,k}(\eta \,|\, \tau)$. As a consequence, given a pair (E, ξ) consisting of a complex elliptic curve E and a point $\xi \in E$, one may unambiguously define $Q_{n,k}(E, \xi) := Q_{n,k}(\eta \,|\, \tau)$ where $\tau \in \mathbb{H}$ is any point such that $\mathbb{C}/\mathbb{Z} + \mathbb{Z}\tau \cong E$ and $\eta \in \mathbb{C}$ is any point whose image in E is ξ. This justifies Feigin and Odesskii's notation $Q_{n,k}(E, \xi)$ for their algebras. The algebras $Q_{n,1}(\eta \,|\, \tau)$ are commonly known as Sklyanin algebras in honor of Sklyanin's discovery of $Q_{4,1}(\eta \,|\, \tau)$.

1. Introduction

We will use the notation $e(z) := e^{2\pi i z}$.

Always, τ denotes a point in the upper-half complex plane,

$$\mathbb{H} \ := \ \{x + iy \,|\, x, y \in \mathbb{R},\ y > 0\}.$$

Some readers might prefer to start with the "big picture" by reading subsections 1.3 and 1.4 before subsections 1.1 and 1.2.

1.1. The main result. Let

$$\vartheta(z \,|\, \tau) \ := \ \sum_{n \in \mathbb{Z}} e\big(nz + \tfrac{1}{2} n^2 \tau\big). \tag{1.1}$$

In 1828, C.G.J. Jacobi [**13**] proved the remarkable identity

$$\vartheta\left(\frac{z}{\tau} \,\middle|\, -\frac{1}{\tau}\right) \ = \ \sqrt{-i\tau}\, e\left(\frac{z^2}{2\tau}\right) \vartheta(z \,|\, \tau), \tag{1.2}$$

2020 *Mathematics Subject Classification.* Primary 16S38; Secondary 16S37, 16W50, 14H52.

The first author was partially supported by NSF grant DMS-2001128. The second author was supported by JSPS KAKENHI Grant Numbers JP16H06337, JP17K14164, JP20K14288, and JP21H04994, Leading Initiative for Excellent Young Researchers, MEXT, Japan, and Osaka Central Advanced Mathematical Institute: MEXT Joint Usage/Research Center on Mathematics and Theoretical Physics JPMXP0619217849.

where $\sqrt{-i\tau}$ is the square root of $-i\tau$ having non-negative real part.[1] Jacobi's identity is a special case of the following functional equation: if

$$\begin{pmatrix} a & b \\ c & d \end{pmatrix} \in \mathrm{SL}(2,\mathbb{Z}) \tag{1.3}$$

is such that ab and cd are even, then there is an 8th root of unity ζ such that

$$\vartheta\left(\frac{z}{c\tau+d} \,\middle|\, \frac{a\tau+b}{c\tau+d}\right) = \zeta\sqrt{c\tau+d}\, e\left(\frac{cz^2}{2(c\tau+d)}\right)\vartheta(z\,|\,\tau). \tag{1.4}$$

The precise value of ζ, which is not important in this paper, can be found at [**15**, Thm. 7.1, p. 32], for example.

In 1989, Feigin and Odesskii defined a remarkable family of graded associative $\mathbb{C}$-algebras which we denote here by $Q_{n,k}(\eta\,|\,\tau)$. These are the elliptic algebras of the title. (When $k=1$, they are called Sklyanin algebras: see §1.4.)

They depend on a pair of relatively prime integers $n > k \geq 1$ and a point $(\eta\,|\,\tau) \in \mathbb{C}\times\mathbb{H}$. We define $Q_{n,k}(\eta\,|\,\tau)$ in subsection 1.2.

The main result in this paper is the following.

THEOREM 1.1 (Theorem 4.5). *If* (1.3) *holds, then there is an isomorphism of graded* $\mathbb{C}$*-algebras*

$$Q_{n,k}\left(\frac{\eta}{c\tau+d} \,\middle|\, \frac{a\tau+b}{c\tau+d}\right) \cong Q_{n,k}(\eta\,|\,\tau). \tag{1.5}$$

In subsection 4.5 we explain how this result can be interpreted as saying that $Q_{n,k}(\,\cdot\,|\,\cdot\,)$ is a "weakly modular function of weight -1 taking values in the category of graded $\mathbb{C}$-algebras".

1.2. The definition of $Q_{n,k}(\eta\,|\,\tau)$. Fix the data (n,k,η,τ) as above; i.e., $n > k \geq 1$ is a pair of relatively prime positive integers and $(\eta\,|\,\tau)$ is a point in $\mathbb{C}\times\mathbb{H}$.

Write $\Lambda_\tau := \mathbb{Z}+\mathbb{Z}\tau$ and $E_\tau := \mathbb{C}/\Lambda_\tau$.

Fix a complex vector space V with basis $x_0,\ldots,x_{n-1}$ indexed by $\mathbb{Z}_n$, the cyclic group of order n. In [**6**, Prop. 2.6], we defined a particular basis $\theta_0,\ldots,\theta_{n-1}$, also indexed by $\mathbb{Z}_n$, for the vector space $\Theta_n(\Lambda_\tau)$ which, by definition, consists of all holomorphic functions $f:\mathbb{C}\to\mathbb{C}$ such that

$$\begin{aligned} f(z+1) &= f(z) \qquad \text{and} \\ f(z+\tau) &= -e(-nz)f(z) \end{aligned}$$

for all $z\in\mathbb{C}$. Copying Feigin and Odesskii, we then defined for each $\eta\in\mathbb{C}-\frac{1}{n}\Lambda_\tau$ a family of holomorphic linear operators

$$R_{n,k}(z,\eta\,|\,\tau): V\otimes V \longrightarrow V\otimes V, \qquad z\in\mathbb{C},$$

given by the formula

$$R_{n,k}(z,\eta\,|\,\tau)(x_i\otimes x_j) := \frac{\theta_0(-z)\cdots\theta_{n-1}(-z)}{\theta_1(0)\cdots\theta_{n-1}(0)} \sum_{r\in\mathbb{Z}_n} \frac{\theta_{j-i+r(k-1)}(-z+\eta)}{\theta_{j-i-r}(-z)\theta_{kr}(\eta)} x_{j-r}\otimes x_{i+r} \tag{1.6}$$

[1] Jacobi did not use this notation, nor did he provide a proof in [**13**]. Jacobi's identity is equivalent to a similar identity for the function $\vartheta_1(z\,|\,\tau)$ that appears in Chapter XXI of Whittaker-Watson [**28**, pp. 475–476]. On the other hand, (1.2) is equivalent to a similar identity for Whittaker and Watson's function $\vartheta_3(z\,|\,\tau)$ since $\vartheta(z\,|\,\tau) = \vartheta_3(\pi z\,|\tau)$ and, as Whittaker and Watson indicate, their identities for $\vartheta_1(z\,|\,\tau)$ and $\vartheta_3(z\,|\,\tau)$ are equivalent.

for all $(i,j) \in \mathbb{Z}_n^2$. Although this formula does not make sense when $\eta \in \frac{1}{n}\Lambda$, because in that case $\theta_{kr}(\eta) = 0$ for some $r \in \mathbb{Z}_n$, the function

$$R_{n,k}(\eta,\eta \,|\, \tau) : \mathbb{C} - \tfrac{1}{n}\Lambda_\tau \longrightarrow \operatorname{End}_{\mathbb{C}}(V^{\otimes 2})$$

extends in a unique way to a holomorphic function on the entire complex plane [**6**, Lem. 3.13]—we will write $R_{n,k}(\eta,\eta \,|\, \tau)$ for the extension. This allows us to define, for all $\eta \in \mathbb{C}$, the algebra

$$Q_{n,k}(\eta \,|\, \tau) \;:=\; \frac{TV}{(\text{the image of } R_{n,k}(\eta,\eta \,|\, \tau))} \tag{1.7}$$

where TV denotes the tensor algebra on V.

Although the formula for $R_{n,k}(z,\eta \,|\, \tau)$ is not very illuminating, by building on results and ideas of Richey and Tracy [**22**] we were able to show in [**5**] that $R(z) = R_{n,k}(z,\eta \,|\, \tau)$ satisfies the quantum Yang-Baxter equation: for all $u, v \in \mathbb{C}$,

$$R(u)_{12}R(u+v)_{23}R(v)_{12} \;=\; R(v)_{23}R(u+v)_{12}R(u)_{23}$$

as operators on $V^{\otimes 3}$, where $R(z)_{12} := R(z) \otimes I$, $R(z)_{23} := I \otimes R(z)$, and I is the identity operator on V. This identity plays a key role in the proof of several results in [**5**] which showed, for suitable η, that $Q_{n,k}(\eta \,|\, \tau)$ has many of the good properties enjoyed by the polynomial ring on n variables; for example, when the image of η in E_τ is not a torsion point, $Q_{n,k}(\eta \,|\, \tau)$ has the same Hilbert series as that polynomial ring.

The work of Feigin and Odesskii in [**9, 18–20**], and our later work in [**4–7**], shows that $Q_{n,k}(\eta \,|\, \tau)$ has many beautiful properties. Roughly speaking, the algebras $Q_{n,k}(\eta \,|\, \tau)$ are a step beyond enveloping algebras and quantized enveloping algebras in the same way as elliptic and theta functions are a step beyond rational and trigonometric functions.

1.3. Feigin and Odesskii's elliptic algebras $Q_{n,k}(E,\xi)$. Let X denote a compact Riemann surface of genus one. A pair (X,p) consisting of a compact Riemann surface X of genus one and a point p on it will be called a complex elliptic curve. We usually omit the adjective "complex". There is a unique way to make X an algebraic (or Lie) group with p as its identity—see, e.g., [**12**, Cor. 1.11]. We will always view (X,p) as a group in this way.

Given an elliptic curve $E = (X,p)$ and a point $\xi \in E$, i.e., a second point $\xi \in X$, which is allowed to be p, Feigin and Odesskii defined an algebra they denoted by $Q_{n,k}(E,\xi)$—see [**19**], and [**6**] for a definition when ξ is one of the points that Feigin and Odesskii disallow.[2]

At [**19**, §1.2], Feigin and Odesskii define $Q_{n,k}(E,\xi)$ by the formula on the right-hand side of (1.7) where $(\eta \,|\, \tau) \in \mathbb{C} \times \mathbb{H}$ is such that E is isomorphic to E_τ and η is a point whose image in $\mathbb{C}/\mathbb{Z} + \mathbb{Z}\tau$ is ξ.

There are many points $(\eta \,|\, \tau) \in \mathbb{C} \times \mathbb{H}$ with the property in the previous sentence. It is easy to see that $Q_{n,k}(\eta \,|\, \tau) = Q_{n,k}(\eta' \,|\, \tau)$ if η and η' have the same image in E_τ so Feigin and Odesskii's definition of $Q_{n,k}(E,\xi)$ does not depend on the choice of η. However, Feigin and Odesskii do not show that the algebra they denote by $Q_{n,k}(E,\xi)$ stays the same when τ is replaced by another point $\tau' \in \mathbb{H}$ for which E is isomorphic to $E_{\tau'}$. We also failed to address this issue in our earlier papers [**4–7**]. Theorem 1.1 remedies this oversight and, as we will now explain,

[2]In truth, Feigin and Odesskii use the notation $Q_{n,k}(E,\tau)$ where τ denotes a point on E—see subsections 1.3.1 and 1.3.2 below.

allows us to conclude that $Q_{n,k}(E,\xi)$ is well-defined up to isomorphism when we define it as

$$Q_{n,k}(E,\xi) \;:=\; Q_{n,k}(\eta\,|\,\tau)$$

where $\tau \in \mathbb{H}$ is any point such that $E_\tau \cong E$ and the image of η under this isomorphism is ξ.

There is a notion of isomorphism between such pairs $(E,\xi) = (X,p,\xi)$, and it follows from Theorem 1.1 that $Q_{n,k}(E,\xi)$ is isomorphic to $Q_{n,k}(E',\xi')$ when (E,ξ) is isomorphic to (E',ξ'). More precisely:

THEOREM 1.2 (Corollary 4.8). *If $(E,\xi) = (E_\tau, \eta+\Lambda_\tau)$ and $(E',\xi') = (E_{\tau'}, \eta'+\Lambda_{\tau'})$ and $\mu : E \to E'$ is an isomorphism of algebraic groups such that the diagram*

$$\begin{array}{ccc} E & \xrightarrow{\ \mu\ } & E' \\ {\scriptstyle x\mapsto x+\xi}\big\downarrow & & \big\downarrow{\scriptstyle x'\mapsto x'+\xi'} \\ E & \xrightarrow[\ \mu\ ]{} & E' \end{array}$$

commutes, then

$$Q_{n,k}(\eta\,|\,\tau) \;\cong\; Q_{n,k}(\eta'\,|\,\tau').$$

All the results about $Q_{n,k}(E,\xi)$ in Feigin and Odesskii's papers and in our earlier papers are, in fact, results about $Q_{n,k}(\eta\,|\,\tau)$.

1.3.1. *Changing the roles of the symbols τ and η.* In our papers [**4**–**7**] we used η to denote a point in $\mathbb{H}$ and τ to denote a point in $\mathbb{C}$. In this paper we switch that notation in order to agree with the common convention in the literature on elliptic functions, theta functions, and modular forms, that τ denotes a point in $\mathbb{H}$.

1.3.2. *Warning.* In [**9**, **19**], Feigin and Odesskii use the symbol τ for both a point in $\mathbb{C}$ and a point on E—in [**19**, §1.1] the symbol τ denotes a point on E but in [**19**, §1.2] the notation $\theta_\alpha(\tau)$ only makes sense when $\tau \in \mathbb{C}$.

1.4. The case when $k = 1$. The algebras $Q_{n,1}(\eta\,|\,\tau)$ are called Sklyanin algebras in honor of E.K. Sklyanin who discovered $Q_{4,1}(E,\eta)$ in 1982 [**25**, **26**].

When $k = 1$, Theorem 1.2 was proved by Tate and Van den Bergh [**27**]. In fact, they proved more because they work with elliptic curves over arbitrary fields.

In [**27**, §4.1], Tate and Van den Bergh define a graded $\mathbb{F}$-algebra $A(E,\sigma,\mathcal{L})$ when E is an elliptic curve over an arbitrary field $\mathbb{F}$, σ is a translation automorphism of E, and $\mathcal{L}$ is an invertible $\mathcal{O}_E$-module of degree $n \geq 3$. They then prove in Proposition 4.1.1 the following result: if $\mu : E' \to E$ is an isomorphism between elliptic curves, then there is a canonical isomorphism $A(E,\sigma,\mathcal{L}) \to A(E',\mu^{-1}\sigma\mu,\mu^*\mathcal{L})$ sending $x \in A(E,\sigma,\mathcal{L})_1 = H^0(E,\mathcal{L})$ to $\mu^* x \in H^0(E',\mu^*\mathcal{L})$.

When E is a complex elliptic curve, σ is translation by ξ, and $\deg\mathcal{L} = n$, then

$$A(E,\sigma,\mathcal{L}) \;\cong\; Q_{n,1}(E,\xi) \;=\; Q_{n,1}(\eta\,|\,\tau)$$

where $\tau \in \mathbb{H}$ is any point such that $E \cong E_\tau$ and ξ is the image of η under an isomorphism $E \to E_\tau$.[3] Thus, [**27**, Prop. 4.1.1] proves Theorem 1.2 when $k = 1$.

[3] A proof that $A(E,\sigma,\mathcal{L})$ is isomorphic to $Q_{n,1}(\eta\,|\,\tau)$ is given in [**6**, §3.2.6].

1.5. Remarks on the proof of the main theorem. The main theorem is an immediate consequence of the "functional equation" (1.8) below. Let us explain.

Let $M = \begin{pmatrix} a & b \\ c & d \end{pmatrix} \in \mathrm{SL}(2,\mathbb{Z})$ and let $(\eta' \,|\, \tau') = M \triangleright (\eta \,|\, \tau) := \left(\frac{\eta}{c\tau+d} \,\middle|\, \frac{a\tau+b}{c\tau+d}\right)$. Since the space of defining relations for $Q_{n,k}(\eta \,|\, \tau)$ is the image of the map

$$R_{n,k}(\eta, \eta \,|\, \tau) : V^{\otimes 2} \to V^{\otimes 2},$$

to show that $Q_{n,k}(\eta' \,|\, \tau')$ is isomorphic to $Q_{n,k}(\eta \,|\, \tau)$ we must show there is an automorphism $\psi \in \mathrm{GL}(V)$, depending on M, such that $\psi^{\otimes 2}$ sends the image of $R_{n,k}(\eta, \eta \,|\, \tau)$ to the image of $R_{n,k}(\eta', \eta' \,|\, \tau')$; i.e., such that

$$(\psi^{\otimes 2}) \circ R_{n,k}(\eta, \eta \,|\, \tau) = R_{n,k}(\eta', \eta' \,|\, \tau') \circ \phi$$

for some $\phi \in \mathrm{GL}(V^{\otimes 2})$; (1.8) is a stronger and more precise result.

To state (1.8) we extend the action of $\mathrm{SL}(2,\mathbb{Z})$ on $(\eta \,|\, \tau) \in \mathbb{C} \times \mathbb{H}$ to an action on $(z, \eta \,|\, \tau) \in \mathbb{C} \times \mathbb{C} \times \mathbb{H}$,

$$M \triangleright (z, \eta \,|\, \tau) \;:=\; \left(\frac{z}{c\tau+d}, \frac{\eta}{c\tau+d} \,\middle|\, \frac{a\tau+b}{c\tau+d}\right).$$

Theorem 4.4 shows there is a $\psi \in \mathrm{GL}(V)$ and a nowhere-vanishing holomorphic function $f(z)$, both depending on $M \in \mathrm{SL}(2,\mathbb{Z})$, such that

$$R_{n,k}\big(M \triangleright (z, \eta \,|\, \tau)\big) \;=\; f(z)\,\big(\psi^{\otimes 2}\big) \circ R_{n,k}(z, \eta \,|\, \tau) \circ \big(\psi^{\otimes 2}\big)^{-1}. \tag{1.8}$$

1.6. The contents of this paper. The equality (1.8) and the main theorem are proved in Section 4. Section 2, which concerns certain normalizations $w_{(u,v)}(z, \eta \,|\, \tau)$ of theta functions with characteristics, and Section 3, which is about a finite Heisenberg group H_n of order n^3 and its extension $\widetilde{H}_n$ when n is even, are essential preliminaries.

The key to proving (1.8) is to use a presentation of the R-matrix defined in (1.6) that has more structure. This is done in several steps.

In Section 3, we make V, the degree-one component of $Q_{n,k}(\eta \,|\tau)$, into an irreducible representation of H_n; as a consequence there is a surjective homomorphism $\rho : \mathbb{C}H_n \to \mathrm{End}_{\mathbb{C}}(V)$ so we can, and do, choose a basis for $\mathrm{End}_{\mathbb{C}}(V)$ consisting of images of certain elements $J_{(u,v)} \in H_n$—this works as stated when n is odd but needs a small adjustment when n is even; we then write $R_{n,k}(z, \eta \,|\tau)$ in terms of this new basis and the functions $w_{(u,v)}(z, \eta \,|\, \tau)$—the essential part of this rewriting involves the operator $T_k(z, \eta \,|\, \tau) : V^{\otimes 2} \to V^{\otimes 2}$ which is defined in (4.2)—its relation to $R_{n,k}(z, \eta \,|\tau)$ is given in (4.3).

The "matrix coefficients" for $T_k(z, \eta \,|\, \tau)$ involve the functions $w_{(u,v)}(z, \eta \,|\, \tau)$ indexed by $(u, v) \in \mathbb{Z}^2$, which are introduced in Section 2. The main result in Section 2 is that for each $M \in \mathrm{SL}(2,\mathbb{Z})$ there is a nowhere-vanishing holomorphic function $g(z)$, depending on M but not on (u, v), such that

$$w_{(u,v)}\big(M \triangleright (z, \eta \,|\, \tau)\big) \;=\; g(z)\, w_{(u,v)M}(z, \eta \,|\, \tau).$$

So, as far as $w_{(u,v)}(z, \eta \,|\, \tau)$ is concerned, the modular action of $\mathrm{SL}(2,\mathbb{Z})$ on $\mathbb{C}\times\mathbb{C}\times\mathbb{H}$ can be replaced by the natural right action of $\mathrm{SL}(2,\mathbb{Z})$ on $\mathbb{Z}^2$ or $\mathbb{Z}_n^2$.

2. Theta functions with characteristics and the functions $w_{(u,v)}(z)$

We will use the notation $\omega := e(\frac{1}{n})$.

When $\tau \in \mathbb{H}$ we will write $\Lambda_\tau := \mathbb{Z} + \mathbb{Z}\tau$ and $E_\tau := \mathbb{C}/\Lambda_\tau$.

We write M^t for the transpose of a matrix M and M^{-t} for the inverse of M^t when it exists.

2.1. Definition of $\theta_{u,v}$.

As in [**5**, §2.5], we will use theta functions with characteristics, namely the functions

$$\theta_{u,v}(z\,|\,\tau) \;=\; \vartheta\begin{bmatrix}u\\v\end{bmatrix}(z\,|\,\tau) \;:=\; e\left(u(z+v)+\tfrac{1}{2}u^2\tau\right)\vartheta(z+u\tau+v\,|\,\tau)$$

where $\vartheta(z\,|\,\tau)$ is the function defined in (1.1). In particular, $\theta_{0,0} = \vartheta$.

PROPOSITION 2.1. *Let $s, t \in \mathbb{Z}$.*

(1) $\theta_{u,v}(z+s\,|\,\tau) = \theta_{u,v+s}(z\,|\,\tau)$.
(2) $\theta_{u,v}(z+s\tau+t\,|\,\tau) = e\big(-s(z+v) - \frac{s^2\tau}{2} + tu\big)\,\theta_{u,v}(z\,|\,\tau)$.
(3) $\theta_{u,v}(z\,|\,\tau) = 0$ *if and only if* $z \in \frac{1}{2}(\tau+1) - (u\tau+v) + \Lambda_\tau$.
(4) *Each of these zeros is a simple zero.*

PROOF. Parts (1), (2), and (3) were stated in [**5**, the introductory remarks in §2.5, and Prop. 2.8], with the roles of η and τ interchanged. For (3), see also [**22**, p. 315, (iv)] and the discussion before Proposition 2.8 in [**5**]. By [**5**, Lem. 2.5], $\theta_{u,v}(z\,|\,\tau)$ has a single zero in each fundamental parallelogram for Λ_τ located at $\frac{1}{2}(1+\tau) - (u\tau+v)$ modulo Λ_τ. □

2.2. Definition of $w_{(u,v)}$.

As in [**5**, §3], we define

$$w_{(u,v)}(z) \;:=\; \frac{\theta_{\frac{u}{n},\frac{v}{n}}(z+\zeta\,|\,\tau)}{\theta_{\frac{u}{n},\frac{v}{n}}(\zeta\,|\,\tau)}, \tag{2.1}$$

where

$$\zeta \;:=\; \eta + \tfrac{1}{2}(\tau+1). \tag{2.2}$$

We denoted $\eta + \frac{1}{2}(\tau+1)$ by ξ in [**5**], but in this paper ξ always denotes an arbitrary point on E.

The next result follows immediately from Proposition 2.1.

PROPOSITION 2.2. *The functions $w_{(u,v)}(z)$ have the following properties:*

(1) $w_{(u,v)}(0) = 1$;
(2) *if $s, t \in \mathbb{Z}$, then*

$$\frac{w_{(u,v)}(z+s\tau+t)}{w_{(u,v)}(z)} \;=\; e\left(-s(z+\eta) - \tfrac{s}{2}(s\tau+\tau+1) + \tfrac{1}{n}(tu-sv)\right);$$

(3) $w_{(u,v)}(z) = 0$ *if and only if* $z = -\eta - \frac{1}{n}(u\tau+v)$ *modulo* Λ_τ;
(4) *each of these zeros is a simple zero.*

In particular,

$$\begin{aligned} w_{(u,v)}(z+1) &= e\big(\tfrac{u}{n}\big)w_{(u,v)}(z), \\ w_{(u,v)}(z+\tau) &= e\big(-z-\eta-\tau-\tfrac{1}{2}-\tfrac{1}{n}v\big)w_{(u,v)}(z). \end{aligned}$$

2.2.1. The notation $\theta_{u,v}$ and $w_{(u,v)}$ makes sense for arbitrary complex numbers u and v. As explained in the discussion preceding Theorem 3.1 in [**5**], when u and v are integers $w_{(u,v)}(z)$ depends only on their images in $\mathbb{Z}_n$ so, in such a case, we will think of the subscript in $w_{(u,v)}(z)$ as an element of $\mathbb{Z}_n^2$.

2.3. The action of $\mathrm{SL}(2,\mathbb{Z})$ **on** $\mathbb{C}\times\mathbb{C}\times\mathbb{H}$**.** If $M = \left(\begin{smallmatrix} a & b \\ c & d\end{smallmatrix}\right) \in \mathrm{GL}(2,\mathbb{R})$ and $\tau\in\mathbb{H}$ we define

$$M \triangleright \tau \;:=\; \frac{a\tau+b}{c\tau+d}.$$

If $\Im(z)$ denotes the imaginary part of a complex number z, then

$$\Im(M\triangleright\tau) \;=\; \det(M)\,\frac{\Im(\tau)}{|c\tau+d|^2}\,.$$

Hence the group $\mathrm{GL}^+(2,\mathbb{R})$ of 2×2 real matrices with positive determinant acts on $\mathbb{H}$. That action extends to actions on $\mathbb{C}\times\mathbb{H}$, given by the formula

$$\begin{pmatrix} a & b \\ c & d\end{pmatrix} \triangleright (z\,|\,\tau) \;:=\; \left(\frac{z}{c\tau+d}\;\middle|\;\frac{a\tau+b}{c\tau+d}\right), \tag{2.3}$$

and on $\mathbb{C}\times\mathbb{C}\times\mathbb{H}$, given by the formula

$$\begin{pmatrix} a & b \\ c & d\end{pmatrix} \triangleright (z,\eta\,|\,\tau) \;:=\; \left(\frac{z}{c\tau+d},\,\frac{\eta}{c\tau+d}\;\middle|\;\frac{a\tau+b}{c\tau+d}\right). \tag{2.4}$$

In particular, the modular group $\mathrm{SL}(2,\mathbb{Z})$ acts on $\mathbb{C}\times\mathbb{H}$ and $\mathbb{C}\times\mathbb{C}\times\mathbb{H}$. The action on $\mathbb{C}\times\mathbb{H}$ is the same as the action in [**22**, (2.12)], which is a source we often refer to, both here and in [**5**].

2.4. Equivariance properties of $w_{(u,v)}(z,\eta\,|\,\tau)$**.** To state the next two results, we make explicit the dependence of $w_{(u,v)}$ on all parameters by writing

$$w_{(u,v)}(z,\eta\,|\,\tau) \;:=\; w_{(u,v)}(z). \tag{2.5}$$

PROPOSITION 2.3. *Let* $M = \left(\begin{smallmatrix} a & b \\ c & d\end{smallmatrix}\right) \in \mathrm{SL}(2,\mathbb{Z})$ *act on the triples* $(z,\eta\,|\,\tau)$ *via* (2.4)*. The functions*

$$z \;\mapsto\; w_{(u,v)M}(z,\eta\,|\,\tau) \tag{2.6}$$

and

$$z \;\mapsto\; w_{(u,v)}\big(M\triangleright(z,\eta\,|\,\tau)\big) \tag{2.7}$$

have the same zeros with the same multiplicities.

PROOF. Let

$$(z',\eta'\,|\,\tau') \;:=\; M\triangleright(z,\eta\,|\,\tau) \;=\; \left(\frac{z}{c\tau+d},\,\frac{\eta}{c\tau+d}\;\middle|\;\frac{a\tau+b}{c\tau+d}\right).$$

By Proposition 2.2(3), $w_{(u,v)M}(z,\eta\,|\,\tau)=0$ if and only if

$$z+\eta+\tfrac{1}{n}\big((au+cv)\tau+(bu+dv)\big) \in \Lambda_\tau \;=\; \mathbb{Z}(a\tau+b)+\mathbb{Z}(c\tau+d),$$

where the equality uses the fact that $M\in\mathrm{SL}(2,\mathbb{Z})$. After dividing by $c\tau+d$, this condition becomes

$$z'+\eta'+\tfrac{1}{c\tau+d}\,\tfrac{1}{n}\,\big(u(a\tau+b)+v(c\tau+d)\big) \in \mathbb{Z}\tau'+\mathbb{Z} \;=\; \Lambda_{\tau'}.$$

Therefore $w_{(u,v)M}(z,\eta\,|\,\tau)=0$ if and only if

$$z'+\eta'+\tfrac{1}{n}\big(u\tau'+v\big) \in \Lambda_{\tau'}.$$

By Proposition 2.2(3), this happens if and only if $w_{(u,v)}(z', \eta' \mid \tau') = 0$; i.e., if and only if $w_{(u,v)}\big(M \triangleright (z, \eta \mid \tau)\big) = 0$. Thus the zeros of the functions are the same. By Proposition 2.2(4), all zeros are simple zeros so their multiplicities are the same. □

We will now improve Proposition 2.3. First, we recall some terminology.

2.4.1. *Quasi-periodicity.* Suppose that $\{\omega_1, \omega_2\}$ is an $\mathbb{R}$-basis for $\mathbb{C}$, and let $\Lambda := \mathbb{Z}\omega_1 + \mathbb{Z}\omega_2$. We say that a function $f : \mathbb{C} \to \mathbb{C}$ is quasi-periodic with respect to Λ if there are elements $a, b, c, d \in \mathbb{C}$ such that

$$\begin{aligned} f(z + \omega_1) &= e^{2\pi i(az+b)} f(z) \qquad \text{and} \\ f(z + \omega_2) &= e^{2\pi i(cz+d)} f(z) \end{aligned}$$

for all $z \in \mathbb{C}$.

2.4.2. *Periodicity Factors.* A periodicity factor for Λ_τ is a function

$$\Lambda_\tau \times \mathbb{C} \longrightarrow \mathbb{C}^\times, \qquad (\lambda, z) \mapsto e_\lambda(z),$$

such that $z \mapsto e_\lambda(z)$ is a nowhere-vanishing holomorphic function on $\mathbb{C}$ such that

$$e_{\lambda+\mu}(z) \;=\; e_\lambda(z + \mu) e_\mu(z)$$

for all $\lambda, \mu \in \Lambda_\tau$; i.e., the function $\lambda \mapsto e_\lambda(-)$ is a 1-cocycle on Λ_τ taking values in the group of nowhere-vanishing holomorphic functions on $\mathbb{C}$, where the latter is a representation of Λ_τ via the action $(\lambda \cdot f)(z) := f(z + \lambda)$.

In Beauville's [**1**, §2.3], $\{e_\lambda \mid \lambda \in \Lambda\}$ is called a *system of multipliers.*

If $f(z \mid \tau)$ is quasi-periodic in z with respect to $\mathbb{Z}+\mathbb{Z}\tau$, then its quasi-periodicity properties imply that the function

$$(s\tau + t, z) \;\mapsto\; e_{s\tau+t}(z) \;:=\; \frac{f(z + s\tau + t \mid \tau)}{f(z \mid \tau)} \qquad\qquad (s, t \in \mathbb{Z})$$

is a periodicity factor for $\mathbb{Z} + \mathbb{Z}\tau$. We call it the periodicity factor for $f(z)$. For example, Proposition 2.2(2) says that the periodicity factor for $w_{(u,v)}(z, \eta \mid \tau)$ is

$$e\left(-s(z+\eta) - \tfrac{s}{2}(s\tau + \tau + 1) + \tfrac{1}{n}(u, v)\begin{pmatrix} t \\ -s \end{pmatrix}\right).$$

2.4.3. The following fact will be used in the next proof: if $f(z \mid \tau)$ and $g(z \mid \tau)$ are quasi-periodic with respect to $\mathbb{Z} + \mathbb{Z}\tau$ having the same periodicity factors and the same zeros (counted with multiplicity), then there is a constant c such that $f(z \mid \tau) = c\, g(z \mid \tau)$ for all $z \in \mathbb{C}$. The reason is that $f(z \mid \tau)/g(z \mid \tau)$ is an elliptic function with respect to Λ_τ having no zeros and no poles, and hence constant.

2.4.4. *Remark.* The relation between periodicity factors, theta functions, and sections of line bundles on complex abelian varieties (in our case line bundles on $E_\tau := \mathbb{C}/\Lambda_\tau$) is spelled out in Mumford's [**16**, §2], in Kempf's [**14**], and in Beauville's [**1**].

THEOREM 2.4. *Let $M \in \mathrm{SL}(2, \mathbb{Z})$. There is a nowhere-vanishing holomorphic function $f(z) = f_M(z)$, that does not depend on (u, v), such that*

$$w_{(u,v)M}(z, \eta \mid \tau) \;=\; f(z)\, w_{(u,v)}\big(M \triangleright (z, \eta \mid \tau)\big).$$

PROOF. If the theorem holds for M and N, then it holds for MN with $f_{MN}(z) = f_M(z) f_N(z)$ and for M^{-1} with $f_{M^{-1}}(z) = f_M(z)^{-1}$. It therefore suffices to prove the theorem for each of the two generators ([**23**, §1.5.3])

$$X := \begin{pmatrix} 0 & -1 \\ 1 & 0 \end{pmatrix} \qquad \text{and} \qquad Y := \begin{pmatrix} 1 & 1 \\ 0 & 1 \end{pmatrix}$$

of $\mathrm{SL}(2,\mathbb{Z})$. To prove the theorem it suffices to show that $w_{(u,v)M}(z,\eta\,|\,\tau)$ and $w_{(u,v)}\big(M \triangleright (z,\eta\,|\,\tau)\big)$ have the same quasi-periodicity properties with respect to Λ_τ, and the same zeros. We already showed in Proposition 2.3 that they have the same zeros, so it remains to show that they have the same periodicity factors with respect to Λ_τ.

Let $M = \left(\begin{smallmatrix} a & b \\ c & d \end{smallmatrix}\right) \in \mathrm{SL}(2,\mathbb{Z})$. As in the previous proof, we adopt the notation

$$(z',\eta'\,|\,\tau') \;:=\; M \triangleright (z,\eta\,|\,\tau) \;=\; \left(\frac{z}{c\tau+d}, \frac{\eta}{c\tau+d} \;\middle|\; \frac{a\tau+b}{c\tau+d}\right).$$

Claim. There are unique integers s', t' such that

$$(z' + s'\tau' + t', \eta'\,|\,\tau') \;=\; M \triangleright (z + s\tau + t, \eta\,|\,\tau),$$

namely $s' := ds - ct$ and $t' := at - bs$, i.e., $\left(\begin{smallmatrix} t' \\ -s' \end{smallmatrix}\right) = M\left(\begin{smallmatrix} t \\ -s \end{smallmatrix}\right)$.

Proof. By definition,

$$M \triangleright (z + s\tau + t, \eta\,|\,\tau) \;=\; \left(\frac{z+s\tau+t}{c\tau+d}, \frac{\eta}{c\tau+d} \;\middle|\; \frac{a\tau+b}{c\tau+d}\right).$$

Since $z' = \frac{z}{c\tau+d}$ it suffices to show there are unique integers s', t' such that

$$s'\tau' + t' \;=\; \frac{s\tau+t}{c\tau+d}.$$

Certainly, if such an $(s',t') \in \mathbb{Z}^2$ exists it is unique because $\{1,\tau'\}$ is linearly independent over $\mathbb{Z}$—even over $\mathbb{R}$. Since $\tau' = \frac{a\tau+b}{c\tau+d}$, a simple calculation shows that $(ds-ct)\tau' + (at-bs) \;=\; \frac{s\tau+t}{c\tau+d}$: just multiply both sides by $c\tau+d$ and use the fact that $ad - bc = 1$.[4] ◊

By Proposition 2.2(2), the periodicity factor for $w_{(u,v)M}(z\,|\,\tau)$ is

$$e\left(-s(z+\eta) - \tfrac{s}{2}(s\tau+\tau+1) + \tfrac{1}{n}\,(u,v)\,M\begin{pmatrix} t \\ -s \end{pmatrix}\right). \tag{2.8}$$

By Proposition 2.2(2), the periodicity factor for $w_{(u,v)}\big(M \triangleright (z,\eta\,|\,\tau)\big)$, i.e., the ratio

$$\frac{w_{(u,v)}\left(M \triangleright (z+s\tau+t,\eta\,|\,\tau)\right)}{w_{(u,v)}\left(M \triangleright (z,\eta\,|\,\tau)\right)} \;=\; \frac{w_{(u,v)}(z'+s'\tau'+t',\eta'\,|\,\tau')}{w_{(u,v)}(z',\eta'\,|\,\tau')},$$

is

$$e\left(-s'(z'+\eta') - \tfrac{s'}{2}(s'\tau'+\tau'+1) + \tfrac{1}{n}(u,v)\begin{pmatrix} t' \\ -s' \end{pmatrix}\right). \tag{2.9}$$

But the terms involving (u,v) in (2.8) and (2.9) are equal, so

$$\frac{(2.8)}{(2.9)} \;=\; \frac{e\left(-s(z+\eta) - \frac{s}{2}(s\tau+\tau+1)\right)}{e\left(-s'(z'+\eta') - \frac{s'}{2}(s'\tau'+\tau'+1)\right)}. \tag{2.10}$$

[4] Alternatively, if $(s',t') \in \mathbb{R}^2$ is such that $(c\tau+d)(s',t')\left(\begin{smallmatrix} \tau' \\ 1 \end{smallmatrix}\right) = (s,t)\left(\begin{smallmatrix} \tau \\ 1 \end{smallmatrix}\right)$, then $(s,t)\left(\begin{smallmatrix} \tau \\ 1 \end{smallmatrix}\right) = (s',t')\left(\begin{smallmatrix} a\tau+b \\ c\tau+d \end{smallmatrix}\right) = (s',t')\,M\left(\begin{smallmatrix} \tau \\ 1 \end{smallmatrix}\right)$ and, since $\{1,\tau\}$ is linearly independent over $\mathbb{R}$, $(s',t') = (s,t)M^{-1} = (s,t)\left(\begin{smallmatrix} d & -b \\ -c & a \end{smallmatrix}\right) = (ds-ct, at-bs)$.

(1) (Proof for $M = Y$.) In this case, $(z', \eta' \,|\, \tau') = (z, \eta \,|\, \tau + 1)$ and

$$(t', -s') = (t, -s)Y^t = (t - s, -s).$$

Thus, when $M = Y$, the denominator in (2.10) is

$$e\left(-s(z+\eta) - \tfrac{s}{2}(s\tau + s + \tau + 1 + 1)\right),$$

which is equal to the numerator in (2.10) because $e\big(-\frac{s}{2}(s+1)\big) = 1$. Hence (2.8)=(2.9) when $M = Y$ so, by the remark in subsection 2.4.3,

$$w_{(u,u+v)}(z, \eta \,|\, \tau) \;=\; c\, w_{(u,v)}(z, \eta \,|\, \tau + 1)$$

for some constant c. But $w_{(u,u+v)}(0, \eta \,|\, \tau) = 1 = w_{(u,v)}(0, \eta \,|\, \tau + 1)$ so $c = 1$, and we conclude that the theorem holds for Y with $f_Y(z) = 1$ for all z.

(2) (Proof for $M = X$.) In this case, $(u, v)M = (v, -u)$,

$$(z', \eta' \,|\, \tau') \;=\; (z/\tau,\, \eta/\tau \,|\, -1/\tau),$$

and $(s', t') = (-t, s)$. Thus, when $M = X$,

$$\frac{(2.8)}{(2.9)} \;=\; e\left(-s(z+\eta) - \tfrac{s}{2}(s\tau + \tau + 1) - \tfrac{t}{2}(\tfrac{t}{\tau} - \tfrac{1}{\tau} + 1) - \tfrac{t}{\tau}(z + \eta)\right). \tag{2.11}$$

If we replaced the term $w_{(u,v)M}(z, \eta \,|\, \tau)$ by $g(z)\, w_{(u,v)M}(z, \eta \,|\, \tau)$ where $g(z) = e(Az^2 + Bz)$, then (2.11) would be multiplied by

$$\begin{aligned}\frac{g(z + s\tau + t)}{g(z)} &= \frac{e(A(z + s\tau + t)^2 + B(z + s\tau + t))}{e(Az^2 + Bz)} \\ &= e\big(A(2z + s\tau + t)(s\tau + t) + B(s\tau + t)\big).\end{aligned}$$

Hence if $A = \frac{1}{2\tau}$ and $B = \frac{1}{2} - \frac{1}{2\tau} + \frac{\eta}{\tau}$, then (2.11) would be multiplied by

$$e\big(\tfrac{1}{2\tau}(2z + s\tau + t)(s\tau + t) + \big(\tfrac{1}{2} - \tfrac{1}{2\tau} + \tfrac{\eta}{\tau}\big)(s\tau + t)\big)$$

which equals

$$e\big(sz + \tfrac{zt}{\tau} + \tfrac{1}{2\tau}(s^2\tau^2 + 2st\tau + t^2) + \tfrac{s\tau}{2} - \tfrac{s}{2} + s\eta + \tfrac{t}{2} - \tfrac{t}{2\tau} + \tfrac{t\eta}{\tau}\big).$$

A straightforward calculation shows that the product of this with (2.11) equals 1, so we conclude that $g(z)\, w_{(u,v)X}(z, \eta \,|\, \tau)$ and $w_{(u,v)}\big(X \triangleright (z, \eta \,|\, \tau)\big)$ have the same periodicity factors with respect to Λ_τ. Since $g(z) \neq 0$ for all z, they also have the same zeros by Proposition 2.3, and both take the value 1 when $z = 0$, so we conclude that

$$w_{(u,v)X}(z, \eta \,|\, \tau) \;=\; e\big(-\tfrac{1}{2\tau}z^2 + \big(\tfrac{1}{2\tau} - \tfrac{1}{2} - \tfrac{\eta}{\tau}\big)z\big)\, w_{(u,v)}\big(X \triangleright (z, \eta \,|\, \tau)\big).$$

The proof is complete. □

3. An extension of the finite Heisenberg group

Heisenberg groups play an important organizational role in the theory of theta functions and abelian varieties—see, for example, [**2**, Ch. 6], [**15**, Ch. I, §3], [**17**], and [**21**]. In the present setting, the relevant Heisenberg group is an extension $1 \to \mu_n \to H_n \to \mathbb{Z}_n^2 \to 0$ where μ_n is the group of complex n^{th} roots of unity. Below we will make V, the degree-one component of $Q_{n,k}(\eta \,|\, \tau)$, an irreducible representation for H_n. As Feigin and Odesskii first noticed, that action lifts to an action of H_n as automorphisms of $Q_{n,k}(\eta \,|\, \tau)$ (see [**6**, §§2.3 and 3.5] for details).

Proposition 3.1 shows that when n is odd the natural action of each $M \in \mathrm{SL}(2, \mathbb{Z})$ on $\mathbb{Z}_n^2 = H_n/\langle\epsilon\rangle$ lifts to an action of M as an automorphism of H_n. (But this does not yield a homomorphism $\mathrm{SL}(2,\mathbb{Z}) \to \mathrm{Aut}(H_n)$.) Such a lifting does not exist when n is even. For this reason we introduce an index-two extension $\widetilde{H}_n$ of H_n for which such a lifting does exist. The utility of H_n and $\widetilde{H}_n$ becomes apparent in Section 4 when we introduce a new presentation for the linear operator $R_{n,k}(z,\eta\,|\,\tau) : V^{\otimes 2} \to V^{\otimes 2}$ in terms of a new linear operator

$$T_k(z,\eta\,|\,\tau) : V^{\otimes 2} \to V^{\otimes 2}$$

that transforms nicely with respect to the actions of, first, $\mathrm{SL}(2,\mathbb{Z})$ on $(z,\eta\,|\,\tau)$ and the subscripts $(u,v) \in \mathbb{Z}_n^2$ in $w_{(u,v)}(z)$, and, second, $\widetilde{H}_n$ on $V^{\otimes 2}$, and, third, an action of $\mathrm{SL}(2,\mathbb{Z})$ on $\widetilde{H}_n$.

3.1. The Heisenberg group H_n.

The Heisenberg group of order n^3 is

$$H_n \;:=\; \langle S,T,\epsilon \mid S^n = T^n = \epsilon^n = 1,\; [S,\epsilon] = [T,\epsilon] = 1,\; [S,T] = \epsilon\rangle .$$

The following equalities are useful when computing in H_n:

$$\begin{aligned}
S^aT^b &= T^bS^a\epsilon^{ab},\\
S^{a_1}T^{b_1}S^{a_2}T^{b_2}S^{a_3}T^{b_3}\cdots &= T^BS^A\epsilon^C,\\
&\quad \text{where } A = a_1+a_2+\cdots, \quad B = b_1+b_2+\cdots,\\
&\quad C = a_1b_1+(a_1+a_2)b_2+(a_1+a_2+a_3)b_3+\cdots,\\
(S^aT^b)^m &= T^{bm}S^{am}\epsilon^X, \quad \text{where } X = \tfrac{1}{2}m(m+1)ab,\\
(T^bS^a)^m &= T^{bm}S^{am}\epsilon^Y, \quad \text{where } Y = \tfrac{1}{2}m(m-1)ab.
\end{aligned}$$

PROPOSITION 3.1. *Let $M = \left(\begin{smallmatrix} a & b\\ c & d\end{smallmatrix}\right) \in \mathrm{GL}(2,\mathbb{Z})$. If n is odd, there is an automorphism $\Psi'_M : H_n \to H_n$ such that*

$$\Psi'_M : \qquad T \mapsto T^aS^c, \qquad S \mapsto T^bS^d, \qquad \epsilon \mapsto \epsilon^{\det M}. \tag{3.1}$$

PROOF. The expressions T^aS^c and T^bS^d make sense when $a,b,c,d \in \mathbb{Z}_n$ because T and S have order n. The result follows from the fact that T^aS^c and T^bS^d have order n, and

$$\left(T^bS^d\right)\left(T^aS^c\right)\left(T^bS^d\right)^{-1}\left(T^aS^c\right)^{-1} \;=\; \epsilon^{ad-bc}.$$

The calculations contain no surprises. □

The map $\mathrm{GL}(2,\mathbb{Z}) \to \mathrm{Aut}(H_n)$, $M \mapsto \Psi'_M$, is *not* a group homomorphism. For example, if $M = \left(\begin{smallmatrix} 0 & -1\\ 1 & 1\end{smallmatrix}\right)$, then $\Psi'_M(T) = S$, $\Psi'_M(S) = T^{-1}S$, and $\Psi'_M\Psi'_M(S) = S^{-1}T^{-1}S = \epsilon T^{-1}$, but $\Psi'_{M^2}(S) = T^{-1}$. Similarly, if $M = \left(\begin{smallmatrix} 0 & 1\\ 1 & 0\end{smallmatrix}\right)$ and $N = \left(\begin{smallmatrix} 1 & 0\\ 1 & 1\end{smallmatrix}\right)$, then $\Psi'_M\Psi'_N(T) = \Psi'_M(TS) = ST \neq TS = \Psi'_{MN}(T)$.

Because $M \mapsto \Psi'_M$ is not a group homomorphism we will make no use of Ψ'_M.

When n is even, an additional problem arises: the map Ψ'_M in (3.1) does not extend to an automorphism of H_n because T^aS^c and T^bS^d need not have order n. For example, TS has order $2n$.[5] In order to address this problem we introduce a new group when n is even, the group $\widetilde{H}_n$ defined below.

[5] If b and c are even residues modulo n, however, both T^aS^c and T^bS^d have order n.

3.2. The extension $\widetilde{H}_n$ of H_n. When n is even we define

$$\widetilde{H}_n := \Big\langle S, T, \epsilon^{1/2} \;\Big|\; S^n = T^n = \epsilon^n = 1,\ [S,\epsilon^{1/2}] = [T,\epsilon^{1/2}] = 1, \\ [S,T] = \epsilon = (\epsilon^{1/2})^2 \Big\rangle.$$

Here $\epsilon^{1/2}$ is a symbol that behaves like a square root of ϵ. We note that $\langle S, T, \epsilon\rangle$ is a normal subgroup of index two in $\widetilde{H}_n$ and is isomorphic to H_n.

3.2.1. *The notation $\widetilde{H}_n$ when n is odd.* For convenience, we define $\widetilde{H}_n := H_n$ when n is odd, with the convention that $\epsilon^{1/2} := \epsilon^{(n+1)/2}$. Thus, when n is odd $\big(\epsilon^{(n+1)/2}\big)^2 = \epsilon$, so $\langle \epsilon^{(n+1)/2}\rangle = \langle\epsilon\rangle$, and $(\epsilon^{1/2})^{-1} = \epsilon^{(n-1)/2}$.

3.2.2. *The action of $\widetilde{H}_n$ as automorphisms of $Q_{n,k}(\eta\,|\,\tau)$.* Feigin and Odesskii observed that H_n acts as automorphisms of $Q_{n,k}(\eta\,|\,\tau)$ [**18**, p. 1143] via the actions

$$S\cdot x_i = \omega^i x_i, \qquad T\cdot x_i = x_{i+1}, \qquad \epsilon\cdot x_i = e\big(\tfrac{1}{n}\big)x_i.$$

A proof of this can be found at [**6**, Prop. 3.23]. It is easy to see that this extends to an action of $\widetilde{H}_n$ as automorphisms of $Q_{n,k}(\eta\,|\,\tau)$ when n is even.

PROPOSITION 3.2. *The group $\widetilde{H}_n$ acts as degree-preserving $\mathbb{C}$-algebra automorphisms of $Q_{n,k}(\eta\,|\,\tau)$ by*

$$S\cdot x_i = \omega^i x_i, \qquad T\cdot x_i = x_{i+1}, \qquad \epsilon^{1/2}\cdot x_i = -e\big(\tfrac{1}{2n}\big)x_i = e\big(\tfrac{n+1}{2n}\big)x_i. \tag{3.2}$$

3.2.3. The reason we chose the action $\epsilon^{1/2}\cdot x_i = -e\big(\frac{1}{2n}\big)x_i$ rather than the action $\epsilon^{1/2}\cdot x_i = e\big(\frac{1}{2n}\big)x_i$ in (3.2) is so the formula works for both even and odd n; i.e., the choice we have made for $\epsilon^{1/2}\cdot x_i$ is compatible with the convention in subsection 3.2.1 that $\epsilon^{1/2} = \epsilon^{(n+1)/2}$ when n is odd, but the other choice is incompatible.

3.3. The action of $\mathrm{SL}(2,\mathbb{Z})$ as automorphisms of $\widetilde{H}_n$. There are many actions of $\mathrm{SL}(2,\mathbb{Z})$ as automorphisms of $\widetilde{H}_n$, but we single out one particular action in Proposition 3.4 that has the virtue of simplicity.

LEMMA 3.3. *Let $\nu = \epsilon^{1/2}$ when n is even and let $\nu = \epsilon^{(n+1)/2}$ when n is odd. If $M = \left(\begin{smallmatrix} a & b\\ c & d\end{smallmatrix}\right) \in \mathrm{GL}(2,\mathbb{Z})$, then there is an automorphism Ψ_M of $\widetilde{H}_n$ such that*

$$\begin{aligned} \Psi_M :\quad & T \mapsto T^aS^c\nu^{ac}, \qquad S \mapsto T^bS^d\nu^{bd}, \qquad \nu\mapsto \nu^{ad-bc}, \\ & T^mS^r \mapsto T^{am+br}S^{cm+dr}\nu^Z, \end{aligned}$$

where $Z = acm^2 + bdr^2 + 2bcmr$.

PROOF. Since

$$\begin{aligned} \Psi_M(S)\Psi_M(T) &= T^bS^d\nu^{bd}T^aS^c\nu^{ac} = T^{a+b}S^{c+d}\epsilon^{ad}\nu^{bd+ac} \quad\text{and} \\ \Psi_M(T)\Psi_M(S) &= T^aS^c\nu^{ac}T^bS^d\nu^{bd} = T^{a+b}S^{c+d}\epsilon^{bc}\nu^{bd+ac}, \end{aligned}$$

we have $\Psi_M(S)\Psi_M(T) = \Psi_M(T)\Psi_M(S)\epsilon^{ad-bc} = \Psi_M(T)\Psi_M(S)\Psi_M(\epsilon)$.

Furthermore,

$$\Psi_M(T)^n = (T^aS^c\nu^{ac})^n = T^{an}S^{cn}\epsilon^D\nu^{acn}$$

where $D = \frac{1}{2}n(n-1)ac$. If n is odd, then D is an integer multiple of n so $\epsilon^D = 1$, and $\nu^{acn} = \epsilon^{acn(n+1)/2} = 1$ so $\Psi_M(T)^n = 1$. If n is even, then $D = m(n-1)ac$ where $m = n/2$ and $\nu^{acn} = \epsilon^{acm}$, whence $\epsilon^D\nu^{acn} = \epsilon^{mnac} = 1$; therefore $\Psi_M(T)^n = 1$

when n is even. In conclusion, $\Psi_M(T)^n = 1$ when n is odd and when n is even. Similar calculations show that $\Psi_M(S)^n = 1$ when n is odd and when n is even.

Hence Ψ_M is a group homomorphism.

Clearly, $\ker(\Psi_M) \cap \langle \nu \rangle = \{1\}$. However, every non-trivial normal subgroup of $\widetilde{H}_n$ has non-trivial intersection with $\langle \nu \rangle$ so we conclude that $\ker(\Psi_M) = \{1\}$. Hence Ψ_M is an automorphism of $\widetilde{H}_n$.

We have

$$\begin{aligned} \Psi_M(T^m S^r) &= (T^a S^c \nu^{ac})^m (T^b S^d \nu^{bd})^r \\ &= T^{am} S^{cm} \epsilon^X T^{br} S^{dr} \epsilon^Y \nu^{acm+bdr} \\ &= T^{am+br} S^{cm+dr} \epsilon^{X+Y+bcmr} \nu^{acm+bdr} \end{aligned}$$

where $X = \frac{1}{2}m(m-1)ac$ and $Y = \frac{1}{2}r(r-1)bd$. Since $\epsilon = \nu^2$, the last part of the lemma holds with $Z = m(m-1)ac + r(r-1)bd + 2bcmr + acm + bdr$. Hence the result. □

PROPOSITION 3.4. *Let $\nu = \epsilon^{1/2}$ when n is even and let $\nu = \epsilon^{(n+1)/2}$ when n is odd. The map $\Psi : \mathrm{SL}(2,\mathbb{Z}) \to \mathrm{Aut}(\widetilde{H}_n)$ that sends $M = \left(\begin{smallmatrix} a & b \\ c & d \end{smallmatrix}\right) \in \mathrm{SL}(2,\mathbb{Z})$ to the automorphism*

$$\Psi_M : \qquad T \mapsto T^a S^c \nu^{ac}, \qquad S \mapsto T^b S^d \nu^{bd}, \qquad \nu \mapsto \nu,$$

is a group homomorphism.

PROOF. Let $N = \left(\begin{smallmatrix} a' & b' \\ c' & d' \end{smallmatrix}\right) \in \mathrm{SL}(2,\mathbb{Z})$. Then

$$\begin{aligned} (\Psi_N \Psi_M)(T) &= (T^{a'} S^{c'} \nu^{a'c'})^a (T^{b'} S^{d'} \nu^{b'd'})^c \nu^{ac} \\ &= T^{a'a} S^{c'a} \epsilon^A \nu^{a'c'a} T^{b'c} S^{d'c} \epsilon^B \nu^{b'd'c} \nu^{ac} \\ &= T^{a'a+b'c} S^{c'a+d'c} \epsilon^{c'ab'c} \epsilon^{A+B} \nu^{a'c'a+b'd'c+ac} \qquad \text{and} \\ \Psi_{NM}(T) &= T^{a'a+b'c} S^{c'a+d'c} \nu^{(aa'+b'c)(c'a+d'c)} \end{aligned}$$

where $A = \frac{1}{2}a(a-1)a'c'$ and $B = \frac{1}{2}c(c-1)b'd'$. We have

$$\nu^{(aa'+b'c)(c'a+d'c)} \nu^{-a'c'a-b'd'c-ac} = \nu^D$$

where

$$\begin{aligned} D &= a'c'a^2 + a'd'ac + b'c'ac + b'd'c^2 - a'c'a - b'd'c - ac \\ &= ac(a'd' + b'c' - 1) + a'c'(a^2 - a) + b'd'(c^2 - c) \\ &= 2acb'c' + a(a-1)a'c' + c(c-1)b'd' \\ &= 2(acb'c' + A + B). \end{aligned}$$

Since $\nu^2 = \epsilon$, $\nu^D = \epsilon^{acb'c'+A+B}$. Hence $(\Psi_N\Psi_M)(T) = \Psi_{NM}(T)$. A similar calculation shows that $(\Psi_N\Psi_M)(S) = \Psi_{NM}(S)$. Hence $\Psi_N\Psi_M = \Psi_{NM}$, so Ψ is a group homomorphism. □

3.3.1. *Remark.* There is another way to prove the existence of the homomorphism $\Psi : \mathrm{SL}(2,\mathbb{Z}) \to \mathrm{Aut}(\widetilde{H}_n)$.

By [**23**, §1.5.3] or [**3**, Prop. 2.1], for example, $\mathrm{SL}(2,\mathbb{Z})$ is the amalgamated free product $\mathbb{Z}_4 *_{\mathbb{Z}_2} \mathbb{Z}_6$ with $\mathbb{Z}_4$ and $\mathbb{Z}_6$ generated by

$$X := \begin{pmatrix} 0 & 1 \\ -1 & 0 \end{pmatrix} \qquad \text{and} \qquad Y := \begin{pmatrix} 1 & 1 \\ -1 & 0 \end{pmatrix}, \tag{3.3}$$

respectively. Independently of Lemma 3.3, it is easy to check that the formulas

$$\Psi_X : \quad T \mapsto S^{-1}, \quad S \mapsto T, \quad \nu \mapsto \nu,$$
$$\Psi_Y : \quad T \mapsto TS^{-1}\nu^{-1}, \quad S \mapsto T, \quad \nu \mapsto \nu,$$

(where $\nu = \epsilon^{1/2}$) extend to automorphisms of $\widetilde{H}_n$. Since

$$\Psi_X^2 = \Psi_Y^3 : \quad T \mapsto T^{-1}, \quad S \mapsto S^{-1}, \quad \nu \mapsto \nu,$$

we obtain $\Psi_X^4 = \Psi_Y^6 = \mathrm{id}$. It follows that there is a unique group homomorphism $\Psi : \mathbb{Z}_4 *_{\mathbb{Z}_2} \mathbb{Z}_6 = \langle X \rangle *_{\mathbb{Z}_2} \langle Y \rangle \to \operatorname{Aut} H_n$ with the property that $\Psi(X) = \Psi_X$ and $\Psi(Y) = \Psi_Y$.

The drawback to this alternative approach is that some additional calculation is needed to produce a formula for $\Psi(M)$ for all $M \in \mathrm{SL}(2,\mathbb{Z})$.

3.3.2. *Remark.* In Proposition 3.4, the automorphism Ψ_M only depends on the image of M in $\mathrm{SL}(2,\mathbb{Z}_n)$ because T^m and S^m only depend on m modulo n.

3.3.3. *Notation.* If $M \in \mathrm{SL}(2,\mathbb{Z})$ and $x \in \widetilde{H}_n$ we will use the notation

$$M \triangleright x \;:=\; \Psi_M(x).$$

3.3.4. A nice feature of the action of $\mathrm{SL}(2,\mathbb{Z})$ on $\widetilde{H}_n$ in Proposition 3.4 is the following: if we define $J_{(u,v)} := T^u S^v \in \widetilde{H}_n$, then in $\widetilde{H}_n/\langle \epsilon^{1/2} \rangle \cong \mathbb{Z}_n^2$ we have $M \triangleright J_{(u,v)} = J_{(u,v)M^t}$.

3.4. Extending representations of H_n to representations of $\widetilde{H}_n$. It is well-known that the n-dimensional irreducible representations of H_n are determined up to isomorphism by their central characters; see, for example, [**11**, §2].

For a moment, let $\rho : H_n \to \mathrm{GL}(n,\mathbb{C})$ be any irreducible representation of H_n on $\mathbb{C}^n$. Because ϵ has order n it acts on $\mathbb{C}^n$ as multiplication by an nth root of unity, ζ say.

When n is even ρ can be extended to a representation $\widetilde{\rho}$ of $\widetilde{H}_n$ by having $\epsilon^{1/2}$ act by a chosen square root of ζ. By [**10**, Prop. 5.1], for example, there are exactly two irreducible $\widetilde{H}_n$-representations $\widetilde{\rho}$ such that $\widetilde{\rho}|_{H_n} \cong \rho$, namely $\widetilde{\rho}$ and $\widetilde{\rho} \otimes \varepsilon$, where ε is the non-trivial 1-dimensional character of $\widetilde{H}_n$ that sends H_n to 1. In particular, such a representation of $\widetilde{H}_n$ is also determined by its central character, i.e., the scalar by which the generator $\epsilon^{1/2}$ of the center acts. The next statement follows.

PROPOSITION 3.5. *Let $\rho : H_n \to \mathrm{GL}(n,\mathbb{C})$ be any n-dimensional irreducible representation of H_n on which ϵ acts as multiplication by ζ. Fix a square root, $\zeta^{1/2}$, of ζ. Extend ρ to a group homomorphism $\widetilde{\rho} : \widetilde{H}_n \to \mathrm{GL}(n,\mathbb{C})$ by declaring that*

$$\widetilde{\rho}(T^a S^b \epsilon^{c/2}) \;:=\; \rho(T)^a \rho(S)^b (\zeta^{1/2})^c.$$

If $\phi \in \operatorname{Aut}(\widetilde{H}_n)$ acts as the identity on $\epsilon^{1/2}$, then the representations $\widetilde{\rho}\phi$ and $\widetilde{\rho}$ are isomorphic.

COROLLARY 3.6. *Let $\rho : H_n \to \mathrm{GL}(n,\mathbb{C})$ be any n-dimensional irreducible representation of H_n and $\widetilde{\rho} : H_n \to \mathrm{GL}(n,\mathbb{C})$ an extension of it to $\widetilde{H}_n$. For $M \in \mathrm{SL}(2,\mathbb{Z})$, let $\Psi_M \in \operatorname{Aut}(\widetilde{H}_n)$ be the automorphism in Proposition 3.4. There is a set map—not a group homomorphism, in general—*

$$\psi : \mathrm{SL}(2,\mathbb{Z}) \to \mathrm{GL}(V) \tag{3.4}$$

such that

$$\widetilde{\rho}(\Psi_M(x)) \;=\; \psi(M)\,\widetilde{\rho}(x)\,\psi(M)^{-1} \tag{3.5}$$

for all $x \in \widetilde{H}_n$.

PROOF. This follows from the previous proposition because $\Psi_M(\epsilon^{1/2}) = \epsilon^{1/2}$. □

3.4.1. *The notation* ρ. From now on we will use the notation $\rho : \widetilde{H}_n \to \mathrm{GL}(V)$ for the representation of $\widetilde{H}_n$ in Proposition 3.2, and also for its linear extension $\mathbb{C}\widetilde{H}_n \to \mathrm{End}_{\mathbb{C}}(V)$ to the group algebra. We will therefore write (3.5) as

$$\rho(M \triangleright x) \;=\; \psi(M)\,\rho(x)\,\psi(M)^{-1}. \tag{3.6}$$

3.4.2. *Remark.* In analogy with the remark in subsection 3.3.2, the automorphism $\psi(M) : V \to V$ depends only on the image of M in $\mathrm{SL}(2,\mathbb{Z}_n)$. For example, if the image of M in $\mathrm{SL}(2,\mathbb{Z}_n)$ is the identity, then $\Psi_M = \mathrm{id}_{\widetilde{H}_n}$ so, in Corollary 3.6, we can take $\psi(M) = \mathrm{id}_V$.

3.5. The algebra A_n and the action of $\mathrm{SL}(2,\mathbb{Z})$ on it. Let

$$A_n \;:=\; \mathbb{C}\widetilde{H}_n \otimes_{\mathbb{C}\langle \epsilon^{1/2}\rangle} \mathbb{C}\widetilde{H}_n,$$

the tensor square of the group algebra $\mathbb{C}\widetilde{H}_n$ over the group algebra of its center, $\langle \epsilon^{1/2}\rangle$. The action of $\mathrm{SL}(2,\mathbb{Z})$ on $\widetilde{H}_n$ extends to an action of $\mathrm{SL}(2,\mathbb{Z})$ on $\mathbb{C}\widetilde{H}_n$ and, because elements of $\mathrm{SL}(2,\mathbb{Z})$ act as the identity on the center of $\widetilde{H}_n$, there is an induced action of $\mathrm{SL}(2,\mathbb{Z})$ on A_n, namely $M \triangleright (x \otimes y) = (M \triangleright x) \otimes (M \triangleright y)$.

4. The main results

4.1. Introduction. Let $g, h \in \mathrm{GL}(V)$ be the linear operators

$$g \cdot x_i \;:=\; \omega^i x_i \qquad\qquad h \cdot x_i \;:=\; x_{i-1}$$

where $\omega := e\big(\frac{1}{n}\big)$. The map $S \mapsto g$, $T \mapsto h$, extends to an irreducible representation of H_n on V (essentially because if m is an integer, then g^m and h^m depend only on m modulo n) in which ϵ acts as multiplication by ω^{-1}. For $(a,b) \in \mathbb{Z}_n^2$, we define the linear operator

$$I_{a,b} \;:=\; h^a g^b \,: V \to V,$$

and its "universal" version

$$J_{a,b} \;:=\; J_{(a,b)} \;:=\; T^a S^b \,\in\, \mathbb{C}\widetilde{H}_n.$$

4.1.1. We note that $M \triangleright J_{(a,b)} = c\, J_{(a,b)M^t}$ where c is some element in the center of $\widetilde{H}_n$. Thus, in A_n we have the useful equality

$$M \triangleright \big(J_{(a,b)} \otimes J_{(a,b)}^{-1}\big) \;=\; J_{(a,b)M^t} \otimes J_{(a,b)M^t}^{-1}. \tag{4.1}$$

This is the reason we treat $J_{(a,b)} \otimes J_{(a,b)}^{-1}$ as an element in A_n rather than an element in $\mathbb{C}\widetilde{H}_n \otimes_{\mathbb{C}} \mathbb{C}\widetilde{H}_n$ when we define $L_k(z,\eta\,|\,\tau)$ in (4.4) below.

4.1.2. Let $T_k(z,\eta\,|\,\tau) : V^{\otimes 2} \to V^{\otimes 2}$ be the linear operator

$$T_k(z,\eta\,|\,\tau) := \sum_{(u,v)\in\mathbb{Z}_n^2} w_{(u,v)}(-nz,\eta\,|\,\tau) I_{-k'u,v} \otimes I_{-k'u,v}^{-1} \tag{4.2}$$

where $k' :=$ the unique integer such that $n > k' \geq 1$ and $kk' = 1 \bmod n$.

4.1.3. Let $P : V \otimes V \to V \otimes V$ be the linear map $P(x \otimes y) = y \otimes x$.

THEOREM 4.1. [**5**, Prop. 3.5] *The images of* $R_{n,k}(z,\eta\,|\,\tau)$ *and* $PT_k(z,\eta\,|\,\tau)$ *are the same because*

$$R_{n,k}(z,\eta\,|\,\tau) \;=\; \tfrac{1}{n}e(-\tfrac{1}{2}n(n+1)z)\,P\,T_k(z,\eta\,|\,\tau). \tag{4.3}$$

Hence

$$Q_{n,k}(\eta\,|\,\tau) \;=\; \frac{TV}{(\textit{the image of } P\,T_k(\eta,\eta\,|\,\tau))}.$$

The only difference between the operator $R_{n,k}(z,\eta\,|\,\tau)$ defined in (1.6) and $\frac{1}{n}e(-\frac{1}{2}n(n+1)z)PT_k(z,\eta\,|\,\tau)$ defined via (4.2) is that they are the matrix representations of the same linear operator with respect to different bases for $\operatorname{End}_{\mathbb{C}}(V)$. The advantage of $T_k(z,\eta\,|\,\tau)$ over $R_{n,k}(z,\eta\,|\,\tau)$ is that we can exploit the way in which the actions of $\widetilde{H}_n$ and A_n on $V^{\otimes 2}$ interact with the actions of $\mathrm{SL}(2,\mathbb{Z})$ on $\widetilde{H}_n$ and the functions $w_{(u,v)}(z,\eta\,|\,\tau)$.[6] To determine how $Q_{n,k}(\eta\,|\,\tau)$ transforms when $\mathrm{SL}(2,\mathbb{Z})$ acts on $(\eta\,|\,\tau)$ we will determine how $T_k(z,\eta\,|\,\tau)$ transforms under the action of $\mathrm{SL}(2,\mathbb{Z})$ on $(z,\eta\,|\,\tau) \in \mathbb{C}\times\mathbb{C}\times\mathbb{H}$.

4.2. $\mathrm{SL}(2,\mathbb{Z})$**-equivariance properties of** $T_k(z,\eta\,|\,\tau)$ **and** $L_k(z,\eta\,|\,\tau)$**.** Following [**5**, Thm. 3.2] and the discussion before it, we will determine how $T_k(z,\eta\,|\,\tau)$ transforms under the action of $\mathrm{SL}(2,\mathbb{Z})$ by examining the action of $\mathrm{SL}(2,\mathbb{Z})$ on the "universal" version of $T_k(z,\eta\,|\,\tau)$ which belongs to the algebra A_n. That universal version is

$$L_k(z,\eta\,|\,\tau) \;:=\; \sum_{(u,v)\in\mathbb{Z}_n^2} w_{(u,v)}(-nz,\eta\,|\,\tau)J_{-k'u,v}\otimes J_{-k'u,v}^{-1} \;\in\; A_n. \tag{4.4}$$

The next "result" formalizes the statement that $L_k(z,\eta\,|\,\tau)$ is the universal version of $T_k(z,\eta\,|\,\tau)$.

LEMMA 4.2. *If* $\rho : \mathbb{C}\widetilde{H}_n \to \operatorname{End}(V)$ *is the representation in Proposition* 3.2, *then*

$$T_k(z,\eta\,|\,\tau) \;=\; (\rho\otimes\rho)\big(L_k(z,\eta\,|\,\tau)\big).$$

The main result in this section, Theorem 4.3, determines the "equivariance" properties of the map

$$L_k : \mathbb{C}\times\mathbb{C}\times\mathbb{H} \longrightarrow A_n, \qquad (z,\eta\,|\,\tau) \mapsto L_k(z,\eta\,|\,\tau) \tag{4.5}$$

with respect to the actions of $\mathrm{SL}(2,\mathbb{Z})$ on its domain and co-domain. To state it we introduce the matrix

$$D \;:=\; \begin{pmatrix} -k & 0 \\ 0 & 1 \end{pmatrix} \;\in\; \mathrm{GL}(2,\mathbb{R}) \tag{4.6}$$

and, for $M \in \mathrm{SL}(2,\mathbb{Z})$, define

$$M' \;:=\; D^{-1}M^{-t}D.$$

Note that $D^{-t} = D^{-1}$ and $M'' = (M')' = M$.

THEOREM 4.3. *Fix the data* (n,k,η,τ) *and* $M \in \mathrm{SL}(2,\mathbb{Z})$. *There is a nowhere-vanishing holomorphic function* $f(z)$ *such that*

$$L_k\big(M \triangleright (z,\eta\,|\,\tau)\big) \;=\; f(z)\,M' \triangleright L_k(z,\eta\,|\,\tau), \tag{4.7}$$

where $M'\triangleright$ *denotes the action of* $M' \in \mathrm{SL}(2,\mathbb{Z})$ *on* A_n.

[6] We learned this from Richey and Tracy's paper [**22**].

PROOF. Note that $(-k'u,v)=(u,v)D^{-1}$ for all $(u,v)\in\mathbb{Z}_n^2$.

Since $M''=M$, (4.7) is equivalent to the statement that there is a nowhere-vanishing holomorphic function $f(z)$ such that

$$L_k\big(M'\triangleright(z,\eta\,|\,\tau)\big)=f(z)\,M\triangleright L_k(z,\eta\,|\,\tau).$$

That is what we will prove.

By the definition of $L_k(z,\eta\,|\,\tau)$ in (4.4),

$$\begin{aligned}M\triangleright L_k(z,\eta\,|\,\tau) &= \sum_{u,v} w_{(u,v)}(-nz,\eta\,|\,\tau)\,J_{(u,v)D^{-1}M^t}\otimes J^{-1}_{(u,v)D^{-1}M^t}\\ &= \sum_{u,v} w_{(u,v)M^{-t}D}(-nz,\eta\,|\,\tau)\,J_{u,v}\otimes J^{-1}_{u,v}\\ &= \sum_{u,v} w_{(u,v)D(D^{-1}M^{-t}D)}(-nz,\eta\,|\,\tau)\,J_{u,v}\otimes J^{-1}_{u,v}.\end{aligned}$$

(The first equality in the previous computation uses the fact that $L_k(z,\eta\,|\,\tau)$ is, by definition, an element of A_n, not an element of $\mathbb{C}\widetilde{H}_n\otimes_{\mathbb{C}}\mathbb{C}\widetilde{H}_n$.) By Theorem 2.4, there is a nowhere-vanishing holomorphic function $g(z)$ such that this equals

$$g(z)\sum_{u,v} w_{(u,v)D}\left(D^{-1}M^{-t}D\triangleright(-nz,\eta\,|\,\tau)\right)J_{u,v}\otimes J^{-1}_{u,v},$$

which is, as we see after re-indexing the summation over (u,v), the same thing as

$$g(z)\sum_{u,v} w_{(u,v)}\left(M'\triangleright(-nz,\eta\,|\,\tau)\right)J_{-k'u,v}\otimes J^{-1}_{-k'u,v}\;=\;g(z)\,L\left(M'\triangleright(z,\eta\,|\,\tau)\right).$$

This finishes the proof. □

If $M=\left(\begin{smallmatrix}a&b\\c&d\end{smallmatrix}\right)\in\mathrm{SL}(2,\mathbb{Z})$, then the left-hand side of (4.7) is

$$L_k\left(\frac{z}{c\tau+d},\,\frac{\eta}{c\tau+d}\;\middle|\;\frac{a\tau+b}{c\tau+d}\right).$$

4.3. R-matrix and elliptic-algebra equivariance, and proof of the main theorem. We now translate (4.7), which is a statement about how $L_k(z,\eta\,|\,\tau)$ transforms under the action of $\mathrm{SL}(2,\mathbb{Z})$, into a statement about how $R_{n,k}(z,\eta\,|\,\tau)$ transforms under the action of $\mathrm{SL}(2,\mathbb{Z})$. To do that we make use of the set map $\psi:\mathrm{SL}(2,\mathbb{Z})\to\mathrm{GL}(V)$ in Corollary 3.6.

THEOREM 4.4. *Fix the data (n,k,η,τ). Let $M\in\mathrm{SL}(2,\mathbb{Z})$ and let $\psi(M')$ be the automorphism of V identified in Corollary 3.6. There is a nowhere-vanishing holomorphic function $f(z)$ such that*

$$R_{n,k}\big(M\triangleright(z,\eta\,|\,\tau)\big)\;=\;f(z)\,\psi(M')^{\otimes 2}\cdot R_{n,k}(z,\eta\,|\,\tau)\cdot\big(\psi(M')^{\otimes 2}\big)^{-1}.$$

PROOF. Let $\rho:\mathbb{C}\widetilde{H}_n\to\mathrm{End}_{\mathbb{C}}(V)$ denote the representation in Proposition 3.2. Since there is a nowhere-vanishing holomorphic function $f(z)$ such that

$$R_{n,k}(z,\eta\,|\,\tau)\;=\;f(z)\,PT_k(z,\eta\,|\,\tau)\;=\;f(z)\,P(\rho\otimes\rho)(L_k(z,\eta\,|\,\tau)),$$

it suffices to show that

$$(\rho\otimes\rho)\big(L_k(M\triangleright(z,\eta\,|\,\tau))\big)\;=\;g(z)\,\psi(M')^{\otimes 2}\cdot(\rho\otimes\rho)\big(L_k(z,\eta\,|\,\tau)\big)\cdot\big(\psi(M')^{\otimes 2}\big)^{-1}$$

for some nowhere-vanishing holomorphic function $g(z)$. However, it follows from Theorem 4.3 that there is a nowhere-vanishing holomorphic function $g(z)$ such that

$$(\rho\otimes\rho)\big(L_k(M\triangleright(z,\eta\,|\,\tau))\big) \;=\; g(z)\,(\rho\otimes\rho)\big(M'\triangleright L_k(z,\eta\,|\,\tau)\big),$$

which is equal to

$$g(z)\,\psi(M')^{\otimes 2}\cdot(\rho\otimes\rho)\big(L_k(z,\eta\,|\,\tau)\big)\cdot\big(\psi(M')^{\otimes 2}\big)^{-1}$$

because, by (3.6), $\rho(M'\triangleright x)=\psi(M')\rho(x)\psi(M')^{-1}$ for all $x\in\widetilde{H}_n$, and hence for all $x\in\mathbb{C}\widetilde{H}_n$. The proof is complete. □

THEOREM 4.5 (Theorem 1.1). *Let $M\in\mathrm{SL}(2,\mathbb{Z})$. The linear isomorphism $\psi(M'):V\to V$ identified in Corollary 3.6 extends to a graded $\mathbb{C}$-algebra isomorphism*

$$\psi(M'):Q_{n,k}(\eta\,|\,\tau)\xrightarrow{\;\sim\;}Q_{n,k}\big(M\triangleright(\eta\,|\,\tau)\big). \tag{4.8}$$

In particular, if $M=\left(\begin{smallmatrix}a&b\\c&d\end{smallmatrix}\right)\in\mathrm{SL}(2,\mathbb{Z})$, then

$$Q_{n,k}\left(\frac{\eta}{c\tau+d}\,\middle|\,\frac{a\tau+b}{c\tau+d}\right)\;\cong\;Q_{n,k}(\eta\,|\,\tau). \tag{4.9}$$

PROOF. By definition, $Q_{n,k}\big(M\triangleright(\eta\,|\,\tau)\big)=TV/I$ where I is the ideal generated by the image of

$$R_{n,k}(M\triangleright(\eta,\eta\,|\,\tau))\;:=\;\lim_{z\to\eta}R_{n,k}(M\triangleright(z,\eta\,|\,\tau))$$

By Theorem 4.4, this is the same as the image of $\psi(M')^{\otimes 2}\cdot R_{n,k}(\eta,\eta\,|\,\tau)$. The conclusion follows. □

4.3.1. *$\psi(M')$ is $\widetilde{H}_n$-equivariant if and only if M is the identity.* If $\psi:R\to R'$ is an isomorphism of rings there is an induced isomorphism of automorphism groups $\widehat{\psi}:\mathrm{Aut}(R)\to\mathrm{Aut}(R')$, $\alpha\mapsto\psi\alpha\psi^{-1}$. Since $\widetilde{H}_n$ acts as automorphisms of $Q_{n,k}(\eta\,|\,\tau)$ and $Q_{n,k}(\eta'\,|\,\tau')$ it is natural to ask whether $\widehat{\psi(M')}$ sends the copy of $\widetilde{H}_n$ in $\mathrm{Aut}(Q_{n,k}(\eta\,|\,\tau))$ to the copy of $\widetilde{H}_n$ in $\mathrm{Aut}(Q_{n,k}(\eta'\,|\,\tau'))$. It does: if $x\in\widetilde{H}_n$, then $\psi(M')\rho(x)\psi(M')^{-1}=\rho(M'\triangleright x)$; i.e., $\widehat{\psi(M')}$ restricts to an isomorphism between the two copies of $\widetilde{H}_n$ and that isomorphism is $x\mapsto M'\triangleright x$. In particular, if M is not the identity, then $\psi(M')$ is not $\widetilde{H}_n$-equivariant.

4.3.2. The next result involves an equality, not just an isomorphism.

COROLLARY 4.6. *If $m\in\mathbb{Z}$, then $Q_{n,k}\big(\frac{\eta}{mn\tau+1}\,\big|\,\frac{\tau}{mn\tau+1}\big)\;=\;Q_{n,k}(\eta\,|\,\tau)$.*

PROOF. As we remarked in subsections 3.3.2 and 3.4.2, the automorphisms Ψ_M of $\widetilde{H}_n$ and $\psi(M)$ of V defined for $M\in\mathrm{SL}(2,\mathbb{Z})$ only depend on the image of M in $\mathrm{SL}(2,\mathbb{Z}_n)$; in particular, if the image of M' in $\mathrm{SL}(2,\mathbb{Z}_n)$ is the identity, then we can take $\psi(M')=\mathrm{id}_V$. If $M=\left(\begin{smallmatrix}1&0\\mn&1\end{smallmatrix}\right)$, then the image of M' in $\mathrm{SL}(2,\mathbb{Z}_n)$ is the identity so we can take $\psi(M')=\mathrm{id}_V$, whence Theorem 4.5 gives the result. □

4.4. The proof of Theorem 1.2. Theorem 1.2 is a consequence of the following lemma.

LEMMA 4.7. *Let Λ_1 and Λ_2 be lattices in $\mathbb{C}$ and $f:\mathbb{C}/\Lambda_1\to\mathbb{C}/\Lambda_2$ a morphism of algebraic groups.*

(1) *There is $u\in\mathbb{C}^\times$ such that $f(z+\Lambda_1)\;=\;uz+\Lambda_2$ for all $z\in\mathbb{C}$.*

(2) *If f is an isomorphism of algebraic groups, then the u in* (1) *is unique.*

(3) *Assume $\Lambda_1 = \Lambda_{\tau_1}$ and $\Lambda_2 = \Lambda_{\tau_2}$ for some $\tau_1, \tau_2 \in \mathbb{H}$. If f is an isomorphism of algebraic groups, then there is a unique*

$$\begin{pmatrix} a & b \\ c & d \end{pmatrix} \in \mathrm{SL}(2,\mathbb{Z})$$

such that $f(z+\Lambda_1) = \frac{z}{c\tau_1+d} + \Lambda_2$ for all $z \in \mathbb{C}$ and $\tau_2 = \frac{a\tau_1+b}{c\tau_1+d}$.

PROOF. (1) This is [**12**, Lem. 1.9].

(2) If $u_1 z + \Lambda_2 = u_2 z + \Lambda_2$ for all $z \in \mathbb{C}$, then $(u_1 - u_2)z \in \Lambda_2$ for all $z \in \mathbb{C}$, which implies that $u_1 - u_2 = 0$. Hence the u in (1) is unique.

(3) (Existence.) Since $f(z+\Lambda_1) = uz + \Lambda_2$ for all $z \in \mathbb{C}$, $u\Lambda_1 = \Lambda_2$. Hence $\tau_2 = u(a\tau_1 + b)$ and $1 = u(c\tau_1 + d)$ for some $a,b,c,d \in \mathbb{Z}$. Since $u\Lambda_1 = \Lambda_2$, $\{a\tau_1 + b, c\tau_1 + d\}$ is a $\mathbb{Z}$-basis for Λ_{τ_1}. Hence $\left(\begin{smallmatrix} a & b \\ c & d \end{smallmatrix}\right) \in \mathrm{GL}(2,\mathbb{Z})$. Since $\tau_2 = \frac{a\tau_1+b}{c\tau_1+d}$, the imaginary part of τ_2 is

$$\Im(\tau_2) \;=\; \det\begin{pmatrix} a & b \\ c & d \end{pmatrix} \frac{\Im(\tau_1)}{|c\tau_1+d|^2}.$$

Since the imaginary parts of τ_1 and τ_2 are positive, $\det\left(\begin{smallmatrix} a & b \\ c & d \end{smallmatrix}\right) > 0$. Hence $\left(\begin{smallmatrix} a & b \\ c & d \end{smallmatrix}\right)$ is in $\mathrm{SL}(2,\mathbb{Z})$.

(Uniqueness.) Since $\tau_1 \notin \mathbb{Q}$, the integers c and d in the equality $1 = u(c\tau_1 + d)$ are unique. Likewise, since $\tau_1 \notin \mathbb{Q}$, the integers a and b in the equality $a\tau_1 + b = u^{-1}\tau_2$ are unique. $\square$

COROLLARY 4.8 (Theorem 1.2). *If $f : E_\tau \to E_{\tau'}$ is an isomorphism of algebraic groups such that $f(\eta + \Lambda_\tau) = \eta' + \Lambda_{\tau'}$, then*

$$Q_{n,k}(\eta \,|\, \tau) \;\cong\; Q_{n,k}(\eta' \,|\, \tau').$$

PROOF. By Lemma 4.7(3), there is an $M = \left(\begin{smallmatrix} a & b \\ c & d \end{smallmatrix}\right) \in \mathrm{SL}(2,\mathbb{Z})$ such that

$$f(z+\Lambda_\tau) \;=\; \frac{z}{c\tau+d} + \Lambda_{\tau'}$$

for all $z \in \mathbb{C}$ and $\tau' = \frac{a\tau+b}{c\tau+d}$. In particular, $\eta' + \Lambda_{\tau'} = \frac{\eta}{c\tau+d} + \Lambda_{\tau'}$. By Theorem 4.5,

$$Q_{n,k}(\eta \,|\, \tau) \;\cong\; Q_{n,k}\big(\tfrac{\eta}{c\tau+d} \,\big|\, \tfrac{a\tau+b}{c\tau+d}\big) \;=\; Q_{n,k}(\eta' \,|\, \tau')$$

where the last equality holds since $\frac{\eta}{c\tau+d}$ and η' have the same image in $E_{\tau'}$. $\square$

4.5. Modularity. The action of the modular group $\mathrm{SL}(2,\mathbb{Z})$ on $\mathbb{C} \times \mathbb{H}$ is the starting point for the theory of modular forms and modular functions. For example, (see, e.g., [**8**, Defn. 1.1.1] and [**24**, §I.3]) a meromorphic function $f : \mathbb{H} \to \mathbb{C}$ is called a weakly modular function of weight k if

$$f\left(\frac{a\tau+b}{c\tau+d}\right) \;=\; (c\tau+d)^k f(\tau) \tag{4.10}$$

for all $\tau \in \mathbb{H}$ and all

$$\begin{pmatrix} a & b \\ c & d \end{pmatrix} \in \mathrm{SL}(2,\mathbb{Z}).$$

We will now restate (4.10) in a way that resembles the isomorphism in (4.9).

First, we associate to a function $f : \mathbb{H} \to \mathbb{C}$ its (backward) graph

$$\Gamma_f \;:=\; \{(f(\tau) \,|\, \tau) \mid \tau \in \mathbb{H}\} \subseteq \mathbb{C} \times \mathbb{H}.$$

Subsets of a given set can be viewed as $\{0,1\}$-valued functions on that set. Here such a procedure produces the function

$$F:\mathbb{C}\times\mathbb{H}\to\{0,1\},\qquad F(z\,|\,\tau)\;:=\;\begin{cases}1 & \text{if } f(\tau)=z,\\ 0 & \text{if } f(\tau)\neq z.\end{cases}$$

With this notation, f satisfies (4.10) if and only if

$$F\biggl((c\tau+d)^k z\;\bigg|\;\frac{a\tau+b}{c\tau+d}\biggr)\;=\;F(z\,|\,\tau).$$

In particular, $f(\tau)$ is a weakly modular function of weight -1 if and only if the $\{0,1\}$-valued function $F(z\,|\,\tau)$ satisfies the identity

$$F\biggl(\frac{z}{c\tau+d}\;\bigg|\;\frac{a\tau+b}{c\tau+d}\biggr)\;=\;F(z\,|\,\tau),$$

which resembles the isomorphism in (4.9).

If we express this identity by saying that F is a "weakly modular function of weight -1 taking values in $\{0,1\}$", and extend this terminology in the obvious way, then Theorem 4.5 says that $Q_{n,k}$ is a "weakly modular function of weight -1 taking values in the category of graded $\mathbb{C}$-algebras".

References

[1] Arnaud Beauville, *Theta functions, old and new*, Open problems and surveys of contemporary mathematics, Surv. Mod. Math., vol. 6, Int. Press, Somerville, MA, 2013, pp. 99–132. MR3204388

[2] Christina Birkenhake and Herbert Lange, *Complex abelian varieties*, 2nd ed., Grundlehren der mathematischen Wissenschaften [Fundamental Principles of Mathematical Sciences], vol. 302, Springer-Verlag, Berlin, 2004, DOI 10.1007/978-3-662-06307-1. MR2062673

[3] Filippo Callegaro, Frederick R. Cohen, and Mario Salvetti, *The cohomology of the braid group B_3 and of $SL_2(\mathbb{Z})$ with coefficients in a geometric representation*, Configuration spaces, CRM Series, vol. 14, Ed. Norm., Pisa, 2012, pp. 193–211, DOI 10.1007/978-88-7642-431-1_8. MR3203639

[4] Alex Chirvasitu, Ryo Kanda, and S. Paul Smith, *The characteristic variety for Feigin and Odesskii's elliptic algebras*, arXiv e-prints (2019), `arXiv:1903.11798`.

[5] Alex Chirvasitu, Ryo Kanda, and S. Paul Smith, *Elliptic R-matrices and Feigin and Odesskii's elliptic algebras*, Selecta Math. (N.S.) **29** (2023), no. 2, Paper No. 31, 81, DOI 10.1007/s00029-023-00827-0. MR4563694

[6] Alex Chirvasitu, Ryo Kanda, and S. Paul Smith, *Feigin and Odesskii's elliptic algebras*, J. Algebra **581** (2021), 173–225, DOI 10.1016/j.jalgebra.2021.04.009. MR4249759

[7] Alex Chirvasitu, Ryo Kanda, and S. Paul Smith, *Maps from Feigin and Odesskii's elliptic algebras to twisted homogeneous coordinate rings*, Forum Math. Sigma **9** (2021), Paper No. e4, 50, DOI 10.1017/fms.2020.60. MR4202489

[8] Fred Diamond and Jerry Shurman, *A first course in modular forms*, Graduate Texts in Mathematics, vol. 228, Springer-Verlag, New York, 2005. MR2112196

[9] B. L. Feigin and A. V. Odesskii, *Sklyanin algebras associated with an elliptic curve*, Preprint deposited with Institute of Theoretical Physics of the Academy of Sciences of the Ukrainian SSR (1989), 33 pages.

[10] William Fulton and Joe Harris, *Representation theory*, Graduate Texts in Mathematics, vol. 129, Springer-Verlag, New York, 1991. A first course; Readings in Mathematics, DOI 10.1007/978-1-4612-0979-9. MR1153249

[11] Johannes Grassberger and Günther Hörmann, *A note on representations of the finite Heisenberg group and sums of greatest common divisors*, Discrete Math. Theor. Comput. Sci. **4** (2001), no. 2, 91–100. MR1835586

[12] Richard Hain, *Lectures on moduli spaces of elliptic curves*, Transformation groups and moduli spaces of curves, Adv. Lect. Math. (ALM), vol. 16, Int. Press, Somerville, MA, 2011, pp. 95–166. MR2883686

[13] C. G. J. Jacobi, *Suite des notices sur les fonctions elliptiques* (French), J. Reine Angew. Math. **3** (1828), 403–404, DOI 10.1515/crll.1828.3.403. MR1577711

[14] George R. Kempf, *Complex abelian varieties and theta functions*, Universitext, Springer-Verlag, Berlin, 1991, DOI 10.1007/978-3-642-76079-2. MR1109495

[15] David Mumford, *Tata lectures on theta. I*, Modern Birkhäuser Classics, Birkhäuser Boston, Inc., Boston, MA, 2007. With the collaboration of C. Musili, M. Nori, E. Previato and M. Stillman; Reprint of the 1983 edition, DOI 10.1007/978-0-8176-4578-6. MR2352717

[16] David Mumford, *Abelian varieties*, Tata Institute of Fundamental Research Studies in Mathematics, vol. 5, Published for the Tata Institute of Fundamental Research, Bombay; by Hindustan Book Agency, New Delhi, 2008. With appendices by C. P. Ramanujam and Yuri Manin; Corrected reprint of the second (1974) edition. MR2514037

[17] David Mumford, *Tata lectures on theta. III*, Progress in Mathematics, vol. 97, Birkhäuser Boston, Inc., Boston, MA, 1991. With the collaboration of Madhav Nori and Peter Norman, DOI 10.1007/978-0-8176-4579-3. MR1116553

[18] A. V. Odesskiĭ, *Elliptic algebras* (Russian, with Russian summary), Uspekhi Mat. Nauk **57** (2002), no. 6(348), 87–122, DOI 10.1070/RM2002v057n06ABEH000573; English transl., Russian Math. Surveys **57** (2002), no. 6, 1127–1162. MR1991863

[19] A. V. Odesskiĭ and B. L. Feĭgin, *Sklyanin's elliptic algebras* (Russian), Funktsional. Anal. i Prilozhen. **23** (1989), no. 3, 45–54, 96, DOI 10.1007/BF01079526; English transl., Funct. Anal. Appl. **23** (1989), no. 3, 207–214 (1990). MR1026987

[20] A. V. Odesskiĭ and B. L. Feĭgin, *Sklyanin's elliptic algebras. The case of a point of finite order* (Russian, with Russian summary), Funktsional. Anal. i Prilozhen. **29** (1995), no. 2, 9–21, 95, DOI 10.1007/BF01080006; English transl., Funct. Anal. Appl. **29** (1995), no. 2, 81–90. MR1340300

[21] Alexander Polishchuk, *Abelian varieties, theta functions and the Fourier transform*, Cambridge Tracts in Mathematics, vol. 153, Cambridge University Press, Cambridge, 2003, DOI 10.1017/CBO9780511546532. MR1987784

[22] Matthew P. Richey and Craig A. Tracy, *Z_n Baxter model: symmetries and the Belavin parametrization*, J. Statist. Phys. **42** (1986), no. 3-4, 311–348, DOI 10.1007/BF01127715. MR833020

[23] Jean-Pierre Serre, *Trees*, Springer-Verlag, Berlin-New York, 1980. Translated from the French by John Stillwell. MR607504

[24] Joseph H. Silverman, *Advanced topics in the arithmetic of elliptic curves*, Graduate Texts in Mathematics, vol. 151, Springer-Verlag, New York, 1994, DOI 10.1007/978-1-4612-0851-8. MR1312368

[25] E. K. Sklyanin, *Some algebraic structures connected with the Yang-Baxter equation* (Russian), Funktsional. Anal. i Prilozhen. **16** (1982), no. 4, 27–34, 96. MR684124

[26] E. K. Sklyanin, *Some algebraic structures connected with the Yang-Baxter equation. Representations of a quantum algebra* (Russian), Funktsional. Anal. i Prilozhen. **17** (1983), no. 4, 34–48. MR725414

[27] John Tate and Michel van den Bergh, *Homological properties of Sklyanin algebras*, Invent. Math. **124** (1996), no. 1-3, 619–647, DOI 10.1007/s002220050065. MR1369430

[28] E. T. Whittaker and G. N. Watson, *A course of modern analysis. An introduction to the general theory of infinite processes and of analytic functions: with an account of the principal transcendental functions*, Cambridge University Press, New York, 1962. Fourth edition. Reprinted. MR178117

DEPARTMENT OF MATHEMATICS, UNIVERSITY AT BUFFALO, BUFFALO, NEW YORK 14260-2900.
Email address: achirvas@buffalo.edu

DEPARTMENT OF MATHEMATICS, GRADUATE SCHOOL OF SCIENCE, OSAKA METROPOLITAN UNIVERSITY, 3-3-138, SUGIMOTO, SUMIYOSHI, OSAKA, 558-8585, JAPAN.
Email address: ryo.kanda.math@gmail.com

DEPARTMENT OF MATHEMATICS, BOX 354350, UNIVERSITY OF WASHINGTON, SEATTLE, WASHINGTON 98195.
Email address: smith@math.washington.edu

Contemporary Mathematics
Volume **801**, 2024
https://doi.org/10.1090/conm/801/16083

Spectra of quantum algebras

K. R. Goodearl

ABSTRACT. This is a survey of what is known and/or conjectured about the prime and primitive spectra of quantum algebras, of quantized coordinate rings in particular. The topological structure of these spectra, their relations to classical affine algebraic varieties, and their relations to each other are discussed.

1. Introduction

Suppose that an algebra A is a quantized coordinate ring of an affine algebraic variety V over an algebraically closed field. The prime and primitive spectra of A are equipped with natural Zariski topologies, and the question arises of how these spaces are related to the corresponding classical spaces, namely the prime spectrum of the classical coordinate ring $\mathcal{O}(V)$ (i.e., the space underlying the scheme structure of V) and the maximal ideal spectrum of $\mathcal{O}(V)$ (which is homeomorphic to V itself). We survey results and conjectures in several directions:

(1) Are the spectra of A topological quotients of the spectra of $\mathcal{O}(V)$?
(2) Are the spectra of A homeomorphic to Poisson spectra of $\mathcal{O}(V)$, where $\mathcal{O}(V)$ is equipped with a Poisson structure from a semiclassical limit process?
(3) Can the topologies on the spectra of A be described via piecewise classical data, from finite collections of classical spectra and connections among them?

Positive answers to (1) are conjectured in general, while for (2), positive answers are only conjectured for generic cases in characteristic zero. Under an additional hypothesis that holds for many quantized coordinate rings, any positive answer to (2) implies a positive answer to (1). Positive answers to (3), finally, are conjectured for generic cases in arbitrary characteristic. In characteristic zero, a positive answer to (3), coupled with a matching positive answer to the corresponding question about the Poisson spectra of $\mathcal{O}(V)$, implies a positive answer to (2).

1.1. Thoughout the paper, we work over an algebraically closed base field k.

1.2. If R is the coordinate ring of a classical affine variety V over k, then the space of maximal ideals of R is homeomorphic to V, and the space of prime ideals of R is homeomorphic to the scheme of irreducible subvarieties of V. In the

2020 *Mathematics Subject Classification.* Primary 16T20; Secondary 20G42.
This research was supported by US National Science Foundation grant DMS-1601184.

noncommutative world, primitive ideals, rather than just maximal ideals, are the appropriate ideals to relate to points, so it is the primitive spectrum rather than the maximal ideal spectrum which takes over the role of the variety.

For a noncommutative algebra A, set

$$\operatorname{Prim} A := \{ \text{ primitive ideals of } A \}, \qquad \operatorname{Spec} A := \{ \text{ prime ideals of } A \},$$

both equipped with the natural Zariski topologies. (In case there is asymmetry of primitivity, let us interpret "primitive" as "left primitive".)

2. Three examples

To illustrate prototypical patterns, we exhibit the spectra of three basic examples – the quantum plane, quantum SL_2, and quantum GL_2, meaning quantized coordinate rings of the plane k^2 and the algebraic groups $SL_2(k)$ and $GL_2(k)$. We just give the standard (single parameter) versions, with parameter $q \in k^\times$.

EXAMPLE 2.1. The quantum plane is the algebra

$$A = \mathcal{O}_q(k^2) := k\langle x, y \mid xy = qyx \rangle.$$

In the *generic* case, i.e., if q is not a root of unity, $\operatorname{Spec} A$ may be drawn as in Figure 1, where dashed lines indicate inclusions (e.g., [**3**, Example II.1.2]).

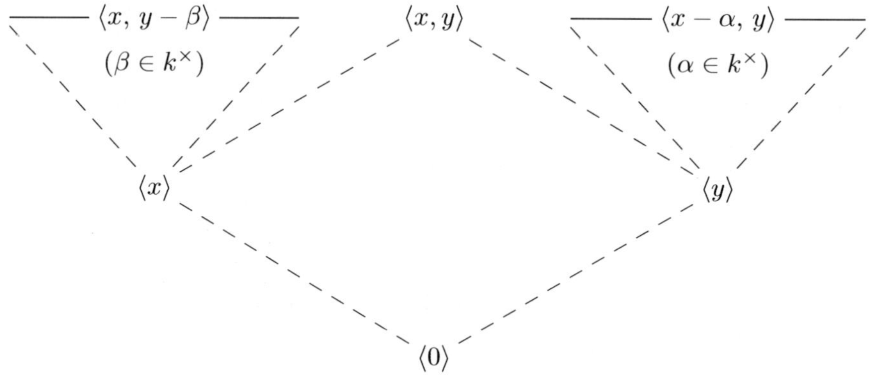

FIGURE 1. $\operatorname{Spec} \mathcal{O}_q(k^2)$

The primitive spectrum of A consists of the maximal ideals together with the zero ideal (e.g., [**3**, Example II.7.2]). We picture this spectrum as in Figure 2 – the x- and y-axes of the classical plane, together with a fat point, which is dense. There is an obvious surjection $\pi : k^2 \to \operatorname{Prim} A$, where $\pi(\alpha, 0) = \langle x - \alpha,\, y \rangle$ and $\pi(0, \beta) = \langle x,\, y - \beta \rangle$ for $\alpha, \beta \in k$ while $\pi(\alpha, \beta) = \langle 0 \rangle$ for $(\alpha, \beta) \in (k^\times)^2$. It is easily checked that π is a topological quotient map. In other words, the topology on $\operatorname{Prim} \mathcal{O}_q(k^2)$ is the quotient topology that results from π.

Quantum SL_2 and quantum GL_2 are obtained from the quantum 2×2 matrix algebra $\mathcal{O}_q(M_2(k))$, which we present via generators a, b, c, d and the following relations:

$$ab = qba \qquad ac = qca \qquad bc = cb$$
$$bd = qdb \qquad cd = qdc \qquad ad - da = (q - q^{-1})bc.$$

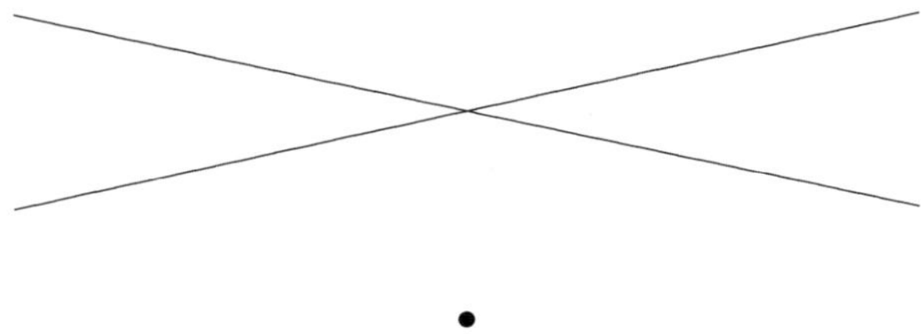

FIGURE 2. Prim $\mathcal{O}_q(k^2)$

The element $D_q := ad - qbc$ in this algebra is a central element known as the (2×2) *quantum determinant.* In parallel with the classical commutative case, quantized coordinate rings of $SL_2(k)$ and $GL_2(k)$ are defined as follows:

$$\mathcal{O}_q(SL_2(k)) := \mathcal{O}_q(M_2(k))/\langle D_q - 1\rangle \qquad \mathcal{O}_q(GL_2(k)) := \mathcal{O}_q(M_2(k))[D_q^{-1}].$$

It is convenient to use the same symbols a, b, c, d for the cosets $a + \langle D_q - 1\rangle, \dots$ in $\mathcal{O}_q(SL_2(k))$.

EXAMPLE 2.2. Consider $A = \mathcal{O}_q(SL_2(k))$ in the generic case, $q \neq \sqrt[\bullet]{1}$. We may picture Spec A as in Figure 3, where the dashed lines again indicate inclusions (e.g., [**3**, Example II.1.3]). Here the primitive ideals consist of the maximal ideals

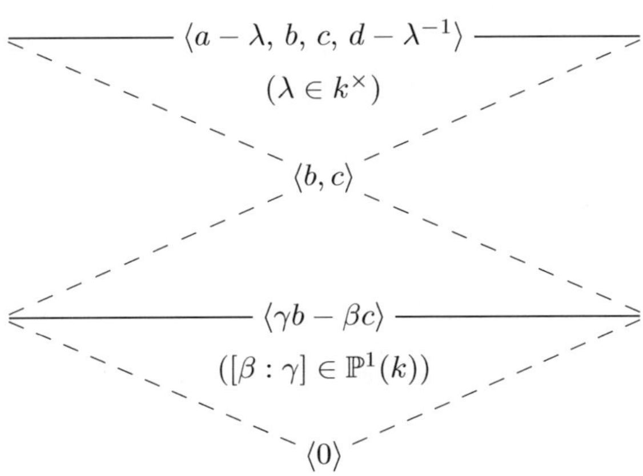

FIGURE 3. Spec $\mathcal{O}_q(SL_2(k))$

together with the height 1 prime ideals $\langle \gamma b - \beta c\rangle$ (e.g., [**3**, Example II.8.6]). Prim A is pictured in Figure 4. The punctured line at the top of the figure represents $k \setminus \{0\}$, while the bottom line stands for $\mathbb{P}^1(k)$. The bullets are sample points, and the inclusions represented by the dashed lines indicate that the closure of each point on the projective line consists of that point together with the full punctured affine line.

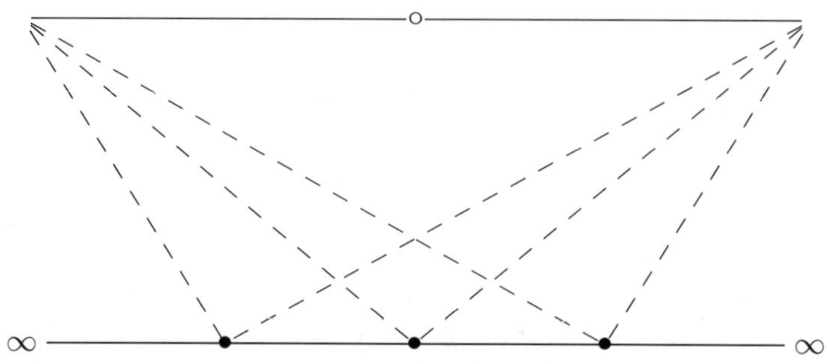

FIGURE 4. Prim $\mathcal{O}_q(SL_2(k))$

As in Example 2.1, there is a topological quotient map $\pi : SL_2(k) \to \operatorname{Prim} A$, this time given by

$$\pi\left(\begin{bmatrix} \alpha & \beta \\ \gamma & \delta \end{bmatrix}\right) = \begin{cases} \langle a-\alpha,\ b,\ c,\ d-\delta\rangle & (\beta=\gamma=0) \\ \langle \gamma b - \beta c\rangle & (\beta,\gamma \text{ not both } = 0). \end{cases} \tag{2.1}$$

EXAMPLE 2.3. Finally, consider $A = \mathcal{O}_q(GL_2(k))$, $q \neq \sqrt[\bullet]{1}$. We display Spec A without indicating all inclusions. As a set, it is the disjoint union of 4 subsets as drawn in Figure 5 (e.g., [**3**, Example II.8.7]). In the upper left subdiagram,

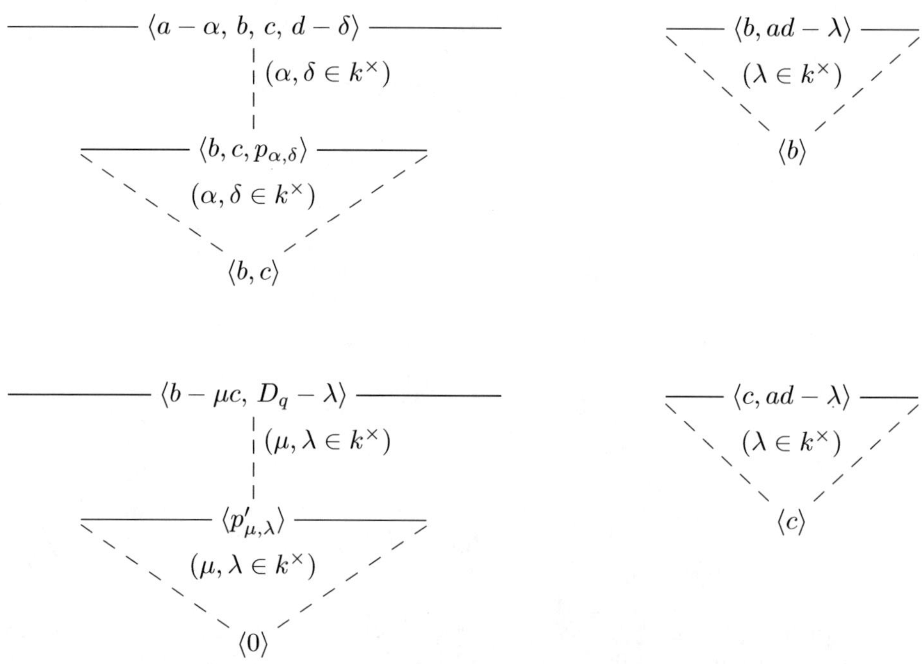

FIGURE 5. Spec $\mathcal{O}_q(GL_2(k))$

$p_{\alpha,\delta}$ denotes an irreducible polynomial in $k[a,d]$ such that $p_{\alpha,\delta}(\alpha,\delta) = 0$, while in the lower left subdiagram, $p'_{\mu,\lambda}$ denotes a polynomial in $k[b, c, D_q]$ of the form $f(bc^{-1}, D_q)c^m$ where f is an irreducible polynomial in $k[s,t]$ of s-degree m with $f(\mu, \lambda) = 0$.

In this algebra, the primitive ideals are the prime ideals appearing at the top levels of the 4 diagrams in Figure 5 (e.g., [**3**, Example II.8.7]; cf. Theorem 5.1). The primes $\langle a - \alpha,\ b,\ c,\ d - \delta \rangle$ are maximal ideals, while the remaining primitive ideals are primes of height 2. The latter may be grouped together as

$$\{\langle \gamma b - \beta c,\ D_q - \lambda \rangle \mid [\beta : \gamma] \in \mathbb{P}^1(k),\ \lambda \in k^\times\}.$$

The only inclusions between primitive ideals in different subsets of $\operatorname{Prim} A$ are

$$\langle \gamma b - \beta c,\ D_q - \lambda \rangle \subseteq \langle a - \alpha,\ b,\ c,\ d - \alpha^{-1}\lambda \rangle \qquad \forall\ [\beta : \gamma] \in \mathbb{P}^1(k),\ \alpha, \lambda \in k^\times.$$

This leads to a rough picture of $\operatorname{Prim} A$ as in Figure 6, where inclusions between primes in the lower and upper layers are not indicated.

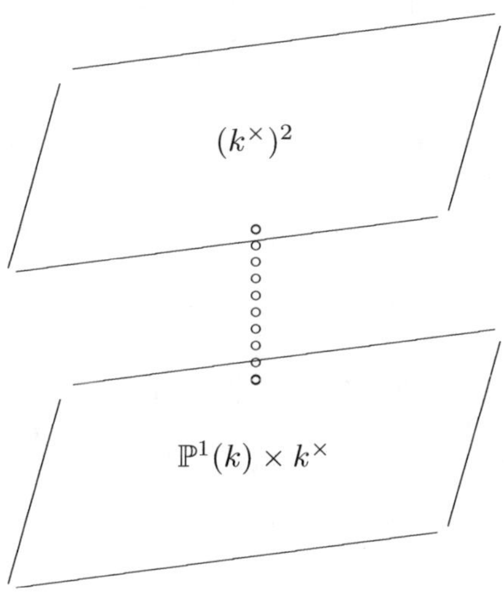

FIGURE 6. $\operatorname{Prim} \mathcal{O}_q(GL_2(k))$

The topology of the space in this picture may be described in terms of the Zariski topologies on $\mathbb{P}^1(k) \times k^\times$ and $(k^\times)^2$, together with the map

$$\phi : \{\text{closed subsets of } \mathbb{P}^1(k) \times k^\times\} \longrightarrow \{\text{closed subsets of } (k^\times)^2\}$$
$$X \longmapsto \bigcup_{([\nu:\mu],\lambda)\in X} \{(x,y) \in (k^\times)^2 \mid xy = \lambda\}.$$

(That the union in the display is closed in $(k^\times)^2$ follows from the fact that the projection $\mathbb{P}^1(k) \times k^\times \to k^\times$ is a closed map.) Namely, a set $X \sqcup Y$, where $X \subseteq \mathbb{P}^1(k) \times k^\times$ and $Y \subseteq (k^\times)^2$, is closed in $\operatorname{Prim} A$ if and only if

- X is closed in $\mathbb{P}^1(k) \times k^\times$;
- Y is closed in $(k^\times)^2$;
- $\phi(X) \subseteq Y$.

See Section 5 for a description of a general framework into which this example fits.

There is a topological quotient map $GL_2(k) \to \operatorname{Prim} A$, parallel to (2.1).

3. Topological quotient properties

As observed in Example 2.1, the primitive ideal space of the quantum plane, $\operatorname{Prim} \mathcal{O}_q(k^2)$, is a topological quotient of the classical plane, k^2, for $q \neq \sqrt[\bullet]{1}$. Modulo a minor technical condition, this quotient picture carries over to quantum affine spaces in general, including the multiparameter algebras

$$\mathcal{O}_{\mathbf{q}}(k^n) := k\langle x_1, \ldots, x_n \mid x_i x_j = q_{ij} x_j x_i \ \forall\, i,j \rangle, \tag{3.1}$$

where $\mathbf{q} = (q_{ij})$ is a multiplicatively skew-symmetric matrix in $M_n(k^\times)$.

THEOREM 3.1 ([**12**, Theorems 4.11, 6.3]). *Let $A = \mathcal{O}_{\mathbf{q}}(k^n)$, and assume either $-1 \notin \langle q_{ij} \rangle$ or* $\operatorname{char} k = 2$. *Then* $\operatorname{Prim} A$ *is a topological quotient of k^n and* $\operatorname{Spec} A$ *is a (compatible) topological quotient of* $\operatorname{Spec} \mathcal{O}(k^n)$. *Compatibility means that there is a topological quotient map $\pi : \operatorname{Spec} \mathcal{O}(k^n) \to \operatorname{Spec} A$ such that the restriction of π to $\max \mathcal{O}(k^n) \approx k^n$ is a topological quotient map onto* $\operatorname{Prim} A$.

These statements hold more generally for cocycle twists of commutative affine algebras graded by torsionfree abelian groups, and hence for quantum affine toric varieties.

The technical condition in Theorem 3.1 excludes elements of order 2 from the subgroup $\langle q_{ij} \rangle$ of $k^\times$. It is not known whether this condition is necessary.

Some other cases following this pattern are known – e.g., as noted in Example 2.2, the primitive spectrum of $\mathcal{O}_q(SL_2(k))$, $q \neq \sqrt[\bullet]{1}$, is a topological quotient of the variety $SL_2(k)$ [**8**, Example 1.1].

We conjecture that such topological quotient results in fact hold much more widely among quantized coordinate rings:

CONJECTURE 3.2 ([**12**, Introduction], [**8**, Conjecture 1.6]). *Let A be a quantized coordinate ring of an affine variety V. Then* $\operatorname{Spec} A$ *and* $\operatorname{Prim} A$ *are compatible topological quotients of* $\operatorname{Spec} \mathcal{O}(V)$ *and* $\max \mathcal{O}(V) \approx V$.

At this point, there is no accepted or even proposed definition of what it should mean for an algebra to be counted as a quantized coordinate ring. To clarify: the conjecture above is to apply at least to currently accepted families of quantized coordinate rings, such as quantum matrices, quantum semisimple groups and quantum Borel subgroups thereof, as well as quantum symplectic and euclidean spaces.

Beyond Theorem 3.1, Conjecture 3.2 has been confirmed for quantum symplectic and euclidean spaces in [**25**, Corollaries 5.9, 5.10, 5.11], [**27**, Corollary 10]. A related conjecture which we discuss below (Conjecture 4.7) leads to additional cases. More precisely, whenever Conjecture 4.7 and an additional hypothesis hold, Conjecture 3.2 follows (see Remark 4.8). Confirmed cases include $\mathcal{O}_q(SL_2(k))$, $\mathcal{O}_q(GL_2(k))$ and $\mathcal{O}_q(SL_3(k))$, as noted below.

4. Poisson relations

One approach to confirming Conjecture 3.2 is to identify a space $X_s(V)$ and a subspace $X_p(V)$ such that (1) $X_s(V)$ and $X_p(V)$ are compatible topological quotients of $\operatorname{Spec} \mathcal{O}(V)$ and $\max \mathcal{O}(V)$; and (2) there are compatible homeomorphisms $X_s(V) \approx \operatorname{Spec} A$ and $X_p(V) \approx \operatorname{Prim} A$. Knowledge of the structures of $X_s(V)$ and $X_p(V)$ then also provides knowledge about the structures of $\operatorname{Spec} A$ and $\operatorname{Prim} A$.

The main instance of this approach will be outlined in the present section. It involves (a) constructing a *semiclassical limit* of a family of quantum algebras;

(b) imposing a *Poisson structure* that arises from the construction process; (c) investigating *Poisson-prime* and *Poisson-primitive spectra* of such a semiclassical limit; (d) relating the spaces obtained in (c) to prime and primitive spectra of *generic* members of the original family. Although semiclassical limits and their Poisson structures can be obtained for general families, the relations implementing (d) do not work well in positive characteristic. Further, clarification of the attribute "generic" in (d) is required. We address these concepts and points separately.

4.1. Poisson algebras.

DEFINITION 4.1. A *Poisson algebra* is an algebra R equipped with a Lie algebra bracket $\{-,-\} : R \times R \to R$ such that

$$\{a, xy\} = \{a, x\}y + x\{a, y\} \quad \forall\, a, x, y \in R.$$

Due to the antisymmetry of the bracket, $\{xy, a\} = x\{y, a\} + \{x, a\}y$ for all $a, x, y \in R$ as well.

A *Poisson ideal* in R is any ideal I such that $\{R, I\} \subseteq I$.

A *Poisson-prime ideal* in R is any proper Poisson ideal P such that

$$IJ \subseteq P \implies I \subseteq P \text{ or } J \subseteq P, \quad \forall \text{ Poisson ideals } I, J \subseteq R.$$

In case R is noetherian and $\operatorname{char} k = 0$, the Poisson-prime ideals of R are precisely those Poisson ideals which are also prime ideals (e.g., [**9**, Lemma 1.1]; commutativity is not required).

A *Poisson-primitive ideal* of R is any ideal P which is the largest of the Poisson ideals contained in some maximal ideal of R. Such ideals are also Poisson-prime.

The *Poisson-prime* and *Poisson-primitive spectra* of R are

$$\operatorname{Pspec} R := \{ \text{ Poisson-prime ideals of } R \}$$
$$\operatorname{Pprim} R := \{ \text{ Poisson-primitive ideals of } R \} \subseteq \operatorname{Pspec} R.$$

Both of these sets are equipped with Zariski topologies.

These Poisson spectra form natural topological quotients of classical spectra in many cases:

THEOREM 4.2 ([**9**, Theorems 4.1, 1.5], [**10**, Lemma 9.3]). *Let R be a commutative noetherian Poisson k-algebra, with* $\operatorname{char} k = 0$. *Then* $\operatorname{Pspec} R$ *is a topological quotient of* $\operatorname{Spec} R$*, via the map*

$$\pi : P \longmapsto (\text{ largest Poisson ideal } \subseteq P\ ).$$

Now assume in addition that R is affine over k and satisfies the following portion of the Poisson Dixmier-Moeglin Equivalence: all $P \in$ $\operatorname{Pprim} R$ *are locally closed in* $\operatorname{Pspec} R$. *Then* $\operatorname{Pprim} R$ *is a topological quotient of* $\max R$*, via π.*

See, e.g., [**2, 5, 9, 11, 17, 18, 20, 23–25**] for some situations where the Poisson Dixmier-Moeglin Equivalence has been established.

4.2. Semiclassical limits. These "limits" arise from parametrized families of algebras. We review the process of forming semiclassical limits in the one-parameter case (a discussion of multiparameter cases is given in [**10**]).

DEFINITION 4.3. Let $\mathcal{A}$ be a torsionfree algebra over a Laurent polynomial ring $k[t^{\pm 1}]$ such that $\mathcal{A}/(t-1)\mathcal{A}$ is commutative. The family $\big(\mathcal{A}_q := \mathcal{A}/(t-q)\mathcal{A}\big)_{q\in k^\times}$ is the corresponding *flat family* of k-algebras, and the commutative algebra $\mathcal{A}_1$ is the algebra underlying the *semiclassical limit* of this family.

This semiclassical limit preserves a trace of the (possible) noncommutativity in the family $\big(\mathcal{A}_q\big)$, expressed via an induced Poisson bracket. Namely, since all additive commutators $[a,b]$ $(= ab - ba)$ in $\mathcal{A}$ are divisible by $t-1$, and uniquely so due to the torsionfreeness assumption on $\mathcal{A}$, there is a bilinear bracket $\frac{1}{t-1}[-,-]$ on $\mathcal{A}$. It is easily checked that $\frac{1}{t-1}[-,-]$ is a Poisson bracket on $\mathcal{A}$, and that it induces a Poisson bracket $\{-,-\}$ on $\mathcal{A}_1$.

More precisely, it is the Poisson algebra $\big(\mathcal{A}_1,\, \{-,-\}\big)$ which is the *semiclassical limit* of the family $\big(\mathcal{A}_q\big)_{q\in k^\times}$.

EXAMPLE 4.4. The Poisson structure on $\mathcal{O}(SL_2(k))$, which is the semiclassical limit of the family $\big(\mathcal{O}_q(SL_2(k))\big)_{q\in k^\times}$ corresponding to the algebra $\mathcal{A} = \mathcal{O}_t(SL_2(k[t^{\pm 1}]))$, is determined by the following data:

$$\{a,b\} = ab \qquad \{a,c\} = ac \qquad \{a,d\} = 2bc$$
$$\{b,c\} = 0 \qquad \{b,d\} = bd \qquad \{c,d\} = cd\,.$$

4.3. The qualifier "generic". *Question*: Which members of a family of quantum algebras should be labelled "generic"? To our knowledge, no rigorous answer to this question has been given. The underlying idea is to apply the label "generic" to those members which exhibit the "most general behavior" within the family – in particular, the "most general noncommutativity" (i.e., the "least commutativity"). The choices that respect this principle are relatively obvious in many cases, much less so in others.

The most common usage of the term "generic" occurs within the "standard" families of quantized coordinate rings, families indexed by a single parameter. These include the examples displayed in Section 2, which are members of the families $\big(\mathcal{O}_q(k^2)\big)_{q\in k^\times}$, $\big(\mathcal{O}_q(SL_2(k))\big)_{q\in k^\times}$ and $\big(\mathcal{O}_q(GL_2(k))\big)_{q\in k^\times}$ respectively. Among the "standard quantized coordinate rings" are the algebras named (for short) quantum affine spaces, quantum tori, quantum matrices, quantum semisimple groups, quantum general linear groups, quantum Borel subgroups of semisimple groups, quantum symplectic and euclidean spaces. (See, e.g., [**3, 7**] for surveys.) They are typically denoted $\mathcal{O}_q(V)$, $k_q[V]$ or $R_q[V]$, where V is an affine variety such as k^n, $(k^\times)^n$, $M_{m,n}(k)$, a semisimple algebraic group G or a Borel subgroup $B^\pm$ of G, a general linear group $GL_n(k)$, etc. Here is the common usage within these families:

DEFINITION 4.5. Let $\mathcal{O}_q(V)$ be a "standard quantized coordinate ring". Within this setting, the parameter q and the algebra $\mathcal{O}_q(V)$ are said to be *generic* when q is not a root of unity.

For an immediate example, observe that in $\mathcal{O}_q(k^2)$, powers x^i and y^j of the given generators commute if and only if $q^{ij} = 1$. Thus, if q is generic, x^i and y^j commute only when one of them equals 1. In contrast, when q is an l-th root of unity, x^i and y^j commute whenever l divides ij.

The properties of the generic members of a family typically differ vastly from those of the non-generic members. For instance, the generic quantum planes $\mathcal{O}_q(k^2)$ have center k (as is easily checked) and are primitive (e.g., [**14**, Exercise 3ZA]).

Since these algebras are infinite dimensional over their centers, Kaplansky's Theorem (e.g., [**22**, Theorem 13.3.8]) shows that they cannot satisfy any polynomial identity. On the other hand, all the non-generic quantum planes do satisfy polynomial identities, because these algebras are finitely generated modules over their centers. (If q is a primitive l-th root of unity, the center of $\mathcal{O}_q(k^2)$ is $k[x^l, y^l]$.)

For another concrete instance, consider the quantum tori of rank two,

$$\mathcal{O}_q((k^\times)^2) := k\langle x^{\pm 1}, y^{\pm 1} \mid xy = qyx\rangle,$$

for $q \in k^\times$. If q is generic, $\mathcal{O}_q((k^\times)^2)$ is a central simple k-algebra with Krull and global dimension 1 ([**16**, Corollary 1.5, Theorems 2.1, 2.4], [**19**, Lemma 1.1], [**21**, Proposition 1.3, Corollary 3.10]). If q is not generic, one finds that $\mathcal{O}_q((k^\times)^2)$ has infinitely many ideals, and its Krull and global dimensions are 2 (e.g., [**16**, Theorems 2.3, 2.4]).

REMARK 4.6. That the generic members of the standard families are the ones indexed by non-roots of unity is an artifact of the way the parameter q is used in the definitions of these families. For example, in the standard quantum plane $\mathcal{O}_q(k^2)$, the scalar q is the coefficient of the (non)commutativity constraint $xy = qyx$, but we can get a similar constraint on replacing q by any scalar-valued function. For simplicity, suppose we build a family $(A_q)_{q\in k}$ where $A_q := k\langle x, y \mid xy = (3q-2)yx\rangle$. Here one should define the generic members of the family to be those A_q such that $3q - 2$ is nonzero and not a root of unity, rather than those for which q is not a root of unity. Going to an extreme, one could even argue that for $\operatorname{char} k \neq 3$, only the single member $A_{2/3}$ should be called generic, since it is a non-noetherian ring with zero-divisors, whereas every other A_q is a noetherian domain.

A number of multiparameter families of quantized coordinate rings occur in the literature (see, e.g., [**3, 7, 11, 15**]). However, the question of which members of such families should be considered "generic" has not even been addressed. For instance, consider the multiparameter quantum affine spaces $\mathcal{O}_{\mathbf{q}}(k^n)$ as in (3.1) and the corresponding quantum tori

$$\mathcal{O}_{\mathbf{q}}((k^\times)^n) := k\langle x_1^{\pm 1}, \dots, x_n^{\pm 1} \mid x_i x_j = q_{ij} x_j x_i \ \forall\ i, j\rangle.$$

At first glance, taking the one-parameter case as a guide, one might consider as "generic" the algebras $\mathcal{O}_{\mathbf{q}}(k^n)$ and $\mathcal{O}_{\mathbf{q}}((k^\times)^n)$ where none of the q_{ij} for $i \neq j$ is a root of unity. However, this still leaves room for "non-general" relations among the q_{ij}, such as $q_{12} = q_{13}^{-1}$, which lets x_1 commute with x_2x_3. If also $q_{12} = q_{13}^{-1} = q_{23}$, then $x_1x_2x_3$ is central in $\mathcal{O}_{\mathbf{q}}(k^3)$ and $\mathcal{O}_{\mathbf{q}}((k^\times)^3)$.

The "most general" multiplicatively skew-symmetric matrices $\mathbf{q} \in M_n(k^\times)$ are those for which the abelian group $\langle q_{ij}\rangle \leq k^\times$ is free of rank $n(n-1)/2$ (in which case the q_{ij} with $i < j$ form a basis). Let us tentatively call these $\mathbf{q}$, as well as the corresponding algebras $\mathcal{O}_{\mathbf{q}}(k^n)$ and $\mathcal{O}_{\mathbf{q}}((k^\times)^n)$, *generic.*

To buttress this choice, note that the non-generic quantum tori $\mathcal{O}_{\mathbf{q}}((k^\times)^n)$ always have "more commutativity" than the generic ones. For example, let $n = 3$ and assume $\mathbf{q}$ is not generic, whence $q_{12}^a q_{13}^b q_{23}^c = 1$ for some nonzero integer triple (a, b, c). If $a = 0$, then $x_1^b x_2^c$ commutes with x_3, and similar nontrivial relations occur if b or c is zero. In any case, $x_1^b x_2^c$ commutes with $x_2^a x_3^b$, which gives a nontrivial relation if at most one of a, b, c is zero. In all these non-generic cases, the group of units of $\mathcal{O}_{\mathbf{q}}((k^\times)^3)$ contains a non-cyclic abelian subgroup G such that $G \cap k^\times = \{1\}$.

In the generic case, the center of $\mathcal{O}_{\mathbf{q}}((k^\times)^n)$ is k, and $\mathcal{O}_{\mathbf{q}}((k^\times)^n)$ is a simple algebra with Krull and global dimension 1 [**21**, Proposition 1.3, Corollary 3.10], although these properties also extend to certain non-generic cases (e.g., [**21**, §3.11]).

4.4. Primary semiclassical limit conjecture. A prerequisite for a useful confirmation or refutal of the following conjecture is to pin down a specification of the term "generic". (For standard families of quantized coordinate rings, Definition 4.5 is to be used.) To emphasize this point, we adopt the phrase *suitably generic* in formulating this and other conjectures.

CONJECTURE 4.7 ([**10**, Conjecture 9.1]). *Assume that* $\operatorname{char} k = 0$. *Let* A *be a suitably generic member of a flat family of quantized coordinate rings for an affine variety* V *over* k, *with semiclassical limit* $\mathcal{O}(V)$. *Then there exist compatible homeomorphisms* $\operatorname{Spec} A \longrightarrow \operatorname{Pspec} \mathcal{O}(V)$ *and* $\operatorname{Prim} A \longrightarrow \operatorname{Pprim} \mathcal{O}(V)$.

We list a number of cases in which Conjecture 4.7 is known to hold:

- Quantum affine spaces and quantum affine toric varieties [**26**, Theorem 3.5, Corollary 3.6], [**13**, Theorems 3.2, 4.1, 5.1];
- Quantum symplectic and euclidean spaces [**25**, Theorem 5.8], [**27**, Corollary 9];
- $\mathcal{O}_q(SL_2(k))$ and $\mathcal{O}_q(GL_2(K))$ [**10**, Example 9.7, Evidence 9.8(a)], [**6**, Corollary 1.3];
- $\mathcal{O}_q(SL_3(K))$ [**6**, Theorem 1.2].

REMARK 4.8. Conjecture 4.7 implies Conjecture 3.2 in the following setting. Suppose k, A, V are as in the above conjecture and that, additionally, $\mathcal{O}(V)$ satisfies the Poisson Dixmier-Moeglin Equivalence. In view of Theorem 4.2, we find: If Conjecture 4.7 holds, then Conjecture 3.2 holds as well.

REMARK 4.9. We briefly illustrate how Conjecture 4.7 may fail in positive characteristic.

(a) Let $A = \mathcal{O}_q(k^2)$, where $q \neq \sqrt[\bullet]{1}$ but $\operatorname{char} k = 2$, and $R = \mathcal{O}(k^2) = k[x, y]$. Due to the characteristic, the elements x^2 and y^2 in R are Poisson-central, meaning that $\{x^2, R\} = \{y^2, R\} = 0$.

As one can check, the ideals $\langle x^2 + \lambda^2 y^2 \rangle$ of R, for $\lambda \in k^\times$, are Poisson-prime but not prime. Since k is infinite, $\bigcap_{\lambda \in k^\times} \langle x^2 + \lambda^2 y^2 \rangle = \langle 0 \rangle$, from which it follows that $\{\langle 0 \rangle\}$ is not open in $\operatorname{Pspec} R$.

In both $\operatorname{Spec} A$ and $\operatorname{Pspec} R$, the zero ideal is the unique dense point. However, $\langle 0 \rangle$ is open in $\operatorname{Spec} A$ (as is clear from Figure 1). Therefore $\operatorname{Spec} A$ is not homeomorphic to $\operatorname{Pspec} R$ in this case.

(b) The example in (a) can be "rescued" by replacing the space $\operatorname{Pspec} R$ by the space $\operatorname{Spec}_{\mathrm{p}} R$ of prime Poisson ideals of R. The set $\operatorname{Spec}_{\mathrm{p}} R$ consists of those prime ideals of R which have generating sets matching those displayed in Figure 1, and so there is an obvious homeomorphism $\operatorname{Spec} A \longrightarrow \operatorname{Spec}_{\mathrm{p}} R$.

(c) Keep $\operatorname{char} k = 2$, but let A be a generic member of the family $\big(\mathcal{O}_{q^2}(k^2)\big)_{q \in k^\times}$. In this case, the semiclassical limit is $R = k[x, y]$ with zero Poisson bracket, so $\operatorname{Pspec} R = \operatorname{Spec} R$, which is not homeomorphic to $\operatorname{Spec} A$.

4.5. A secondary semiclassical limit conjecture. Whenever Conjecture 4.7 holds, all of the considered generic quantized coordinate rings of V must have homeomorphic prime and primitive spectra. Thus, we are led to a related conjecture:

CONJECTURE 4.10. *Assume that* $\operatorname{char} k = 0$. *Let* A *and* A' *be suitably generic members of some flat family of quantized coordinate rings for an affine variety* V *with semiclassical limit* $\mathcal{O}(V)$. *Then there exist compatible homeomorphisms* $\operatorname{Spec} A \longrightarrow \operatorname{Spec} A'$ *and* $\operatorname{Prim} A \longrightarrow \operatorname{Prim} A'$.

Evidently, Conjecture 4.10 holds in all the cases where Conjecture 4.7 has been verified. Another positive scenario is known, although the proof involves a shift of the base field. In this case, we take "suitably generic" to mean "q transcendental over a suitable subfield of k".

OBSERVATION 4.11. Let $(A_q)_{q\in k^\times} = (\mathcal{A}/(t-q)\mathcal{A})_{q\in k^\times}$ be a flat family of quantized coordinate rings for an affine variety V, where $\mathcal{A}$ is again a torsionfree algebra over a Laurent polynomial ring $k[t^{\pm 1}]$ and $A_1 \cong \mathcal{O}(V)$.

Further, assume that $\mathcal{A}$ is defined over a subfield $k_0 \subset k$, that is, $\mathcal{A} = \mathcal{A}_0 \otimes_{k_0} k$ for a torsionfree $k_0[t^{\pm 1}]$-algebra $\mathcal{A}_0$. We claim that if $p, q \in k^\times$ are transcendental over k_0, then $\operatorname{Spec} A_p \approx \operatorname{Spec} A_q$ and $\operatorname{Prim} A_p \approx \operatorname{Prim} A_q$.

To see this, observe that there is a k_0-automorphism $\phi \in \operatorname{Aut} k$ such that $\phi(p) = q$. Then ϕ extends to a $k_0[t^{\pm 1}]$-algebra automorphism $\widehat{\phi} = \mathrm{id}_{\mathcal{A}_0} \otimes \phi$ of $\mathcal{A}$, and $\widehat{\phi}$ induces an isomorphism $A_p \longrightarrow A_q$ of k_0-algebras. This isomorphism, in turn, induces the claimed homeomorphisms.

Let us consider these homeomorphisms in a particular instance, namely when $\mathcal{A} := k[t^{\pm 1}]\langle x, y \mid xy = tyx\rangle$. This algebra determines a flat family $(\mathcal{O}_q(k^2))_{q\in k^\times}$ and is defined over the prime field k_0 of k. Consequently, the quantum planes $\mathcal{O}_p(k^2)$ and $\mathcal{O}_q(k^2)$ are isomorphic as rings for any $p, q \in k^\times$ which are transcendental over k_0. On the other hand, these rings are not isomorphic as k-algebras unless $p = q^{\pm 1}$, as follows from [**1**, Proposition 3.2, Corollaire 3.11(c)].

On the side, we point out that the homeomorphisms arising from $\widehat{\phi}$ in this instance are not entirely geometric, in that their restrictions to classical portions of the given spectra are not always morphisms of classical varieties. The homeomorphism $\operatorname{Prim} A_p \longrightarrow \operatorname{Prim} A_q$ induced by $\widehat{\phi}$ sends

$$\langle x-\alpha,\, y\rangle A_p \quad\longmapsto\quad \langle x-\phi(\alpha),\, y\rangle A_q$$

for $\alpha \in k^\times$. In other words, $k^\times$ is being mapped to itself via ϕ, and the latter map is not a morphism of k-varieties unless $p = q$ and $\phi = \mathrm{id}_k$.

5. Piecewise classical structure of spectra

The prime and primitive spectra of many quantum algebras, particularly generic quantized coordinate rings, are known to have a "piecewise classical" structure, in that they possess finite stratifications in which each stratum is homeomorphic to a classical affine scheme or variety. We outline the relevant partitions, raise the question of fully describing these spectra in terms of classical data, and set up a conjectural framework for achieving an integrated picture.

THEOREM 5.1 ([**3**, Theorems II.2.13, II.8.4, Proposition II.8.3]). *Let* A *be a noetherian* k*-algebra and* H *a torus* $(k^\times)^r$ *acting rationally on* A.

(a) *There are partitions*

$$\operatorname{Spec} A = \bigsqcup_{J\in H\text{-}\operatorname{Spec} A} \operatorname{Spec}_J A \quad \textit{and} \quad \operatorname{Prim} A = \bigsqcup_{J\in H\text{-}\operatorname{Spec} A} \operatorname{Prim}_J A,$$

where H-$\operatorname{Spec} A$ *is the set of* H*-stable prime ideals of* A *and*

$$\operatorname{Spec}_J A := \{P \in \operatorname{Spec} A \mid \bigcap_{h\in H} h(P) = J\}, \qquad \operatorname{Prim}_J A := \operatorname{Prim} A \cap \operatorname{Spec}_J A.$$

(b) *For each* $J \in H$-$\operatorname{Spec} A$*, there are homeomorphisms* $\operatorname{Spec}_J A \approx \operatorname{Spec} Z_J$ *and* $\operatorname{Prim}_J A \approx \max Z_J$*, where* Z_J *is the center of a localization of* A/J *and is isomorphic to a Laurent polynomial ring over* k*.*

(c) *Now assume additionally that* H-$\operatorname{Spec} A$ *is finite, and that* A *satisfies the noncommutative Nullstellensatz over* k*. Then*

$$\operatorname{Prim}_J A = \{\text{ maximal elements of } \operatorname{Spec}_J A\}$$

for all $J \in H$-$\operatorname{Spec} A$*.*

The above partition of $\operatorname{Spec} A$ is a *stratification* in the sense that the closure of any stratum $\operatorname{Spec}_J A$ is a union of strata, and likewise for the partition of $\operatorname{Prim} A$. For details on quantized coordinate rings that satisfy the hypotheses of Theorem 5.1, see [**3**, Corollary II.8.5 and proof].

The topologies on these spectra are determined by the topologies on the strata together with relations between the strata, such as conditions on relative closures. In order to describe these topologies, we write $\mathrm{CL}(T)$ for the collection of closed subsets of a topological space T, and we use overbars to denote closures in $\operatorname{Spec} A$ and $\operatorname{Prim} A$.

LEMMA 5.2 ([**4**, Lemma 2.3, Observation 3.4]). *Let* A *and* H *be as in Theorem* 5.1. *For* $J \subset J'$ *in* H-$\operatorname{Spec} A$*, define*

$$\phi^s_{JJ'} : \mathrm{CL}(\operatorname{Spec}_J A) \to \mathrm{CL}(\operatorname{Spec}_{J'} A), \qquad \phi^s_{JJ'}(Y) = \overline{Y} \cap \operatorname{Spec}_{J'} A$$
$$\phi^p_{JJ'} : \mathrm{CL}(\operatorname{Prim}_J A) \to \mathrm{CL}(\operatorname{Prim}_{J'} A), \qquad \phi^p_{JJ'}(Y) = \overline{Y} \cap \operatorname{Prim}_{J'} A.$$

Then the closed subsets of $\operatorname{Spec} A$ *are the subsets* X *such that*

- $X \cap \operatorname{Spec}_J A \in \mathrm{CL}(\operatorname{Spec}_J A)$ *for all* J*;*
- $\phi^s_{JJ'}(X \cap \operatorname{Spec}_J A) \subseteq X \cap \operatorname{Spec}_{J'} A$ *for all* $J \subseteq J'$*.*

Similarly, the closed subsets of $\operatorname{Prim} A$ *are the subsets* X *such that*

- $X \cap \operatorname{Prim}_J A \in \mathrm{CL}(\operatorname{Prim}_J A)$ *for all* J*;*
- $\phi^p_{JJ'}(X \cap \operatorname{Prim}_J A) \subseteq X \cap \operatorname{Prim}_{J'} A$ *for all* $J \subseteq J'$*.*

Since the strata of $\operatorname{Spec} A$, respectively $\operatorname{Prim} A$, are classical schemes, respectively varieties, one would like to also describe the maps $\phi^\bullet_{JJ'}$ in terms of classical geometric data. The most obvious possibilities fail, however. Immediate examples show that these maps cannot always be given by inverse images under morphisms, nor by closures of images under morphisms (cf. [**4**, Example 3.5] and Example 5.5 below). Potentially, the $\phi^\bullet_{JJ'}$ can be described in terms of morphisms connecting the relevant spectra to auxiliary schemes or varieties as follows, where we identify $\operatorname{Prim}_J A$ and $\operatorname{Spec}_J A$ with the spaces $\max Z_J$ and $\operatorname{Spec} Z_J$ given in Theorem 5.1(b).

CONJECTURE 5.3. [**4**, Conjecture 3.11] *Let* A *and* H *be as in Theorem* 5.1(c).

For $J \subset J'$ in H-$\operatorname{Spec} A$, there exist an affine variety $V_{JJ'}$, a corresponding scheme $\widetilde{V}_{JJ'}$, and morphisms

$$\begin{array}{ccc} \operatorname{Prim}_{J'} A & & \\ & \searrow^{f_{JJ'}} & \\ \operatorname{Prim}_J A & \xrightarrow{\;g_{JJ'}\;} & V_{JJ'} \end{array} \qquad \begin{array}{ccc} \operatorname{Spec}_{J'} A & & \\ & \searrow^{\widetilde{f}_{JJ'}} & \\ \operatorname{Spec}_J A & \xrightarrow{\;\widetilde{g}_{JJ'}\;} & \widetilde{V}_{JJ'} \end{array}$$

such that

$$\phi^p_{JJ'}(Y) = f^{-1}_{JJ'}(\overline{g_{JJ'}(Y)}) \text{ for } Y \in \mathrm{CL}(\operatorname{Prim}_J A);$$

$$\phi^s_{JJ'}(Z) = \widetilde{f}^{-1}_{JJ'}(\overline{\widetilde{g}_{JJ'}(Z)}) \text{ for } Z \in \mathrm{CL}(\operatorname{Spec}_J A).$$

Conjecture 5.3 was confirmed for $\mathcal{O}_q(GL_2(k))$, $\mathcal{O}_q(SL_3(k))$ and $\mathcal{O}_q(M_2(k))$ in [**4**, Example 4.3, Theorems 5.4, 7.5]; Poisson analogs for $\mathcal{O}(GL_2(k))$ and $\mathcal{O}(SL_3(k))$ were established in [**6**, Corollaries 4.5, 4.6].

REMARK 5.4. Suppose that k, A, V are as in the hypotheses of Conjecture 4.7, and that the hypotheses of Theorem 5.1(c) hold for a suitable torus H. In view of Theorem 5.1 and Lemma 5.2, we see that if Conjecture 5.3 is validated, then the spaces $\operatorname{Spec} A$ and $\operatorname{Prim} A$ are determined by the strata $\operatorname{Spec}_J A$ and $\operatorname{Prim}_J A$ together with the auxiliary data from the latter conjecture.

This leads to further bridges among the conjectures. Indeed, if one assumes Conjecture 5.3 as well as a matching Poisson analogue for $\mathcal{O}(V)$, one obtains homeomorphisms $\operatorname{Pspec} \mathcal{O}(V) \approx \operatorname{Spec} A$ and $\operatorname{Pprim} \mathcal{O}(V) \approx \operatorname{Prim} A$. This approach was used to verify Conjecture 4.7 for $\mathcal{O}_q(SL_3(k))$ and $\mathcal{O}_q(GL_2(k))$ in [**6**, Theorem 5.21, Corollaries 5.22, 5.23].

EXAMPLE 5.5. To illustrate the auxiliary varieties and morphisms that may appear in the context of Conjecture 5.3, consider the algebra $A = \mathcal{O}_q(GL_2(k))$, $q \neq \sqrt[\bullet]{1}$, and the auxiliary data that have been used to confirm the conjecture in this case [**4**, Example 4.3].

There is a natural action of the torus $H = (k^\times)^4$ on A, under which A has exactly 4 H-primes; the poset H-$\operatorname{Spec} A$ is drawn on the left in Figure 7. The strata $\operatorname{Prim}_J A$ are homeomorphic to the varieties shown on the right. We identify the $\operatorname{Prim}_J A$ with these varieties.

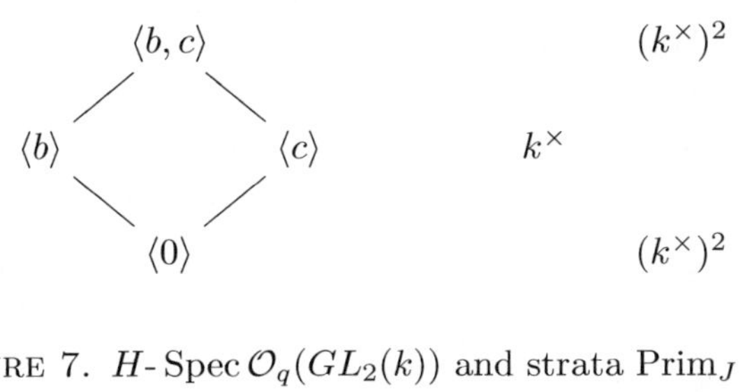

FIGURE 7. H-$\operatorname{Spec} \mathcal{O}_q(GL_2(k))$ and strata $\operatorname{Prim}_J \mathcal{O}_q(GL_2(k))$

The auxiliary data used for $\operatorname{Prim} A$ [**4**, Figure 1] are displayed in Figure 8, where the varieties $V_{JJ'}$ are marked with square brackets. (The data for $\phi^p_{\langle b\rangle,\langle b,c\rangle}$ and $\phi^p_{\langle c\rangle,\langle b,c\rangle}$ are the same, so we display only one set.)

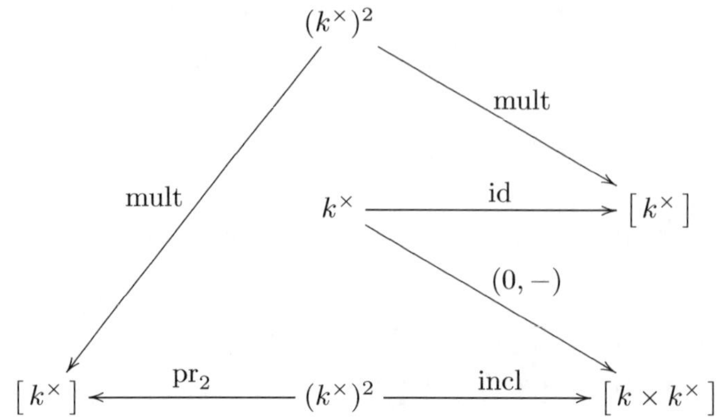

FIGURE 8. Auxiliary data for $\operatorname{Prim} \mathcal{O}_q(GL_2(k))$

Obviously the map $\phi^p_{\langle b\rangle,\langle b,c\rangle}$ can be expressed in the form $Y \mapsto h_b^{-1}(Y)$, where $h_b : (k^\times)^2 \to k^\times$ is the multiplication map. Describing the maps $\phi^p_{\langle 0\rangle,\langle b\rangle}$ and $\phi^p_{\langle 0\rangle,\langle b,c\rangle}$ in similar terms is a different story, however. For instance, $\phi^p_{\langle 0\rangle,\langle b\rangle}$ sends all finite subsets of $(k^\times)^2$ to the empty set. Hence, it cannot be given by inverse images under any map $k^\times \to (k^\times)^2$, nor can it be given by closures of images under any map $(k^\times)^2 \to k^\times$.

References

[1] J. Alev and F. Dumas, *Sur le corps des fractions de certaines algèbres quantiques* (French, with English summary), J. Algebra **170** (1994), no. 1, 229–265, DOI 10.1006/jabr.1994.1336. MR1302839

[2] Jason Bell, Stéphane Launois, Omar León Sánchez, and Rahim Moosa, *Poisson algebras via model theory and differential-algebraic geometry*, J. Eur. Math. Soc. (JEMS) **19** (2017), no. 7, 2019–2049, DOI 10.4171/JEMS/712. MR3656478

[3] Ken A. Brown and Ken R. Goodearl, *Lectures on algebraic quantum groups*, Advanced Courses in Mathematics. CRM Barcelona, Birkhäuser Verlag, Basel, 2002, DOI 10.1007/978-3-0348-8205-7. MR1898492

[4] Kenneth A. Brown and Kenneth R. Goodearl, *Zariski topologies on stratified spectra of quantum algebras*, Commutative algebra and noncommutative algebraic geometry. Vol. II, Math. Sci. Res. Inst. Publ., vol. 68, Cambridge Univ. Press, New York, 2015, pp. 63–91. MR3496861

[5] Kenneth A. Brown and Iain Gordon, *Poisson orders, symplectic reflection algebras and representation theory*, J. Reine Angew. Math. **559** (2003), 193–216, DOI 10.1515/crll.2003.048. MR1989650

[6] Siân Fryer, *The prime spectrum of quantum SL_3 and the Poisson prime spectrum of its semiclassical limit*, Trans. London Math. Soc. **4** (2017), no. 1, 1–29, DOI 10.1112/tlm3.12004. MR3653053

[7] K. R. Goodearl, *Prime spectra of quantized coordinate rings*, Interactions between ring theory and representations of algebras (Murcia), Lecture Notes in Pure and Appl. Math., vol. 210, Dekker, New York, 2000, pp. 205–237. MR1759846

[8] K. R. Goodearl, *Quantized primitive ideal spaces as quotients of affine algebraic varieties*, Quantum Groups and Lie Theory (Durham 1999) (A. Pressley, Ed.), London Math. Soc. Lecture Notes 290, Cambridge (2001) Cambridge Univ. Press, pp. 130–148.

[9] K. R. Goodearl, *A Dixmier-Moeglin equivalence for Poisson algebras with torus actions*, Algebra and its applications, Contemp. Math., vol. 419, Amer. Math. Soc., Providence, RI, 2006, pp. 131–154, DOI 10.1090/conm/419/08001. MR2279114

[10] K. R. Goodearl, *Semiclassical limits of quantized coordinate rings*, Advances in Ring Theory (Athens, Ohio 2008) (D.V. Huynh and S.R. López-Permouth, Eds.), Basel (2010) Birkhäuser/Springer, pp. 165–204.

[11] K. R. Goodearl and S. Launois, *The Dixmier-Moeglin equivalence and a Gel′fand-Kirillov problem for Poisson polynomial algebras* (English, with English and French summaries), Bull. Soc. Math. France **139** (2011), no. 1, 1–39, DOI 10.24033/bsmf.2598. MR2815026

[12] K. R. Goodearl and E. S. Letzter, *Quantum n-space as a quotient of classical n-space*, Trans. Amer. Math. Soc. **352** (2000), no. 12, 5855–5876, DOI 10.1090/S0002-9947-00-02639-8. MR1781280

[13] K. R. Goodearl and E. S. Letzter, *Semiclassical limits of quantum affine spaces*, Proc. Edinburgh Math. Soc. (2) **52** (2009), 387–407.

[14] K. R. Goodearl and R. B. Warfield Jr., *An introduction to noncommutative Noetherian rings*, 2nd ed., London Mathematical Society Student Texts, vol. 61, Cambridge University Press, Cambridge, 2004, DOI 10.1017/CBO9780511841699. MR2080008

[15] Karen L. Horton, *The prime and primitive spectra of multiparameter quantum symplectic and Euclidean spaces*, Comm. Algebra **31** (2003), no. 10, 4713–4743, DOI 10.1081/AGB-120023129. MR1998025

[16] Vasanti A. Jategaonkar, *A multiplicative analog of the Weyl algebra*, Comm. Algebra **12** (1984), no. 13-14, 1669–1688, DOI 10.1080/00927878408823074. MR743310

[17] David A. Jordan and Sei-Qwon Oh, *Poisson spectra in polynomial algebras*, J. Algebra **400** (2014), 56–71, DOI 10.1016/j.jalgebra.2013.11.004. MR3147362

[18] Stéphane Launois and Omar León Sánchez, *On the Dixmier-Moeglin equivalence for Poisson-Hopf algebras*, Adv. Math. **346** (2019), 48–69, DOI 10.1016/j.aim.2019.01.036. MR3904282

[19] Martin Lorenz, *Group rings and division rings*, Methods in ring theory (Antwerp, 1983), NATO Adv. Sci. Inst. Ser. C: Math. Phys. Sci., vol. 129, Reidel, Dordrecht, 1984, pp. 265–280. MR0770597

[20] Juan Luo, Xingting Wang, and Quanshui Wu, *Poisson Dixmier-Moeglin equivalence from a topological point of view*, Israel J. Math. **243** (2021), no. 1, 103–139, DOI 10.1007/s11856-021-2154-9. MR4299143

[21] J. C. McConnell and J. J. Pettit, *Crossed products and multiplicative analogues of Weyl algebras*, J. London Math. Soc. (2) **38** (1988), no. 1, 47–55, DOI 10.1112/jlms/s2-38.1.47. MR949080

[22] J. C. McConnell and J. C. Robson, *Noncommutative Noetherian rings*, Revised edition, Graduate Studies in Mathematics, vol. 30, American Mathematical Society, Providence, RI, 2001. With the cooperation of L. W. Small, DOI 10.1090/gsm/030. MR1811901

[23] Sei-Qwon Oh, *Symplectic ideals of Poisson algebras and the Poisson structure associated to quantum matrices*, Comm. Algebra **27** (1999), no. 5, 2163–2180, DOI 10.1080/00927879908826555. MR1683857

[24] Sei-Qwon Oh, *Poisson polynomial rings*, Comm. Algebra **34** (2006), no. 4, 1265–1277, DOI 10.1080/00927870500454463. MR2220812

[25] Sei-Qwon Oh, *Quantum and Poisson structures of multi-parameter symplectic and Euclidean spaces*, J. Algebra **319** (2008), no. 11, 4485–4535, DOI 10.1016/j.jalgebra.2008.03.010. MR2416732

[26] Sei-Qwon Oh, Chun-Gil Park, and Yong-Yeon Shin, *Quantum n-space and Poisson n-space*, Comm. Algebra **30** (2002), no. 9, 4197–4209, DOI 10.1081/AGB-120013313. MR1936465

[27] Sei-Qwon Oh and Mi-Yeon Park, *Relationship between quantum and Poisson structures of odd dimensional Euclidean spaces*, Comm. Algebra **38** (2010), no. 9, 3333–3346, DOI 10.1080/00927870903114987. MR2724222

Department of Mathematics, University of California, Santa Barbara, California 93106

Email address: goodearl@math.ucsb.edu

Contemporary Mathematics
Volume **801**, 2024
https://doi.org/10.1090/conm/801/16038

Introduction to Belavin-Drinfeld Quantum Groups

Timothy J. Hodges

ABSTRACT. In this article we discuss the Belavin-Drinfeld quantum groups, one of the most fascinating and elusive classes of non-standard quantum groups. Belavin and Drinfeld classified all non-skewsymmetric classical r-matrices for simple Lie algebras $\mathfrak{g}$. Etingof, Schedler and Schiffmann constructed explicit formal quantizations of the associated Poisson groups. Yakimov has shown that the same construction can be used to construct algebraic quantizations. This opens up a potentially extremely interesting field of study. This article describes the construction of the associated quantized algebra of functions and the analog of the quantized universal enveloping algebra. A number of conjectures and open questions are described.

1. Introduction

One of the most interesting problems in the theory of non-commutative algebra and geometry is the description of the primitive spectrum of well-structured algebras and the comparison of this spectrum with the symplectic leaves of an associated Poisson algebra. This approach essentially began with the study of the enveloping algebra $U(\mathfrak{g})$ of a finite dimensional complex Lie algebra $\mathfrak{g}$ in the 1970's and 80's. With the discovery of quantum groups in the late 1980's, it was natural to ask for a description of the primitive ideals of the quantized function algebras $\mathbb{C}_q[G]$ and to relate this description to that of the symplectic leaves of the associated Poisson structure on G. Building on work of the author and Levasseur [**12, 13**], a complete description of the primitive spectrum of $\mathbb{C}_q[G]$ was given by Joseph [**20, 21**]. The maximal torus, H, of G acts naturally on Prim and the orbits are indexed by elements of $W \times W$ where W is the Weyl group of G. This kind of behavior was found to occur in a number of related algebras and much work has been done in this direction over the last 25 years.

Joseph's work was extended to the slightly more general case of multi-parameter quantized function algebras in [**14**]. This naturally raised the question of whether there existed similar results for more general quantized function algebras - those which arose in some way as quantizations of more general Poisson structures on G.

Poisson structures on an algebraic group G are given by Lie bialgebra structures on its Lie algebra $\mathfrak{g}$. An important family of such structures on simple groups was described by Belavin and Drinfeld in 1984 [**3**] and it was expected that quantizations of these groups would also exist. Some specific examples of such quantizations

2020 *Mathematics Subject Classification.* Primary 20G42, 16T20, 17B37; Secondary 16T25, 17B38, 17B63, 81R60, 53D17.

This work was partially supported by the National Science Foundation.

occurred naturally and were studied in an ad-hoc fashion (see, for instance, [**1**, **2**, **7**, **15**, **16**]). However it was not until 2001 that an explicit general construction of these non-standard quantum groups was achieved by Etingof, Schedler and Schiffman [**10**]. This construction was only given in the formal context (as a Hopf algebra over $\mathbb{C}[[h]]$). However, Yakimov has shown [**18**] that the ESS construction can be adapted to yield complex quantum groups $\mathbb{C}_{\pi,q}[G]$ (that is, Hopf algebras which are finitely generated over $\mathbb{C}$ and depend on a parameter $q \in \mathbb{C}$; the meaning of the parameter π will be explained in Section 2). Yakimov also gave a detailed description of the symplectic leaves of these Poisson groups [**30**]. This naturally leads us to the following question.

PROBLEM 1. Describe the primitive ideals of the Belavin-Drinfeld quantized function algebras $\mathbb{C}_{\pi,q}[G]$ for G a complex semi-simple algebraic group and $q \in \mathbb{C}$ not a root of unity.

Since the standard quantized enveloping algebras $U_q(\mathfrak{g})$ occur as homomorphic images of certain $\mathbb{C}_{\pi,q}[G]$, a complete solution to this problem will necessarily involve as special cases the description of the primitive spectrum of both of these algebras. Thus one does not expect that an answer to Problem 1 will be easy. However, as in the standard case, one expects that the classification of the symplectic leaves of the associated Poisson group G will be a useful guide to the description of the completely prime primitive spectrum.

The object of this article is to sketch out the groundwork of such a research program. We describe the construction of the general quantum function algebra $\mathbb{C}_{\pi,q}[G]$ and of the corresponding quantized enveloping algebra $U_{\pi,q}(\mathfrak{g})$. There are a large number of interesting questions about these algebras. As an example we discuss in some detail the analog of the triangular decomposition for $U_{\pi,q}(\mathfrak{g})$.

We begin in Section 3, with a brief discussion of Belavin-Drinfeld triples and solutions of the classical Yang-Baxter equation. In the next section we give some explicit examples of Belavin-Drinfeld quantum groups, the double quantum groups and the Cremmer-Gervais quantum groups, which we shall use to illustrate the later discussion and to provide evidence for some of the conjectures. In Section 7 we outline Yakimov's version of the Etingof-Schedler-Schiffman construction for complex quantum groups. The final section describes the construction of the FRT duals $U_{\pi,q}(\mathfrak{g})$ of Belavin-Drinfeld quantum groups and outlines some important structural conjectures.

The goal of this brief survey is to give a broad overview of the subject, without getting bogged down in too many of the technical details which are available elsewhere. We give enough background review to establish the necessary notation, while trying to avoid the technicalities that can make this subject inaccessible to the non-expert. At the same time we refer frequently to our two illustrative examples, double quantum groups and Cremmer-Gervais quantum groups in order to give the reader a concrete feel of the subject.

2. Standard quantum groups

We begin with a brief review of the standard notation that we shall be using for the quantized enveloping algebra $U_q(\mathfrak{g})$ and quantized function algebra $\mathbb{C}_q[G]$. The reader familiar with this material can safely skip this section or scan quickly over it. For simplicity we follow the approach of [**14**] with some minor notational changes. Any missing details can be found in that paper.

Let $\mathfrak{g}$ be a complex semi-simple Lie algebra. Let $\mathfrak{h}$ be a Cartan subalgebra of $\mathfrak{g}$, Δ the associated root system, $\Gamma = \{\alpha_1, \ldots, \alpha_n\}$ a base of Δ, Δ^+ the set of positive roots and W the Weyl group. We denote by $\mathcal{P}$ and $\mathcal{Q}$ the lattices of weights and roots respectively. Let κ be the Killing form on $\mathfrak{g}$; this will identify $\mathfrak{g}$, resp. $\mathfrak{h}$, with its dual $\mathfrak{g}^*$, resp. $\mathfrak{h}^*$ and induce a bilinear form on $\mathfrak{h}^*$ denoted by $(\cdot,\cdot)$. We also set as usual

$$\mathfrak{n}^{\pm} = \bigoplus_{\alpha \in \Delta^+} \mathfrak{g}_{\pm\alpha}, \quad \mathfrak{b}^{\pm} = \mathfrak{h} \oplus \mathfrak{n}^{\pm}$$

Fix a lattice $\mathcal{L}$ such that $\mathcal{Q} \subset \mathcal{L} \subset \mathcal{P}$ and let G be a connected semi-simple Lie group such that $\mathrm{Lie}(G) = \mathfrak{g}$ and such that G contains a maximal torus H such that the character group of H is isomorphic to $\mathcal{L}$.

Let $q \in \mathbb{C}^*$ and assume that q *is not a root of unity.* Let $\hbar \in \mathbb{C}$ such that $q = \exp(-\hbar/2)$ and define rational powers of q by $q^d = \exp(-d\hbar/2)$ for $d \in \mathbb{Q}$. Denote by U^0 the group algebra of $\mathcal{L}$,

$$U^0 = \mathbb{C}[K_\lambda\,;\, \lambda \in \mathcal{L}], \quad K_0 = 1, \quad K_\lambda K_\mu = K_{\lambda+\mu}.$$

The one parameter quantized enveloping algebra associated to this data is the Hopf algebra

$$U_q(\mathfrak{g}) = U^0[E_i, F_i\,;\, 1 \le i \le n]$$

with the usual relations and Hopf algebra structure [**14**, Section 3]. We further define subalgebras of $U_q(\mathfrak{g})$ as follows

$$U_q(\mathfrak{n}^+) = \mathbb{C}[E_i,\,;\, 1 \le i \le n], \quad U_q(\mathfrak{n}^-) = \mathbb{C}[F_i,\,;\, 1 \le i \le n]$$
$$U_q(\mathfrak{b}^+) = U^0[E_i,\,;\, 1 \le i \le n], \quad U_q(\mathfrak{b}^-) = U^0[F_i,\,;\, 1 \le i \le n].$$

For $\Lambda \in \mathcal{P}^+$, denote by $L(\Lambda)$ the simple module of highest weight Λ. Let $\mathcal{C}_\mathcal{L}$ be the subcategory of finite dimensional $U_q(\mathfrak{g})$-modules consisting of finite direct sums of $L(\Lambda)$, $\Lambda \in \mathcal{L}^+$. The category $\mathcal{C}_\mathcal{L}$ is closed under tensor products and the formation of duals.

Let $M \in \mathrm{obj}(\mathcal{C}_\mathcal{L})$. For $f \in M^*$, $v \in M$ we define the coordinate function $c_{f,v} \in U_q(\mathfrak{g})^*$ by

$$\forall u \in U_q(\mathfrak{g}), \quad c_{f,v}(u) = f(uv).$$

Using the standard isomorphism $(M \otimes N)^* \cong N^* \otimes M^*$ one has the following formula for multiplication,

$$c_{f,v} c_{f',v'} = c_{f' \otimes f, v \otimes v'}.$$

DEFINITION 2.1. The *quantized function algebra* $\mathbb{C}_q[G]$ is the restricted dual of $U_q(\mathfrak{g})$ with respect to the category $\mathcal{C}_\mathcal{L}$: that is to say

$$\mathbb{C}_q[G] = \mathbb{C}[c_{f,v}\,;\, v \in M, f \in M^*,\, M \in \mathrm{obj}(\mathcal{C}_\mathcal{L})].$$

The algebra $\mathbb{C}_q[G]$ is a Hopf algebra. If $\{v_1, \ldots, v_s; f_1, \ldots, f_s\}$ is a dual basis for $M \in \mathrm{obj}(\mathcal{C}_\mathcal{L})$ one has

$$\Delta(c_{f,v}) = \sum_i c_{f,v_i} \otimes c_{f_i,v}, \quad \epsilon(c_{f,v}) = f(v), \quad S(c_{f,v}) = c_{v,f}.$$

We set

$$C(M) = \mathbb{C}\langle c_{f,v}\,;\, f \in M^*,\, v \in M\rangle$$

and when $M = L(\Lambda)$ we abbreviate the notation to $C(M) = C(\Lambda)$. It is then classical that

$$\mathbb{C}_q[G] = \bigoplus_{\Lambda \in \mathcal{L}+} C(\Lambda).$$

3. Belavin-Drinfeld triples and non-skew-symmetric solutions of the CYBE

In this section we summarize the description of certain types of Poisson structures on a semi-simple group given by Belavin and Drinfeld (in the simple case) in [**3**]. A Poisson structure on G is given by a Lie bialgebra structure on its Lie algebra $\mathfrak{g}$ and by Whitehead's lemma, all such bialgebra structures are coboundary when $\mathfrak{g}$ is semi-simple. That is, the cocommutator $\delta : \mathfrak{g} \to \mathfrak{g} \otimes \mathfrak{g}$ is given by $\delta(X) = X.r$ for an $r \in \mathfrak{g} \otimes \mathfrak{g}$ satisfying

(*i*) $r_{12} + r_{21}$ is a $\mathfrak{g}$-invariant element of $\mathfrak{g} \otimes \mathfrak{g}$

(*ii*) $[[r, r]] = [r_{12}, r_{13}] + [r_{12}, r_{23}] + [r_{13}, r_{23}]$ is a $\mathfrak{g}$-invariant element of $\mathfrak{g} \otimes \mathfrak{g} \otimes \mathfrak{g}$

The equation $[[r, r]] = 0$ is known as the classical Yang-Baxter equation and solutions to the CYBE are known as *classical r-matrices.* The Lie bialgebra structures arising from classical r-matrices are said to be *quasi-triangular.* If also r is skew-symmetric, this structure is said to be triangular. At the opposite end of the spectrum is the case where $r_{12} + r_{21}$ is non-degenerate. In this case we call $\mathfrak{g}$ a *factorizable* quasi-triangular Lie bialgebra (note that when $\mathfrak{g}$ is simple this is equivalent to r being non-skew-symmetric). The classification of non-skew-symmetric solutions of the CYBE when $\mathfrak{g}$ is simple was given by Belavin and Drinfeld [**3**]. To describe these r-matrices we need two parameters: a discrete parameter known as a Belavin-Drinfeld triple; and a continuous parameter $s \in \wedge^2\mathfrak{h}$ satisfying certain compatibility conditions.

DEFINITION 3.1. A *Belavin-Drinfeld triple* is a triple $(\tau, \Gamma_1, \Gamma_2)$ where $\Gamma_1,\ \Gamma_2 \subset \Gamma$ and $\tau : \Gamma_1 \to \Gamma_2$ is a bijection such that

(*i*) for all $\alpha, \beta \in \Gamma_1$, $(\tau(\alpha), \tau(\beta)) = (\alpha, \beta)$;

(*ii*) for all $\alpha \in \Gamma_1$ there exists a k such that $\tau^k(\alpha) \notin \Gamma_1$.

For any Belavin-Drinfeld triple τ we define $n := n(\tau)$ to be the largest integer such that $\tau^n(\alpha)$ is defined for some $\alpha \in \Gamma_1$. We call this number the *order of the Belavin–Drinfeld triple* τ. For instance, if the triple is disjoint ($\Gamma_1 \cap \Gamma_2 = \emptyset$), then $n(\tau) = 1$.

EXAMPLE 3.2. Let $\Gamma = A_n = \{\alpha_1, \dots, \alpha_n\}$, let $\Gamma_1 = \{\alpha_1, \dots, \alpha_{n-1}\}$, let $\Gamma_2 = \{\alpha_2, \dots, \alpha_n\}$ and let $\tau : \Gamma_1 \to \Gamma_2$ be given by $\tau(\alpha_i) = \alpha_{i+1}$. For instance, when $n = 5$, the triple is:

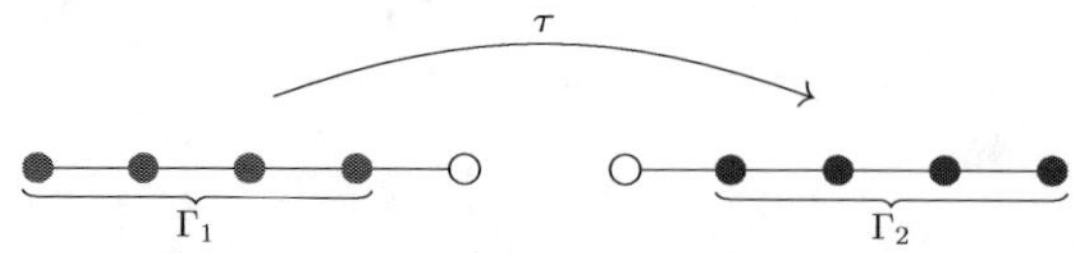

Since the quantization of the associated Poisson group yields the Cremmer-Gervais quantum group, we call this triple the *Cremmer-Gervais triple.* The order of this triple is $n(\tau) = n - 1$

EXAMPLE 3.3. Let G be a semi-simple group and let $D(G) = G \times G$ be the double group. In this case $\Gamma = \Gamma_1 \sqcup \Gamma_2$ where Γ_1 and Γ_2 are identical copies of a base for G. The natural bijection $\tau : \Gamma_1 \to \Gamma_2$ is a Belavin-Drinfeld triple. For instance when G is E_6 the triple can be visualized as:

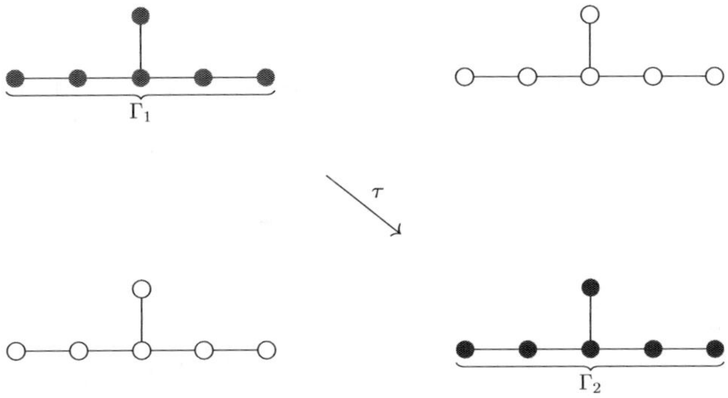

Identify $\mathfrak{h}$ with $\mathfrak{h}^*$ via nondegenerate restriction of the form κ to $\mathfrak{h}$. Set

$$\mathfrak{h}_k := \mathrm{Span}_{\mathbb{C}}\{\alpha \in \Gamma_k\},\ \ k = 1, 2 \quad \text{and} \quad \mathfrak{z} = \mathrm{Span}_{\mathbb{C}}\{\alpha - \tau(\alpha) \mid \alpha \in \Gamma_1\}^{\perp}.$$

By [**10**, Lemma 3.1],

$$\mathfrak{h} = \mathfrak{h}_1 \oplus \mathfrak{z}$$

and by [**10**, Corollary 3.1] τ can be extended to an orthogonal isomorphism

$$\tau\colon \mathfrak{h} \to \mathfrak{h} \quad \text{such that} \quad \tau|_{\mathfrak{z}} = \mathrm{id}.$$

For a positive root $\alpha \in \Delta^+$, choose root vectors e_α and f_α for the roots α and $-\alpha$, normalized by $\kappa(e_\alpha, f_\alpha) = 1$. Let $\mathfrak{n}^+$ be the nilpotent subalgebra of $\mathfrak{g}$ generated by $\{e_\alpha\}_{\alpha\in\Gamma}$ and τ be the unique Lie algebra endomorphism of $\mathfrak{n}_+$ defined by

$$\tau(e_\alpha) = e_{\tau(\alpha)} \text{ for } \alpha \in \Gamma_1 \quad \text{and} \quad \tau(e_\alpha) = 0 \text{ for } \alpha \in \Gamma\backslash\Gamma_1.$$

The standard classical r-matrix for $\mathfrak{g}$ is the element

$$r := \frac{1}{2}\Omega_0 + \sum_{\alpha\in\Delta^+} e_\alpha \otimes f_\alpha \in \mathfrak{g} \otimes \mathfrak{g}$$

where $\Omega_0 \in S^2\mathfrak{h}$ is the Cartan component of Ω, the quadratic Casimir of $\mathfrak{g}$.

THEOREM 3.4 (Belavin–Drinfeld [**3**]). *Let $\mathfrak{g}$ be a simple Lie bialgebra. The orbits of the adjoint group of $\mathfrak{g}$ on the set of non-skewsymmetric classical r-matrices for $\mathfrak{g}$ have a unique representative of the form*

$$r_\pi = r - s + \sum_{\alpha\in\Delta^+} \sum_{k=1}^{n} \tau^k(e_\alpha) \wedge f_\alpha$$

for a pair $\pi = (\tau, s)$ consisting of a triple $(\Gamma_1, \Gamma_2, \tau)$ of order n and an element

$$s \in \wedge^2\mathfrak{h} \quad \textit{such that} \quad 2((\alpha - \tau(\alpha)) \otimes 1)s = ((\alpha + \tau(\alpha)) \otimes 1)\Omega_0, \forall \alpha \in \Gamma_1.$$

This result may be generalized to an analogous result for reductive Lie bialgebras.

DEFINITION 3.5. A pair $\pi = (\tau, s)$ of the form described in the theorem will be called a *Belavin-Drinfeld pair.*

EXAMPLE 3.6. In the case of the trivial triple ($\tilde{\Gamma}_1 = \tilde{\Gamma}_2 = \emptyset$), there is no condition on s and the corresponding Poisson structures are the multi-parameter standard ones whose quantizations are the multi-parameter quantum groups $\mathbb{C}_{q,p}[G]$ (see, for example, [**14**]).

4. Examples of Belavin-Drinfeld quantum groups

Before considering the general construction of Belavin-Drinfeld quantum groups, let us consider a couple of examples which can be constructed in simpler, more straightforward ways and which have consequently been studied in a fair amount of detail.

The first example is the quantum Lorentz group and its generalization, the double quantum groups. These quantum groups have been studied in a variety of different forms for many years. The reader can consult, for instance, [**4–6, 9, 15, 26, 28**].

EXAMPLE 4.1 (Double quantum groups and the quantum Lorentz group). The quantum Lorentz group is a non-standard quantization of the group $SL_2(\mathbb{C}) \times SL_2(\mathbb{C})$. As an algebra it is generated by the elements a, b, c, d and $\tilde{a}, \tilde{b}, \tilde{c}, \tilde{d}$ subject to the relations:

$$
\begin{gathered}
ba = qab,\ ca = qac,\ db = qbd,\ dc = qcd \\
cb = bc, \quad da - ad = (q - q^{-1})bc, \quad da - qbc = 1 \\
\tilde{b}\tilde{a} = q\tilde{a}\tilde{b},\ \tilde{c}\tilde{a} = q\tilde{a}\tilde{c},\ \tilde{d}\tilde{b} = q\tilde{b}\tilde{d},\ \tilde{d}\tilde{c} = q\tilde{c}\tilde{d} \\
\tilde{c}\tilde{b} = \tilde{b}\tilde{c}, \quad \tilde{d}\tilde{a} - \tilde{a}\tilde{d} = (q - q^{-1})\tilde{b}\tilde{c}, \quad \tilde{d}\tilde{a} - q\tilde{b}\tilde{c} = 1 \\
a\tilde{a} = \tilde{a}a,\ qa\tilde{b} = \tilde{b}a,\ a\tilde{c} = q\tilde{c}a - (q - q^{-1})c\tilde{a},\ a\tilde{d} + (q - q^{-1})c\tilde{b} = \tilde{d}a \\
qb\tilde{a} = \tilde{a}b + (q - q^{-1})\tilde{b}a,\ b\tilde{b} = \tilde{b}b,\ b\tilde{c} + (q - q^{-1})d\tilde{a} = (q - q^{-1})\tilde{d}a + \tilde{c}b \\
b\tilde{d} + (q - q^{-1})d\tilde{b} = q\tilde{d}b \\
c\tilde{a} = q\tilde{a}c, \quad c\tilde{b} = \tilde{b}c, \quad c\tilde{c} = \tilde{c}c, \quad qc\tilde{d} = \tilde{d}c \\
d\tilde{a} = \tilde{a}d + (q - q^{-1})\tilde{b}c,\ d\tilde{b} = q\tilde{b}d,\ qd\tilde{c} = \tilde{c}d + (q - q^{-1})\tilde{d}c,\ d\tilde{d} = \tilde{d}d
\end{gathered}
$$

More abstractly, it is the Drinfeld double

$$\mathbb{C}_q[D(SL(2))] = \mathbb{C}_q[SL(2)] \bowtie \mathbb{C}_q[SL(2)].$$

More generally, given any semi-simple group G we can form the Drinfeld double $\mathbb{C}_q[D(G)] = \mathbb{C}_q[G] \bowtie \mathbb{C}_q[G]$. These algebras are quantizations of the Poisson structures on $D(G) = G \times G$ given by the triples in Example 3.3. This approach was used in [**17**] to give a straightforward construction of Belavin-Drinfeld quantum groups in the case where the triple is disjoint ($\Gamma_1 \cap \Gamma_2 = \emptyset$) .

The next example is the Cremmer-Gervais quantum groups which first arose in some work of Cremmer and Gervais on the non-linearly extended Virasoro algebras [**7**]. These constructions were further studied in [**1, 2, 16**].

EXAMPLE 4.2 (The Cremmer-Gervais quantum groups). Let $0 \neq q, p \in \mathbb{C}$ and let n be a positive integer. Denote by e_{ij} the standard matrix units in $M_n(\mathbb{C})$. Set

$$
\begin{aligned}
R = R_n(q,p) = p^{-1} \Bigg(& \sum_{i=1}^{n} q e_{ii} \otimes e_{ii} + q \sum_{i>j} p^{-2(i-j)} e_{ii} \otimes e_{jj} \\
& + q^{-1} \sum_{i<j} p^{-2(i-j)} e_{ii} \otimes e_{jj} \\
& + (q^{-1} - q) \sum_{i<j} \sum_{k=1}^{j-i-1} p^{2k} e_{i,j-k} \otimes e_{j,i+k} \\
& + (q - q^{-1}) \sum_{i>j} \sum_{k=0}^{i-j-1} p^{-2k} e_{i,j+k} \otimes e_{j,i-k} \Bigg)
\end{aligned}
$$

Then R is a solution of the quantum Yang-Baxter equation [**16**, Theorem 4.1]. Define $A(R)$ to be the associated quantum matrix algebra with generators T_i^k and relations

$$\sum_{u,v} R_{ji}^{uv} T_u^k T_v^l = \sum_{u,v} T_i^u T_j^v R_{vu}^{kl}$$

These algebras have a normal, group-like determinant element $\det_q$. We define $\mathbb{C}_R[GL(n)] = A(R)[\det_q^{-1}]$. When $q = p^n$ the element $\det_q$ is central and we define $\mathbb{C}_R[SL(n)] = \mathbb{C}_R[GL(n)]/(\det_q - 1)$. Both $\mathbb{C}_R[GL(n)]$ and $\mathbb{C}_R[SL(n)]$ have natural Hopf algebra structures. The Hopf algebra $\mathbb{C}_R[SL(n)]$ is a quantization of the Poisson structure on $SL(n)$ given by the triple in Example 3.2.

This construction suggests a possible route to constructing quantizations of the Belavin-Drinfeld Poisson groups in the case $G = SL(n)$ - one "quantizes" the classical r-matrix r using a kind of exponentiation to a solution R of the quantum Yang-Baxter equation. One then constructs $A(R)$, proves the existence of a central determinant $\det_q$ and constructs the new quantum group as $A(R)/(\det_q - 1)$. Gerstenhaber, Giaquinto and Schack conjectured a form for R in [**11**]; however this conjecture proved hard to verify directly and was eventually established by Schedler [**29**] using the ESS construction.

5. Braided Hopf algebras

Let us begin by reviewing the relevant definitions and constructions from [**16**]. We begin with the definition of a braided bialgebra. Such algebras are also called *co-quasitriangular* (since they are dual to the notion of quasi-triangular Hopf algebras) but we prefer the term braided because this allows us to talk about a braiding rather than the more long-winded *co-quasi triangular structure.* A braiding is the dual of the notion of universal R-matrix for quasi-triangular Hopf algebras.

DEFINITION 5.1. A *braiding* on a bialgebra A is a bilinear pairing $\langle \mid \rangle$ such that the following conditions hold for all a, b, and c in A

(1) $\sum \langle a_{(1)} \mid b_{(1)} \rangle b_{(2)} a_{(2)} = \sum \langle a_{(2)} \mid b_{(2)} \rangle a_{(1)} b_{(1)}$
(2) $\langle \, , \, \rangle$ is invertible in $(A \otimes A)^*$;
(3) $\langle a \mid bc \rangle = \sum \langle a_{(1)} \mid b \rangle \langle a_{(2)} \mid c \rangle$.
(4) $\langle ab \mid c \rangle = \sum \langle b \mid c_{(1)} \rangle \langle a \mid c_{(2)} \rangle$

Note that in this situation, we also have that $\langle a \mid 1\rangle = \epsilon(a) = \langle 1 \mid a\rangle$ for all $a \in A$.

A *braided bialgebra* is a pair (A, β) where A is a bialgebra and β is a braiding on A. A braided Hopf algebra is a Hopf algebra which is braided as a bialgebra. The antipode of a braided Hopf algebra is always bijective by [**8**, Theorem 1.3]. Moreover, the braiding is S-invariant in the sense that $\langle a \mid b\rangle = \langle S(a) \mid S(b)\rangle$.

Let's recall briefly the braiding on $\mathbb{C}_q[G]$ defined via the quantum R-matrix as constructed in [**23**, Ch. 4]. For $\nu \in \mathcal{Q}^+$, set $m(\nu) := \dim U_q(\mathfrak{n}^+)_\nu = \dim U_q(\mathfrak{n}^-)_{-\nu}$. Let $\{x_{\nu,j}\}_{j=1}^{m(\nu)}$ and $\{x_{-\nu,j}\}_{j=1}^{m(\nu)}$ be two dual bases of $U_q(\mathfrak{n}^+)_\nu$ and $U_q(\mathfrak{n}^-)_{-\nu}$ with respect to the Rosso–Tanisaki form, cf. [**19**, Ch. 6]. Define the quasi R-matrix [**23**, Ch. 4]

$$\Theta = \sum_{\nu \in \mathcal{Q}^+} \sum_{j=1}^{m(\nu)} x_{\nu,j} \otimes x_{-\nu,j} \in U_q(\mathfrak{n}^+) \widehat{\otimes} U_q(\mathfrak{n}^-) = \varprojlim \bigoplus_{\gamma \in \mathcal{Q}^+, \gamma \le \nu} U_q(\mathfrak{n}^+)_\gamma \widehat{\otimes} U_q(\mathfrak{n}^-)_{-\gamma}.$$

Let

$$\omega \colon \mathcal{P} \times \mathcal{P} \to \mathbb{Q}$$

be an additive bicharacter. For $V_1, V_2 \in \mathcal{C}_{\mathcal{P}}$, define the invertible diagonal operator

$$q^\omega \in \operatorname{End}_{\mathbb{C}}(V_1 \otimes V_2) \quad \text{by} \quad q^\omega|_{(V_1)_\lambda \otimes (V_2)_\mu} := q^{\omega(\lambda,\mu)}\mathrm{id} \text{ for } \lambda, \mu \in \mathcal{P}.$$

The quantum R-matrix operator is defined by

$$\mathcal{R}_{V_1,V_2} := q^{\Omega_0}\Theta \in \operatorname{End}_{\mathbb{C}}(V_1 \otimes V_2)$$

where

$$\Omega_0(\lambda, \mu) := \langle \lambda, \mu\rangle.$$

The braiding on $\mathbb{C}_q[G]$ is then defined by

$$\beta(c_{\xi_1,v_1}, c_{\xi_2,v_2}) = (\xi_2 \otimes \xi_1)(\mathcal{R}_{V_1,V_2} v_1 \otimes v_2)$$

6. Cocycle twisting

The construction of the BD quantum groups is performed using the now well-understood technique of twisting by a 2-cocycle [**9, 14, 24, 27**].

DEFINITION 6.1. Let A be a Hopf algebra. A *2-cocycle on* A is an invertible pairing $\sigma : A \otimes A \to k$ such that for all x, y and z in A,

$$\sum \sigma(x_{(1)}, y_{(1)})\sigma(x_{(2)}y_{(2)}, z) = \sum \sigma(y_{(1)}, z_{(1)})\sigma(x, y_{(2)}z_{(2)})$$

and $\sigma(1, 1) = 1$.

Given a 2-cocycle σ on a Hopf algebra A, one can twist the multiplication to get a new Hopf algebra A_σ. The new multiplication is given by

$$x \cdot y = \sum \sigma(x_{(1)}, y_{(1)})x_{(2)}y_{(2)}\sigma^{-1}(x_{(3)}, y_{(3)}).$$

See [**9**] for further details.

THEOREM 6.2. *Let A be a braided bialgebra with braiding β. Let σ be a 2-cocycle on A. Let A_σ be the twisted bialgebra defined above. Then $\sigma\beta(\sigma^\top)^{-1}$ (convolution product) is a braiding on A_σ.*

7. The algebraic quantization of Belavin–Drinfeld r-matrices

Let $(\tau, \Gamma_1, \Gamma_2)$ be a Belavin-Drinfeld triple of order n. For $k \in [2, n]$, let Γ_{2k-1} and Γ_{2k} be the domain and range of τ^k So,

$$\Gamma_1 \supsetneq \Gamma_3 \supsetneq \ldots \supsetneq \Gamma_{2n-1} \quad \text{and} \quad \Gamma_2 \supsetneq \Gamma_4 \supsetneq \ldots \supsetneq \Gamma_{2n},$$

where the strict inclusions follow from the nilpotency property of τ. The map τ^k defines a graph isomorphism

$$\tau^k \colon \Gamma_{2k-1} \to \Gamma_{2k}.$$

In particular, $(\tau^k, \Gamma_{2k-1}, \Gamma_{2k})$ is a Belavin–Drinfeld triple for $\mathfrak{g}$ for all $k \in [1, n]$.

Recall that the braiding on $\mathbb{C}_q[G]$ is given by the standard quasi R-matrix

$$\Theta \in U_q(\mathfrak{n}^+)\widehat{\otimes} U_q(\mathfrak{n}^-)$$

Denote by τ the unique algebra endomorphism of $U_q(\mathfrak{n}^+)$ defined by

$$\tau(E_\alpha) = E_{\tau(\alpha)} \text{ for } \alpha \in \Gamma_1 \quad \text{and} \quad \tau(E_\alpha) = 0 \text{ for } \alpha \in \Gamma \backslash \Gamma_1.$$

For $V_1, V_2 \in \mathcal{C}_{\mathcal{L}}$ and $k \in [1, n]$, the operator $(\tau^k \otimes 1)(\Theta)$ acts on $V_1 \otimes V_2$ because all but finitely many terms in the summation above act by 0 on $V_1 \otimes V_2$. Fix a Belavin-Drinfeld pair $\pi = (\tau, s)$ and define $\mathcal{J}_{V_1,V_2} \in \mathrm{End}_{\mathbb{C}}(V_1 \otimes V_2)$ by

$$\begin{aligned}\mathcal{J}_{V_1,V_2} :=& q^{(\tau\otimes 1)\Omega_0}\big((\tau \otimes 1)\Theta\big) \ldots q^{(\tau^n\otimes 1)\Omega_0}\big((\tau^n \otimes 1)\Theta\big) \\ &\times q^{-(\tau^n\otimes 1)\Omega_0 - \cdots - (\tau\otimes 1)\Omega_0} q^{-s-\Omega_{\mathfrak{z}^\perp}/2}.\end{aligned}$$

THEOREM 7.1 (Yakimov, Etingof-Schedler-Schiffman). *Let $U_q(\mathfrak{g})$ be the quantized universal enveloping algebra of a complex semi-simple Lie algebra $\mathfrak{g}$ and $q \in \mathbb{C}^*$ be a non-root of unity. Let $\pi = (\tau, s)$ be a Belavin-Drinfeld pair for $\mathfrak{g}$. Then, for all $V_1, V_2, V_3 \in \mathcal{C}_{\mathcal{L}}$, the $\mathcal{J}$-operators satisfy the condition*

$$\mathcal{J}_{V_1\otimes V_2, V_3}\mathcal{J}_{V_1,V_2} = \mathcal{J}_{V_1, V_2\otimes V_3}\mathcal{J}_{V_2,V_3}.$$

Define the map $\theta_\pi \colon \mathbb{C}_q[G] \times \mathbb{C}_q[G] \to \mathbb{C}$ by

$$\theta_\pi(c_{\xi_1,v_1}, c_{\xi_2,v_2}) := (\xi_2 \otimes \xi_1)(\mathcal{J}_{V_1,V_2}(v_1 \otimes v_2))$$

for $V_1, V_2 \in \mathcal{C}_{\mathcal{L}}$, $v_i \in V_i$ and $\xi_i \in V_i^*$.

THEOREM 7.2. *The map $\theta_\pi \colon \mathbb{C}_q[G] \times \mathbb{C}_q[G] \to \mathbb{C}$ is a Hopf algebra 2-cocycle on $\mathbb{C}_q[G]$.*

This enables us to define the Belavin-Drinfeld quantum group for the pair π.

DEFINITION 7.3. The Hopf algebra obtained by twisting $\mathbb{C}_q[G]$ by the Hopf 2-cocycle θ_π is denoted

$$\mathbb{C}_{\pi,q}[G] := \mathbb{C}_q[G]_{\theta_\pi}.$$

The Hopf algebra $\mathbb{C}_{\pi,q}[G]$ is an algebraic quantization of the coordinate ring $\mathbb{C}[G]$ of the group G with Poisson structure given by the Belavin–Drinfeld r-matrix r_π. The product in $\mathbb{C}_{\pi,q}[G]$ is given by

$$(c_{\xi_1,v_1}c_{\xi_2,v_2})(x) = (\xi_2 \otimes \xi_1)(\mathcal{J}^{-1}_{V_1,V_2}\Delta(x)\mathcal{J}_{V_1,V_2}(v_1 \otimes v_2))$$

for $v_i \in V_i$ and $\xi_i \in V_i^*$.

8. The FRT dual $U_{\pi,q}(\mathfrak{g})$

It is natural to think that when one twists the multiplication of a quantum group $\mathbb{C}_q[G]$ by a 2-cocyle, then one can obtain the corresponding version of the quantized universal enveloping algebra $U_q(\mathfrak{g})$ by twisting the comultiplication. For instance, this is the situation for the small quantum groups studied by Negron [**25**]. However this approach does not work in this situation and one must instead create the appropriate version of $U_q(\mathfrak{g})$ using the FRT construction. We begin by summarizing this construction in the case of an arbitrary braided Hopf algebra.

Let A be a braided Hopf algebra. The braiding induces a pair of Hopf algebra maps $l^{\pm} : A^{op} \to A^{\circ}$ given by

$$l^{+}(a)(b) = \langle a \mid b \rangle \quad \text{and} \quad l^{-}(a)(b) = \langle b \mid S(a) \rangle$$

Denote by $U^{\pm}$ the images of $l^{\pm}$ respectively. Then $U^{\pm}$ are obviously Hopf subalgebras of A°. The braiding on A induces a Hopf pairing on $U^{+} \otimes (U^{-})^{op}$ given by $\pi(l^{+}(a), l^{-}(b)) = \langle a \mid b \rangle$ using which one can construct the Drinfeld double $U^{+} \bowtie U^{-}$. The multiplication map $u \otimes v \mapsto uv$ is then a Hopf algebra map from $U^{+} \bowtie U^{-}$ to A^{o}.

DEFINITION 8.1. The *FRT dual* of a braided Hopf algebra A is the Hopf algebra $U(A) = U^{+}U^{-}$.

When A is finite dimensional, $U(A_{\sigma})$ can be obtained from $U(A)$ by an analogous twisting of the coalgebra structure on $U(A)$ [**24**]. Thus $U(A_{\sigma})$ and $U(A)$ are isomorphic as algebras. However in the infinite dimensional case, $U(A_{\sigma})$ will generally be quite different from $U(A)$ as an algebra.

DEFINITION 8.2. Let π be a BD-pair. Denote $U(\mathbb{C}_{\pi,q}[G])$ by $U_{\pi,q}(\mathfrak{g})$.

The FRT dual of a braided Hopf algebra is of course most important when A and $U(A)$ form a non-degenerate dual pair. We expect this to happen for all Belavin-Drinfeld quantum groups, but this is still an open question. The problem is essentially equivalent to showing that all simple $\mathbb{C}_{\pi,q}[G]$-comodules are simple as $U_{\pi,q}(\mathfrak{g})$-modules.

CONJECTURE 8.3. *Let π be a Belavin-Drinfeld pair.*

(1) *All finite-dimensional simple $\mathbb{C}_{\pi,q}[G]$-comodules are simple as $U_{\pi,q}(\mathfrak{g})$-modules.*
(2) *$\mathbb{C}_{\pi,q}[G]$ is the restricted dual of $U_{\pi,q}(\mathfrak{g})$ with respect to a subcategory of finite dimensional $U_{\pi,q}(\mathfrak{g})$-modules.*
(3) *The pairing between $\mathbb{C}_{\pi,q}[G]$ and $U_{\pi,q}(\mathfrak{g})$ is non-degenerate.*

EXAMPLE 8.4. These results are proved in [**15**] for the double quantum groups $\mathbb{C}_q[D(G)]$. In [**22**] it is further shown that $U_q(\mathfrak{d}(\mathfrak{g})) \cong U_q(\mathfrak{g}) \bowtie \mathbb{C}_q[G]^{\mathrm{op}}$.

We next want to consider the analog of the triangular decomposition of $U_q(\mathfrak{g})$. In particular, we would like to understand the analog of the Cartan component $U_0 = \mathbb{C}[K_{\lambda}\,;\, \lambda \in \mathcal{L}]$.

For any braided Hopf algebra, we can set

$$U_0(A) = U^{+} \cap U^{-}$$

Then $U_0(A)$ is a Hopf subalgebra of $U(A)$ (when there is no danger of ambiguity we shall denote $U_0(A)$ by U_0). Consider the pairing on U_0 induced from the pairing on A in the following way: if $x = l^+(a)$ and $y = l^-(b)$ are elements of U^0, then

$$(x \mid y) = \langle a \mid S(b) \rangle$$

This pairing is a braiding on U_0 [**16**, Prop 2.1]. Thus the FRT dual $U(A)$ of a braided Hopf algebra A contains a canonical braided Hopf subalgebra $U_0(A)$.

DEFINITION 8.5. Let $(\tau, \Gamma_1, \Gamma_2)$ be a Belavin-Drinfeld triple. Set

$$\tilde{\Gamma} = \Gamma_1, \quad \tilde{\Gamma}_1 = \tau^{-1}(\Gamma_1 \cap \Gamma_2), \quad \tilde{\Gamma}_2 = \Gamma_1 \cap \Gamma_2$$

and

$$\tilde{\tau} = \tau|_{\tilde{\Gamma}} : \tilde{\Gamma}_1 \to \tilde{\Gamma}_2$$

Denote by $\tilde{\mathfrak{g}}$ the semi-simple Lie subalgebra of $\mathfrak{g}$ corresponding to $\tilde{\Gamma}$. Then $(\tilde{\tau}, \tilde{\Gamma}_1, \tilde{\Gamma}_2)$ is a Belavin-Drinfeld triple.

CONJECTURE 8.6. *There is a reductive algebraic group $\hat{G}$ whose Lie algebra has semi-simple component $\tilde{\mathfrak{g}}$ such that $U_0(\mathbb{C}_{\pi,q}[G]) \cong \mathbb{C}_{\tilde{\pi},q}[\hat{G}]$ for some Belavin-Drinfeld pair $\tilde{\pi} = (\tilde{\tau}, \hat{s})$ where $\hat{s} \in \hat{\mathfrak{h}}$.*

Of course, this conjecture assumes an extension of the constructions of Section 7 to the reductive case. This is fairly straightforward and we shall not give the details here.

EXAMPLE 8.7. Consider the case of the double quantum groups described in Section 4.1. In this case $\tilde{\Gamma}$ is the first copy of Γ_0; $\tilde{\Gamma}_1 = \tilde{\Gamma}_2 = \emptyset$, and $\tilde{\tau}$ is the trivial triple. We have an isomorphism $U_q(\mathfrak{g}) \bowtie \mathbb{C}_q[G]^{\mathrm{op}} \to U_q(\mathfrak{d}(\mathfrak{g}))$. It can be shown that U_0 in this case is $U_0(\mathfrak{g}) \bowtie \mathbb{C}_q[G]^{\mathrm{op}}$ which can be thought of as a multi-parameter quantum group $\mathbb{C}_{p,q}[H \times G]$ for an appropriate choice of p.

Next we turn to the dual version of Conjecture 8.6; namely that there is a homomorphic image of $\mathbb{C}_{\pi,q}[G]$ isomorphic to $U_{\tilde{\pi},q}(\hat{\mathfrak{g}})$. We begin with some general observations.

The natural map from A into the dual of $U_0(A)$ yields a homomorphic image of A inside $U_0(A)^o$. In fact, as the following result shows, this image contains the FRT dual of the braided algebra $U_0(A)$.

PROPOSITION 8.8 ([**16**, Theorem 2.2]). *Let A be a braided Hopf algebra and let B be a braided Hopf subalgebra of $U_0(A)$. Let $\phi : A \to B^\circ$ be the natural Hopf algebra map given by $\phi(a)(u) = u(a)$ for $a \in A$ and $u \in B$. Then $\phi(A) \supset U(B)$.*

CONJECTURE 8.9. *The image of $\mathbb{C}_{\pi,q}[G]$ in the dual of $U_0(\mathbb{C}_{\pi,q}[G])$ is isomorphic to $U_{\tilde{\pi},q}(\hat{\mathfrak{g}})$, the FRT dual of $\mathbb{C}_{\tilde{\pi},q}[\hat{G}]$.*

Our usual examples provide good evidence for this conjecture.

EXAMPLE 8.10. In the standard case, $U(\mathbb{C}_q[G]) = U_q(\mathfrak{g})$ and the algebra U_0 is the usual 'Cartan part' U^0 with braiding given by restriction of the Rosso form. So here $\tilde{\mathfrak{g}}$ and $\tilde{\tau}$ are trivial and $\hat{G}$ is just the maximal torus H. The map ϕ : $\mathbb{C}_q[G] \to (U_0)^\circ$ becomes an identification of the undeformed algebra of functions on the maximal torus $\mathbb{C}[H]$ with the FRT dual of U_0.

EXAMPLE 8.11. Continuing with the example of the double quantum group above, the image of $\mathbb{C}_q[D(G)]$ in the dual of $U_0 = U_q(\mathfrak{h}) \bowtie \mathbb{C}_q[G]^{\mathrm{op}}$ is the algebra $U^+ \bowtie U^-$ which can be considered as $U_{p,q}(\mathfrak{g} \oplus \mathfrak{h})$, a multiparameter quantized universal enveloping algebra for the reductive group $\mathfrak{g} \oplus \mathfrak{h}$.

EXAMPLE 8.12. In the Cremmer-Gervais case of $\mathbb{C}_R[SL(n)]$, $\Gamma = A_{n-1}$ and τ is the triple given in Example 3.2. Then $\tilde{\Gamma} = \Gamma_1 = A_{n-2}$ and $\tilde{\tau}$ is just the usual Cremmer-Gervais triple on A_{n-2}. The reductive group $\hat{G}$ is $GL(n-1)$ and [**16**, Section 5] establishes that there is a homomorphism from $\mathbb{C}_{\hat{R}}[GL(n-1)]$ to $U_R(\mathfrak{sl}(n))_0$ for $\hat{R}$ an appropriately chosen multi-parameter Cremmer-Gervais R-matrix. We expect this map to be an isomorphism. This map induces a surjective homomorphism $\phi : \mathbb{C}_R[SL(n)] \to U_{\hat{R}}(\mathfrak{gl}(n-1))$ [**16**, Theorem 5.5] which, we conjecture to be the surjection of Conjecture 8.9.

Notice that this conjecture implies that most or all of our generalized quantized enveloping algebras $U_{\pi,q}(\mathfrak{g})$ occur as homomorphic images of a larger generalized quantized function algebra, so that any descriptions of the algebraic structure of $U_{\pi,q}(\mathfrak{g})$ and $\mathbb{C}_{\pi,q}[G]$ will be intricately intertwined.

The strongest evidence to believe these conjectures is that analogous results are true in the "classical limit"; that is, there is a corresponding result for the associated factorizable Lie bialgebras.

THEOREM 8.13 ([**16**, Theorem 6.4]). *Let $\mathfrak{g}$ be a semi-simple Lie algebra of type Γ. Let $(\tau, \Gamma_1, \Gamma_2)$ be a triple on Γ and let $r \in \mathfrak{g} \otimes \mathfrak{g}$ be a classical r-matrix associated to a Belavin-Drinfeld pair $\pi = (\tau, s)$. Let $\phi_\pm : \mathfrak{g}^* \to \mathfrak{g}$ be the maps give by*

$$\phi_+(\xi) = \sum \xi(r_i) r_i', \quad \phi_-(\xi) = \sum r_i \xi(r_i')$$

and let $\mathfrak{k} = \operatorname{Ker} \phi_+ + \operatorname{Ker} \phi_-$. Then $\mathfrak{k}$ is a Lie biideal of $\mathfrak{g}^*$. The bialgebra $\tilde{\mathfrak{g}} = \mathfrak{g}^*/\mathfrak{k}$ is a reductive Lie bialgebra whose semi-simple part is of type Γ_1 and whose bialgebra structure is given by a Belavin-Drinfeld pair $\tilde{\pi} = (\tilde{\tau}, \hat{s})$ where $\tilde{\tau}$ is the triple given in Definition 8.5.

To emphasize the analog with Conjecture 8.6, note that if $\mathfrak{c}_0 = \operatorname{Im} \phi_+ \cap \operatorname{Im} \phi_-$, then $\mathfrak{c}_0$ is a Lie subbialgebra of $\mathfrak{g}$ whose Lie bialgebra dual is $\mathfrak{g}^*/\mathfrak{k}$. So the quasi-triangular Lie bialgebra $\mathfrak{g}$ contains a canonical co-quasi-triangular subalgebra $\mathfrak{c}_0$.

9. Conclusion

Ultimately one would like to understand the structure of $\mathbb{C}_{\pi,q}[G]$ and $U_{\pi,q}(\mathfrak{g})$ in the same depth as our understanding of the standard quantum groups $\mathbb{C}_q[G]$ and $U_q(\mathfrak{g})$. At the level of finite dimensional representations of $U_{\pi,q}(\mathfrak{g})$ we do not expect anything new. The construction of $\mathbb{C}_{\pi,q}[G]$ by twisting assures that the category of finite dimensional comodules for $\mathbb{C}_{\pi,q}[G]$ is the same as that of $\mathbb{C}_q[G]$; and Conjecture 8.3 implies that this will be the category of finite dimensional representations of $U_{\pi,q}(\mathfrak{g})$. However the similarities are not likely to extend much further. For instance all primitive ideals of $\mathbb{C}_q[G]$ are completely prime. But given Conjecture 8.9, it is likely that this is never true for non-trivial triples. Similarly the primitive ideals of $\mathbb{C}_q[G]$ fall into finitely many orbits under the natural action of the maximal torus H. Again we do not expect any such behavior to hold in the case of a non-trivial triple. Of course, from another point of view this is to be expected because we expect the primitive spectrum to be closely connected to the description of the symplectic leaves of the associated Poisson structure where a similar phenomenon is found [**30**].

Since Joseph's proof of the classification of primitive ideals in $\mathbb{C}_q[G]$ relies heavily on the theory of highest weight modules for $U_q(\mathfrak{g})$, one would like to have an analogous theory for $U_{\pi,q}(\mathfrak{g})$. However this will be significantly more complicated because of the more complex structure of the subalgebra U_0. There are also many interesting questions to ask about the algebraic structure of $U_{\pi,q}(\mathfrak{g})$: What is its center? Are the minimal primitive ideals generated by their intersection with the center? Are all primitive ideals the annihilators of highest weight modules? Do the (completely prime) primitive ideals of $U_{\pi,q}(\mathfrak{g})$ correspond closely to the closures of the symplectic leaves of the dual Poisson group? One might also expect that a more direct construction of $U_{\pi,q}(\mathfrak{g})$ is possible similar to that of $U_q(\mathfrak{g})$ but incorporating the data from the Drinfeld pair π.

In conclusion, we believe that a deeper study of Belavin-Drinfeld quantum groups will reveal an intricate and fascinating structure with many hard but elegant problems which will well reward the work needed to solve them. Our current knowledge of quantum groups should turn out to be but the tip of a more complex iceberg that is noncommutative algebraic group theory.

The author would like to thank Milen Yakimov for many helpful conversations on the topics discussed in the article.

References

[1] J. Balog, L. Dąbrowski, and L. Fehér, *A new quantum deformation of* SL(3), Phys. Lett. B **257** (1991), no. 1-2, 74–78, DOI 10.1016/0370-2693(91)90861-J. MR1100861

[2] J. Balog, L. Dąbrowski, and L. Fehér, *Nonstandard quantum group in Toda and WZNW theories*, Nonperturbative methods in low-dimensional quantum field theories (Debrecen, 1990), World Sci. Publ., River Edge, NJ, 1991, pp. 279–293. MR1191209

[3] A. A. Belavin and V. G. Drinfel′d, *Triangle equations and simple Lie algebras*, Mathematical physics reviews, Vol. 4, Soviet Sci. Rev. Sect. C: Math. Phys. Rev., vol. 4, Harwood Academic Publ., Chur, 1984, pp. 93–165. Translated from the Russian. MR768939

[4] U. Carow-Watamura, M. Schlieker, M. Scholl, and S. Watamura, *A quantum Lorentz group*, Internat. J. Modern Phys. A **6** (1991), no. 17, 3081–3108, DOI 10.1142/S0217751X91001507. MR1113986

[5] Ursula Carow-Watamura and Satoshi Watamura, *Complex quantum group, dual algebra and bicovariant differential calculus*, Comm. Math. Phys. **151** (1993), no. 3, 487–514. MR1207262

[6] Chryssomalis Chryssomalakos, Ralf A. Engeldinger, Branislav Jurčo, Michael Schlieker, and Bruno Zumino, *Complex quantum enveloping algebras as twisted tensor products*, Lett. Math. Phys. **32** (1994), no. 4, 275–281, DOI 10.1007/BF00761138. MR1310292

[7] Eugène Cremmer and Jean-Loup Gervais, *The quantum group structure associated with nonlinearly extended Virasoro algebras*, Comm. Math. Phys. **134** (1990), no. 3, 619–632. MR1086746

[8] Yukio Doi, *Braided bialgebras and quadratic bialgebras*, Comm. Algebra **21** (1993), no. 5, 1731–1749, DOI 10.1080/00927879308824649. MR1213985

[9] Yukio Doi and Mitsuhiro Takeuchi, *Multiplication alteration by two-cocycles—the quantum version*, Comm. Algebra **22** (1994), no. 14, 5715–5732, DOI 10.1080/00927879408825158. MR1298746

[10] Pavel Etingof, Travis Schedler, and Olivier Schiffmann, *Explicit quantization of dynamical r-matrices for finite dimensional semisimple Lie algebras*, J. Amer. Math. Soc. **13** (2000), no. 3, 595–609, DOI 10.1090/S0894-0347-00-00333-7. MR1758755

[11] Murray Gerstenhaber, Anthony Giaquinto, and Samuel D. Schack, *Construction of quantum groups from Belavin-Drinfel′d infinitesimals*, Quantum deformations of algebras and their representations (Ramat-Gan, 1991/1992; Rehovot, 1991/1992), Israel Math. Conf. Proc., vol. 7, Bar-Ilan Univ., Ramat Gan, 1993, pp. 45–64. MR1261900

[12] Timothy J. Hodges and Thierry Levasseur, *Primitive ideals of* $\mathbf{C}_q[\mathrm{SL}(3)]$, Comm. Math. Phys. **156** (1993), no. 3, 581–605. MR1240587

[13] Timothy J. Hodges and Thierry Levasseur, *Primitive ideals of* $\mathbf{C}_q[\mathrm{SL}(n)]$, J. Algebra **168** (1994), no. 2, 455–468, DOI 10.1006/jabr.1994.1239. MR1292775

[14] Timothy J. Hodges, Thierry Levasseur, and Margarita Toro, *Algebraic structure of multiparameter quantum groups*, Adv. Math. **126** (1997), no. 1, 52–92, DOI 10.1006/aima.1996.1612. MR1440253

[15] Timothy J. Hodges, *Double quantum groups and Iwasawa decomposition*, J. Algebra **192** (1997), no. 1, 303–325, DOI 10.1006/jabr.1996.6933. MR1449964

[16] Timothy J. Hodges, *On the Cremmer-Gervais quantizations of* $\mathrm{SL}(n)$, Internat. Math. Res. Notices **10** (1995), 465–481, DOI 10.1155/S107379289500033X. MR1358031

[17] Timothy J. Hodges, *Nonstandard quantum groups associated to certain Belavin-Drinfeld triples*, Perspectives on quantization (South Hadley, MA, 1996), Contemp. Math., vol. 214, Amer. Math. Soc., Providence, RI, 1998, pp. 63–70, DOI 10.1090/conm/214/02905. MR1601221

[18] T. J. Hodges and M. T. Yakimov, *Belavin-Drinfeld quantum groups*, in preparation.

[19] Jens Carsten Jantzen, *Lectures on quantum groups*, Graduate Studies in Mathematics, vol. 6, American Mathematical Society, Providence, RI, 1996, DOI 10.1090/gsm/006. MR1359532

[20] Anthony Joseph, *On the prime and primitive spectra of the algebra of functions on a quantum group*, J. Algebra **169** (1994), no. 2, 441–511, DOI 10.1006/jabr.1994.1294. MR1297159

[21] Anthony Joseph, *Idéaux premiers et primitifs de l'algèbre des fonctions sur un groupe quantique* (French, with English and French summaries), C. R. Acad. Sci. Paris Sér. I Math. **316** (1993), no. 11, 1139–1142. MR1221638

[22] Ulrich Krähmer, *FRT-duals as quantum enveloping algebras*, J. Algebra **264** (2003), no. 1, 68–81, DOI 10.1016/S0021-8693(03)00116-9. MR1980686

[23] George Lusztig, *Introduction to quantum groups*, Progress in Mathematics, vol. 110, Birkhäuser Boston, Inc., Boston, MA, 1993. MR1227098

[24] Shahn Majid, *Tannaka-Kreĭn theorem for quasi-Hopf algebras and other results*, Deformation theory and quantum groups with applications to mathematical physics (Amherst, MA, 1990), Contemp. Math., vol. 134, Amer. Math. Soc., Providence, RI, 1992, pp. 219–232, DOI 10.1090/conm/134/1187289. MR1187289

[25] Cris Negron, *Small quantum groups associated to Belavin-Drinfeld triples*, Trans. Amer. Math. Soc. **371** (2019), no. 8, 5401–5432, DOI 10.1090/tran/7438. MR3937297

[26] P. Podleś and S. L. Woronowicz, *Quantum deformation of Lorentz group*, Comm. Math. Phys. **130** (1990), no. 2, 381–431. MR1059324

[27] N. Reshetikhin, *Multiparameter quantum groups and twisted quasitriangular Hopf algebras*, Lett. Math. Phys. **20** (1990), no. 4, 331–335, DOI 10.1007/BF00626530. MR1077966

[28] N. Yu. Reshetikhin and M. A. Semenov-Tian-Shansky, *Quantum R-matrices and factorization problems*, J. Geom. Phys. **5** (1988), no. 4, 533–550 (1989), DOI 10.1016/0393-0440(88)90018-6. MR1075721

[29] Travis Schedler, *Proof of the GGS conjecture*, Math. Res. Lett. **7** (2000), no. 5-6, 801–826, DOI 10.4310/MRL.2000.v7.n6.a12. MR1809303

[30] Milen Yakimov, *Symplectic leaves of complex reductive Poisson-Lie groups*, Duke Math. J. **112** (2002), no. 3, 453–509, DOI 10.1215/S0012-9074-02-11233-2. MR1896471

Department of Mathematical Sciences, University of Cincinnati, Cincinnati, Ohio 45221-0025

Email address: `hodgest@ucmail.uc.edu`

Contemporary Mathematics
Volume **801**, 2024
https://doi.org/10.1090/conm/801/16085

A survey on Zariski cancellation problems for noncommutative and Poisson algebras

Hongdi Huang, Xin Tang, and Xingting Wang

Abstract. In this article, we discuss some recent developments of the Zariski Cancellation Problem in the setting of noncommutative algebras and Poisson algebras.

1. Introduction

As remarked by Kraft in his 1995 survey [**32**] that "*... there is no doubt that complex affine n-space* $\mathbb{A}^n = \mathbb{A}^n_{\mathbb{C}}$ *is one of the basic objects in algebraic geometry. It is therefore surprising how little is known about its geometry and its symmetries. Although there has been some remarkable progress in the last few years, many basic problems remain open.*"

Among these open questions, features the "Zariski Cancellation Problem". A birational version of the cancellation problem was first raised by Zariski during the 1949 Paris Colloquium on Algebra and Theory of Numbers [**55**]. Throughout this paper, we work over a base field $\Bbbk$. All the algebras are over $\Bbbk$ and we reserve $\otimes$ to mean $\otimes_{\Bbbk}$ unless stated otherwise. Let K and K' be two finitely generated fields over $\Bbbk$. Denote by $K(x)$ and $K'(x)$ the simple transcendental extensions of K and K', respectively. Zariski asked whether the isomorphism $K(x) \cong K'(x)$ implies $K \cong K'$? Over the years, a 'biregular' version of the cancellation problem has attracted a great deal of research attention. More precisely, let $\mathbb{A}^n = \Bbbk^n$ be the affine space. One has the following "Zariski Cancellation Problem" (abbreviated as **ZCP**).

Question 1.1 (Zariski Cancellation Problem). Does an isomorphism

$$Y \times \mathbb{A}^1 \cong \mathbb{A}^{n+1}$$

imply an isomorphism $Y \cong \mathbb{A}^n$, for any affine variety Y?

Algebraically, the Zariski Cancellation Problem asks whether an algebra isomorphism $A[x_1] \cong \Bbbk[x_1, \cdots, x_{n+1}]$ implies that A is isomorphic to $\Bbbk[x_1, \cdots, x_n]$ as $\Bbbk$-algebras. The Zariski Cancellation Problem has been solved for $A = \Bbbk[x]$ in [**1**] and for $A = \Bbbk[x, y]$ in [**21, 49, 52**]. Recently, the Zariski Cancellation Problem has been settled negatively in [**27, 28**] for $\Bbbk[x_1, \cdots, x_n]$ where $n \geq 3$ and $\mathrm{char}(\Bbbk) > 0$.

2020 *Mathematics Subject Classification.* Primary 16P99, 16W99.

The third author was partially supported by Simons collaboration grant #688403 and AFOSR grant FA9550-22-1-0272.

The Zariski Cancellation Problem remains open for $A = \Bbbk[x_1, \cdots, x_n]$ for $n \geq 3$ in the zero characteristic case. The Zariski Cancellation Problem and its analogs have been further studied for commutative domains, see [**16, 17, 18, 20, 21, 31, 46**] for a sample of references. We refer the reader to the excellent survey papers [**29, 30**] for a complete account in the setting of commutative domains.

In this survey article, we would like to discuss some recent developments in the Zariski Cancellation Problem and its generalizations in the setting of noncommutative algebras and Poisson algebras. Note that the study of cancellation properties for noncommutative rings or algebras dates back to the early 1970s, see for example [**1, 4, 10, 15, 19**]. There is a vast literature on the cancellation problem for noncommutative algebras and Poisson algebras; see for instance the incomplete list [**5, 8, 7, 9, 23, 24, 38, 39, 57, 58**].

In general, it is natural to identify noncommutative algebras A with similar cancellation properties. If A is a finitely generated $\Bbbk$-algebra, then we will simply call A an *affine algebra* over $\Bbbk$. We denote by $Z(A)$ the center of A. The following notions are formulated in [**8**].

DEFINITION 1.2 ([**8**, Definition 1.1]). Let A be a $\Bbbk$-algebra.

(1) A is said to be *cancellative* if any $\Bbbk$-algebra isomorphism $A[t] \cong B[t]$ for some $\Bbbk$-algebra B implies that $A \cong B$.
(2) A is said to be *strongly cancellative* if, for each $n \geq 1$, any $\Bbbk$-algebra isomorphism $A[t_1, \cdots, t_n] \cong B[t_1, \cdots, t_n]$ for some algebra B implies that $A \cong B$.
(3) A is said to be *universally cancellative* if, for every affine commutative $\Bbbk$-domain R such that the natural map $\Bbbk \to R \to R/I$ is an isomorphism for some ideal $I \subset R$ and every $\Bbbk$-algebra B, $A \otimes R \cong B \otimes R$ implies that $A \cong B$.

It follows from the definition that "A is universally cancellative" implies "A is strongly cancellative" and "A is strongly cancellative" implies "A is cancellative". Moreover, the size of $Z(A)$ seems to play a significant role in the cancellation problem as shown below.

THEOREM 1.3 ([**8**, Proposition 1.3]). *Let A be a $\Bbbk$-algebra with center being $\Bbbk$. Then A is universally cancellative.*

This is a useful result since there are many noncommutative algebras that have their centers equal to the base field $\Bbbk$ and are thus universally cancellative. We have the following typical example from [**8**]. We refer the reader to [**37**] for the calculation of centers for many classes of noncommutative algebras.

EXAMPLE 1.4 ([**8**, Example 1.4]). (1) Let $\Bbbk$ be a field of characteristic zero and A_n be the nth Weyl algebra over $\Bbbk$. Then $Z(A_n) = \Bbbk$. So A_n is universally cancellative.

(2) Let $\Bbbk$ be a field and $q \in \Bbbk^{\times}$. Let $A = \Bbbk_q[x_1, \cdots, x_n]$ be the skew polynomial ring generated by $x_1, \cdots, x_n$ subject to the relations $x_j x_i = q x_i x_j$ for all $1 \leq i < j \leq n$. If $n \geq 2$ and q is not a root of unity, then $Z(A) = \Bbbk$. So A is universally cancellative.

On the other hand, the centers of many PI noncommutative algebras are strictly larger than the base field $\Bbbk$. Thus novel ideas are needed to address the cancellation problem for noncommutative algebras. We will discuss more about the cancellation problem for noncommutative algebras, with a focus on many important families of Artin–Schelter regular algebras in Section 2, where we will also address the cancellation problem for non-domains. In Section 3, we will discuss certain generalized cancellation problems such as Morita and skew cancellations. Section 4 is devoted to the Zariski Cancellation Problem for Poisson algebras. We will employ the notion of the *Gelfand–Kirillov (GK) dimension* instead of the classical Krull dimension. We refer the reader to [**33**] for the definition and basic properties of the GK-dimension. It is well known that the GK-dimension of any affine commutative $\Bbbk$-algebra is always equal to its Krull dimension.

2. Zariski cancellation for noncommutative algebras

In this section, we will discuss the cancellative properties for noncommutative algebras. Among them is an important family called Artin–Schelter (AS) regular algebras, which are considered noncommutative analogs of polynomial algebras. We refer the reader to [**3**] for the definition and basic properties of AS-regular algebras. AS-regular algebras should serve as important testing examples for the Zariski Cancellation Problem.

There are two crucial ingredients in the treatment of ZCP for noncommutative algebras, namely the *Makar-Limanov invariant* and the *discriminant*. The Makar-Limanov invariant was originally introduced by Makar-Limanov [**46**] and has become a very useful invariant in commutative algebra. It was first used in [**8**] to solve rigidity, automorphism, isomorphism, and embedding problems for several classes of noncommutative algebras. In the following three subsections, we state the cancellativity results according to the GK-dimension of an algebra.

2.1. Definitions. In this subsection, we list some basic definitions and terminologies. We recall the definition of the Makar-Limanov invariant and its variations from [**8**], see also [**47**]. Let A be a $\Bbbk$-algebra and $\mathrm{Der}(A)$ be the collection of all $\Bbbk$-linear derivations of A. Denote by $\mathrm{LND}(A) = \{\delta \in \mathrm{Der}(A) \mid \delta \text{ is locally nilpotent}\}$. When $\mathrm{char}(\Bbbk) > 0$, one needs to use higher derivations. We denote by $\mathrm{LND}^H(A)$ the set of all locally nilpotent higher derivations of A and $\mathrm{LND}^I(A)$ the set of locally nilpotent iterative higher derivations of A. We refer the reader to [**8, 53, 34**] for more details on higher derivations. We have the following definition of rigidity in terms of derivations. Let * be either blank, ^{H}, or ^{I}.

DEFINITION 2.1 ([**8**, Definition 2.3]). Let A be an algebra over $\Bbbk$.

(1) The *Makar-Limanov* invariant* of A is defined to be
$$\mathrm{ML}^*(A) = \bigcap_{\delta \in \mathrm{LND}^*(A)} \ker(\delta)$$
This means that we have the original $\mathrm{ML}(A)$, as well as, $\mathrm{ML}^H(A)$ and $\mathrm{ML}^I(A)$.

(2) We say that A is LND^*-rigid if $\mathrm{ML}^*(A) = A$, or $\mathrm{LND}^*(A) = \{0\}$.

(3) A is called *strongly* LND^*-rigid if $\mathrm{ML}^*(A[t_1, \cdots, t_n]) = A$, for all $n \geq 1$.

The Makar-Limanov invariant of an algebra A plays a critical role in controlling the cancellation property for A. Indeed, we have the following result from [**8**].

THEOREM 2.2 ([**8**, Theorem 3.3]). *Let A be an affine $\Bbbk$-domain of finite GK-dimension. When $*$ is blank we further assume that A contains $\mathbb{Z}$.*

(a) *If A is strongly* LND*-rigid, then A is strongly cancellative.*
(b) *If* $\mathrm{ML}^*(A[t]) = A$, *then A is cancellative.*

Therefore, to establish the cancellation property for an algebra A, it suffices to prove that the algebra A is LND*-rigid. In order to achieve that goal, another invariant called *discriminant* is introduced for any noncommutative algebra that is module-finite over its center.

DEFINITION 2.3 ([**12**, Definition 1.3]). Let A be a $\Bbbk$-algebra that is a finitely generated free module over its center Z of rank w. Let $\{z_1, \ldots, z_w\}$ be a Z-basis of A and let $\mathrm{tr} : A \to Z$ be a trace map. Then the *w-discriminant* of A over Z is defined to be

$$d_w(A/Z) :=_{Z^\times} \det(\mathrm{tr}(z_i z_j))_{w\times w}.$$

It is shown in [**8**] that certain properties such as being dominating and effective of the discriminant control the LND*-rigidity. We refer the reader to [**8, 12, 13**] for the definition of the notions of effectiveness and dominance of an element in an algebra and their relations to automorphism groups.

EXAMPLE 2.4 ([**12**, Example 5.6]). Suppose $\mathrm{char}(\Bbbk) \neq 2$. Consider the following algebra

$$A = \Bbbk\langle x, y\rangle/(x^2y - yx^2, xy^2 - y^2x, x^6 - y^2).$$

A computation shows that the center Z of A is the polynomial ring generated by x^2 and $xy + yx$ and its 4-discriminant is given by $d_4(A/Z) = (xy - yx)^4$ which is both dominating and effective.

2.2. Algebras of GK-dimension one or two. It is well known that the coordinate ring of an affine curve is cancellative [**1**]. In light of this result, one is led to consider the cancellativity for every affine prime $\Bbbk$-algebra of GK-dimension one. Many nice properties of algebras of GK-dimension one are investigated in [**56**], which are useful to characterize the cancellativity. We remark that cancellativity is prevalent in GK-dimensions 1 and 2, while many specific families of algebras are known to be cancellative in higher dimensions [**8, 36**].

For an algebra A of GK-dimension one, some affirmative answers to the Zariski cancellation problem from [**38**] and [**5**] are listed in the following theorem.

THEOREM 2.5. *Let A be an affine $\Bbbk$-algebra of GK-dimension one.*

(a) [**38**, Theorem 0.6] *If $\Bbbk$ is algebraically closed and A is prime, then A is cancellative.*
(b) [**5**, Theorem 1.1] *If $\Bbbk$ is of characteristic zero and A is a domain, then A is cancellative.*

Notice that in the above theorem, for part (a) the authors invoke Tsen's theorem which requires the base field $\Bbbk$ to be algebraically closed. Part (b) is somewhat orthogonal to part (a) since domains of GK-dimension one over algebraically closed fields are commutative by an application of Tsen's theorem to a result of Small and Warfield [**56**]. Hence the only case of part (b) covered by (a) is when the algebra is commutative, which was previously known from the result of Abhyankar-Eakin-Heinzer [**1**].

We remark that the base field $\Bbbk$ plays an important role here, indeed. As mentioned before, $\Bbbk[x_1, \cdots, x_n]$ with $n \geq 3$ is not cancellative when $\mathrm{char}(\Bbbk) > 0$ by [**27, 28**]. When it comes to a noncommutative $\Bbbk$-algebra A, where $\Bbbk$ has positive characteristic and is not algebraically closed, a family of non-cancellative algebras is constructed in [**5**] as below.

EXAMPLE 2.6. Let p be a prime, and let $K = \mathbb{F}_p(x_1, \ldots, x_{p^2-1})$. Let $k = \mathbb{F}_p(x_1^p, \ldots, x_{p^2-1}^p)$ and we let δ be the k-linear derivation of K given by $\delta(x_i) = x_{i+1}$ for $i = 1, \ldots, p^2 - 1$, where we take $x_{p^2} = x_1$. Since k has characteristic $p > 0$, δ^{p^i} is a k-linear derivation for every $i \geq 0$, and since $\delta^{p^2}(x_i) = \delta(x_i) = x_{i+1}$ for $i = 1, \ldots, p^2 - 1$, $\delta^{p^{j+2}} = \delta^{p^j}$ for every $j \geq 0$. We let $\delta' := \delta^p$, which as we have just remarked is a k-linear derivation of K. We let $A = K[x; \delta]$ and $B = K[x'; \delta']$. Since $\mathrm{ad}_u^p = \mathrm{ad}_{u^p}$ for any element u in a ring of characteristic p, we have $z := x^{p^2} - x$ and $z' := (x')^{p^2} - x'$ are central by the above remarks. One can check that A and B have Gelfand-Kirillov dimension one and $A[t] \cong B[t']$, but $A \not\cong B$.

It is worth pointing out that the above examples indicate the necessity of the condition that $\Bbbk$ is algebraically closed in the positive characteristic case. Next, we summarize the results of cancellativity of an algebra having GK-dimension two.

THEOREM 2.7 ([**8**, Theorem 0.5]). *Let $\Bbbk$ be an algebraically closed field of characteristic zero. Let A be an affine $\Bbbk$-domain of GK-dimension two. If A is not commutative, then A is cancellative.*

Note that, indeed, there are examples of commutative affine domains of GK-dimension two that are not cancellative [**18, 20, 21, 31**]. We quote some counterexamples from [**18**] below.

EXAMPLE 2.8.

(a) Let $n \geq 1$ and let B_n be the coordinate ring of the surface $x^n y = z^2 - 1$ over $\mathbb{C}$. Then $B_i \not\cong B_j$ if $i \neq j$, but $B_i[t] \cong B_j[t]$ for all $i, j \geq 1$. Therefore, all the B_n's are not cancellative.
(b) Let A and B be the affine coordinate rings of the surfaces $xy - z^2 + 1 = 0$ and $x^2 y - z^2 + 1 = 0$. We have $A \not\cong B$ while $A[t] \cong B[t]$.

2.3. Algebras with 'small' center or higher GK-dimension. As shown in Theorem 1.3, the center $Z(A)$ of an algebra A provides some control on the cancellation property of A. Intuitively, if A has a smaller center, then A is closer to be cancellative. So it is natural to ask if the cancellativity of the center can imply that of the algebra itself.

QUESTION 2.9 ([**38**, Question 0.1][**5**, Conjecture 2.1]). Let A be an affine noetherian domain. Suppose that $Z(A)$ is affine and cancellative. Is A cancellative? In particular, if the center of A is finite-dimensional over $\Bbbk$, is A cancellative?

In [**38**], cancellation property was proved for non-domain noncommutative algebras using a generalization of ideas from [**8**] relating Makar-Limanov invariant and LND-rigidity to their 'small' centers, that is, the ML_Z-invariant and the notion of LND_Z-rigidity.

THEOREM 2.10. *The following algebras A are cancellative:*

(a) ([**38**, Theorem 0.2]) *A is strongly Hopfian with artinian center.*
(b) ([**38**, Corollary 0.3]) *A is left (or right) artinian; in particular, A is finite-dimensional over a base field $\Bbbk$.*
(c) ([**38**, Corollary 0.3]) *A is the path algebra of a finite quiver.*

Recall that an algebra A is called *Hopfian* if any $\Bbbk$-algebra epimorphism from A to itself is an automorphism. Moreover, we say A is *strongly Hopfian* if $A[t_1, \cdots, t_n]$ is Hopfian for any $n \geq 0$. (see [**38**, Definition 3.4]). As a consequence, the following three large classes of strongly Hopfian algebras are Zariski cancellative once their centers are artinian.

EXAMPLE 2.11 ([**38**, Lemma 3.5]). The following algebras are strongly Hopfian.

(1) Left or right noetherian algebras.
(2) Locally finite $\mathbb{N}$-graded affine $\Bbbk$-algebras.
(3) Prime affine $\Bbbk$-algebras satisfying a polynomial identity.

One may also consider the cancellation problem in the category of graded $\Bbbk$-algebras. An isomorphism lemma is established for graded algebras in [**7**] stating that for any two connected graded $\Bbbk$-algebras that are finitely generated in degree one, if they are isomorphic as ungraded algebras, then they must be isomorphic as graded algebras. As an application, we have the following result.

THEOREM 2.12 ([**7**]). *Let Ω be the class of connected graded $\Bbbk$-algebras finitely generated in degree one. If $A \in \Omega$ satisfies $Z(A) \cap A_1 = \{0\}$, then A is cancellative within Ω, where A_1 is the set of elements of degree one in A.*

The following results are proved in [**10**], which indicate how the center of an algebra affects its cancellativity. The property of *(strongly) retractable* and its weaker version of *(strongly) detectable* are introduced in [**38**] for non-domain noncommutative algebras parallel to the discriminant computation method. Recall that *the nilpotent radical* of a ring R is defined to be the intersection of all prime ideals in R.

THEOREM 2.13. *Let R be a ring with center Z and N be the nilpotent radical of R. In each of the following cases, R is strongly cancellative:*

(a) *Z/N is strongly retractable.*
(b) *Z has Krull dimension zero,*
(c) *Z is a finite direct sum of local rings.*

For arbitrary affine $\Bbbk$-domain of finite GK-dimension, Theorem 2.2 provides a general approach to the Zariski cancellation problem using LND-rigidity. Discriminant computation so far is the most practical way to achieve that goal in the PI case.

THEOREM 2.14 ([**8**, Theorem 5.2]). *Let A be a PI $\Bbbk$-domain. Suppose the w-discriminant of A over its center is effective for some w. Then A is strongly* LND^H*-rigid. As a consequence, A is strongly cancellative.*

The above theorem can be applied to several important families of noncommutative PI-algebras.

EXAMPLE 2.15 ([**8**, Theorem 0.8]). Let A be a finite tensor product of the skew polynomial rings $\Bbbk_q[x_1, \cdots, x_n]$ where n is even and $q \in \Bbbk\setminus\{0, 1\}$ is a root of unity. Then A is LND-rigid. As a consequence, A is cancellative.

In the example of skew polynomial rings $\Bbbk_q[x_1, \cdots, x_n]$, we ask the following question when n is odd.

QUESTION 2.16 ([**57**, Question 0.8]). Let $q \in \Bbbk\setminus\{0, 1\}$ be a root of unity. Is the skew polynomial ring $\Bbbk_q[x_1, \cdots, x_n]$ cancellative for n odd and $n \geq 3$?

Calculating the discriminants is usually difficult. Even if the discriminant is computable, it often lacks the dominating or effective property. Therefore, it is important to search for other means to prove the LND-rigidity for a given algebra. Via the approach of *Nakayama automorphisms*, the cancellation problem is considered for certain special classes of non-PI AS-regular algebras of GK-dimension 3 in [**40**]. In particular, many examples of cancellative algebras A of GK-dimension no greater than 3 are given in [**40**, Corollary 0.9].

The cancellation problem is further considered for noetherian connected graded AS-regular algebras of GK-dimension 3 or higher in [**57**]. The following more general results are proved in [**57**].

THEOREM 2.17 ([**57**, Theorem 0.2 and 0.3]). *Suppose* $\mathrm{char}(\Bbbk) = 0$. *Let A be a noetherian AS-regular algebra generated in degree one. If either of the following sets of hypotheses holds, then A is cancellative:*

(a) $\mathrm{gl.dim}(A) = 3$ *and A is not PI.*
(b) $\mathrm{GKdim}(Z(A)) \leq 1$, *and* $\mathrm{gl.dim}(A/(t)) = \infty$ *for every homogeneous central element $t \in Z(A)$ of positive degree.*

The following conjecture is presented in [**57**].

CONJECTURE 2.18 ([**57**, Conjecture 0.4]). *Suppose* $\mathrm{char}(\Bbbk) = 0$. *Let A be a noetherian finitely generated prime algebra.*

(1) *If* $\mathrm{GKdim}(Z(A)) \leq 1$, *then A is cancellative.*
(2) *If* $\mathrm{GKdim}(A) = 3$ *and A is not PI, then A is cancellative.*

Regarding Theorem 2.17, Conjecture 2.18 is just claiming that the global dimension hypotheses can be dropped if A is finitely generated prime or replaced by $\mathrm{GKdim}(A) = 3$ in part (2). This further leads to the following natural question from the perspective of the global or Krull dimension.

QUESTION 2.19 ([**38**, Question 0.8]). Let A be a $\Bbbk$-algebra of global dimension one (respectively, Krull dimension one). Is then A cancellative?

It is important to mention that, under some natural assumptions, the PI quantized Weyl algebras are shown to be LND^H-rigid and thus strongly cancellative using Poisson geometry [**36**].

3. Morita and skew cancellations

It is natural to consider other generalizations of the Zariski cancellation problem. Indeed, one may consider the categorical versions of the ZCP. The notion of cancellation for abelian categories and derived categories are initiated in [**39**]. For any algebra A, let $M(A)$ denote the category of the right A-modules and $D(A)$ denote the derived category of right A-modules. The following definitions are proposed in [**39**].

DEFINITION 3.1. Let A be any algebra.

(1) [**39**, Definitions 0.1] We say A is *Morita cancellative* if the statement that $M(A[t])$ is equivalent to $M(B[t])$ for another algebra B implies that $M(A)$ is equivalent to $M(B)$.
(2) [**39**, Definitions 0.2] We say A is *derived cancellative* if the statement that $D(A[t])$ is triangulated equivalent to $D(B[t])$ for an algebra B implies that $D(A)$ is triangulated equivalent to $D(B)$.

From Morita theory, any two Morita equivalent algebras must have isomorphic centers. This suggests that many techniques used in the Zariski cancellation problem for non-domain noncommutative algebras based on their centers can be applied to the Morita cancellation problem. For example, according to [**39**, Theorem 0.7] a commutative domain is cancellative if and only if it is Morita cancellative if and only if Z is derived cancellative.

Moreover, a representation-theoretical version of discriminant was introduced in [**39**] in terms of the module category which is well-behaved under Morita equivalence. As a consequence, relative ML_Z-invariant and the notion of LND_Z-rigidity with respect to the center Z was extended to element-wise versions with respect to some representation-theoretical discriminant. Using these new ideas, we have the Morita cancellation property for the following noncommutative algebras

THEOREM 3.2. *The following algebras are Morita cancellative.*

(1) *(*[**39**, Theorem 0.3]*) Every strongly Hopfian algebra with artinian center.*
(2) *(*[**39**, Theorem 0.5]*) The path algebra $\Bbbk Q$ for every finite quiver Q.*
(3) *(*[**39**, Theorem 0.6]*) Every affine prime $\Bbbk$-algebra of GK-dimension one over an algebraically closed field $\Bbbk$.*
(4) *(*[**39**, Corollary 0.8]*) Every non-PI Sklyanin algebra of global dimension three.*

The Morita cancellation is further studied in [**58**], where the Morita version of the universally cancellative property is proposed in [**58**, Definition 0.3] as a natural combination of being Morita cancellative and universally cancellative. As a Morita version of [**8**, Proposition 1.3], one has the following result, which once again demonstrates the control of the center of A over its cancellation property.

THEOREM 3.3 ([**58**, Theorem 0.4]). *Let A be an algebra with center $Z(A)$ being the base field $\Bbbk$. Then A is universally Morita cancellative.*

As both a Morita version and a strengthened version of a partial combination of [**38**, Theorem 4.1] with [**38**, Theorem 4.2], the following result is proved in [**58**]. Note that *strongly Morita cancellation* can be defined analogously as strongly cancellation. We denote the nilradical of an algebra A by $N(A)$.

THEOREM 3.4. [**58**, Theorem 0.5] *Let A be an algebra with center Z such that either Z or $Z/N(Z)$ is strongly retractable (respectively, strongly detectable), then A is strongly Morita cancellative.*

We remark that the hypotheses of being "strongly Hopfian" in the previous results such as [**38**, Theorem 0.2, Theorem 4.2] and [**39**, Theorem 0.3, Lemma 3.6, Theorem 4.2(2), Corollary 4.3, Corollary 7.3] are not needed. So the above theorem provides a method to study the Morita Zariski cancellation problem for noncommutative algebras via their centers (e.g., [**38**, Question 0.1]). Recall that a commutative algebra is called *von Neumann regular* if it is reduced and has Krull dimension zero. One has the following result.

COROLLARY 3.5 ([**58**, Corollary 0.6]). *Let A be an algebra with center Z.*

(1) *If $Z/N(Z)$ is generated by a set of units of $Z/N(Z)$, then Z and A are strongly cancellative and strongly Morita cancellative.*
(2) *If $Z/N(Z)$ is a von Neumann regular algebra, then Z and A are strongly cancellative and strongly Morita cancellative.*

(3) *If Z is a finite direct sum of local algebras, then Z and A are strongly cancellative and strongly Morita cancellative.*

The problem of skew cancellations was considered in [**2**] and has been recently revisited in [**5**, **9**]. Let A be a $\Bbbk$-algebra. Let σ be a $\Bbbk$-algebra automorphism of A and δ be a σ-derivation of A. Then one can form the Ore extension, denoted by $A[t;\sigma,\delta]$, which shares many nice properties with the polynomial extension $A[t]$. An iterated Ore extension of an algebra A is of the form

$$A[t_1;\sigma_1,\delta_1][t_2;\sigma_2,\delta]\cdots[t_n;\sigma_n,\delta_n],$$

where σ_i is an algebra automorphism of $A_{i-1} := A[t_1;\sigma_1,\delta_1]\cdots[t_{i-1};\sigma_{i-1},\delta_{i-1}]$ and δ_i is a σ_i-derivation of A_{i-1}. The reader is referred to [**48**, Chapter 1] for more details, see also [**26**].

DEFINITION 3.6 ([**58**, Definition 0.7 and 0.8]). Let A be an algebra.

(1) We say A is *skew cancellative* if any isomorphism of algebras

$$A[t;\sigma,\delta] \cong A'[t';\sigma',\delta'] \tag{3.6.1}$$

for another algebra A', implies an isomorphism of algebras

$$A \cong A'.$$

(2) We say A is *σ-cancellative* if it is skew cancellative for Ore extensions (3.6.1) with $\delta = 0$ and $\delta' = 0$.

(3) We say A is *δ-cancellative* if it is skew cancellative for Ore extensions (3.6.1) with $\sigma = \mathrm{Id}_A$ and $\sigma' = \mathrm{Id}_{A'}$.

(4) We say A is *σ-algebraically cancellative* if it is skew cancellative for Ore extensions (3.6.1) with locally algebraic σ and σ'.

Moreover, we can define A to be *strongly skew cancellative* if

$$A[t_1;\sigma_1,\delta_1][t_2;\sigma_2,\delta_2]\cdots[t_n;\sigma_n,\delta_n] \cong A'[t_1';\sigma_1',\delta_1'][t_2';\sigma_2',\delta_2']\cdots[t_n';\sigma_n',\delta_n']$$

implies that $A \cong A'$ for any $n \geq 1$.

Skew cancellation has appeared earlier in different notions. Given two Ore extensions $A[t;\delta_1]$ and $B[t;\delta_2]$, A is said to be *Ore invariant* if the isomorphism $A[t;\delta_1] \cong B[t;\delta_2]$ implies $A \cong B$. When, moreover, the isomorphism between the Ore extensions carries A onto B, then A is said to be *strongly Ore invariant* [**2**]. In [**2**], it is proved that certain abelian regular rings are strongly Ore invariant, and regular self-injective PI rings with no $\mathbb{Z}$-torsion are Ore invariant.

In [**9**], it is proved that if the skew polynomial ring $R[t;\delta_1]$ and $\Bbbk[x][t;\delta_2]$ are isomorphic and $\delta_2(x) \in \Bbbk[x]$ has degree at least one, then $R \cong \Bbbk[x]$. This result is generalized in [**5**] as follows.

THEOREM 3.7 ([**5**, Theorem 1.2]). *Let A be an affine commutative $\Bbbk$-domain of Krull dimension one. Then we have the following.*

(1) *A is σ-cancellative.*

(2) *If $\Bbbk$ has characteristic zero, then A is δ-cancellative.*

However, the question of whether the cancellation property holds for any skew polynomial extensions of mixed type with coefficient rings being domains of Krull dimension one remains as an unresolved one in [**5**].

QUESTION 3.8 ([**5**, Question 5.7]). Let R be an affine commutative $\Bbbk$-domain of Krull dimension one. Is any Ore extension $R[x;\sigma,\delta]$ skew cancellative?

Skew cancellation is further studied in [**58**] in terms of iterated Ore extensions and Ore extensions of special types. Moreover, an automorphism σ of A is called *locally algebraic* if every finite-dimensional subspace of A is contained in a σ-stable finite dimensional subspace of A [**61**]. It is obvious that the identity map is locally algebraic.

There are two new tools used in the skew cancellation problem. The first tool is the notion of divisor subalgebra, which was introduced in [**14**] to solve Zariski cancellation and automorphism problems for certain noncommutative algebras. Let A be a $\Bbbk$-domain, and F be any subset of A. We denote by $Sw(F)$ the set of $g \in A$ such that $f = agb$ for some $a, b \in A$ and $0 \neq f \in F$. That is, $Sw(F)$ is the set consisting of all the subwords of the elements in F. Let us set $D_0(F) = F$ and inductively define $D_n(F)$ for $n \geq 1$ as the $\Bbbk$-subalgebra of A generated by $Sw(D_{n-1}(F))$. The subalgebra $\mathbb{D}(F) = \bigcup_{n\geq 0} D_n(F)$ is called the *divisor subalgebra* of A generated by F. Any nonzero element $f \in A$ is called *controlling* if $\mathbb{D}(\{f\}) = A$; see [**58**, Examples] for controlling elements for various examples of noncommutative affine domains.

THEOREM 3.9 ([**58**, Theorem 0.9]). *Let A be an affine $\Bbbk$-domain of finite GK-dimension. Suppose the unit element 1 is controlling. Then A is strongly σ-algebraically cancellative. As a consequence, A is strongly δ-cancellative.*

The second tool relies on a structural result of division algebras. Recall that a simple artinian ring S is called *stratiform* over $\Bbbk$ [**54**] if there is a chain of simple artinian rings

$$S = S_n \supseteq S_{n-1} \supseteq \cdots \supseteq S_1 \supseteq S_0 = \Bbbk$$

where, for every i, either

(i) S_{i+1} is finite over S_i on both sides; or
(ii) S_{i+1} is equal to the quotient ring of the Ore extension $S_i[t_i; \sigma_i, \delta_i]$ for an automorphism σ_i of S_i and σ_i-derivation δ_i of S_i.

Such a chain of simple artinian rings is called a stratification of S. The *stratiform length* of S is the number of steps in the chain that are of type (ii). A crucial result proved in [**54**] is that the stratiform length is an invariant of S. Moreover, a Goldie prime ring A is called *stratiform* if its quotient division ring, denoted by $Q(A)$, is stratiform. Many stratiform noetherian domains are proved to be skew cancellative.

THEOREM 3.10 ([**58**, Theorem 0.10]). *Let A be a noetherian domain that is stratiform. Suppose the unit element 1 is controlling. Then A is strongly skew cancellative in the category of noetherian stratiform domains.*

At the end of [**58**], motivated by the results [**5**, Lemma 4.3, Proposition 5.6], it is established that every LND-rigid algebra is δ-cancellative. The following is the detailed statement.

THEOREM 3.11 ([**58**, Theorem 5.4]). *Suppose that $\Bbbk$ is a field of characteristic zero. Let A be an affine $\Bbbk$-domain of finite GK-dimension. Suppose that* $\mathrm{ML}(A) = A$. *Then A is δ-cancellative.*

We end this section by stating the following question from [**58**].

QUESTION 3.12 ([**58**, Question 5.6]). Suppose that A is an affine domain of finite GK-dimension over a base field $\Bbbk$ of characteristic zero. Suppose either $\mathrm{ML}(A) = A$ or A is strongly LND-rigid. Is A strongly δ-cancellative?

4. Zariski cancellation for Poisson algebras

Zariski cancellation problems can be asked for different types of associative and non-associative algebras. In this section, we focus on the Zariski cancellation problem for Poisson algebras.

The notion of the Poisson bracket, first introduced by Siméon Denis Poisson, arises naturally in Hamiltonian mechanics and differential geometry.

DEFINITION 4.1. A *Poisson algebra* is a commutative $\Bbbk$-algebra A together with a bilinear form $\{-,-\} : A \times A \to A$ that is both a Lie bracket and a biderivation.

Poisson algebras have recently been studied intensively by many researchers with topics related to noncommutative discriminant [**11**, **50**], representation theory of PI Sklyanin algebras [**59**, **60**], Poisson Dixmier-Moeglin equivalences [**6**, **35**, **45**], Poisson enveloping algebras [**41**, **43**, **42**, **44**], invariant theory [**22**] and so on. Moreover, the following question was asked in [**23**] regarding the Zariski cancellation property.

QUESTION 4.2 (Zariski cancellation problem for Poisson algebras). When is a Poisson algebra A cancellative? That is, does an isomorphism of Poisson algebras $A[t] \cong B[t]$ for another Poisson algebra B imply an isomorphism $A \cong B$ as Poisson algebras?

We first restrict our attention to graded Poisson algebras. The main result below can be viewed as a Poisson version of [**7**, Theorem 3.1]. Recall that any Poisson algebra A, the *Poisson center* of A is denoted by $\mathcal{Z}_P(A) := \{a \in A \,|\, \{a,-\} = 0\}$.

THEOREM 4.3 ([**23**, Theorem 4.5]). *Let A and B be two connected graded Poisson algebras finitely generated in degree one. Suppose either $\mathcal{Z}_P(A)$ or $\mathcal{Z}_P(B)$ is generated in degree at least 2. If $A[t_1, \dots, t_n] \cong B[t_1, \dots, t_n]$ as ungraded Poisson algebras for some $n \geq 1$, then $A \cong B$ as connected graded Poisson algebras.*

The following result shows a deep connection between the Poisson center and Poisson cancellation.

THEOREM 4.4. *Let A be a Poisson algebra.*

(a) *([**23**, Corollary 5.4]) If A is noetherian with artinian Poisson center, then A is Poisson cancellative.*
(b) *([**23**, Theorem 5.5]) If A has trivial Poisson center, then A is Poisson cancellative.*

Theorem 4.4 (b) can be applied to show that Poisson integral domains of Krull dimension two which have nontrivial Poisson brackets are Poisson cancellative ([**23**, Corollary 5.6]). Here the non-triviality of the Poisson bracket plays an essential role since by [**18**, **20**] there are commutative domains of Krull dimension two that are not cancellative.

It is important to point out that many theories developed in the (associative) noncommutative setting [**8**, **7**] can be adapted to the Poisson setting. Hence the main work of [**23**] is to modify the ideas of the noncommutative discriminant [**12**, **13**] and the Makar-Limanov invariant [**47**] into the setting of Poisson algebras.

DEFINITION 4.5. Let A be a Poisson algebra. A derivation δ of A is called a *Poisson derivation* if

$$\delta(\{a,b\}) \;=\; \{\delta(a),b\} + \{a,\delta(b)\} \text{ for all } a,b \in A.$$

The *Poisson Makar-Limanov invariant* of A is defined to be

$$\mathrm{PML}(A) = \bigcap_{\delta \in \mathrm{PLND}(A)} \ker(\delta),$$

where $\mathrm{PLND}(A)$ denotes the space of all locally nilpotent Poisson derivations of A.

THEOREM 4.6 ([**23**, Theorem 6.12]). *Assume $\Bbbk$ is a field of characteristic zero. Let A be an affine Poisson domain over $\Bbbk$ with finite Krull dimension. If A has no nontrivial locally nilpotent Poisson derivations, then A is Poisson cancellative.*

In positive characteristic, we can prove a similar result by replacing Poisson derivations by higher Poisson derivations introduced in [**34**].

It is important to mention that for Poisson algebras in characteristic zero, their Poisson centers are usually not large enough for us to emulate the definition of discriminant for noncommutative algebras by simply replacing the algebraic center with Poisson center. The following definition used the idea in [**39**, §2] to introduce the notion of Poisson discriminant from a representation-theoretic point of view.

DEFINITION 4.7 ([**23**, Definition 7.1]). Let A be a Poisson algebra and let $\mathcal{Z}_P = \mathcal{Z}_P(A)$ be its Poisson center. Let $\mathcal{P}$ be a property defined for Poisson algebras that is invariant under isomorphism classes of Poisson algebras. We define the following terms for sets/ideals in A.

(a) ($\mathcal{P}$-locus) $L_{\mathcal{P}}(A) := \{\mathfrak{m} \in \mathrm{Maxspec}(\mathcal{Z}_P) : A/\mathfrak{m}A \text{ has property } \mathcal{P}\}$.
(b) ($\mathcal{P}$-discriminant set) $D_{\mathcal{P}}(A) := \mathrm{Maxspec}(\mathcal{Z}_P) \backslash L_{\mathcal{P}}(A)$.
(c) ($\mathcal{P}$-discriminant ideal) $I_{\mathcal{P}}(A) := \bigcap_{\mathfrak{m} \in D_{\mathcal{P}}(A)} \mathfrak{m} \subset \mathcal{Z}_P$.

In the case that $I_{\mathcal{P}}(A)$ is a principal ideal, generated by $d \in \mathcal{Z}_P$, then d is called the *$\mathcal{P}$-discriminant* of A, denoted by $d_{\mathcal{P}}(A)$. Observe that, if $\mathcal{Z}_P$ is a domain, $d_{\mathcal{P}}(A)$ is unique up to an element of $\mathcal{Z}_P^{\times}$.

There is also a notion of effectiveness for Poisson discriminant. As in the noncommutative setting, effectiveness controls the locally nilpotent Poisson derivations and hence yields Poisson cancellation.

THEOREM 4.8 ([**23**, Theorem 7.16]). *Let A be an affine Poisson domain with affine Poisson center. If the Poisson discriminant exists and is effective either in A or its Poisson center, then A is Poisson cancellative.*

Next, we turn to the cancellation of quadratic Poisson polynomial algebras in three variables in comparison with their quantizations as AS-regular algebras of dimension three. By analyzing the Poisson center of an arbitrarily connected graded Poisson domain, it was shown that Poisson polynomial algebras in three variables with Poisson bracket either being quadratic or derived from a Lie algebra are cancellative.

THEOREM 4.9. *Let $\Bbbk$ be a base field that is algebraically closed of characteristic zero.*

(a) *(*[**24**, Theorem 3.11] *and* [**24**, Corollary 3.14]*) Let $A = \Bbbk[x, y, z]$ be a quadratic Poisson algebra with nontrivial Poisson bracket. Then the dth Veronese Poisson subalgebra $A^{(d)}$ is Poisson cancellative for any $d \geq 1$. In particular, A is Poisson cancellative.*
(b) *(*[**24**, Theorem 3.15]*) Let $\mathfrak{g}$ be a non-abelian Lie algebra of dimension ≤ 3. Then the symmetric algebra $S(\mathfrak{g})$ on $\mathfrak{g}$ together with the Konstant-Kirillov bracket is Poisson cancellative.*

As the Ore extensions of rings, there is a Poisson version of the extension of an arbitrary Poisson algebra, called Poisson-Ore extension, using a Poisson derivation α and a Poisson α-derivation δ of the base Poisson algebra. In [**24**], the *skew Zariski cancellation problem* was introduced for Poisson algebras in terms of any Poisson-Ore extension. That is, it asks whether or not an isomorphism between the Poisson-Ore extensions of two base Poisson algebras implies an isomorphism between the two base Poisson algebras. Like in the noncommutative algebraic setting [**5, 9, 58**], similar invariants including the Poisson Makar-Limanov invariant, the divisor Poisson subalgebra, and the Poisson stratiform length can be defined for Poisson algebras, which play important roles in their skew cancellation problem.

THEOREM 4.10. *Let* $\Bbbk$ *be a base field of characteristic zero, and* A *be a noetherian Poisson domain.*

(a) *(*[**24**, Theorem 4.4]*) If* A *is either Poisson simple of finite Krull dimension with units in* $\Bbbk$ *or affine of Krull dimension one, then* A *is Poisson* α*-cancellative.*
(b) *(*[**24**, Theorem 4.5]*) If the Poisson Makar-Limanov invariant of* A *equals* A*, then* A *is Poisson* δ*-cancellative.*
(c) *(*[**24**, Theorem 4.8]*) If* A *has finite Krull dimension and the* 1*-divisor Poisson subalgebra of* A *equals* A*, then* A *is Poisson skew cancellative.*
(d) *(*[**24**, Theorem 4.15]*) If* A *is Poisson stratiform and the* 1*-divisor Poisson subalgebra of* A *equals* A*, then* A *is Poisson skew cancellative in the category of noetherian Poisson stratiform domains.*

In the last part of this section, we will discuss some open questions regarding the Poisson cancellation problem. (see [**23**, §8] and [**24**, §5] for more questions).

For any Poisson algebra A, there is a notion of Poisson universal enveloping algebra $U(A)$ (see [**51**]), whose representation category is Morita equivalent to the category of Poisson modules over A.

QUESTION 4.11 ([**23**, Question 8.7]). Let A be any affine Poisson algebra, and $U(A)$ be its Poisson universal enveloping algebra. What is the relationship between $U(A)$ being cancellative and A being Poisson cancellative?

In practice, many Poisson structures can be derived from the process of semi-classical limits of quantized coordinate rings, for instance, see [**25**].

QUESTION 4.12 ([**23**, Question 8.8]). What is the relation between the cancellation property of a quantized coordinate ring and the Poisson cancellation property of its semiclassical limit?

Recall that, in [**39**, Definition 0.2], a notion of *derived cancellation* is further introduced for any associative algebra by replacing the respective module category with its derived category. We are interested in its Poisson version.

QUESTION 4.13 ([**24**, Question 5.4]). Can one define a suitable notion of *derived Poisson cancellation*? Under what conditions does a Poisson algebra satisfy this condition?

Acknowledgments

The authors thank the referee for carefully reading the paper and providing many thoughtful comments and also thank Milen Yakimov for valuable correspondence on the subject. Part of this paper was formulated during the conference "Recent Advances and New Directions in the Interplay of Noncommutative Algebra and Geometry" in June 2022 and the BIRS Workshop on "Noncommutative Geometry and Noncommutative Invariant Theory" in September 2022. The authors want to thank the University of Washington and Banff International Research Station for their hospitality and support.

References

[1] Shreeram S. Abhyankar, William Heinzer, and Paul Eakin, *On the uniqueness of the coefficient ring in a polynomial ring*, J. Algebra **23** (1972), 310–342, DOI 10.1016/0021-8693(72)90134-2. MR306173

[2] Efraim P. Armendariz, Hyeng Keun Koo, and Jae Keol Park, *Isomorphic Ore extensions*, Comm. Algebra **15** (1987), no. 12, 2633–2652, DOI 10.1080/00927878708823556. MR917758

[3] Michael Artin and William F. Schelter, *Graded algebras of global dimension* 3, Adv. in Math. **66** (1987), no. 2, 171–216, DOI 10.1016/0001-8708(87)90034-X. MR917738

[4] Teruo Asanuma, *On strongly invariant coefficient rings*, Osaka Math. J. **11** (1974), 587–593. MR396541

[5] Jason Bell, Maryam Hamidizadeh, Hongdi Huang, and Helbert Venegas, *Noncommutative analogues of a cancellation theorem of Abhyankar, Eakin, and Heinzer*, Beitr. Algebra Geom. **62** (2021), no. 2, 295–315, DOI 10.1007/s13366-020-00518-3. MR4254621

[6] Jason Bell, Stéphane Launois, Omar León Sánchez, and Rahim Moosa, *Poisson algebras via model theory and differential-algebraic geometry*, J. Eur. Math. Soc. (JEMS) **19** (2017), no. 7, 2019–2049, DOI 10.4171/JEMS/712. MR3656478

[7] Jason Bell and James J. Zhang, *An isomorphism lemma for graded rings*, Proc. Amer. Math. Soc. **145** (2017), no. 3, 989–994, DOI 10.1090/proc/13276. MR3589298

[8] Jason Bell and James J. Zhang, *Zariski cancellation problem for noncommutative algebras*, Selecta Math. (N.S.) **23** (2017), no. 3, 1709–1737, DOI 10.1007/s00029-017-0317-7. MR3663593

[9] Jeffrey Bergen, *Cancellation in skew polynomial rings*, Comm. Algebra **46** (2018), no. 2, 705–707, DOI 10.1080/00927872.2017.1327060. MR3764890

[10] J. W. Brewer and E. A. Rutter, *Isomorphic polynomial rings*, Arch. Math. (Basel) **23** (1972), 484–488, DOI 10.1007/BF01304919. MR320068

[11] K. A. Brown and M. T. Yakimov, *Azumaya loci and discriminant ideals of PI algebras*, Adv. Math. **340** (2018), 1219–1255, DOI 10.1016/j.aim.2018.10.024. MR3886192

[12] S. Ceken, J. H. Palmieri, Y.-H. Wang, and J. J. Zhang, *The discriminant controls automorphism groups of noncommutative algebras*, Adv. Math. **269** (2015), 551–584, DOI 10.1016/j.aim.2014.10.018. MR3281142

[13] S. Ceken, J. H. Palmieri, Y.-H. Wang, and J. J. Zhang, *The discriminant criterion and automorphism groups of quantized algebras*, Adv. Math. **286** (2016), 754–801, DOI 10.1016/j.aim.2015.09.024. MR3415697

[14] Kenneth Chan, Alexander A. Young, and James J. Zhang, *Discriminant formulas and applications*, Algebra Number Theory **10** (2016), no. 3, 557–596, DOI 10.2140/ant.2016.10.557. MR3513131

[15] D. B. Coleman and E. E. Enochs, *Isomorphic polynomial rings*, Proc. Amer. Math. Soc. **27** (1971), 247–252, DOI 10.2307/2036302. MR272805

[16] Anthony J. Crachiola, *Cancellation for two-dimensional unique factorization domains*, J. Pure Appl. Algebra **213** (2009), no. 9, 1735–1738, DOI 10.1016/j.jpaa.2009.01.003. MR2518172

[17] A. Crachiola and L. Makar-Limanov, *On the rigidity of small domains*, J. Algebra **284** (2005), no. 1, 1–12, DOI 10.1016/j.jalgebra.2004.09.015. MR2115001

[18] W. Danielewski, *On the cancellation problem and automorphism groups of affine algebraic varieties*, preprint, Warsaw, 1989.

[19] Paul Eakin and William Heinzer, *A cancellation problem for rings*, Conference on Commutative Algebra (Univ. Kansas, Lawrence, Kan., 1972), Lecture Notes in Math., Vol. 311, Springer, Berlin-New York, 1973, pp. 61–77. MR349664

[20] Karl-Heinz Fieseler, *On complex affine surfaces with* $\mathbf{C}^+$*-action*, Comment. Math. Helv. **69** (1994), no. 1, 5–27, DOI 10.1007/BF02564471. MR1259603

[21] Takao Fujita, *On Zariski problem*, Proc. Japan Acad. Ser. A Math. Sci. **55** (1979), no. 3, 106–110. MR531454

[22] Jason Gaddis, Padmini Veerapen, and Xingting Wang, *Reflection groups and rigidity of quadratic Poisson algebras*, Algebr. Represent. Theory **26** (2023), no. 2, 329–358, DOI 10.1007/s10468-021-10096-0. MR4568848

[23] Jason Gaddis and Xingting Wang, *The Zariski cancellation problem for Poisson algebras*, J. Lond. Math. Soc. (2) **101** (2020), no. 3, 1250–1279, DOI 10.1112/jlms.12303. MR4111940

[24] Jason Gaddis, Xingting Wang, and Daniel Yee, *Cancellation and skew cancellation for Poisson algebras*, Math. Z. **301** (2022), no. 4, 3503–3523, DOI 10.1007/s00209-022-03026-3. MR4449718

[25] K. R. Goodearl, *Semiclassical limits of quantized coordinate rings*, Advances in ring theory, Trends Math., Birkhäuser/Springer Basel AG, Basel, 2010, pp. 165–204, DOI 10.1007/978-3-0346-0286-0_12. MR2664671

[26] K. R. Goodearl and R. B. Warfield Jr., *An introduction to noncommutative Noetherian rings*, 2nd ed., London Mathematical Society Student Texts, vol. 61, Cambridge University Press, Cambridge, 2004, DOI 10.1017/CBO9780511841699. MR2080008

[27] Neena Gupta, *On the cancellation problem for the affine space* $\mathbb{A}^3$ *in characteristic* p, Invent. Math. **195** (2014), no. 1, 279–288, DOI 10.1007/s00222-013-0455-2. MR3148104

[28] Neena Gupta, *On Zariski's cancellation problem in positive characteristic*, Adv. Math. **264** (2014), 296–307, DOI 10.1016/j.aim.2014.07.012. MR3250286

[29] Neena Gupta, *A survey on Zariski cancellation problem*, Indian J. Pure Appl. Math. **46** (2015), no. 6, 865–877, DOI 10.1007/s13226-015-0154-3. MR3438041

[30] N. Gupta, *The Zariski Cancellation Problem and related problems in Affine Algebraic Geometry*, To appear in *Proceedings ICM 2022*, `arXiv:2208.14736`, 2022.

[31] M. Hochster, *Nonuniqueness of coefficient rings in a polynomial ring*, Proc. Amer. Math. Soc. **34** (1972), 81–82, DOI 10.2307/2037901. MR294325

[32] Hanspeter Kraft, *Challenging problems on affine n-space*, Astérisque **237** (1996), Exp. No. 802, 5, 295–317. Séminaire Bourbaki, Vol. 1994/95. MR1423629

[33] G. R. Krause and T. H. Lenagan, *Growth of algebras and Gel′fand-Kirillov dimension*, Research Notes in Mathematics, vol. 116, Pitman (Advanced Publishing Program), Boston, MA, 1985. MR781129

[34] Stéphane Launois and César Lecoutre, *A quadratic Poisson Gel'fand-Kirillov problem in prime characteristic*, Trans. Amer. Math. Soc. **368** (2016), no. 2, 755–785, DOI 10.1090/tran/6352. MR3430349

[35] Stéphane Launois and Omar León Sánchez, *On the Dixmier-Moeglin equivalence for Poisson-Hopf algebras*, Adv. Math. **346** (2019), 48–69, DOI 10.1016/j.aim.2019.01.036. MR3904282

[36] Jesse Levitt and Milen Yakimov, *Quantized Weyl algebras at roots of unity*, Israel J. Math. **225** (2018), no. 2, 681–719, DOI 10.1007/s11856-018-1675-3. MR3805662

[37] Oswaldo Lezama and Helbert Venegas, *Center of skew PBW extensions*, Internat. J. Algebra Comput. **30** (2020), no. 8, 1625–1650, DOI 10.1142/S0218196720500575. MR4177582

[38] O. Lezama, Y.-H. Wang, and J. J. Zhang, *Zariski cancellation problem for non-domain noncommutative algebras*, Math. Z. **292** (2019), no. 3-4, 1269–1290, DOI 10.1007/s00209-018-2153-7. MR3980292

[39] D.-M. Lu, Q.-S. Wu, and J. J. Zhang, *A Morita cancellation problem*, Canad. J. Math. **72** (2020), no. 3, 708–731, DOI 10.4153/s0008414x1900004x. MR4098598

[40] J.-F. Lü, X.-F. Mao, and J. J. Zhang, *Nakayama automorphism and applications*, Trans. Amer. Math. Soc. **369** (2017), no. 4, 2425–2460, DOI 10.1090/tran/6718. MR3592516

[41] Jiafeng Lü, Sei-Qwon Oh, Xingting Wang, and Xiaolan Yu, *Enveloping algebras of double Poisson-Ore extensions*, Comm. Algebra **46** (2018), no. 11, 4891–4904, DOI 10.1080/00927872.2018.1459643. MR3864270

[42] Jiafeng Lü, Xingting Wang, and Guangbin Zhuang, *Universal enveloping algebras of Poisson Hopf algebras*, J. Algebra **426** (2015), 92–136, DOI 10.1016/j.jalgebra.2014.12.010. MR3301903

[43] Jiafeng Lü, Xingting Wang, and Guangbin Zhuang, *Universal enveloping algebras of Poisson Ore extensions*, Proc. Amer. Math. Soc. **143** (2015), no. 11, 4633–4645, DOI 10.1090/S0002-9939-2015-12631-7. MR3391023

[44] Jiafeng Lü, Xingting Wang, and Guangbin Zhuang, *Homological unimodularity and Calabi-Yau condition for Poisson algebras*, Lett. Math. Phys. **107** (2017), no. 9, 1715–1740, DOI 10.1007/s11005-017-0967-6. MR3687261

[45] Juan Luo, Xingting Wang, and Quanshui Wu, *Poisson Dixmier-Moeglin equivalence from a topological point of view*, Israel J. Math. **243** (2021), no. 1, 103–139, DOI 10.1007/s11856-021-2154-9. MR4299143

[46] L. Makar-Limanov, *On the hypersurface $x + x^2y + z^2 + t^3 = 0$ in $\mathbf{C}^4$ or a $\mathbf{C}^3$-like threefold which is not $\mathbf{C}^3$*. part B, Israel J. Math. **96** (1996), no. part B, 419–429, DOI 10.1007/BF02937314. MR1433698

[47] L. Makar-Limanov, *Locally nilpotent derivations, a new ring invariant and applications*, preprint, pages 1–60, 1998.

[48] J. C. McConnell and J. C. Robson, *Noncommutative Noetherian rings*, Revised edition, Graduate Studies in Mathematics, vol. 30, American Mathematical Society, Providence, RI, 2001. With the cooperation of L. W. Small, DOI 10.1090/gsm/030. MR1811901

[49] Masayoshi Miyanishi and Tohru Sugie, *Affine surfaces containing cylinderlike open sets*, J. Math. Kyoto Univ. **20** (1980), no. 1, 11–42, DOI 10.1215/kjm/1250522319. MR564667

[50] Bach Nguyen, Kurt Trampel, and Milen Yakimov, *Noncommutative discriminants via Poisson primes*, Adv. Math. **322** (2017), 269–307, DOI 10.1016/j.aim.2017.10.018. MR3720799

[51] Sei-Qwon Oh, *Poisson enveloping algebras*, Comm. Algebra **27** (1999), no. 5, 2181–2186, DOI 10.1080/00927879908826556. MR1683858

[52] Peter Russell, *On affine-ruled rational surfaces*, Math. Ann. **255** (1981), no. 3, 287–302, DOI 10.1007/BF01450704. MR615851

[53] F. K. Schmidt and H. Hasse, *Noch eine Begründung der Theorie der höheren Differentialquotienten in einem algebraischen Funktionenkörper einer Unbestimmten. (Nach einer brieflichen Mitteilung von F.K. Schmidt in Jena)* (German), J. Reine Angew. Math. **177** (1937), 215–237, DOI 10.1515/crll.1937.177.215. MR1581557

[54] A. H. Schofield, *Stratiform simple Artinian rings*, Proc. London Math. Soc. (3) **53** (1986), no. 2, 267–287, DOI 10.1112/plms/s3-53.2.267. MR850221

[55] B. Segre, *Sur un probléme de M. Zariski*, Algébre et Théorie des Nombres, Colloques Internationaux du Centre National de la Recherche Scientifique, no. 24, pages 135–138. Centre National de la Recherche Scientifique, Paris, 1950.

[56] L. W. Small and R. B. Warfield Jr., *Prime affine algebras of Gel′fand-Kirillov dimension one*, J. Algebra **91** (1984), no. 2, 386–389, DOI 10.1016/0021-8693(84)90110-8. MR769581

[57] Xin Tang, Helbert J. Venegas Ramírez, and James J. Zhang, *Cancellation problem for AS-regular algebras of dimension three*, Pacific J. Math. **312** (2021), no. 1, 233–255, DOI 10.2140/pjm.2021.312.233. MR4298801

[58] Xin Tang, James J. Zhang, and Xiangui Zhao, *Cancellation of Morita and skew types*, Israel J. Math. **244** (2021), no. 1, 467–500, DOI 10.1007/s11856-021-2199-9. MR4344034

[59] Chelsea Walton, Xingting Wang, and Milen Yakimov, *Poisson geometry of PI three-dimensional Sklyanin algebras*, Proc. Lond. Math. Soc. (3) **118** (2019), no. 6, 1471–1500, DOI 10.1112/plms.12220. MR3957827

[60] Chelsea Walton, Xingting Wang, and Milen Yakimov, *Poisson geometry and representations of PI 4-dimensional Sklyanin algebras*, Selecta Math. (N.S.) **27** (2021), no. 5, Paper No. 99, 60, DOI 10.1007/s00029-021-00713-7. MR4323328

[61] James J. Zhang, *A note on GK dimension of skew polynomial extensions*, Proc. Amer. Math. Soc. **125** (1997), no. 2, 363–373, DOI 10.1090/S0002-9939-97-03602-2. MR1350966

Department of Mathematics, Rice University, Houston, Texas 77005
Email address: h237huan@rice.edu

Department of Mathematics & Computer Science, Fayetteville State University, Fayetteville, North Carolina 28301
Email address: xtang@uncfsu.edu

Department of Mathematics, Louisiana State University, Baton Rouge, Louisiana 70803
Email address: xingtingwang@math.lsu.edu

Contemporary Mathematics
Volume **801**, 2024
https://doi.org/10.1090/conm/801/16010

Log centres of noncommutative crepant resolutions are Kawamata log terminal: Remarks on a paper of Stafford and Van den Bergh

Colin Ingalls and Takehiko Yasuda

Abstract. We show that if a finitely generated prime algebra Δ is a finitely generated maximal Cohen-Macaulay module over its centre Z, and has global dimension equal to $\dim Z$, then the pair given by its centre and ramification divisor is Kawamata log terminal.

1. Introduction

We will work over an algebraically closed field of characteristic zero. Noncommutative crepant resolutions (NCCRs), as discussed in [**2**], indicate an interplay between noncommutative algebra and algebraic geometry. It is not clear when a commutative algebra admits an NCCR, and various necessary conditions have been found. In [**18**], it is shown that any commutative algebra admitting an NCCR has rational singularities. In particular, they show that if an affine prime algebra Δ that is finite over its centre is homologically homogenous, or equivalently Δ is Cohen-Macaulay over its centre with finite global dimension, then its centre has rational singularities. We use standard results about the singularities of the minimal model program [**11**], and the argument of [**18**] to show that the singularities of the log pair given by the centre and the ramification divisor is Kawamata log terminal. This shows in particular that the pair is $\mathbb{Q}$-Gorenstein. We now review the contents of this paper.

In Section 2, we discussion the singularities of the minimal model program and we recall the definitions and properties of log canonical covers. In Section 3, we compute the centralizers of the canonical bimodule ω for a hereditary order over a DVR. We also quote relevant results from [**11**]. This will be used in the proof of the main result. In Section 4, we recall the properties of homologically homogeneous algebras that are finite over their centres. This material is from [**18**]. In Section 5, we prove the main result. Given Δ, a prime affine algebra that finite over its centre and homologically homogeneous, with centre Z and discriminant D, we show that the centre of the canonical cover Γ of the order Δ is the log canonical cover of the pair (Z, Δ). Lastly, in Section 6, we list algebras to which the main theorem applies, and present some relevant examples.

The first author was supported by an NSERC Discovery Grant. The second author was supported by JSPS KAKENHI Grant Numbers JP22740020, JP18H01112, JP21H04994.

There are many cases of singularities which are known to not admit a NCCR, for example see [**7**].

The results of this paper suggest the following question.

QUESTION 1.1. Let Z be a commutative local ring. Under what conditions is Z the centre of a prime affine homologically homogeneous ring A? In other words, if we allow the flexibility of adding a boundary divisor D, and allowing A to have a non-trivial generic Brauer class, can we obtain a twisted ramified NCCR of the pair $(\operatorname{Spec} Z, D)$?

See 4.2 for the definitions of the relevant terms. The results and a sketch of the proofs of this paper were presented in [**9**]. The paper [**8**] works in the setting of NCCRs and Theorem 1.1(a) and Corollary 1.7 of [**8**] prove our main results in that special case.

2. The singularities of birational geometry

In this section we review some material about the singularities that arise in the study of birational geometry and the minimal model program. Standard references include [**6, 11, 13**].

2.1. Kawamata log terminal, log terminal and canonical singularities.

Let X be a normal variety. A $\mathbb{Q}$-*divisor* in X is a linear combination of codimension one subvarieties of X with rational coefficients. A $\mathbb{Q}$-divisor is *effective* if all its coefficients are non-negative. A $\mathbb{Q}$-divisor D is $\mathbb{Q}$-*Cartier* if there is an integer m so that mD is a Cartier divisor. A *log variety* is a pair (X, D) consisting of a normal variety X with a $\mathbb{Q}$-divisor D.

Let (X, D) be a log variety such that $K_X + D$ is $\mathbb{Q}$-Cartier. In this case we say that (X, D) is $\mathbb{Q}$-*Gorenstein.* Then for a proper birational morphism $f : Y \to X$ with Y smooth, we write

$$K_Y = f^*(K_X + D) + \sum_E a_E E, \; a_E \in \mathbb{Q}$$

where E runs over exceptional divisors and strict transforms of prime divisors appearing in D. The coefficient a_E is independent of the morphism f and depends only on the divisorial valuation of the function field associated to E. The map $f : Y \to (X, D)$ is called *crepant* if all $a_E = 0$, even in the case where Y is only normal and not necessarily smooth. We say that (X, D) has *Kawamata log terminal (klt) singularities* if $a_E > -1$ for every E. When (X, D) is $\mathbb{Q}$-Gorenstein and $D = 0$, then X is $\mathbb{Q}$-Gorenstein. In this case, we say that X has *log terminal (resp. canonical) singularities* if $a_E > -1$ (resp. ≥ 0) for every E. Note that K_X is Cartier if and only if X is Gorenstein. We recall the following well known results.

PROPOSITION 2.1 ([**11**, Cor. 5.24]). *Suppose that K_X is Cartier. Then X has rational singularities if and only if it has canonical singularities.*

PROPOSITION 2.2 ([**11**, Th. 5.22]). *If (X, D) has Kawamata log terminal singularities with D effective, then X has rational singularities.*

2.2. Log canonical covers. Next we recall the standard technique of log canonical covers (see [**10**, §8.5][**14**, §1.3]).

Suppose now that a $\mathbb{Q}$-Gorenstein log variety (X, D) has its boundary divisor of the form

$$D = \sum_i \frac{e_i - 1}{e_i} D_i, \ e_i \in \mathbb{Z}_{>0}$$

with D_i prime divisors. We say that D has *standard coefficients.* Let m be the *index* of (X, D), that is, the least positive integer such that $m(K_X + D)$ is Cartier. Given a coherent sheaf $\mathcal{F}$ on X we write $\mathcal{F}^{(n)} := (\mathcal{F}^{\otimes n})^{**}$, the reflexive hull of the tensor power. Locally on X, fixing an isomorphism $\omega_X^{(m)}(mD) \cong \mathcal{O}_X$, we obtain an $\mathcal{O}_X$-algebra

$$\mathcal{A} := \bigoplus_{i=0}^{m-1} \omega_X^{(i)} \left(\lfloor iD \rfloor\right). \tag{2.1}$$

(In [**10**,**14**], a slightly different description of the sheaf, $\bigoplus_{i=0}^{m-1} \mathcal{O}_X \left(\lfloor -i(K_X + D) \rfloor\right)$, is used, though it is isomorphic to $\mathcal{A}$.) The associated finite morphism

$$\phi : \widetilde{X} := \mathcal{S}\mathrm{pec}_X\, \mathcal{A} \to X$$

is called a log canonical cover and has the following properties:

PROPOSITION 2.3 ([**10**, §8.5][**14**, §1.3]). *Let (X, D) be a $\mathbb{Q}$-Gorenstein log variety with log canonical cover $\widetilde{X}$. Then $\widetilde{X}$ is an irreducible normal variety, $K_{\widetilde{X}}$ is Cartier, and*

$$K_{\widetilde{X}} = \phi^*(K_X + D).$$

Note that the map $\psi : \widetilde{X} \to (X, D)$ is crepant by definition.

PROPOSITION 2.4. [**11**, Prop. 5.20] *With the notation as above, $\widetilde{X}$ has canonical singularities if and only if (X, D) has Kawamata log terminal singularities.*

3. Centralizers of powers of ω

Let R be a DVR with parameter t. We write $(t^i) = t^i R$ for $i \in \mathbb{Z}$, the fractional ideal generated by t^i and $(t^0) = (1) = R$. Let Δ be a *standard hereditary order* over R with ramification index e, defined to be

$$\Delta = \begin{pmatrix} (1) & \cdots & \cdots & (1) \\ (t) & \ddots & \ddots & \vdots \\ \vdots & \ddots & \ddots & \vdots \\ (t) & \cdots & (t) & (1) \end{pmatrix} \subseteq R^{e \times e}.$$

The dualizing module of Δ is described as a bimodule by $\omega = \mathrm{Hom}_R(\Delta, \omega_R)$ as described in [**18**, Prop. 2.7]. We compute this bimodule by forming the dual of Δ into R component wise and taking the transpose. Let J be the Jacobson radical of Δ, we have:

$$\omega \simeq \begin{pmatrix} (1) & (t^{-1}) & \cdots & (t^{-1}) \\ \vdots & \ddots & \ddots & \vdots \\ \vdots & \ddots & \ddots & (t^{-1}) \\ (1) & \cdots & \cdots & (1) \end{pmatrix} \text{ and } J = \begin{pmatrix} (t) & (1) & \cdots & (1) \\ \vdots & \ddots & \ddots & \vdots \\ \vdots & \ddots & \ddots & (1) \\ (t) & \cdots & \cdots & (t) \end{pmatrix}.$$

Let

$$y = \begin{pmatrix} 0 & 1 & 0 & \cdots & 0 \\ \vdots & \ddots & \ddots & \ddots & \vdots \\ \vdots & \ddots & \ddots & \ddots & 0 \\ 0 & \ddots & \ddots & \ddots & 1 \\ t & 0 & \cdots & \cdots & 0 \end{pmatrix}$$

Note that y is a normal element in Δ and

$$J = y\Delta = \Delta y \qquad y^e = t \qquad \omega = y^{1-e}\Delta = yt^{-1}\Delta.$$

An induction computation shows that

$$J^i = y^i\Delta = \begin{pmatrix} \left(t^{-\lfloor \frac{-i}{e} \rfloor}\right) & \cdots & \left(t^{-\lfloor \frac{e-1-i}{e} \rfloor}\right) \\ \vdots & \ddots & \vdots \\ \left(t^{-\lfloor \frac{-e+1-i}{e} \rfloor}\right) & \cdots & \ddots \end{pmatrix}$$

We note that

$$\omega^i = y^{-i(e-1)}\Delta = \begin{pmatrix} \left(t^{-\lfloor \frac{i(e-1)}{e} \rfloor}\right) & \cdots & \left(t^{-\lfloor \frac{(i+1)(e-1)}{e} \rfloor}\right) \\ \vdots & \ddots & \vdots \\ \left(t^{-\lfloor \frac{(i-1)(e-1)}{e} \rfloor}\right) & \cdots & \ddots \end{pmatrix}$$

So the centralizer of the bimodule ω^i is

$$Z(\omega^i) = \left(t^{-\lfloor \frac{i(e-1)}{e} \rfloor}\right).$$

Write $a^{m\times n}$ for the $m \times n$ matrix with entries all a. We will use this notation for a being an ideal or an element of a ring.

Given a sequence of natural numbers $[N] = [n_1, \dots, n_e]$, we let Δ be a *block hereditary order* given by

$$\Delta = \begin{pmatrix} (1)^{n_1\times n_1} & \cdots & \cdots & (1)^{n_1\times n_e} \\ (t)^{n_2\times n_1} & \ddots & \ddots & \vdots \\ \vdots & \ddots & \ddots & \vdots \\ (t)^{n_e\times n_1} & \cdots & (t)^{n_e\times n_{e-1}} & (1)^{n_e\times n_e} \end{pmatrix}$$

which we abbreviate as

$$\Delta \simeq \begin{pmatrix} (1) & \cdots & \cdots & (1) \\ (t) & \ddots & \ddots & \vdots \\ \vdots & \ddots & \ddots & \vdots \\ (t) & \cdots & (t) & (1) \end{pmatrix}^{[N]}$$

We note that Δ is Morita equivalent to the standard hereditary R-order with ramification index e. Since we can multiply matrices by blocks, we have the following equations

$$J = (y\Delta)^{[N]} = (\Delta y)^{[N]} \qquad (y^e)^{[N]} = t \qquad \omega = (y^{1-e}\Delta)^{[N]} = (yt^{-1}\Delta)^{[N]}.$$

As above we get the following lemma.

LEMMA 3.1. *Let R be a DVR with parameter t. Let Δ be a block hereditary order over with ramification index e. Then the centralizer of powers of the dualizing module ω^i are given by*

$$Z(\omega^i) = \left(t^{-\lfloor \frac{i(e-1)}{e} \rfloor}\right).$$

4. Noncommutative resolutions or algebraic set up

We firstly recall some results and notation from [**18**]. Recall that a k-algebra is *affine* if it is finitely generated as an algebra. We let Δ be an affine prime k-algebra which is finitely generated as a module over its centre $Z := Z(\Delta)$. Note that Z is an integral domain [**12**, §10, Ex. 1], and we will write K for its fraction field. Let A be the simple Artinian ring of fractions of Δ. We have that A is a central simple K-algebra.

We say Δ is a *tame* Z-order if Δ is a finitely generated, reflexive Z module and Δ_p is hereditary for all codimension one primes p of Z. We say that Δ is *homologically homogeneous* if the projective dimensions of all simple Δ modules are equal. We recall the following result, mostly due to Brown-Hajarnavis [**5**], which we cite from [**18**].

PROPOSITION 4.1 ([**18**, Th. 2.3]). *Let Δ be an affine homologically homogeneous order of dimension d with centre Z, then*

(1) *Δ is a Cohen-Macaulay module over Z.*
(2) $\operatorname{gldim}\Delta = \operatorname{GKdim}\Delta = d$.
(3) *Z is an affine Cohen-Macaulay normal domain.*
(4) *Δ is a tame Z-order.*

Note that the third conclusion depends on our base field k having characteristic zero.

We will now define the discriminant of a tame order in codimension one. Let p be a codimension one prime of Z. Since Δ_p is hereditary, there is an étale extension R of the DVR Z_p so that $\Delta_R := \Delta \otimes R$ is isomorphic to

$$\Delta_R \simeq \begin{pmatrix} (1)^{n_1\times n_1} & \cdots & \cdots & (1)^{n_1\times n_{e_p}} \\ (t)^{n_2\times n_1} & \ddots & \ddots & \vdots \\ \vdots & \ddots & \ddots & \vdots \\ (t)^{n_{e_p}\times n_1} & \cdots & (t)^{n_{e_p}\times n_{e_p-1}} & (1)^{n_{e_p}\times n_{e_p}} \end{pmatrix}$$

where t is a parameter of R [**1,16**]. We will write $[N] = [n_1, \ldots, n_{e_p}]$ and abbreviate as

$$\Delta_R \simeq \begin{pmatrix} (1) & \cdots & \cdots & (1) \\ (t) & \ddots & \ddots & \vdots \\ \vdots & \ddots & \ddots & \vdots \\ (t) & \cdots & (t) & (1) \end{pmatrix}^{[N]}$$

We call e_p the *ramification index* of Δ at p. We define the *discriminant* of Δ to be the $\mathbb{Q}$-divisor

$$D = \sum_{\substack{p\subset Z \\ \operatorname{cod} p=1}} \left(\frac{e_p - 1}{e_p}\right) D_p.$$

Note that this sum is finite since $\lceil D \rceil$ is the classical discriminant of the order Δ which is defined via the reducued trace pairing [**16**, §25]. We will combine the centre of Δ with its discriminant into the *log centre,* $\mathrm{logZ}(\Delta) = (\operatorname{Spec} Z(\Delta), D)$.

Suppose Δ is an affine prime algebra that is a finite module over its centre. We have [**18**, Lemma 2.4] that Δ is homologically homogeneous if and only if Δ is a Cohen-Macaulay R-module and $\operatorname{GKdim} \Delta = \operatorname{gldim} \Delta$.

DEFINITION 4.2. Let Δ be a prime, affine k-algebra, module finite over its centre. If Δ is homologically homogeneous, or equivalently by [**18**, Lemma 2.4], the algebra Δ is a Cohen-Macaulay with $\operatorname{GKdim} \Delta = \operatorname{gldim} \Delta$, then we say that Δ is a *twisted ramified noncommutative crepant resolution (NCCR) of the pair* $\mathrm{logZ}(\Delta) = (X, D)$. If furthermore $D = 0$ then we say the NCCR is *unramified.* Let K be the fraction field of the centre Z. The twist of a twisted NCCR Δ is the class of $\Delta \otimes_Z K$ in $\operatorname{Br} K$. If this is trivial, then we say Δ is an untwisted NCCR of the pair $\mathrm{logZ}(\Delta)$. A *classical* NCCR is untwisted and unramified, and this the usual defintion of an NCCR in the literature.

Note that in the case that $Z(\Delta)$ is Gorenstein, our definition of a classical NCCR is equivalent to the definition of an NCCR in [**2**, Defn. 2.2] and our definition of twisted unramified NCCR is equivalent to the definition of a twisted NCCR in [**2**, Defn. 2.2].

The next statement is well known.

PROPOSITION 4.3. *If Δ is a maximal order and a classical NCCR of the pair (Z, D) then $D = 0$ and there is a reflexive Z-module M so that $\Delta \simeq \operatorname{End}_Z(M)$.*

PROOF. Since the generic Brauer class of Δ is trivial and Z is a normal integral domain, the above statement is true in codimension one, and so taking the reflexive hull gives us the result. □

5. Noncommutative crepant resolutions and canonical covers

PROPOSITION 5.1. *Let $\Gamma = \Delta \oplus \omega_\Delta \oplus \cdots \oplus \omega_\Delta^{(r-1)}$ be the canonical cover of Δ and put $\widetilde{Z} := Z(\Gamma)$ and $\widetilde{X} := \operatorname{Spec} \widetilde{Z}$. Then the morphism $\widetilde{X} \to X$ is a log canonical cover of (X, D).*

PROOF. Let p be a codimension one prime of Z. By [**1**], there is an étale extension R of Δ_p so that $\Delta_R := \Delta \otimes R$ is isomorphic to

$$\Delta_R \simeq \begin{pmatrix} R^{n_1 \times n_1} & \cdots & \cdots & R^{n_1 \times n_k} \\ (tR)^{n_2 \times n_1} & \ddots & \ddots & \vdots \\ \vdots & \ddots & \ddots & \vdots \\ (tR)^{n_k \times n_1} & \cdots & (tR)^{n_{k-1} \times n_k} & R^{n_k \times n_k} \end{pmatrix}$$

where t is a parameter of the DVR R. Now Lemma 3.1 shows that

$$Z(\omega^i_{\Delta_R}) = \left(t^{-\lfloor \frac{i(e-1)}{e} \rfloor}\right)$$

and so

$$Z(\omega_\Delta^{(i)}) = \omega_X^{(i)}(\lfloor iD \rfloor).$$

étale locally in codimension one. Both sides are naturally isomorphic at the generic point, so we obtain this isomorphism globally since both sides are reflexive. It follows immediately that

$$\widetilde{Z} = Z(\Gamma) = \bigoplus_{i=0}^{m-1} Z(\omega_\Delta^{(i)}) = \bigoplus_{i=0}^{m-1} \omega_X^{(i)}(\lfloor iD \rfloor)$$

where m is the index of $K_X + D$.

□

Let Δ be an affine prime k-algebra that is module finite over its centre. Suppose further that Δ is homologically homogeneous. Let $Z := Z(\Delta)$ its centre and $X := \operatorname{Spec} Z$ its spectrum. Stafford and Van den Bergh proved an important result about the possible singularities of the centre:

THEOREM 5.2 ([**18**]). *X has only rational singularities.*

In fact, we can prove the following along the same line, which is a refinement of their result because of Proposition 2.2.

THEOREM 5.3. *Let D be the discriminant divisor on X associated to Δ. Then the log variety (X, D) has Kawamata log terminal singularities. Moreover, if X is $\mathbb{Q}$ -Gorenstein, then X has log terminal singularities.*

PROOF. Since the proof is local on X, we may suppose that if m is the index of (X, D), then $\omega_X^{(m)}(mD)$ is isomorphic to Z. Then putting $\widetilde{Z} := \bigoplus \omega_X^{(i)}(\lfloor iD \rfloor)$, we can construct a log canonical cover of (X, D),

$$\phi : \widetilde{X} = \operatorname{Spec} \widetilde{Z} \to X = \operatorname{Spec} Z$$

as in (2.1). On the other hand, the isomorphism $\omega_X^{(m)}(mD) \cong Z$ defines an isomorphism $\omega_\Delta^{(m)} \cong \Delta$ as Δ-bimodules and we have a canonical cover of Δ,

$$\Gamma := \Delta \oplus \omega_\Delta \oplus \cdots \oplus \omega_\Delta^{(m-1)}.$$

We show in 5.1 that $Z(\omega_\Delta^{(i)}) = \omega_X^{(i)}(\lfloor iD \rfloor)$ and $Z(\Gamma) = \widetilde{Z}$.

Since [**18**, Prop. 3.1(1)] shows that Γ is also homologically homogeneous, so by Theorem 5.2, $\widetilde{Z}$ has rational singularities. Now, from Propositions 2.1 and 2.4, the pair (X, D) has Kawamata log terminal singularities. The last assertion is straightforward. □

We note the following corollaries.

COROLLARY 5.4. *Let A be an unramified NCCR of a commutative algebra R. Then R is log terminal and $\mathbb{Q}$-Gorenstein.*

COROLLARY 5.5. *Let $\mathcal{A}$ be a (twisted, ramified) NCCR of the pair (X, D). Then the pair is Kawamata log terminal and $\mathbb{Q}$-Gorenstein.*

We present an alternate proof of the theorem above. First note that:

PROPOSITION 5.6. *Let Δ be an affine, prime, finite over centre, homologically homogeneous k-algebra. Let Γ be the canonical cover of Δ as in 5.1. Let (X, D) the the log centre of Δ and let $\widetilde{X} = \operatorname{Spec} Z(\Gamma)$. Then we have that $K_{\widetilde{X}} = \pi^*(K_X + D)$.*

PROOF. Let m be the index of $K_X + D$ and let $\hat{\otimes}$ denote the reflexive hull of the tensor product. We have the following isomorphisms:

$$\Gamma \hat{\otimes}_\Delta \omega_\Delta \simeq \omega_\Gamma$$
$$\omega_\Gamma \simeq \omega_{\widetilde{X}} \hat{\otimes}_{\widetilde{X}} \Gamma$$
$$\omega_\Delta^{(m)} = \omega_X^{(m)}(mD) \otimes_X \Delta.$$

In other words if we consider the following diagram:

$$\begin{array}{ccc} \operatorname{Spec}\Delta & \longleftarrow & \operatorname{Spec}\Gamma \\ \downarrow & & \downarrow \\ (X,D) & \longleftarrow & \widetilde{X} \end{array}$$

We know that all maps except the bottom one are crepant. Hence the bottom map is crepant as well. □

Using Prop. 5.6 we obtain another proof of Theorem 5.3.

PROOF. Since Γ is homologically homogeneous with $Z(\Gamma) = \widetilde{Z}$, by Thm. 5.2, we know that $\operatorname{Spec}\widetilde{Z} = \widetilde{X}$ is has rational singularities. Therefore $\widetilde{X}$ is canonical by Prop. 2.1 and hence log terminal. By Prop. 5.6 we have that $K_{\widetilde{X}} = \pi^*(K_X + D)$. So by Prop. 2.4, we have that (X, D) is Kawamata log terminal. In particular if X is $\mathbb{Q}-$Gorenstein then it is log terminal. □

6. Corollaries and examples

Some algebras are well known to be homologically homogeneous and so provide examples of twisted and ramified NCCRs. We would like to thank Kenneth Brown for the references for the following statement.

PROPOSITION 6.1. *Suppose that Δ is finite over its centre and Δ is one of*

(1) *a Hopf algebra over an algebraically closed field.*
(2) *a connected graded affine noetherian algebra of finite global dimension.*
(3) *a not necessarily connected Artin-Schelter regular algebra.*

then Δ is a (twisted, ramified) NCCR of the pair $(Z, D) = \mathrm{logZ}(\Delta)$ which is Kawamata log terminal and $\mathbb{Q}$-Goreinstein.

PROOF. The first case is [**4**, Prop. 1.6]. The second case is in [**19**]. The third case is in [**17**]. □

COROLLARY 6.2. *The log centres of the following algebras are Kawamata log terminal.*

- *Quantised function algebras of semisimple groups $\mathcal{O}_q(G)$ when q is a root of unity.*
- *Quantised enveloping algebras of semisimple Lie algebras $U_q(\mathfrak{g})$ when q is a root of unity.*

Next, we produce an example of a variety X which is not $\mathbb{Q}$-Gorenstein, but we add an appropriate boundary divisor D, we obtain a $\mathbb{Q}$-Goreinstein pair (X, D) which is $\mathbb{Q}$-Goreinstein and admits a ramified NCCR.

EXAMPLE 6.3. Consider the quiver with vertices $Q_0 = \{0, 1\}$ with two arrows u, v from 0 to 1 and two arrows $\bar{u}, \bar{v}$ from 1 to 0. We let A be the path algebra of this quiver with relations

$$u\bar{u}v = v\bar{u}u \quad u\bar{v}v = v\bar{v}u$$
$$\bar{u}u\bar{v} = \bar{v}u\bar{u}, \quad \bar{u}v\bar{v} = \bar{v}v\bar{u}.$$

Consider the commutative ring and fractional ideals

$$R = k[a, b, c, d]/(ad - bc)$$
$$I = aR + cR \qquad I^* = R + \frac{b}{a}R.$$

Note also that

$$A \simeq \begin{pmatrix} R & I \\ I^* & R \end{pmatrix}$$

under the isomorphism

$$u \mapsto \begin{pmatrix} 0 & a \\ 0 & 0 \end{pmatrix} \qquad v \mapsto \begin{pmatrix} 0 & c \\ 0 & 0 \end{pmatrix}$$
$$\bar{u} \mapsto \begin{pmatrix} 0 & 0 \\ 1 & 0 \end{pmatrix} \qquad \bar{v} \mapsto \begin{pmatrix} 0 & 0 \\ \frac{b}{a} & 0 \end{pmatrix}.$$

It is well known that A is Azumaya except at the origin, and is an NCCR of R. We consider an action of $\mathbb{Z}/2$ on A that fixes the vertices and acts as $-1, 1, 1, 1$ on the arrows $u, v, \bar{u}, \bar{v}$ respectively. Note that the skew group ring

$$A \rtimes \mathbb{Z}/2$$

has global dimension three and is homologically homogeneous. Hence it is a ramified NCCR of its log centre.

We compute

$$Z(A) = R = k[a, b, c, d](ad - bc)$$
$$S = Z(A \rtimes \mathbb{Z}/2) = Z(A)^{\mathbb{Z}/2} = k[a^2, ab, b^2, c, d] \subset R.$$

The referee noted $S \simeq k[x, y, z, c, d]/(xz - y^2, xd - yc, yd - zc)$ which is the base of the Francia flip. So there is a small resolution $f : Y \to \operatorname{Spec} S$ given by blowing up the ideal (c, d). So K_Y must be ample or anti-ample, and $K_Y = f^*K_S$. For more details see [**3**, (2·1),Lem. 2] or [**15**, (3.8)]. If S is $\mathbb{Q}$-Gorenstein then $nK_S = 0$ for some n, giving a contradiction. We also give a toric computation below which shows that S is not $\mathbb{Q}$-Gorenstein.

We write $N = \mathbb{Z}^3$ and $N' = \mathbb{Z}^2 \oplus \frac{1}{2}\mathbb{Z}$ and let $M = N^\vee, M' = N'^\vee$ be the dual lattices. Let σ is the cone generated by the columns of the matrix:

$$\begin{pmatrix} 0 & 0 & 1 & 1 \\ 0 & 1 & 0 & 1 \\ 1 & 1 & 1 & 1 \end{pmatrix}.$$

We will see that the map of toric varieties determined by $\sigma \subset N$ to $\sigma \subset N'$ is the map $\operatorname{Spec} R \to \operatorname{Spec} S$. We write

$$k[M] = k[x^{\pm 1}, y^{\pm 1}, z^{\pm 1}] \subset k[x^{\pm 1}, y^{\pm 1}, z^{\pm 2}] = k[M'].$$

We have

$$R = k[a,b,c,d] = k[zx^{-1}, zy^{-1}, y, x] = k[\sigma^\vee] \subset k[M'].$$
$$S = k[a^2, ab, b^2, c, d] = k[z^2x^{-2}, z^2x^{-1}y^{-1}, z^2y^{-2}, y, x] = k[\sigma^\vee] \subset k[M'].$$

We compute that $A \rtimes \mathbb{Z}/2$ is ramified with index two on the toric divisor D_ρ corresponding to the ray generated by $\rho = (0,0,1/2) \in \sigma \subset N'$. Hence the pair $(\operatorname{Spec} S, \frac{1}{2}D_\rho)$ is $\mathbb{Q}$-Gorenstein and admits a ramified NCCR. If we write the primitive generators of the rays of σ in terms of the basis $e_1, e_2, \frac{1}{2}e_3$ of N', we obtain the columns of the following matrix

$$C = \begin{pmatrix} 0 & 0 & 1 & 1 \\ 0 & 1 & 0 & 1 \\ 1 & 2 & 2 & 2 \end{pmatrix}.$$

Let $D_1, \ldots, D_4$ be the toric Weil divisors corresponding to the rays. A toric Weil divisor $n_1D_1 + n_2D_2 + n_3D_3 + n_4D_4$ is $\mathbb{Q}$-Cartier if and only if $(n_1, \ldots, n_4)$ is in the $\mathbb{Q}$ span of the rows of the matrix B. It is clear that $K_{\operatorname{Spec} S} = -D_1 - D_2 - D_3 - D_4$ is not in the row span of C. However as above we have $\frac{1}{2}D_\rho = \frac{1}{2}D_1$ then

$$K + B = -\frac{1}{2}D_1 - D_2 - D_3 - D_4$$

is in the row span of C, verifying that the pair $(\operatorname{Spec} S, \frac{1}{2}D_\rho)$ is $\mathbb{Q}$-Gorenstein. So we see that the toric variety $\operatorname{Spec} S$ is not $\mathbb{Q}$-Goreinstein and does not admit an unramified NCCR, and there is a ramified NCCR of the pair $(\operatorname{Spec} S, \frac{1}{2}D_\rho)$.

In the next example, the singularity X is well known to not have a classical NCCR [**7**, Thm. 3.1,Rmk. 3.2], but with an appropriate divisor D, we obtain a twisted ramified NCCR of the pair (X, D). This shows that the pair is Kawamata log terminal and $\mathbb{Q}$-Gorenstein. We would like to thank Michael Wemyss for providing us with this example.

EXAMPLE 6.4. Let A be the k-algebra generated by a, b, c with relations

$$ac + ca, \quad bc + cb, \quad ab - ba - 2c^3.$$

The centre $Z(A)$ is generated by

$$x := a^2, \quad y := b^2, \quad z := ab + ba, \quad t := c^2.$$

One may verify that

$$z^2 - 4xy = (ab - ba)^2 = 4c^6 = 4t^3.$$

So the centre has the compound Du Val singularity with a quadric cone in its resolution [**13**, Fig. 8-3-5]. This singularity does not have a noncommutative crepant resolution [**7**, Thm. 3.1,Rmk. 3.2].

The resolution

$$0 \to A \xrightarrow{\begin{pmatrix} b & a & c \end{pmatrix}} A^3 \xrightarrow{\begin{pmatrix} -c & 0 & -a \\ 0 & c & b \\ -b & a & -2c^2 \end{pmatrix}} A^3 \xrightarrow{\begin{pmatrix} a \\ b \\ c \end{pmatrix}} A \to k \to 0$$

shows that A is Artin-Schelter regular and so is homologically homogeneous. Lastly, we compute that A is ramified on the divisor B defined by $t = 0$. So if we let $X = \operatorname{Spec} Z(A)$ and $D = \frac{1}{2}B$ we have that (X, D) has a twisted noncommutative crepant resolution and so is Kawamata log terminal and $\mathbb{Q}$-Gorenstein.

We also point out that A can be viewed as a Clifford algebra. Let

$$Z = k[x,y,z,t]/(z^2 - 4xy - 4t^3)$$

and define a quadratic form Q on Z given by the Gram matrix

$$\begin{pmatrix} 2x & z & 0 \\ z & 2y & 0 \\ 0 & 0 & 2t \end{pmatrix}.$$

Write $\mathrm{Cl}(Q)$ for the Clifford algebra of this form, generated by a, b, c as an Z-algebra with relations

$$2a^2 = 2x, \quad ab + ba = z, \quad 2b^2 = 2y, \quad ac + ca = bc + cb = 0, \quad 2c^2 = 2t.$$

Note that in this algebra we have $(ab - ba)^2 = (2c^3)^2$ giving a prime ideal generated by $(ab - ba - 2c^3)$ so we have

$$A = \mathrm{Cl}(Q)/(ab - ba - 2c^3).$$

This description can be used to show that A is non-trivially twisted.

Acknowledgments

The first author would like to thank Kenneth Brown and Michael Wemyss for helpful conversations. We would also like to thank an anonymous referee for helpful comments.

References

[1] M. Artin, *Left ideals in maximal orders*, Brauer groups in ring theory and algebraic geometry (Wilrijk, 1981), Lecture Notes in Math., vol. 917, Springer, Berlin-New York, 1982, pp. 182–193. MR657429

[2] Michel Van den Bergh, *Non-commutative crepant resolutions, an overview*, 2022.

[3] Gavin Brown, *Flips arising as quotients of hypersurfaces*, Math. Proc. Cambridge Philos. Soc. **127** (1999), no. 1, 13–31, DOI 10.1017/S0305004198003351. MR1692523

[4] K. A. Brown and K. R. Goodearl, *Homological aspects of Noetherian PI Hopf algebras of irreducible modules and maximal dimension*, J. Algebra **198** (1997), no. 1, 240–265, DOI 10.1006/jabr.1997.7109. MR1482982

[5] K. A. Brown and C. R. Hajarnavis, *Homologically homogeneous rings*, Trans. Amer. Math. Soc. **281** (1984), no. 1, 197–208, DOI 10.2307/1999529. MR719665

[6] Herbert Clemens, János Kollár, and Shigefumi Mori, *Higher-dimensional complex geometry* (English, with French summary), Astérisque **166** (1988), 144 pp. (1989). MR1004926

[7] Hailong Dao, *Remarks on non-commutative crepant resolutions of complete intersections*, Adv. Math. **224** (2010), no. 3, 1021–1030, DOI 10.1016/j.aim.2009.12.016. MR2628801

[8] Hailong Dao, Osamu Iyama, Ryo Takahashi, and Michael Wemyss, *Gorenstein modifications and* $\mathbb{Q}$*-Gorenstein rings*, J. Algebraic Geom. **29** (2020), no. 4, 729–751, DOI 10.1090/jag/760. MR4158464

[9] Colin Ingalls, *Centres of nccrs are klt*, `https://www.msri.org/workshops/643/schedules/16756`, A talk at the workshop "Interactions Between Noncommutative Algebra, Representation Theory, And Algebraic Geometry", MSRI, 2013.

[10] Yujiro Kawamata, *Log canonical models of algebraic 3-folds*, Internat. J. Math. **3** (1992), no. 3, 351–357, DOI 10.1142/S0129167X9200014X. MR1163730

[11] János Kollár and Shigefumi Mori, *Birational geometry of algebraic varieties*, Cambridge Tracts in Mathematics, vol. 134, Cambridge University Press, Cambridge, 1998. With the collaboration of C. H. Clemens and A. Corti; Translated from the 1998 Japanese original, DOI 10.1017/CBO9780511662560. MR1658959

[12] T. Y. Lam, *A first course in noncommutative rings*, 2nd ed., Graduate Texts in Mathematics, vol. 131, Springer-Verlag, New York, 2001, DOI 10.1007/978-1-4419-8616-0. MR1838439

[13] Kenji Matsuki, *Introduction to the Mori program*, Universitext, Springer-Verlag, New York, 2002, DOI 10.1007/978-1-4757-5602-9. MR1875410

[14] Yuri G. Prokhorov, *Lectures on complements on log surfaces*, MSJ Memoirs, vol. 10, Mathematical Society of Japan, Tokyo, 2001. MR1830440

[15] Miles Reid, *Decomposition of toric morphisms*, Arithmetic and geometry, Vol. II, Progr. Math., vol. 36, Birkhäuser Boston, Boston, MA, 1983, pp. 395–418. MR717617

[16] I. Reiner, *Maximal orders*, London Mathematical Society Monographs. New Series, vol. 28, The Clarendon Press, Oxford University Press, Oxford, 2003. Corrected reprint of the 1975 original; With a foreword by M. J. Taylor. MR1972204

[17] Manuel L. Reyes and Daniel Rogalski, *Graded twisted Calabi-Yau algebras are generalized Artin-Schelter regular*, Nagoya Math. J. **245** (2022), 100–153, DOI 10.1017/nmj.2020.32. MR4413364

[18] J. T. Stafford and M. Van den Bergh, *Noncommutative resolutions and rational singularities*, Michigan Math. J. **57** (2008), 659–674, DOI 10.1307/mmj/1220879430. Special volume in honor of Melvin Hochster. MR2492474

[19] J. T. Stafford and J. J. Zhang, *Homological properties of (graded) Noetherian* PI *rings*, J. Algebra **168** (1994), no. 3, 988–1026, DOI 10.1006/jabr.1994.1267. MR1293638

School of Mathematics and Statistics, Carleton University, Ottawa, Ontario K1S 5B6, Canada

Email address: cingalls@math.carleton.ca

Department of Mathematics, Graduate School of Science, Osaka University, Toyonaka, Osaka 560-0043, Japan

Email address: yasuda.takehiko.sci@osaka-u.ac.jp

Contemporary Mathematics
Volume **801**, 2024
https://doi.org/10.1090/conm/801/16039

On the geometry of some algebras related to the Weyl groupoid.

Ian M. Musson

Abstract. Let $\mathtt{k}$ be an algebraically closed field of characteristic zero. Let $\mathfrak{g}$ be a finite dimensional classical simple Lie superalgebra over $\mathtt{k}$ or $\mathfrak{gl}(m,n)$. In the case that $\mathfrak{g}$ is a Kac-Moody algebra of finite type with set of roots Δ, Sergeev and Veselov introduced the Weyl groupoid $\mathfrak{W} = \mathfrak{W}(\Delta)$, which has significant connections with the representation theory of $\mathfrak{g}$. Let $\mathfrak{h}$, W and $Z(\mathfrak{g})$ be a Cartan subalgebra of $\mathfrak{g}_0$, the Weyl group of $\mathfrak{g}_0$ and the center of $U(\mathfrak{g})$ respectively. Also let G be a Lie supergroup with Lie $G = \mathfrak{g}$. There are several important commutative algebras related to $\mathfrak{W}$. Namely

- The image $I(\mathfrak{h})$ of the injective Harish-Chandra map $Z(\mathfrak{g}) \longrightarrow S(\mathfrak{h})^W$.
- The supercharacter $\mathbb{Z}$-algebras $J(\mathfrak{g})$ and $J(G)$ of finite dimensional representations of $\mathfrak{g}$ and G.

Let $\mathcal{A} = \mathcal{A}(\mathfrak{g})$ be denote either $I(\mathfrak{h})$ or $J(G) \otimes_{\mathbb{Z}} \mathtt{k}$. The purpose of this paper is to investigate the algebraic geometry of $\mathcal{A}$. In many cases, the algebra $\mathcal{A}$ satisfies the Nullstellensatz. This gives a bijection between radical ideals in $\mathcal{A}$ and superalgebraic sets (zero loci of such ideals). Any superalgebraic set is uniquely a finite union of irreducible superalgebraic components. In the Kac-Moody case, we describe the smallest superalgebraic set containing a given (Zariski) closed set, and show that the superalgebraic sets are exactly the closed sets that are unions of groupoid orbits.

Contents

1. Introduction

1.1. Background.

1.1.1. *The classical Lie superalgebras.* We are interested in the following *classical* Lie superalgebras $\mathfrak{g}$. All have $\mathfrak{g}_0$ reductive and $\mathfrak{g}_0 \neq \mathfrak{g}$.

(a) The KM[1] algebras $\mathfrak{gl}(m|n)$, $\mathfrak{osp}(r|2n)$, $\mathfrak{sl}(m|n)$ with $m \neq n$, $D(2|1;\alpha)$, $F(4)$, $G(3)$,

(b) The strange Lie superalgebras $\mathfrak{p}(n)$, $\mathfrak{q}(n)$ and their simple relatives $\mathfrak{sp}(n)$, $\mathfrak{psq}(n)$.

There are some conditions on $r, m, n \in \mathbb{N}$ and $\alpha \in \mathtt{k}$ which we omit. The last three algebras listed in (a) are called *exceptional.* The algebras listed in (a) with $\mathfrak{gl}(m|n)$ replaced by $\mathfrak{psl}(n|n)$, together with the simple algebras from (b) are precisely the finite dimensional simple Lie superalgebras $\mathfrak{g}$ with $\mathfrak{g}_0$ reductive and $\mathfrak{g}_0 \neq \mathfrak{g}$.

1.1.2. *Algebras related to the Weyl groupoid.* The Weyl groupoid $\mathfrak{W}(\Delta)$ of Sergeev and Veselov, [**SV11**] has become a fundamental tool in the representation theory of a KM algebra $\mathfrak{g}$. Here Δ is the root system of $\mathfrak{g}$. This groupoid was introduced in relation to the algebras $J(\mathfrak{g})$ and $J(G)$ listed in the abstract. The Lie supergroups G we consider are defined in Lemma 3.4. In particular the supergroups $P(n)$ and $Q(n)$ have Lie $P(n) = \mathfrak{p}(n)$ and Lie $Q(n) = \mathfrak{q}(n)$. Weyl groupoids for $P(n)$ and $Q(n)$ were defined in [**IRS21**] Section 5.4 and [**Rei23**], Section 5 respectively.

For a KM algebra $\mathfrak{g}$, an important role is played by a non-degenerate, symmetric bilinear form $(\, , \,)$ on $\mathfrak{g}$. This can be constructed in a uniform way, [**Mus12**] Theorem 5.4.1, Remark 5.4.2. The restriction of this form to $\mathfrak{h}$ remains non-degenerate and this yields an isomorphism $\mathfrak{h} \longrightarrow \mathfrak{h}^*$, denoted $h_\alpha \mapsto \alpha$. Thus we obtain a non-degenerate, symmetric bilinear form on $\mathfrak{h}^*$ which we also denote by $(\, , \,)$. This latter form is invariant under the Weyl group W of $\mathfrak{g}_0$, and we can normalize the whole construction by requiring that its Gram matrix is the (symmetrized) Cartan matrix (α_i, α_j), where $\{\alpha_i\}$ is the distinguished basis of simple roots for $\mathfrak{g}$, [**Mus12**] Section 3.2. The issue with $\mathfrak{psl}(n|n)$ is that any set of simple roots $\{\alpha_i\}$ is linearly dependent, and the matrix $A = (\alpha_i, \alpha_j)$ is singular. Thus there is no bilinear form as above. The usual solution to this problem is to take a minimal realization of A, leading to the construction of $\mathfrak{gl}(n|n)$, which has $\mathfrak{psl}(n|n)$ as a subquotient. For $\alpha \in \Delta_{iso}$, the set of isotropic roots of $\mathfrak{g}$, set

$$\Pi_\alpha = \{\lambda \in \mathfrak{h}^* | (\lambda, \alpha) = 0\}. \tag{1.1}$$

The other algebra of interest $I(\mathfrak{h})$ is isomorphic to $Z(\mathfrak{g})$ via the Harish-Chandra map. This algebra may be defined as

$$I(\mathfrak{h}) = \{f \in S(\mathfrak{h})^W | f(\lambda) = f(\lambda + t\alpha) \text{ for all } \alpha \in \Delta_{iso}, \lambda \in \Pi_\alpha \text{ and } t \in \mathtt{k}\}. \tag{1.2}$$

Several other definitions of $I(\mathfrak{h})$ can be given based on [**Mus12**], Lemma 12.1.1. By work of Gorelik and Kac [**Gor04**], [**Kac84**], the Harish-Chandra map yields an isomorphism $Z(\mathfrak{g}) \longrightarrow I(\mathfrak{h})$, see also [**Mus12**], Theorem 13.1.1. For a description of $Z(\mathfrak{g})$ when $\mathfrak{g} = \mathfrak{p}(n)$ and $\mathfrak{q}(n)$, see [**Gor01**], [**Gor06**] respectively.

There is a description of $J(G)$ parallel to (1.2). Let $\mathbb{T}$ be a maximal torus in G_0, and $X(\mathbb{T}) = \mathrm{Hom}(\mathbb{T}, \mathtt{k}^*)$ (resp. $Y(\mathbb{T}) = \mathrm{Hom}(\mathtt{k}^*, \mathbb{T})$) the group of characters (resp. one-parameter subgroups) of $\mathbb{T}$. Let D_α be the derivation of $\mathbb{Z}[X(\mathbb{T})]$ given by

$$D_\alpha(\mathsf{e}^\beta) = (\alpha, \beta)\mathsf{e}^\beta.$$

Then in the KM case, we have by [**SV11**] Equation (1),

$$J(G) = \{f \in \mathbb{Z}[X(\mathbb{T})]^W | D_\alpha f \in (\mathsf{e}^\alpha - 1) \text{ for all } \alpha \in \Delta_{iso}\}. \tag{1.3}$$

---

[1] Kac-Moody Lie superalgebras of finite type. These algebras are also known as contragredient Lie superalgebras, [**Kac77**], [**Mus12**] Chapter 5.

Various conditions equivalent to that defining $J(G)$ are given in [**Mus22**] Section 3.

1.1.3. *Continuous Weyl groupoids.* In the theory of algebraic groups acting on varieties, a key role is played by closed orbits. However the orbits of $\mathfrak{W}$ on $\mathfrak{h}^*$ or $\mathbb{T}$ are not closed unless they are finite, see Remark 5.4, and for this reason in Section 5 we introduce the continuous Weyl groupoids $\mathfrak{W}^c$ and $\mathfrak{W}^c_*$. The action of a groupoid $\mathfrak{G}$ on an affine variety X and the invariant ring $\mathcal{O}(X)^{\mathfrak{G}}$ are defined in Section 5. In Proposition 5.9, we show there are actions of $\mathfrak{W}^c$ and $\mathfrak{W}^c_*$ on $\mathfrak{h}^*$ and $\mathbb{T}$ respectively such that

$$I(\mathfrak{h}) = S(\mathfrak{h})^{\mathfrak{W}^c} \text{ and } J(G) \otimes_{\mathbb{Z}} \mathtt{k} = \mathcal{O}(\mathbb{T})^{\mathfrak{W}^c_*}.$$

NOTATION 1.1. We assume one of the following holds

(i) $X = \mathfrak{h}^*$, $\mathfrak{G} = \mathfrak{W}^c$, $\mathcal{B} = \mathcal{O}(X)^W = S(\mathfrak{h})^W$ and $\mathcal{A} = I(\mathfrak{h})$.

(ii) $X = \mathbb{T}$, $\mathfrak{G} = \mathfrak{W}^c_*$, $\mathcal{B} = \mathcal{O}(X)^W$ and $\mathcal{A} = J(G) \otimes_{\mathbb{Z}} \mathtt{k}$.

EXAMPLE 1.2. If $\mathfrak{g} = \mathfrak{gl}(m|n)$ the algebras $I(\mathfrak{h})$ and $J(G) \otimes_{\mathbb{Z}} \mathtt{k}$ are the algebras of supersymmetric polynomials and Laurent supersymmetric polynomials over $\mathtt{k}$ respectively. These algebras have very beautiful combinatorics, see [**Mus12**] Chapter12, [**PT92**] and [**Ser19**]. Suprisingly though, the algebras of (Laurent) supersymmetric polynomials have received little attention from a geometric point of view, perhaps because they are not Noetherian.

In the simplest case where $m = n = 1$, $\mathfrak{W}$ is the smallest groupoid that is not a disjoint union of groups. Thus $\mathfrak{W}$ has two objects, and two non-identity morphisms. To describe the action of $\mathfrak{W}$ on the plane $\mathfrak{h}^* = \mathtt{k}^{1|1}$, take the objects to be $\pm a$ where $a \neq 0$ is a point in $\mathfrak{h}^*$. The functor sends $\pm a$ to the line L they span, and the non-identity morphisms add or subtract a to a point on L. The invariant ring $S(\mathfrak{h})^{\mathfrak{W}}$ consists of all polynomial functions that are constant on L. So $S(\mathfrak{h})^{\mathfrak{W}} = \mathtt{k} + TS(\mathfrak{h})$ where $S(\mathfrak{h}) = \mathtt{k}[S,T]$ and L is the line defined by T. The algebra $S(\mathfrak{h})^{\mathfrak{W}}$ is a well-known example of a non-Noetherian domain of Krull dimension two. This answers a question of S. Paul Smith.

1.1.4. *The Duflo-Serganova functor.* We recall the definition of the Duflo-Serganova functor DS_x, in the case that x is a root vector such that $[x,x] = 0$. For M as above, set $M_x = \text{Ker}\ {}_M x/xM$. Then M_x is a module for $\mathfrak{g}_x$ where $\mathfrak{g}_x = \text{Ker ad}\ x/\text{Im ad}\ x$, which is a Lie superalgebra of smaller rank than $\mathfrak{g}$. Although $\mathfrak{g}_x$ is a quotient of a subalgebra of $\mathfrak{g}$, it was shown in [**DS05**] Lemma 6.3, that it embeds in $\mathfrak{g}$ at least in the KM case. We follow the description of [**HR18**] section 2C in the case that x is a root vector corresponding to the isotropic root β.

LEMMA 1.3. *Let x be a root vector corresponding to the isotropic root β. We can identify $\mathfrak{g}_x$ with a subalgebra of $\mathfrak{g}$ such that the centralizer $\mathfrak{g}^x$ of x in $\mathfrak{g}$ is a semidirect sum $\mathfrak{g}^x = \mathfrak{g}_x \ltimes [x, \mathfrak{g}]$. The subalgebra $\mathfrak{g}_x$ is given by*

$$\mathfrak{g}_x = \mathfrak{h}(\beta) \oplus \bigoplus_{\alpha \in \Delta(\beta)} \mathfrak{g}^\alpha, \tag{1.4}$$

where $\Delta(\beta) = \{\alpha \in \Delta | (\alpha, \beta) = 0, \alpha \neq \pm\beta\}$, and $\mathfrak{h}(\beta) = \text{span}\ \{h_\alpha | \alpha \in \Delta(\beta)\}$. We have $\mathfrak{h}(\beta) = \mathfrak{g}_x \cap \mathfrak{h}$.

LEMMA 1.4. *The restriction of $(\ ,\ )$ to* span $\Delta(\beta)$ *is non-degenerate.*

PROOF. Let $\{\alpha_1, \ldots, \alpha_r\}$ be a basis for span $\Delta(\beta)$ and suppose $\sigma \in \mathfrak{h}^*$ with $(\sigma, \beta) = 1$. For $i \in [r]$, let $\alpha_i' = \alpha_i - (\sigma, \alpha_i)$. The Gram matrix of $(\ ,\ )$ using the basis $\{\alpha_1, \dots \alpha_r, \beta, \sigma\}$ for $\mathfrak{h}^*$ has the same determinant as the Gram matrix using the basis where each α_i is replaced by α_i'. Since the latter matrix is block diagonal and $(\alpha_i, \alpha_j) = (\alpha_i', \alpha_j')$, the result follows. □

Under the isomorphism $\mathfrak{h}^* \longrightarrow \mathfrak{h}^*$, span $\Delta(\beta)$ maps to span $\{h_\alpha | \alpha \in \Delta(\beta)\}$ which we identify with $\mathfrak{h}(\beta)$. We have span $\Delta(\beta) = \mathfrak{h}(\beta)^*$. If $(\sigma(\beta), (\beta)) = 1$, we have an orthogonal direct sum

$$\mathfrak{h}^* = \mathfrak{h}(\beta)^* \oplus \text{span}\,\{\beta, \sigma(\beta)\}. \tag{1.5}$$

In the table below, taken from [**GHSS22**] Section 4.5, the first column lists the Lie superalgebras we consider, x is a root vector, and the second column describes $\mathfrak{g}_x$ up to isomorphism. If $\mathfrak{g} = \mathfrak{p}(n)$, we assume that $x \in \mathfrak{g}^{\epsilon_1+\epsilon_2}$ in the notation of [**Mus12**], 2.4.1.

$\mathfrak{g}$	$\mathfrak{g}_x$
$\mathfrak{gl}(m\|n)$	$\mathfrak{gl}(m-1\|n-1)$
$\mathfrak{sl}(m\|n)$, $m \neq n$	$\mathfrak{sl}(m-1\|n-1)$
$\mathfrak{osp}(m\|2n)$	$\mathfrak{osp}(m-2\|2n-2)$
$D(2\|1;\alpha)$	$\mathtt{k}$
$F(4)$	$\mathfrak{sl}(3)$
$G(3)$	$\mathfrak{sl}(2)$
$\mathfrak{q}(n)$	$\mathfrak{q}(n-2)$
$\mathfrak{p}(n)$	$\mathfrak{p}(n-2)$

Since we use induction on the rank as a proof technique, we make some brief remarks about the base cases. These occur when $\mathfrak{g}_x$ has no isotropic roots. Then the Weyl groupoid for $\mathfrak{g}_x$ is just the Weyl group. The desired results hold trivially in these cases. By convention $\mathfrak{gl}(0|k) = \mathfrak{gl}(k|0) = \mathfrak{gl}(k)$, and other low dimensional entries in the column for $\mathfrak{g}_x$ should be interpreted similarly.

The main use we make of the functor DS_x is to construct certain algebra maps ds_x in Subsections 3.1 and 3.2. These algebra maps can be realized in two equivalent ways, and both realizations can be found in the literature, at least implicitly. Both methods can be easily adapted to obtain maps on suitable overrings as in the diagram preceding Lemma 2.7. In the situation of Lemma 1.3, $\mathfrak{h}^*$ has been identified with a subspace of $\mathfrak{h}^*$, so we have a resriction map res : $S(\mathfrak{h}) \longrightarrow S(\mathfrak{h}_x), f \mapsto f|_{\mathfrak{h}^*}$. This map descends to $S(\mathfrak{h})^W$ and $I(\mathfrak{h})$ and we have res $= ds_x : I(\mathfrak{h}) \longrightarrow I(\mathfrak{h}_x)$. On the other hand the algebra $I(\mathfrak{h})$ can be defined in terms of polynomials satisfying certain partial evaluation conditions, and then it is natural to view ds_x as evaluation ev : $I(\mathfrak{h}) \longrightarrow I(\mathfrak{h}_x)$. Similar remarks apply to $ds_x : J(G) \longrightarrow J(G_x)$.

1.1.5. *Harish-Chandra pairs and representations of Lie supergroups.* An (algebraic) Lie supergroup G can be defined using a Hopf superalgebra $\mathcal{O}(G)$ and the functor of points. Denote the category of supercommutative $\mathtt{k}$-algebras **Alg**. Then for any supercommutative algebra A, the A-points of G are given by

$$G(A) = \text{Mor}_{\mathbf{Alg}}(\mathcal{O}(G), A).$$

The Hopf superalgebra structure on $\mathcal{O}(G)$ induces a natural group structure on $G(A)$. Another useful approach (due to Kostant [**Kos77**] and Koszul [**Kos83**]) to G uses Harish-Chandra pairs. Following [**CCF11**] Definition 7.4.1, a *Harish-Chandra pair* (HCP) is a pair $(G_0, \mathfrak{g})$ where consisting of an algebraic group G_0 and a Lie superalgebra $\mathfrak{g}$ such that

(a) $\mathfrak{g}_0 = \text{Lie}\, G_0$,

(b) There is a representation ρ of G_0 on $\mathfrak{g}$ such that $\rho(G_0)|_{\mathfrak{g}_0} = \mathrm{Ad}$ and the differential of ρ acts on $\mathfrak{g}$ as the adjoint representation: for $X \in \mathfrak{g}_0, Y \in \mathfrak{g}$,

$$d\rho(X)Y = [X, Y].$$

We write $(G_0, \mathfrak{g}, \rho)$ if we want to stress the role of ρ. Harish-Chandra pairs form a category, for morphisms see [**CCF11**] Definition 7.4.2. Given a Lie supergroup G, we obtain a HCP $(G_0, \mathrm{Lie}\, G, \mathrm{Ad})$. Conversely given a HCP, we can put a Hopf superalgebra structure on

$$\mathcal{O}(G) := \mathrm{Hom}_{U(\mathfrak{g}_0)}(U(\mathfrak{g}), \mathcal{O}(G_0)), \tag{1.6}$$

and thus obtain a Lie supergroup G. In (1.6), $U(\mathfrak{g})$ is naturally a left $U(\mathfrak{g}_0)$-module, and the action of $x \in \mathfrak{g}_0$ on $\mathcal{O}(G_0)$ is given by a left-invariant differential operator, [**CCF11**] page 125. This is the basis for the following result, see [**CCF11**] Theorem 7.4.5.

THEOREM 1.5. *There is an equivalence of categories between the category of Lie supergroups and the category of Harish-Chandra pairs.*

The connection between representations of Harish-Chandra pairs and the corresponding Lie supergroup was explained in [**CC11**]. A *representation* of a HCP $(G_0, \mathfrak{g})$ consists of the following data

(a) A representation $\Sigma : G_0 \longrightarrow GL(V)$,
(b) A representation $\sigma : \mathfrak{g} \longrightarrow \mathfrak{gl}(V)$, such that $d\Sigma = \sigma|_{\mathfrak{g}_0}$ and a further compatabilty condition given in [**CC11**] Definition 6 (2) holds.

THEOREM 1.6. *There is a bijection between representations of the HCP $(G_0, \mathfrak{g})$ and representations of G on V.*

PROOF. See [**CC11**] Proposition 6. □

Let $\mathcal{F}_\mathfrak{g}$ (resp. $\mathcal{F}_G$) be the category of finite dimensional $\mathbb{Z}_2$-graded $\mathfrak{g}$-modules (resp. G-modules), and Π the parity change functor on $\mathcal{F}_\mathfrak{g}$. The extra condition mentioned above essentially says that the representations of G that arise correspond to representations of $\mathfrak{g}_0$ that are integrable as G_0-modules. For example, suppose that $\mathfrak{g}_0$ has a non-trivial center. If $\mathfrak{a}$ is an abelian Lie algebra and T is a torus with $\mathrm{Lie}\, T = \mathfrak{a}$, then integrable representations of $\mathfrak{a}$ correspond to elements of the character group $X(T)$. Hence $\mathfrak{g}_0$ has many more finite dimensional modules than G_0, and these modules can be induced to give graded modules for $\mathfrak{g}$. Thus the category $\mathcal{F}_G$ is more manageable than $\mathcal{F}_\mathfrak{g}$. In addition, the results we quote from [**HR18**], [**IRS21**] and [**Rei23**] work best for supergroups. Let $\mathcal{F} = \mathcal{F}_G$ or $\mathcal{F}_\mathfrak{g}$. The Grothendieck group $K(\mathcal{F})$ of $\mathcal{F}$ is the free abelian group on the symbols $[M]$, where M is an object in $\mathcal{F}$ modulo the relations that $[M] = [L] + [N]$ whenever there is an exact sequence $0 \to L \to M \to N \to 0$ in $\mathcal{F}$. Since $\mathcal{F}$ is closed under tensor product, $K(\mathcal{F})$ has a ring structure determined by $[M][N] = [M \otimes N]$. The *supercharacter ring* (resp. *character ring*) is the factor ring of $K(\mathcal{F}_\mathfrak{g})$ by the ideal generated by all $[\Pi(M)]+[M]$ (resp. $[\Pi(M)]-[M]$) where $M \in \mathcal{F}_\mathfrak{g}$. For the classical Lie superalgebras $\mathfrak{g}$ listed in Subsection 1.1.1, we define $J(\mathfrak{g})$ to be the character ring of $\mathfrak{g}$ if $\mathfrak{g}$ has type $\mathfrak{q}$ and the supercharacter ring in all other cases. The $\mathbb{Z}$-algebra $J(G)$ is the factor ring of $K(\mathcal{F}_G)$ defined in a similar way. The Lie superalgebras $\mathfrak{g}$ listed in 1.1.1 such that $\mathfrak{g}_0$ has a non-trivial center are $\mathfrak{g} = \mathfrak{gl}(m|n), \mathfrak{sl}(m|n)$ and $\mathfrak{g} = \mathfrak{osp}(2|2n)$. In these cases $J(\mathfrak{g})$ is graded by an abelian group, and $J(G)$ is the identity component of this grading, see [**SV11**] Propositions 7.1, 7.2, 7.3 and 7.5.

We mention an immediate consequence of the weak Nullstellensatz for $Z(\mathfrak{g})$, Theorem 4.2. Let $\mathfrak{g}$ be any classical Lie superalgebra. A *central character* of $\mathfrak{g}$ is an algebra map $Z(\mathfrak{g}) \longrightarrow \mathtt{k}$, [**GHSS22**] Section 3. Any central character χ determines, and is determined by the maxmal ideal Ker χ. An element $z \in Z(\mathfrak{g})$ acts on any Verma module $M(\lambda)$ as a scalar $\chi_\lambda(z)$ and χ_λ is called the *central character afforded* by $M(\lambda)$, [**Mus12**] 8.2.4. The next result proves and extends Conjecture 13.5.1 from [**Mus12**].

COROLLARY 1.7. *Any central character of* $\mathfrak{g}$ *is afforded by a Verma module.*

This paper replaces the unpublished preprint [**Mus19**] which dealt only with the cases where $\mathfrak{g} = \mathfrak{gl}(m|n)$ or $\mathfrak{osp}(r|2n)$. I am grateful to Maria Gorelik, Hanspeter Kraft and Shifra Reif for helpful correspondence.

2. General results.

Throughout $[k]$ denotes the set of the first k positive integers and iff means if and only if. We start with some general results on commutative rings.

2.1. Certain ring extensions. In this Subsection we assume the following hypothesis on the ring extension $A \subseteq B$.

HYPOTHESIS 2.1. For some $T \in A$, a non-zero divisor in B, we have $TB \subseteq A$.

If I is an ideal of A, let

$$\text{rad }(I) = \{f \in A | f^n \in I, \text{ for some } n > 0\}.$$

If $I = \text{rad }(I)$ we say I is a *radical ideal.* Suppose R is a commutative ring and $T \in R$, is a non-zero divisor. Denote the localization by R_T.

LEMMA 2.2. *Assume Hypothesis* 2.1.

(a) *We have an equality of localizations* $A_T = B_T$.
(b) *If* P *is a radical ideal of* A *with* $T \in P$, *then* $TB \subseteq P$.

PROOF. We leave (a) as an exercise. Then (b) holds since $(TB)^2 \subseteq AP = P$. □

If L is an ideal of R define the *extension* of L to R_T to be $L^e = L_T$, the localization of L. If M is an ideal of R_T, set $M^c = M \cap R$, the *contraction* of M to R. Then extension and contraction provide order preserving bijections

$$\{P \in \text{Spec } R | T \notin P\} \leftrightarrow \text{Spec } R_T. \tag{2.1}$$

This applies to $R = A$ or B in Lemma 2.2, so we have a bijection

$$\{P \in \text{Spec } B | T \notin P\} \leftrightarrow \{P \in \text{Spec } A | T \notin P\}, \tag{2.2}$$

given by $P \mapsto P \cap A$. From now on we identify the two sets on either side of (2.1).

THEOREM 2.3. *Let* $\phi : A \longrightarrow A/BT = C$ *be the natural map. Then we have a disjoint union*

$$\text{Spec } A = \text{Spec } B_T \cup \phi^{-1}(\text{Spec } C).$$

PROOF. As noted above we have $A_T = B_T$. For a prime ideal Q of A there are two possibilities. If $T \notin Q$, then Q_T is a prime ideal of $A_T = B_T$ such that $Q_T \cap A = Q$. If $T \in Q$, then Q is the inverse image of the prime ideal of Q/BT under ϕ. □

COROLLARY 2.4. *If m is a maximal ideal of A and $T \notin m$, the maximal ideal M of B given by $M = m_T \cap B$ satisfies $m = M \cap A$. Also $mB \neq B$.*

PROOF. This first statement follows from Equation (2.2), and the second is an immediate consequence. □

In Corollary 2.4, if $B = \mathcal{O}(X)$ for an affine variety X, with B a finitely generated $\mathtt{k}$-algebra, then m and M have the same set of zeroes in X, so by the usual Nullstellensatz $M = \text{Rad } mB$. However M can strictly contain mB, see [**Mus22**] Example 2.2.

LEMMA 2.5. *In the situation of Theorem 2.3, suppose that the rings B_T and C satisfy the ascending chain condition on radical ideals. Then so does A.*

PROOF. Let $R^1 \subseteq R^2 \subseteq \ldots$ be an ascending chain of radical ideals in A. Then $R^1_T \subseteq R^2_T \ldots$ is an ascending chain of radical ideals in $A_T = B_T$, so by assumption, there is an m such that $R^m_T = R^i_T$ for all $i \geq m$. Write $R^m_T = p_1 \cap p_2 \cap \ldots \cap p_r$ for some prime ideals of B_T. We can assume that this intersection is irredundant, and all the p_k are minimal over R^m_T. Then if $P_k = p_k \cap A$ and $i \geq m$, the P_k are exactly the minimal primes over R^i that do not contain T. By [**Eis95**] Corollary 2.12 every radical ideal in a commutative ring is the intersection of (a possibly infinite number of) prime ideals. So for $i \geq m$, write R^i as an intersection of prime ideals in A. Say

$$R^i = P_1 \cap \ldots \cap P_r \cap \bigcap_{j \in \Lambda_i} Q_j.$$

where the Q_j are prime ideals that contain TB. Thus $D_i = \bigcap_{j\in\Lambda_i} Q_j$ is a radical ideal in A containing TB, and if $\overline{D}_i = D_i/TB$, then $\overline{D}_1 \subseteq \overline{D}_2 \subseteq \ldots$ is an ascending chain of radical ideals of C. So there is an $n \geq m$, such that $\overline{D}_i = \overline{D}_n$ for all $i > n$ and then $R^i = R^n$. □

COROLLARY 2.6. *With the same assumptions as the Lemma, every radical ideal I of A is a finite intersection of prime ideals. These prime ideals can be taken to be the prime ideals minimal over I.*

PROOF. See [**Kap74**] Theorem 87. □

2.2. Towards the weak Nullstellensatz. Consider a ring extension $A \subseteq B$, and $X \subseteq \text{Max } A$. We say the *Relative Weak Nullstellensatz* (RWN) for X and the pair (A, B), if for every $m \in \text{Max } A$ we have $mB \neq B$. If the RWN holds in the case $X = \text{Max } A$, we say simply that RWN holds for the pair (A, B). If A, B are $\mathtt{k}$-algebras, with B finitely generated and $M \in \text{Max } B$, then $B/M \cong \mathtt{k}$, so if $m = M \cap A$, then

$$\mathtt{k} \hookrightarrow A/m \cong (A + M)/M \subseteq B/M \cong \mathtt{k}. \tag{2.3}$$

Thus $m \in \text{Max } A$ and we have a map

$$\psi : \text{Max } B \longrightarrow \text{Max } A, \quad M \mapsto M \cap A. \tag{2.4}$$

If RWN holds for the pair (A, B), and $m \in \text{Max } A$, then by Zorn's Lemma there is a $M \in \text{Max } B$ containing mB, we have $M \cap A = m$, so the map in (2.4) is surjective.

Consider the following diagram of ring homomorphisms where the vertical maps are inclusions, and ϕ is the restriction of Φ to A.

$$\begin{array}{ccc} A & \xrightarrow{\phi} & A' \\ \downarrow & & \downarrow \\ B & \xrightarrow{\Phi} & B' \end{array}$$

LEMMA 2.7. *Suppose Φ and ϕ are onto and* RWN *holds for the pair (A', B'). Then the RWN holds for the pair (A, B) and the set $X = \phi^{-1}(\text{Max } A)$.*

PROOF. Suppose $m \in X$, that is m is a maximal ideal of A containing Ker ϕ. Then $\phi(m)$ is a maximal ideal of A', so by assumption $J = \phi(m)B'$ is a proper ideal of B'. Since $\Phi(mB) = J$, it follows that mB is a proper ideal of B. □

LEMMA 2.8. *Suppose the finite group G acts on S and assume $|G|$ is a unit in S. If* RWN *holds for the pair (R, S), then the RWN holds for the pair (R^G, S^G).*

PROOF. In this situation we have a Reynolds operator $R \longrightarrow R^G$. Regarded as a G-module R^G is the isotypic component of R corresponding to the trivial module. Let R^+ be the sum of the remaining isotypic components. Thus $R = R^G \oplus R^+$ is a decomposition into R^G-modules. If $m \in \text{Max } R^G$, then $mR = m \oplus mR^+$ is a proper ideal of R, so by assumption mS is a proper ideal of S. The result follows since $mS^G = S^G$ implies $mS = S$. □

RWN does not hold always when $S = \mathtt{k}[x, y]$ is a polynomial algebra in 2 variables: if $R = \mathtt{k}[x] + (xy - 1)S$, then $m = x\mathtt{k}[x] + (xy - 1)S$ is a maximal ideal of R such that $mS = S$.

3. Applications.

In this Section if $\mathcal{A} = I(\mathfrak{h})$ or $J(G) \otimes_{\mathbb{Z}} \mathtt{k}$ we consider a ring extension $\mathcal{A} \subset \mathcal{B}$ and an element $T \in \mathcal{A}$ such that $T\mathcal{B} \subset \mathcal{A}$. Thus Hypothesis 2.1 holds. Because we will need to change the Lie superalgebra, we sometimes write $\mathcal{A} = \mathcal{A}(\mathfrak{g})$. The algebra $\mathcal{B}$ is either $S(\mathfrak{h})^W$ or $\mathcal{O}(\mathbb{T})^W$, so is in particular a finitely generated $\mathtt{k}$-algebra. To prove the weak Nullstellensatz we consider a map with kernel $T\mathcal{B}$,

$$\mathcal{A}(\mathfrak{g}) \longrightarrow \mathcal{A}(\mathfrak{g}_x),$$

where $\mathfrak{g}_x$ is as in the table from Subsection 1.1.4, and use induction on the rank of $\mathfrak{g}$.

3.1. Application to $I(\mathfrak{h})$. By [**GHSS22**] Section 3, the functor DS_x induces an algebra map $Z(\mathfrak{g}) \longrightarrow Z(\mathfrak{g}_x)$. Using the Harish-Chandra map, this gives rise to a map $ds_x : I(\mathfrak{h}) \longrightarrow I(\mathfrak{h}_x)$. Because of (1.4) and the above remarks, this map can be thought of as restriction of functions $\text{res}(f) = f|_{\mathfrak{h}_x^*}$.

3.1.1. *The Kac-Moody case.* In the KM case, if $\mathfrak{g}_1$ is a simple $\mathfrak{g}_0$-module, let $\Omega = \Delta_{iso}$. Otherwise $\mathfrak{g}_1 = \mathfrak{g}_1^+ \oplus \mathfrak{g}_1^-$ is a direct sum of two simple $\mathfrak{g}_0$-modules, and we let $\Omega = \Delta_1^+$ be the set of roots of $\mathfrak{g}_1^+$. Then W acts transitively on Ω. Define $T = \prod_{\beta \in \Omega} h_\beta$. Then T is W-invariant. We describe an automorphism σ on $\mathfrak{g}_x$ which we are going to use below.

(a) For $\mathfrak{g} = \mathfrak{osp}(2m|2n)$, then σ is the diagram automorphism of $\mathfrak{g}_x$ described in [**Mus12**] Lemma 5.5.12. This automorphism preserves $\mathfrak{h}$.

(b) For $\mathfrak{g} = F(4)$, then σ is induced by an involution of the Dynkin diagram for $\mathfrak{g}_x = \mathfrak{sl}(3)$.
(c) for $\mathfrak{g} = D(2|1;\alpha)$ we have $\mathfrak{g}_x = \mathtt{k}$ and $\sigma = -\mathrm{Id}$.
(d) In all other cases $\sigma = \mathrm{Id}$.

THEOREM 3.1. *We have*

(a) $T\mathcal{S}(\mathfrak{h})^W \subset I(\mathfrak{h})$ *and the kernel of* $\mathrm{res} : I(\mathfrak{h}) \to I(\mathfrak{h}_x)$ *equals* $T\mathcal{S}(\mathfrak{h})^W$.
(b) *We have* $\mathrm{res}(I(\mathfrak{h}) = I(\mathfrak{h}_x)^{\sigma_x}$.

PROOF. By construction T is W-invariant and T vanishes on all hyperplanes Π_α. Thus by (1.2), $T\mathcal{S}(\mathfrak{h})^W \subset \mathrm{Ker\,res}$. Suppose $x = e_\beta$ is a root vector and $f \in \mathrm{Ker\,res}$. Then h_β divides f in $\mathcal{S}(\mathfrak{h})$, so by W-invariance, so does T. Write $f = gT$ with $g \in \mathcal{S}(\mathfrak{h})$. Since f, T are W-invariant, so is g proving (a). For (b) see [**Gor20**] Theorem 6.4. $\square$

3.1.2. *Algebras of type* $\mathfrak{q}$. There are four related Lie superalgebras of type $\mathfrak{q}$. We refer to [**Rei23**] 2.1, [**CW12**] for full details, but mention that $\mathfrak{q}(n)$ is constructed as a subalgebra of $\mathfrak{gl}(n|n)$, and has derived subalgebra $\mathfrak{sq}(n)$. We have $\mathfrak{q}(n)_0 \cong \mathfrak{gl}(n)$ and $\mathfrak{sq}(n)_0 \cong \mathfrak{sl}(n)$. The algebras $\mathfrak{q}(n)$, $\mathfrak{sq}(n)$ have factor algebras $\mathfrak{pq}(n)$ and $\mathfrak{psq}(n)$ respectively. The algebra $\mathfrak{psq}(n)$ is simple for $n \geq 3$. By [**Gor06**] Theorem 13.1.

$$Z(\mathfrak{q}(n)) \cong \{f \in S(\mathfrak{h})^W | f|_{x_n=-x_{n-1}=t} \text{ is independent of } t\} := I_n. \tag{3.1}$$

Now (3.1) shows that if $T = \prod_{i<j}(x_i + x_j)$, then $T \in \mathcal{A} = I_n$. Set $\mathcal{B} = S(\mathfrak{h})^W$. Define f $\mathrm{ev} : I_n \to I_{n-2}$ by $\mathrm{ev}(f) = f|_{x_n=-x_{n-1}=0}$.

THEOREM 3.2. *We have*

(a) $T\mathcal{B} \subset I_n$ *and the kernel of* $\mathrm{ev} : I_n \to I_{n-2}$ *equals* $T\mathcal{B}$.
(b) *We have* $\mathrm{ev}(I_n) = I_{n-2}$

PROOF. (a) follows as in the proof of Theorem 3.1 (a). For (b) see [**Gor20**] Proposition 5.8.3. $\square$

To unify notation, we set $I(\mathfrak{h}) = I_n$ if $\mathfrak{g} = \mathfrak{q}(n)$. In this case, by [**Mus12**] Lemma 12.1.1.

$$I(\mathfrak{h}) = \{f \in S(\mathfrak{h})^W | f(\lambda) = f(\lambda + t\alpha) \text{ for all } \alpha \in \Delta^+, \lambda \in \Pi_\alpha \text{ and } t \in \mathtt{k}\} \tag{3.2}$$

3.1.3. *Algebras of type* $\mathfrak{p}$. We are interested in two Lie superalgebras of type $\mathfrak{p}$. We refer to [**CW12**] for details, but recall that $\mathfrak{p}(n)$ is constructed as a subalgebra of $\mathfrak{gl}(n|n)$, and has derived subalgebra $\mathfrak{sp}(n)$. We have $\mathfrak{p}(n)_0 \cong \mathfrak{gl}(n)$ and $\mathfrak{sp}(n)_0 \cong \mathfrak{sl}(n)$. The algebra $\mathfrak{psq}(n)$ is simple for $n \geq 3$. If $\mathfrak{g} = \mathfrak{p}(n)$, then by [**Gor01**] Theorem 4.1, $Z(\mathfrak{g})$ contains an element t such that $t^2 = 0$ and there is a linear isomorphism $Z(\mathfrak{g}) \cong \mathtt{k} \oplus tS(\mathfrak{h})^W$. This means that $Z(\mathfrak{g})$ has a unique prime ideal and the methods of Subsection 2.1 do not apply to the study of $Z(\mathfrak{g})$. For this reason $\mathfrak{p}(n)$ is excluded in some results. Corollary 1.7 holds when $\mathfrak{g} = \mathfrak{p}(n)$, since there is only one central character. The ring $J(G)$ when G is the Lie supergroup $P(n)$ remains of interest.

3.2. Application to $J(G) \otimes_{\mathbb{Z}} \mathtt{k}$.

3.2.1. *Lie supergroups.* We make a basic hypothesis, compare [**GHSS22**] Section 8.

HYPOTHESIS 3.3. Assume one of the following holds

(i) If $\mathfrak{g} = \mathfrak{gl}(m|n)$, set $G_0 = GL(m) \times GL(n)$.
(ii) If $\mathfrak{g} = \mathfrak{osp}(r|2n)$, set $G_0 = SO(r) \times SP(2n)$, where $r = 2m+1$ or $2m$.
In cases (i) and (ii) set

$$P = \sum_{i=1}^{m} \mathbb{Z}\epsilon_i \oplus \sum_{i=1}^{n} \mathbb{Z}\delta_i.$$

(iii) If $\mathfrak{g} = \mathfrak{sl}(m|n)$ $m \neq n$, set $G_0 = \{(A, B) \in GL(m) \times Gl(n) | \det A = \det B\}$ and let P be the root lattice of $\mathfrak{g}$.
(iv) If $\mathfrak{g} = \mathfrak{p}(n)$, set $G_0 = \{(A, A^{-t}) \in GL(n) \times GL(n)\}$.
(v) If $\mathfrak{g} = \mathfrak{q}(n)$, set $G_0 = \{(A, A) \in GL(n) \times GL(n)\}$.

In (iv) A^{-t} is the transpose of A^{-1}. Suppose (iv) or (v) holds. The simple roots are $\alpha_i = \epsilon_i - \epsilon_{i+1}, i \in [n]$. We use the bilinear form on $\mathfrak{h}^*$ given by $(\epsilon_i, \epsilon_j) = \delta_{i,j}$ and set $P = \sum_{i=1}^{n} \mathbb{Z}\epsilon_i$. Recall that $\mathbb{T}$ is a maximal torus in G_0.

LEMMA 3.4.

(a) *There is a Harish-Chandra pair $(G_0, \mathfrak{g})$ and the Lie supergroup determined by (1.6) is $G = GL(m|n), OSP(r|2n), SL(m|n), P(n)$ or $Q(n)$ in cases (i)-(v) respectively.*
(b) *There is an isomorphism $P \longrightarrow X(\mathbb{T})$, which we write as $\alpha \longrightarrow \mathsf{e}^{\alpha}$. Here e is a formal symbol used to convert to multiplicative notation.*

PROOF. The possible weights of G_0-modules (and hence also G-modules) are determined by P. Thus the result follows from the explicit description of the characters or supercharacters of these modules given in [**SV11**], [**IRS21**] and [**Rei23**]. □

If $\mathfrak{g} = \mathfrak{p}(n)$ or $\mathfrak{q}(n)$, and $\alpha = \epsilon_i - \epsilon_j \in \Delta_0^+$, the set of positive even roots, set $\overline{\alpha} = \epsilon_i + \epsilon_j$ and define in place of (1.1),

$$\Pi_\alpha = \{\lambda \in \mathfrak{h}^* | (\lambda, \overline{\alpha}) = 0\}. \tag{3.3}$$

In the KM case, an *iso-set* is a linearly independent set of mutually orthogonal isotropic roots. If $\mathfrak{g} = \mathfrak{p}(n)$ or $\mathfrak{q}(n)$, an *iso-set* is a linearly independent subset A of Δ_0^+ such that $(\alpha, \overline{\beta}) = 0$ for all $\alpha, \beta \in A$. The non-standard convention that $\Delta_{iso} = \Delta_0^+$ in types p and q, allows us to give unified statements of certain results.

Let $P \subset \mathfrak{h}^*$ be as in Hypothesis 3.3. For all $\alpha \in P$, define the subtorus $\mathtt{T}_\alpha$ of $\mathbb{T}$ by

$$\mathtt{T}_\alpha = \text{Ker } \mathsf{e}^{\alpha} \text{ in cases (i)-(iii), or } \mathtt{T}_\alpha = \text{Ker } \mathsf{e}^{\overline{\alpha}} \text{ in cases (iv)-(v).} \tag{3.4}$$

Composition gives a non-degenerate pairing $X(\mathbb{T}) \times Y(\mathbb{T}) \longrightarrow \text{Hom}(\mathtt{k}^*, \mathtt{k}^*) \cong \mathbb{Z}$. The next result relates this pairing to the bilinear form on P.

LEMMA 3.5. *For each root β there is a unique $c_\beta \in Y(\mathbb{T})$ such that*

$$\mathsf{e}^{\alpha} c_\beta(t) = t^{(\alpha,\beta)} \text{ for all roots } \alpha \text{ and } t \in \mathtt{k}^*. \tag{3.5}$$

PROOF. First suppose $G = GL(m|n)$ or $OSP(r|2n)$, with $r = 2m+1$ or $2m$ and let $p = m+n$. We show (3.5) holds for all $\beta \in P$. It is convenient to relabel the ordered basis of P as follows

$$(\alpha_1, \ldots, \alpha_p) := (\epsilon_1, \ldots, \epsilon_m, \delta_1, \ldots, \delta_n).$$

Then the dual basis is

$$(\varpi_1, \ldots, \varpi_p) := (\epsilon_1, \ldots, \epsilon_m, -\delta_1, \ldots, -\delta_n).$$

There are unique $c_1, \ldots, c_p \in Y(T)$ such that $\mathsf{e}^{\alpha_i} c_j(t) = t^{\delta_{i,j}}$. If $\beta \in P$, write $\beta = \sum_{i=1}^p a_j \varpi_j$ with $a_j \in \mathbb{Z}$. Then set $c_\beta = \prod_{i=1}^p c_j^{a_j}$. It is enough to show (3.5) for $\alpha = \alpha_i$, but in this case we have, for $t \in \mathtt{k}^*$ we have

$$\mathsf{e}^{\alpha_i} c_\beta(t) = t^{a_i} = t^{(\alpha_i, \beta)}.$$

If $G = SL(m|n)$ the distinguished set of simple roots is

$$\begin{aligned} \alpha_i &= \epsilon_i - \epsilon_{i+1}, & i \in [m-1] \\ \alpha_m &= \epsilon_m - \delta_1, & \\ \alpha_{m+i} &= \delta_i - \delta_{i+1}, & i \in [m-1] \end{aligned}$$

Set $x_i = \mathsf{e}^{\epsilon_i}, y_i = \mathsf{e}^{\delta_i}$. Then for a simple root α, $\mathsf{e}^\alpha = x_i x_{i+1}^{-1}, x_m y_1^{-1}$ and $y_i y_{i+1}^{-1}$ in the three cases listed above. For a root β define $c_\beta \in Y(\mathbb{T})$, by

$$c_\beta(t) = (t^{(\beta,\epsilon_1)}, \ldots, t^{(\beta,\epsilon_m)}, t^{(\beta,\delta_1)}, \ldots, t^{(\beta,\delta_n)}).$$

It suffices to show (3.5) for α, β for simple roots and this is done with a short computation. □

LEMMA 3.6. *If $G = P(n)$ or $Q(n)$, then for all $\beta \in \Delta_0^{\mathsf{l}}$ there is a unique $c_\beta \in Y(\mathbb{T})$ such that*

$$\mathsf{e}^{\overline{\alpha}} c_\beta(t) = t^{(\overline{\alpha},\beta)} \text{ for all } \alpha \in \Delta_0^+ \text{ and } t \in \mathtt{k}^*. \tag{3.6}$$

PROOF. Define $c_\beta \in Y(\mathbb{T})$ by $c_\beta(t) = (1, \ldots, t, \ldots, t^{-1}, \ldots, 1)$, where t, t^{-1} are in positions i and j. The result holds by a short computation. □

In the KM case, it was shown by Hoyt and Reif, [**HR18**] Proposition 8, that the functor DS_x induces a ring homomorphism $ds_x : J(G) \longrightarrow J(G_x)$. Our next aim is to describe the image and kernel of the map ds_x, and deduce that the inclusion $\mathcal{A} = J(G) \otimes_{\mathbb{Z}} \mathtt{k} \subseteq \mathcal{B} = \mathbb{Z}[X(\mathbb{T})]^W \otimes_{\mathbb{Z}} \mathtt{k}$ satisfy Hypothesis 2.1. Though not strictly necessary for all the results we cite, we extend scalars to $\mathtt{k}$. This brings out the analogy with $I(\mathfrak{h})$. Note that $\mathbb{Z}[X(\mathbb{T})] \otimes_{\mathbb{Z}} \mathtt{k} = \mathcal{O}(\mathbb{T})$.

3.2.2. *Properties of ds_x.* In the the KM case, set $\rho_{\text{iso}} = \frac{1}{2} \sum_{\alpha \in \Delta_{\text{iso}}^+} \alpha$.

THEOREM 3.7.

(a) *There is an element $T \in \mathcal{B}$ such that* $\operatorname{Ker} ds_x = T\mathcal{B} \subset \mathcal{A}$. *In particular Hypothesis* 2.1 *holds.*
(b) *Let G be one of the Lie supergroups $SL(m|n), m \neq n, GL(m|n)$, or $OSP(r|2n)$. Then $ds_x : J(G) \longrightarrow J(G_x)$ is surjective.*

PROOF. (a) Define T by

$$T = \mathsf{e}^{\rho_{\text{iso}}} \prod_{\alpha \in \Delta_{\text{iso}}^+} (1 - \mathsf{e}^{-\alpha}).$$

In the notation of [**HR18**] Lemma 16 we have $T = k(\rho_{\rm iso})$. So the result follows from the cited Lemma and [**HR18**] Theorem 17. Also (b) is [**HR18**] Theorem 20. □

3.2.3. *The case of* $Q(n)$. The case $G = Q(n)$ is studied in [**Rei23**]. Here are some relevant results. In this and the following Subsection, let $\mathbb{T}_n$ be a maximal torus in $Q(n)_0 = GL(n)$. Then [**Rei23**] Proposition 3, gives an explicit description of a ring J_n isomorphic to $J(Q(n))$. For our purposes it is convenient to extend scalars to $\mathbb{Z}[\frac{1}{2}]$. Thus we define

$$J_n = \{f \in \mathbb{Z}[\frac{1}{2}][X(\mathbb{T}_n)]^W | f|_{x_n=-x_{n-1}=t} \text{ is independent of } t\}. \tag{3.7}$$

With this definition $J_n \cong J(Q(n)) \otimes_{\mathbb{Z}} \mathbb{Z}[\frac{1}{2}]$. Using this isomorphism, ds_x identifies with an evaluation map $\mathcal{A} = J_n \longrightarrow J_{n-2} = \mathcal{A}'$, given by $\mathrm{ev}(f) = f|_{x_n=-x_{n-1}=1}$. Note that $\mathcal{A}$ is contained in $\mathcal{B} = \mathbb{Z}[\frac{1}{2}][X(\mathbb{T}_n)]^W$. So we have a diagram as in Subsection 2.2, where the map $\phi = \mathrm{ev}$, and the map $\Phi : \mathcal{B} \longrightarrow \mathcal{B}' = \mathbb{Z}[\frac{1}{2}][X(\mathbb{T}_{n-2})]^{W'}$ is defined by

$$\Phi(f) = f|_{x_n=-x_{n-1}=1}.$$

THEOREM 3.8. (a) *The map* ev : $J_n \longrightarrow J_{n-2}$ *is surjective.*

(b) *Define* $T \in \mathcal{B} = \mathbb{Z}[\frac{1}{2}][X(\mathbb{T}_n)]^W$, *by* $T = \prod_{i<j}(x_i + x_j)$. *Then* $T\mathcal{B} \subset \mathcal{A}$. *That is Hypothesis* 2.1 *holds. Furthermore* Ker ev $= T\mathcal{B}$.

PROOF. For (a) see [**Rei23**] Proposition 7. Since clearly T is fixed by W and $\mathrm{ev}(T) = 0$, we have $T\mathcal{B} \subset \mathcal{A}$ and $T\mathcal{B} \subseteq$ Ker ev. The other inclusion in (b) follows from the proof of [**Rei23**] Proposition 6. Without extending scalars, we would not have $T\mathcal{B} \subset \mathcal{A}$. □

3.2.4. *The case of* $P(n)$. The case of $G = P(n)$ is studied in [**IRS21**]. We define

$$J_n = \{f \in \mathbb{Z}[X(\mathbb{T}_n)]^W | f|_{x_n=x_{n-1}^{-1}=t} \text{ is independent of } t\}. \tag{3.8}$$

By [**IRS21**] Theorem 1.0.1 $J_n \cong J(P(n))$, the super character ring of $P(n)$. As before this allows us to identify ds_x with an evaluation map $\mathcal{A} = J_n \longrightarrow J_{n-2} = \mathcal{A}'$, given by $\mathrm{ev}(f) = f|_{x_n=x_{n-1}^{-1}=1}$. Note that $\mathcal{A}$ is contained in $\mathcal{B} = \mathbb{Z}[X(T_n)]^W$. So we have a diagram as in Subsection 2.2, where the map $\phi = \mathrm{ev}$, and the map $\Phi : \mathcal{B} \longrightarrow \mathcal{B}' = \mathbb{Z}[X(\mathbb{T}_{n-2})]^{W'}$ is defined by $\Phi(f) = f|_{x_n=x_{n-1}^{-1}=1}$.

THEOREM 3.9. (a) *The map* ev : $J_n \longrightarrow J_{n-2}$ *is surjective.*

(b) *Define* $T \in \mathcal{B} = \mathbb{Z}[X(\mathbb{T}_n)]^W$, *by* $T = \prod_{i<j}(1 - x_i x_j)$. *Then such that* $T\mathcal{B} \subset \mathcal{A}$. *That is Hypothesis* 2.1 *holds. Furthermore* Ker ev $= T\mathcal{B}$.

PROOF. For (a) see [**IRS21**] Theorem 4.0.1. Statement (b) follows from the proof of [**IRS21**] Proposition 3.2.1. □

4. The Nullstellensatz.

In this Section we assume one of the following holds, compare Notation 1.1

(i) $\mathfrak{g} \neq \mathfrak{p}(n)$, $\mathcal{A} = I(\mathfrak{h})$, $X = \mathfrak{h}^*$ and $\mathcal{B} = \mathcal{O}(X)^W = S(\mathfrak{h})^W$.

(ii) $\mathcal{A} = J(G) \otimes_{\mathbb{Z}} \mathtt{k}$, $X = \mathbb{T}, \mathcal{B} = \mathcal{O}(X)^W$ and G is not exceptional KM.

4.1. Maximal ideals. First we prove the analog of the weak Nullstellensatz.

THEOREM 4.1. *If m is a maximal ideal of $\mathcal{A}$, then there is a maximal ideal M of $\mathcal{B}$ such that $m = M \cap \mathcal{A}$.*

PROOF. We give the details in case (i). If $T \notin m$ this follows from Corollary 2.4. Assume is KM and let W_x be the Weyl group of $\mathfrak{g}_x$. By induction the conclusion holds with $(\mathcal{A}, \mathcal{B})$ replaced by the pair $(I(\mathfrak{h}_x), S(\mathfrak{h}_x)^{W_x})$. Thus by Lemma 2.8 it holds also for $(I(\mathfrak{h}_x)^\sigma, S(\mathfrak{h}_x)^{W_x,\sigma})$ where $\sigma = \sigma_x$ is as in Theorem 3.1. Thus if $T \in m$, the result holds by Lemma 2.7, taking M to be any maximal ideal containing $m\mathcal{B}$. The same argument works for $\mathfrak{g} = \mathfrak{q}(n)$ using Theorem 3.2. The proof in case (ii) is similar. □

THEOREM 4.2. *If $\lambda \in X$, set*

$$m_\lambda = \{f \in \mathcal{A} | f(\lambda) = 0\}.$$

If m is a maximal ideal of $\mathcal{A}$, then $m = m_\lambda$ for some $\lambda \in X$.

PROOF. This follows from Theorem 4.1. Note that maximal ideals in $\mathcal{B}$ correspond to W-orbits in X. □

For $\lambda \in X$, let $M_\lambda \in \operatorname{Max} \mathcal{O}(X)$ be the ideal of functions vanishing at λ.

COROLLARY 4.3. *We have*

(a) *There are surjective maps*

$$\operatorname{Max} \mathcal{O}(X) \longrightarrow \operatorname{Max} \mathcal{B} \longrightarrow \operatorname{Max} \mathcal{A}.$$

(b) *Denote the composite of the maps in (a) by σ and suppose that $\sim$ is the smallest equivalence relation on X such that*
   (i) *$w\lambda \sim \lambda$ for all $w \in W$.*
   (ii) *If $X = \mathfrak{h}^*$ and $(\lambda, \alpha) = 0$ for $\alpha \in \Delta_{iso}$, then $\lambda \sim \lambda + t\alpha$ for all $t \in \mathtt{k}$.*
   (iii) *If $X = \mathbb{T}$ and $\lambda \in \mathtt{T}_\alpha$ for $\alpha \in \Delta_{iso}$, then $\lambda \sim c_\alpha(t)\lambda$ for all $t \in \mathtt{k}$.*
   Then $M_\mu \in \sigma^{-1}(m_\lambda)$ iff $\lambda \sim \mu$.

PROOF. In (a) the map $\operatorname{Max} \mathcal{B} \longrightarrow \operatorname{Max} \mathcal{A}$ is given by $M \longrightarrow M \cap \mathcal{A}$, and the other map is defined similarly. The first map is surjective since $\mathcal{B}$ is the fixed ring of $\mathcal{O}(X)$ under the action of the finite group W. The second map is surjective by Theorem 4.1. If $X = \mathfrak{h}^*$, part (b) is a reformulation of [**Mus12**] Theorem 13.5.4, and the proof when $X = \mathbb{T}$ is similar. □

4.2. The strong Nullstellensatz. We deduce the strong Nullstellensatz from Theorem 4.1. There is a small problem because we need the analogous result when the pair $(\mathcal{A}, \mathcal{B})$ is replaced by $(\mathcal{A} \otimes_{\mathtt{k}} \mathtt{k}[z], \mathcal{B} \otimes_{\mathtt{k}} \mathtt{k}[z])$. This is easily taken care of by noting that the description of $\mathcal{A}$ given by (1.2) and (1.3) is independent of the algebraically closed field, and hence holds also for any field extension K of $\mathtt{k}$.

COROLLARY 4.4. *If $m \in \operatorname{Max} \mathcal{A} \otimes_{\mathtt{k}} \mathtt{k}[z]$, then $m\mathcal{B} \otimes_{\mathtt{k}} \mathtt{k}[z]$ is a proper ideal of $\mathcal{B} \otimes_{\mathtt{k}} \mathtt{k}[z]$.*

PROOF. Let $S = \mathtt{k}[z] \backslash 0$. If $m \cap S = \emptyset$, then $m_S \in \operatorname{Max} \mathcal{A} \otimes_{\mathtt{k}} K$, where $K = \mathtt{k}(z)$, and the result follows by the above remarks. Otherwise, we have $z - \lambda \in m$ for some $\lambda \in \mathtt{k}$. Now if $C = \mathcal{A}$ or $\mathcal{B}$, set $C_\lambda = C \otimes_{\mathtt{k}} \mathtt{k}[z]/(z - \lambda)$. Then $\overline{m} := m/(z - \lambda) \in \operatorname{Max} \mathcal{A}_\lambda$. Now

$$\mathcal{A} \cong \mathcal{A}_\lambda \subseteq \mathcal{B}_\lambda \cong \mathcal{B},$$

so by Theorem 4.1 $\overline{m}\mathcal{B}_\lambda$ is a proper ideal of $\mathcal{B}_\lambda$. □

If I is a subset of $\mathcal{A}$, let $\mathcal{V}(I) = \{x \in X | f(x) = 0 \text{ for all } f \in I\}$. Such a set is called an *superalgebraic set*. If instead I is a subset of $\mathcal{B}$, we say that $\mathcal{V}(I)$ is an *algebraic set* (or *closed*) in X. Thus any superalgebraic set is algebraic. In addition if V is a subset of X, set

$$\mathcal{I}_{\mathcal{A}}(V) = \{f \in \mathcal{A} | f(x) = 0 \text{ for all } x \in V\}.$$

We will also need

$$\mathcal{I}(V) = \{f \in \mathcal{O}(X) | f(x) = 0 \text{ for all } x \in V\}.$$

THEOREM 4.5. *The maps $\mathcal{I}_{\mathcal{A}}$ and $\mathcal{V}$ are inverse bijections between the set of superalgebraic sets in X, and the set of radical ideals in $\mathcal{A}$. Both maps are order reversing.*

PROOF. The key point is that $\mathcal{I}_{\mathcal{A}}(\mathcal{V}(I)) \subseteq \text{rad}\,(I)$. To show I is not assumed finitely generated, we repeat the well-known "Rabinowitsch trick", [**Eis95**] Theorem 4.19, [**Ful89**] Chapter 1. Equation (4.1) below uses only finitely many elements from I. Suppose $G \in \mathcal{I}_{\mathcal{A}}(\mathcal{V}(I))$, and let J be the ideal of $\mathcal{A} \otimes \mathtt{k}[z]$ generated by I and $zG-1$. Then $\mathcal{V}(J)$ is empty, since G vanishes wherever all polynomials in I vanish. Therefore by Corollary 4.4, $1 \in J$, and we can write

$$1 = \sum_{i=1}^{r} A_i F_i + B(zG - 1), \tag{4.1}$$

where the F_i are in I and $B, A_i \in \mathcal{A} \otimes \mathtt{k}[z]$. Now set $Y = 1/z$ and multiply by a large power of Y to obtain

$$Y^N = \sum_{i=1}^{r} C_i F_i + D(G - Y)$$

where $D, C_i \in \mathcal{A} \otimes \mathtt{k}[Y]$. The result follows by setting $Y = G$. □

PROPOSITION 4.6. *The maps $\mathcal{V}$, and $\mathcal{I}_{\mathcal{A}}$ satisfy the following properties. Suppose that E_λ, V_λ are subsets of $\mathcal{A}$, and X respectively and that $\mathfrak{a}, \mathfrak{b}$ are ideals of $\mathcal{A}$.*

(a) $\mathcal{V}(\cup_{\lambda \in \Lambda} E_\lambda) = \cap_{\lambda \in \Lambda} \mathcal{V}(E_\lambda)$.
(b) $\mathcal{V}(\mathfrak{a} \cap \mathfrak{b}) = \mathcal{V}(\mathfrak{a}\mathfrak{b}) = \mathcal{V}(\mathfrak{a}) \cup \mathcal{V}(\mathfrak{b})$.
(c) $\mathcal{I}_{\mathcal{A}}(\cup_{\lambda \in \Lambda} V_\lambda) = \cap_{\lambda \in \Lambda} \mathcal{I}_{\mathcal{A}}(V_\lambda)$.

PROOF. Left to the reader. □

4.3. Prime ideals and irreducible components. We say that a superalgebraic set is *irreducible* if it cannot be written as the union of two proper superalgebraic sets.

PROPOSITION 4.7. *If I is a radical ideal of $\mathcal{A}$ and $V = \mathcal{V}(I)$ is the corresponding superalgebraic set, then I is prime iff V is irreducible as a superalgebraic set.*

PROOF. The proof is completely analagous to the classical case [**Ful89**] Chapter 1.5, Proposition 1, page 15. □

In general if I is a radical ideal of $\mathcal{A}$, then using Corollary 2.6, we can write I uniquely in the form $I = P_1 \cap \ldots \cap P_r$, where the P_i are the prime ideals of $\mathcal{A}$ which are minimal over I. Then $\mathcal{V}(I) = \bigcup_{i=1}^{r} \mathcal{V}(P_i)$ by Proposition 4.6. We call the superalgebraic sets $\mathcal{V}(P_i)$ the *irreducible superalgebraic components of* $\mathcal{V}(I)$. They are irreducible.

COROLLARY 4.8. *Every superalgebraic set is uniquely a finite union of irreducible superalgebraic components.*

PROOF. This follows from the Nullstellensatz and Proposition 4.6 (b). □

REMARK 4.9. Our approach to the Nullstellensatz was to relate ideals of $\mathcal{A}$ and $\mathcal{B} = \mathcal{O}(X)^W = \mathcal{O}(X/W)$. Therefore it might seem more naural to relate radical ideals in $\mathcal{A}$ to their zero loci in X/W. However the connection between closed sets in X and X/W is well understood, when $\mathcal{O}(X)$ is finitely generated, [**DK02**] Section 2.3.1. If U is a closed W-invariant subset of X, then $\mathcal{I}(U)$ is a W-stable ideal of $\mathcal{O}(X)$. Let $\pi : X \longrightarrow X/W$ be the natural map, corresponding to the inclusion of rings $\mathcal{O}(X)^W \subseteq \mathcal{O}(X)$. If V is closed in X/W, and U is an irreducible component of $\pi^{-1}(V)$, then so is wU for $w \in W$, and $\pi(U) = \pi(wU)$. In Section 6 it will be more convenient to work with closed sets in X, rather than X/W. Several closed subsets of X we consider are identified under π, see Lemma 6.10.

5. Weyl groupoids and their actions.

5.1. Actions of groupoids on varieties. A groupoid $\mathfrak{G}$ can be defined as a small category with all morphisms invertible. We denote the set of objects by $\mathfrak{B}$ which we call the *base*. As in [**SV11**] we use the same notation $\mathfrak{G}$ for the set of morphisms as for the groupoid itself. For a variety X consider the groupoid $\mathfrak{A}(X)$ with base all subvarieties of X, and morphisms all isomorphisms between subvarieties. We say the groupoid $\mathfrak{G}$ acts on X if there is a functor

$$F : \mathfrak{G} \longrightarrow \mathfrak{A}(X).$$

If $g : b \longrightarrow b'$ is a morphism in $\mathfrak{G}$, set $s(g) = b$. Then for $\lambda \in X$, set $\mathfrak{B}_\lambda = \{b \in \mathfrak{B} | \lambda \in F(b)\}$ and $\mathfrak{G}_{(\lambda)} = \{g \in \mathfrak{G} | s(g) \in \mathfrak{B}_\lambda\}$. If $g \in \mathfrak{G}_{(\lambda)}$, we say *$g$ is defined at* λ and often write $g\lambda = F(g)\lambda$. Set $\mathfrak{G}_{(0)}\lambda = \{\lambda\}$ and for $i \geq 0$, define $\mathfrak{G}_{(i+1)}\lambda = \{g\mu | \mu \in \mathfrak{G}_{(i)}\lambda, g \in \mathfrak{G}_{(\mu)}\}$. The *$\mathfrak{G}$-orbit* of λ is $\mathfrak{G}\lambda = \bigcup_{i\geq 0} \mathfrak{G}_{(i)}\lambda$. Informally, $\mathfrak{G}_{(i)}\lambda$ is the set of points in X we can reach from λ using i morphisms from $\mathfrak{G}$. If X is affine, the *invariant ring* for the action of $\mathfrak{G}$ on X is

$$\mathcal{O}(X)^{\mathfrak{G}} = \{f \in \mathcal{O}(X) | f \text{ is constant on } \mathfrak{G}\text{-orbits}\}.$$

Unlike the case of a group action on X, the groupoid $\mathfrak{G}$ does not in general act on functions.

LEMMA 5.1. *The comorphism $\pi : X \longrightarrow \operatorname{Spec} \mathcal{O}(X)^{\mathfrak{G}} = Y$ is constant on $\mathfrak{G}$-orbits.*

PROOF. This follows from the definition of $\mathcal{O}(X)^{\mathfrak{G}}$. □

For continuous Weyl groupoids the map π has some pleasant properties, see Corollary 5.8.

5.2. Continuous Weyl groupoids and their actions. In [**SV11**] Sergeev and Veselov associated a certain groupoid $\mathfrak{W} = \mathfrak{W}(\Delta)$, which they the call Weyl groupoid, to the root system Δ of a KM algebra. They also defined a functor $\mathfrak{W} \longrightarrow \mathfrak{A}(\mathfrak{h}^*)$. We introduce two closely related groupoids $\mathfrak{W}^c = \mathfrak{W}^c(\Delta)$ and $\mathfrak{W}^c_* = \mathfrak{W}^c_*(\Delta)$.

We need a preliminary construction, namely the semi-direct product groupoid $\Gamma \ltimes \mathfrak{G}$. Let $\mathfrak{G}$ be a groupoid and Γ a group acting on $\mathfrak{G}$ by automorphisms of the corresponding category. In particular, Γ acts on the base $\mathfrak{B}$ of $\mathfrak{G}$. Then the *semi-direct product groupoid* $\Gamma \ltimes \mathfrak{G}$ has the same base $\mathfrak{B}$, and is defined in [**SV11**]. To define the continuous Weyl groupoids, we first introduce some auxilliary groupoids $\mathcal{G}^c$ and $\mathcal{G}^c_*$.

5.2.1. *An action on* $\mathfrak{h}^*$. In the KM case consider the following groupoid $\mathfrak{T}^c_{iso}$ with base $\Delta^c_{iso} = \Delta_{iso} \times \mathtt{k}$. The non-identity morphisms are $\tau_{\alpha,t} : (\alpha, t) \to (-\alpha, t)$. The group W acts on $\mathfrak{T}^c_{iso}$ in a natural way: $\alpha \to w(\alpha)$, $\tau_{\alpha,t} \to \tau_{w(\alpha),t}$. Let $\mathcal{G}^c$ be the semi-direct product groupoid $\mathcal{G}^c = W \ltimes \mathfrak{T}^c_{iso}$ with base Δ^c_{iso}. The *continuous Weyl groupoid* $\mathfrak{W}^c$ is defined by

$$\mathfrak{W}^c = W \coprod \mathcal{G}^c, \tag{5.1}$$

the disjoint union of the group W considered as a groupoid with a single point base $[W]$ and $\mathcal{G}^c$. There is an action of $\mathfrak{W}^c$ on $\mathfrak{h}^*$, given by a functor from $\mathfrak{W}^c$ to $\mathfrak{A}(\mathfrak{h}^*)$. Here we adapt [**SV11**], Section 9. The base point $[W]$ maps to the whole space $\mathfrak{h}^*$. The base element (α, t) maps to the hyperplane $\Pi_\alpha = \Pi_{-\alpha}$. The morphism $\tau_{\alpha,t}$ acts via

$$\tau_{\alpha,t}(x) = x + t\alpha,\ x \in \Pi_\alpha. \tag{5.2}$$

If $\mathfrak{g} = \mathfrak{p}(n)$ or $\mathfrak{q}(n)$, define the groupoid $\mathfrak{T}^c$ with base $\mathfrak{B}^c = \{(\pm\overline{\alpha}, t) \mid \alpha \in \Delta^+_0, t \in \mathtt{k}\}$ and non-identity morphisms $\tau_{\alpha,t} : (\overline{\alpha}, t) \longrightarrow (-\overline{\alpha}, t)$ and $\tau_{-\alpha,t} : (-\overline{\alpha}, t) \longrightarrow (\overline{\alpha}, t)$ for $\alpha \in \Delta^+_0$. Next form the semi-direct product $\mathcal{G}^c = W \ltimes \mathfrak{T}^c$ with base $\mathfrak{B}^c$, and define $\mathfrak{W}^c$ as in (5.1). There is an action of $\mathfrak{W}^c$ on $\mathfrak{h}^*$ defined as before, except that in (5.2) for $\alpha \in \Delta^+_0$, Π_α is given by (3.3). The orbits of $\mathfrak{W}^c$ on $\mathfrak{h}^*$ are described in all cases in Theorem 5.7. However, we make no use of this action when $\mathfrak{g} = \mathfrak{p}(n)$, see Subsection 3.1.3.

5.2.2. *An action on* $\mathbb{T}$. We need a variant $\mathfrak{W}^c_*$ of $\mathfrak{W}^c$. In the KM case define $\mathcal{G}^c_*$ in exactly the same as $\mathcal{G}^c$, except that the base of $\mathcal{G}^c_*$ is $\Delta_{iso} \times \mathtt{k}^*$. Then set $\mathfrak{W}^c_* = W \coprod \mathcal{G}^c_*$. There is an action of $\mathfrak{W}^c_*$ on $\mathbb{T}$, corresponding to a functor from $\mathfrak{W}^c_*$ to $\mathfrak{A}(\mathbb{T})$. As before, the base point $[W]$ maps to the whole space $\mathbb{T}$. The base element (α, t) maps to the subtorus $\mathtt{T}_\alpha$ defined in (3.4). The morphism $\tau_{\alpha,t} : \mathtt{T}_\alpha \longrightarrow \mathtt{T}_\alpha$ acts via

$$\tau_{\alpha,t}(x) = c_\alpha(t)x,\ x \in \mathtt{T}_\alpha. \tag{5.3}$$

If $G = P(n)$ or $Q(n)$, The groupoids $\mathcal{G}^c_*$ and $\mathfrak{W}^c_*$ are defined by analogy with the groupoids for $\mathfrak{p}(n)$ or $\mathfrak{q}(n)$ except that the base of $\mathcal{G}^c_*$ is $\mathfrak{B}^c_* = \{(\pm\overline{\alpha}, t) \mid \alpha \in \Delta^+, t \in \mathtt{k}^\times\}$. There is a functor $\mathfrak{W}^c_* \longrightarrow \mathfrak{A}(\mathbb{T})$, such that $W = N_{G_0}(\mathbb{T})/\mathbb{T}$ acts in the obvious way. The base elements $(\pm\overline{\alpha}, t)$ map to $\mathtt{T}_\alpha = \text{Ker } \mathtt{e}^{\overline{\alpha}}$, and for $\alpha \in \Delta^+_0$ $\tau_{\alpha,t}$ acts via (5.3).

LEMMA 5.2. *If* $(\alpha, \overline{\beta}) = 0$, *then* $(\beta, \overline{\alpha}) = 0$.

PROOF. Suppose $\alpha = \epsilon_i - \epsilon_j$, and $\beta = \epsilon_k - \epsilon_\ell$ where $i < j, k < \ell$. If $(\alpha, \overline{\beta}) = 0$, then $0 = \delta_{i,k} + \delta_{i,\ell} - \delta_{j,k} - \delta_{j,\ell}$. If $i = k$, then $j = \ell$ and hence $(\beta, \overline{\alpha}) = \delta_{i,k} - \delta_{i,\ell} + \delta_{j,k} - \delta_{j,\ell} = 0$. Similar arguments for the cases $i = \ell$ and $i \neq k, \ell$ complete the proof. □

Equation (5.3) is a multiplicative analog of (5.2). If $\alpha \in \Delta_{iso}$ is isotropic, $\mathtt{T}_\alpha$ is a codimension one subtorus in $\mathtt{T}_\alpha$, and by (3.5) Im $c_\alpha \subseteq \mathtt{T}_\alpha$. Then $\tau_{\alpha,t}$ translates $x \in \mathtt{T}_\alpha$ using the one-parameter subgroup c_α. In types P and Q, this holds since $(\alpha, \overline{\alpha}) = 0$.

5.2.3. *Orbits.* If $\mathfrak{G}$ is the coutinuous Weyl groupoid as in Notation 1.1, we describe the orbit $\mathfrak{G}\lambda$ of $\lambda \in X$. Define the *degree of atypicality* of λ to be

$$\text{atyp}\,\lambda = \max\{s|(\lambda, A) = 0 \text{ for some iso-set } A \text{ with } |A| = s\} \quad \text{if } X = \mathfrak{h}^*,$$

or

$$\text{atyp}\,\lambda = \max\{s|\lambda \in \bigcap_{\alpha\in A} \mathtt{T}_\alpha \text{ for some iso-set } A \text{ with } |A| = s\} \quad \text{if } X = \mathbb{T}.$$

We say λ is *typical* if atyp $\lambda = 0$. Next set

$$E(\lambda) = \{\alpha \in \Delta_{iso}|\lambda \in \Pi_\alpha\}, \quad \mathtt{E}_\lambda = \bigcup_{w\in W} w[\bigcup_{\alpha\in E(\lambda)} (\lambda + \mathtt{k}\alpha)]$$

or

$$E(\lambda) = \{\alpha \in \Delta_{iso}|\lambda \in \mathtt{T}_\alpha\}, \quad \mathtt{E}_\lambda = \bigcup_{w\in W} w[\bigcup_{\alpha\in E(\lambda)} (\text{Im}\, c_\alpha\lambda)],$$

if $X = \mathfrak{h}^*$ or $X = \mathbb{T}$ respectively. We need a preliminary result.

LEMMA 5.3. *We have* $\mathfrak{G}\lambda = \mathtt{E}_\lambda$.

PROOF. Clearly $\mathtt{E}_\lambda \subseteq \mathfrak{G}\lambda$. We show that for $\mu \in \mathfrak{G}_{(i+1)}\lambda$, we have $\mathtt{E}_\mu = \mathtt{E}_\lambda$. Write $\mu = g\nu$ where $\nu \in \mathfrak{G}_{(i)}\lambda$ and $g \in \mathfrak{G}_{(\nu)}$. By induction $\mathtt{E}_\nu = \mathtt{E}_\lambda$. Thus we can assume that $\nu = \lambda$. Since $\mathtt{E}_\lambda$ is W-invariant, we may further assume $g = \tau_{\alpha,t}$. The action of g is given by (5.2) or (5.3), so since g is defined at λ, $\lambda \in \Pi_\alpha$ or $\lambda \in \mathtt{T}_\alpha$. Hence $\alpha \in E(\lambda)$ and this gives the result. □

REMARK 5.4. If $\lambda \in \mathfrak{h}^*$, the proof shows that $\mathfrak{W}\lambda = \bigcup_{w\in W} w[\bigcup_{\alpha\in E(\lambda)}(\lambda+\mathbb{Z}\alpha)]$, which has closure $\mathfrak{W}^c\lambda$. This explains why our primary interest is in orbits of continuous Weyl groupoids. To give a definitive description of these orbits we need two more Lemmas.

LEMMA 5.5. *Let* $F(\lambda) \subseteq E(\lambda)$ *be an iso-set with* $|F(\lambda)| =$ atyp λ, *and set* $G(\lambda) = F(\lambda) \cup -F(\lambda)$. *If* $\beta \in E(\lambda)\backslash F(\lambda)$, *then for some reflection* $u \in W$ *we have* $u\lambda = \lambda$ *and* $u\beta = \pm\alpha \in G(\lambda)$.

PROOF. See [**Gor20**] Equation (14). For convenience we give a short proof. In the KM case, we can argue as follows. There is some $\alpha \in F(\lambda)$ such that $(\alpha, \beta) \neq 0$. The Jacobi identity applied to the three root vectors $e_{\pm\alpha} \in \mathfrak{g}^{\pm\alpha}$ and $e_\beta \in \mathfrak{g}^\beta$ implies that one of $[e_\beta, e_{\pm\alpha}]$ is non-zero. Replacing α by $-\alpha$ if necessary, we can assume $\gamma = \beta - \alpha$ is an (even) root. Then from the rank two case, [**Mus12**] Table 3.4.1 we see that γ-string through $-\beta$ consists of $-\alpha$ and $-\beta$. Thus if $u = s_\gamma$ is the reflection corresponding to γ, we have $u\beta = \alpha$. Since $(\lambda, \gamma) = 0$, u fixes λ. For types P and Q we adapt [**Gor20**]. We have $(\alpha, \overline{\beta}) \neq 0$ for some $\alpha \in F(\lambda)$. We can assume $\alpha = \epsilon_i - \epsilon_j, \beta = \pm(\epsilon_i - \epsilon_k) \in \Delta_0^+$, where $i < j \neq i, k$. Then

$$(\lambda, \overline{\beta}) = 0 = (\lambda, \overline{\alpha}),$$

and the result holds with $\gamma = \epsilon_j - \epsilon_k$. □

Next set

$$\mathtt{F}_\lambda = \bigcup_{w\in W} w(\lambda + \sum_{\alpha\in F(\lambda)} \mathtt{k}\alpha) \text{ or } \mathtt{F}_\lambda = \bigcup_{w\in W} w\left(\prod_{\alpha\in F(\lambda)} \text{Im}\, c_\alpha\right)\lambda$$

if $X = \mathfrak{h}^*$ or $X = \mathbb{T}$ respectively. The definitions do not change if $F(\lambda)$ is replaced by $G(\lambda)$.

LEMMA 5.6. *We have* $\mathtt{F}_\lambda \subseteq \mathfrak{G}\lambda$.

PROOF. By (5.3) and Lemma 5.2, if $\alpha, \beta \in F(\lambda)$, $s, t \in \mathtt{k}$ and $\lambda \in \mathtt{T}_\alpha \cap \mathtt{T}_\beta$ then $c_\alpha(t)\lambda \in \mathtt{T}_\beta$. Similar remarks apply to (5.2) if $\lambda \in \Pi_\alpha \cap \Pi_\beta$. This implies the statement. □

THEOREM 5.7. *We have*

(a) $\mathfrak{G}\lambda = \mathtt{F}_\lambda$,
(b) $\dim \mathfrak{G}\lambda = \operatorname{atyp} \lambda$,
(c) *Every* $\mathfrak{G}$*-orbit is closed.*
(d) *Let* $\sim$ *be the relation from Corollary* 4.3 *(b). Then* $\mu \sim \lambda$ *iff* $\mathfrak{G}\mu = \mathfrak{G}\lambda$.
(e) *If* $m_\lambda = M_\lambda \cap \mathcal{A}$ *as in Corollary* 4.3, *then* $\mathcal{V}(m_\lambda) = \mathfrak{G}\lambda$.

PROOF. By Lemmas 5.3 and 5.6 $\mathtt{F}_\lambda \subseteq \mathtt{E}_\lambda = \mathfrak{G}\lambda$. We show $\mathtt{E}_\lambda \subseteq \mathtt{F}_\lambda$. If $\beta \in E(\lambda)\backslash F(\lambda)$, let u be as in Lemma 5.5. Then $u(\lambda + \mathtt{k}\beta) = (\lambda + \mathtt{k}\alpha)$ and $u(c_\beta(t)\lambda) = c_\alpha(t)\lambda$ in the two cases under consideration. This shows (a) and (b), (c) are immediate consequences. By definition $\sim$ is the smallest equivalence relation satisfying the conditions in Corollary 4.3 (b). Hence if $\mu \sim \lambda$ we have $\mathfrak{G}\mu = \mathfrak{G}\lambda$. Conversely if $\mathfrak{G}\mu = \mathfrak{G}\lambda$, then $\mu \sim \lambda$ by Lemma 5.3, proving (d). From Corollary 4.3 and (d) we have

$$m_\lambda = m_\mu \text{ iff } \mu \sim \lambda \text{ iff } \mathfrak{G}\mu = \mathfrak{G}\lambda. \tag{5.4}$$

Since $\lambda \in \mathcal{V}(m_\lambda)$ this implies $\mathcal{V}(m_\lambda) \supseteq \mathfrak{G}\lambda$. For the other inclusion, if $\mu \in \mathcal{V}(m_\lambda)$, then $m_\mu = \mathcal{I}_\mathcal{A}(\mu) \supseteq m_\lambda$, and equality must hold. Thus $\mu \in \mathfrak{G}\mu = \mathfrak{G}\lambda$, proving (e). □

We interpret some of our results geometrically. The map π from Lemma 5.1 restricts to the map

$$\psi : \operatorname{Max} \mathcal{O}(X) \longrightarrow \operatorname{Max} A. \tag{5.5}$$

as in (2.4).

COROLLARY 5.8. (a) *The map in* (5.5) *is surjective.*
(b) *Identifying* X *with* $\operatorname{Max} \mathcal{O}(X)$, *the fiber over* m_λ *is* $\mathfrak{G}\lambda$.

PROOF. (a) is a restatement of the Weak Nullstellensatz, Corollary 4.3 (a) and (b) follows from (5.4). □

PROPOSITION 5.9. *Assume Hypothesis* 3.3 *holds. Then*

(a) $\mathcal{O}(\mathbb{T})^{\mathfrak{W}^c_*} = J(G) \otimes_\mathbb{Z} \mathtt{k}$.
(b) $S(\mathfrak{h})^{\mathfrak{W}} = S(\mathfrak{h})^{\mathfrak{W}^c} = I(\mathfrak{h})$ *if* $\mathfrak{g} \neq \mathfrak{p}(n)$.

PROOF. By (1.3), (3.7), (3.8), and [**Mus22**] Lemma 3.1, if $f \in \mathcal{O}(\mathbb{T})$, then $f \in J(G) \otimes_\mathbb{Z} \mathtt{k}$ iff $f(\lambda) = f(c_\alpha(t)\lambda)$ for all relevant roots α, $t \in \mathtt{k}$ and all $\lambda \in \mathtt{T}_\alpha$. Equivalently $f \in \mathcal{O}(\mathbb{T})^{\mathfrak{W}^c_*}$. This shows (a) and the proof of (b) is similar using (1.2). The first equality in (b) holds because any polynomial that is constant on infinitely many points on a line is constant on the whole line. □

6. The geometry of superalgebraic sets.

6.1. Outline. Let $\mathfrak{g}$ be a KM Lie superalgebra and let G be the corresponding supergroup as in Lemma 3.4. We use Notation 1.1. We define the S-topology on Spec $\mathcal{O}(X)$ by declaring that the S-closed sets are the $\mathfrak{G}$-stable (Zariski) closed sets. To give an idea of the geometry of superalgebraic sets, we show that they are are precisely the the S-closed sets. We also compute the S-closure V^S of an arbitrary closed set $V = V_0 = \mathcal{V}(I)$ with I a W-invariant radical ideal in $\mathcal{O}(X)$. From V we obtain a superalgebraic set which we call the *heart* and another closed set $V_1 = \mathcal{V}(I_1)$ with I_1 a W-invariant radical ideal of $\mathcal{O}(X)$. The process can be repeated giving new closed sets V_i and new hearts. Then V^S is the union of all the hearts and also the union of all the V_i. The hearts are determined by the minimal degrees of atypicality in the Zariski irreducible components of the V_i. A somewhat simplified version of this procedure and some examples are given in [**Mus22**]. If the defining equations for V are known, it can be done algorithmically using Groebner bases, see [**Mus22**] Example 2.11. The heart is defined in the next subsection, and then V^S is analyzed using projections onto intersections of hyperplanes corresponding to iso-sets. Let

$$\text{atyp } V = \max\{s| \text{ for some } \lambda \in V, \text{atyp } \lambda = s\}.$$

Now suppose r = atyp V and choose an iso-set $S = \{\beta = \beta_1, \beta_2, \dots, \beta_r\}$ with $(\lambda, S) = 0$ for some $\lambda \in V$. Since we can replace $\alpha \in S$ by $-\alpha$ if necessary, we can assume $S \subseteq \Omega$.

6.2. The heart. Suppose $I = \bigcap_{i=1}^{\ell} P_i$, an irredundant intersection of prime ideals of $\mathcal{B}$, where $T \notin P_i$ iff $i \in [k]$. Equivalently $i \in [k]$ iff $\mathcal{V}(P_i)$ contains typical points. For $i \in [k]$, set $Q_i = P_i \cap \mathcal{A}$. Then set

$$\text{heart } V = \text{heart}_0 V = \mathcal{V}(\bigcap_{i=1}^{k} Q_i), \quad V_{\text{reg}} = \mathcal{V}(\bigcap_{i=1}^{k} P_i).$$

If $k = 0$, set heart $V = \emptyset$. Note that heart V is superalgebraic, since $\bigcap_{i=1}^{k} Q_i$ is an ideal of $\mathcal{A}$. Set $V^c = \mathcal{V}(I, T)$.

LEMMA 6.1. *For $i \in [k]$,*

(a) *$\mathcal{V}(P_i) \subseteq \mathcal{V}(Q_i)$ and $V_{\text{reg}} \subseteq$ heart V.*
(b) *If $\lambda \in \mathcal{V}(Q_i)$, and $T(\lambda) \neq 0$, then $\lambda \in \mathcal{V}(P_i)$.*
(c) *If $\lambda \in$ heart V, and $T(\lambda) \neq 0$, then $\lambda \in V_{\text{reg}}$.*
(d) heart $V \backslash V_{\text{reg}} \subseteq V^c$.
(e) *If $i > k$, then $\mathcal{V}(P_i) \subseteq V^c$.*

PROOF. (a) is clear since the $\mathcal{V}(P_i)$ and $\mathcal{V}(Q_i)$ are the irreducible components of V_{reg} and heart V respectively. Next if $f \in P_i$ we have $T^b f \in Q_i$ for some $b \geq 0$. So if $\lambda \in \mathcal{V}(Q_i)$, and $T(\lambda) \neq 0$ we obtain $f(\lambda) = 0$, proving (b) and (c) is proved similarly. By (c), if $\lambda \in$ heart $V \backslash V_{\text{reg}}$, then $T(\lambda) = 0$ and (d) follows from this. Finally (e) follows since $T \in P_i$ for $i > k$. $\square$

Note that

$$V = V_{\text{reg}} \cup \bigcup_{i>k} \mathcal{V}(P_i). \tag{6.1}$$

COROLLARY 6.2.

$$V = \text{heart } V \cup V^c.$$

PROOF. The inclusion $\subseteq$ follows from (a), (e) in the Lemma and (6.1). The opposite inclusion results from $V^c \subseteq V$, (d) in the Lemma and (6.1). $\square$

6.3. Analysis using projections. If $q : X \longrightarrow Y$ is a morphism of affine varieties, define the comorphism $q^* : \mathcal{O}(Y) \longrightarrow \mathcal{O}(X)$ by $(q^*f)(x) = (fq)(x)$ for all $f \in \mathcal{O}(Y)$ and $x \in X$. It is easy to check the following.

LEMMA 6.3. *Let $i : Y \longrightarrow X$ be an embedding of affine varieties and suppose $q : X \longrightarrow Y$ is a morphism such that $qi = \mathrm{id}_Y$. If $I = \ker i^*$, then*

$$I \cap q^*\mathcal{O}(Y) = 0 \text{ and } I + q^*\mathcal{O}(Y) = \mathcal{O}(X).$$

Furthermore, if V is the closed subset of Y defined by the radical ideal K of $\mathcal{O}(Y)$, then as a closed subset of X we have $V = \mathcal{V}(q^*K)$.

EXAMPLE 6.4. Let $X = \mathfrak{h}^*$ and $X(\beta) = \mathfrak{h}(\beta)^*$. Define $q_\beta : \mathfrak{h}^* \longrightarrow \mathfrak{h}(\beta)^*$ by restriction $q_\beta(\lambda) = \lambda|_{\mathfrak{h}(\beta)}$, and let $i_\beta : \mathfrak{h}(\beta)^* \longrightarrow \mathfrak{h}^*$ be the map induced by the inclusion $\Delta(\beta) \subset \Delta$. The map q_β can also be described as follows. IAN pink notebook Let $\{\epsilon_i\}_{i=1}^p$ and $\{h_i\}_{i=1}^p$ be dual bases for $\mathfrak{h}(\beta)^*, \mathfrak{h}(\beta)$. Extend the h_i to functionals on $\mathfrak{h}^*$ that vanish on the orthogonal complement of $\mathfrak{h}(\beta)^*$ in $\mathfrak{h}^*$, see (1.5). Then for all $\lambda \in \mathfrak{h}^*$ we have $q_\beta(\lambda) = \sum_{i=1}^p h_i(\lambda)\epsilon_i$.

EXAMPLE 6.5. If $X = \mathbb{T}$, let $X(\beta) = \mathbb{T}(\beta)$ be the subtorus of X generated by $\mathrm{Im}\, c_\alpha$ for $\alpha \in \Delta(\beta)$. Let $\{\chi_i\}_{i=1}^p$ and $\{c_i\}_{i=1}^p$ be bases for $X(\mathbb{T}(\beta)), Y(\mathbb{T}(\beta))$ respectively such that $\chi_i c_j(t) = t^{\delta_{i,j}}$. Extend the c_i to $Y(\mathbb{T})$ by analogy with the previous example. Then define $q_\beta : \mathbb{T} \longrightarrow \mathbb{T}(\beta)$ by $q_\beta(t) = \prod_{i=1}^p c_i(\chi_i(t))$.

Now for $X = \mathfrak{h}^*$ or $\mathbb{T}$, and β an isotropic root, define $X(\beta)$ as in the Examples. Define $p_\beta = i_\beta \circ q_\beta$. We have

$$\Pi_\beta = \mathfrak{h}(\beta)^* + \mathtt{k}\beta \text{ and } \mathtt{T}_\beta = \mathbb{T}(\beta) \times \mathrm{Im}\, c_\beta. \tag{6.2}$$

For the torus case, $\mathbb{T}(\beta) \subset \mathtt{T}_\beta$ by Lemma 3.5 and equality holds because $\mathtt{T}_\beta$ is the subtorus generated by $\mathrm{Im}\, c_\alpha$ for $\alpha \in \Delta$ with $(\alpha, \beta) = 0$.

LEMMA 6.6. (a) *If $\lambda \in \Pi_\beta$, then $\lambda - p_\beta(\lambda) \in \mathtt{k}\beta$.*
(b) *If $t \in \mathtt{T}_\beta$, then $tp_\beta(t)^{-1} \in \mathrm{Im}\, c_\beta$.*

PROOF. We prove (a). The proof of (b) is similar. By (6.2) we can write $\lambda = \lambda_0 + c\beta$ with $\lambda_0 \in \mathfrak{h}(\beta)^*$, $c \in \mathtt{k}$. Since $i_\beta \circ q_\beta$ is the identity on $\mathfrak{h}(\beta)^*$ and $q_\beta(\beta) = 0$, we obtain the result. $\square$

The action of W on $\mathcal{O}(X)$ is given by $(wf)(x) = f(w^{-1}x)$ for $w \in W, f \in \mathcal{O}(X)$ and $x \in X$. Then $w\Delta(\beta) = \Delta(w\beta)$, and we have $wX(\beta) = X(w\beta)$. Since W acts transitively on Ω we may assume $\sigma(w\beta) = w\sigma(\beta)$ in (1.5). Then

$$w \circ q_\beta(\lambda) = q_{w\beta}(w\lambda). \tag{6.3}$$

We have an embedding $q_\beta^* : \mathcal{O}(X(\beta)) \longrightarrow \mathcal{O}(X)$ as in Lemma 6.3 with $Y = X(\beta)$ and $q = q_\beta$. In what follows we suppress q_β^*.

Set $X_\beta = \Pi_\beta$ or $\Pi_\beta = \mathtt{T}_\beta$ in case (i) or (ii) of 1.1 holds respectively For $\beta \in \Delta_{iso}$, set $V_\beta = V \cap X_\beta$. Since $T(\lambda) = 0$ for $\lambda \in V^c$, we have

$$V^c = \bigcup_{\beta \in \Omega} V_\beta.$$

Next we study the sets V_β in more detail. If $\beta = \beta_1$, set $U_\beta = q_\beta(V_\beta) \subseteq X(\beta)$ and

$$K_\beta = \mathcal{I}(V_\beta) \cap \mathcal{O}(X(\beta)),$$

We denote the closure of $U \subseteq X$ by $\overline{U}$.

LEMMA 6.7. *For all $\beta \in \Omega$, in case (i) V^S contains the sets*

$$Y_\beta := \overline{U_\beta + \mathtt{k}\beta} = \overline{U_\beta} + \mathtt{k}\beta. \tag{6.4}$$

PROOF. If $\lambda \in V_\beta$, then by Lemma 6.6, $\lambda - p_\beta(\lambda) = a\beta$ for some $a \in \mathtt{k}$. By $\mathfrak{W}^c$-invariance, V^S contains $\lambda + (t-a)\beta = p_\beta(\lambda) + t\beta$ for all $t \in \mathtt{k}$. For the second equality in (6.4), note that $\overline{U_\beta} \cap \mathtt{k}\beta = 0$. □

Similarly in case (ii) V^S contains the sets

$$Y_\beta := \overline{U_\beta \times \operatorname{Im} c_\beta} = \overline{U_\beta} \times \operatorname{Im} c_\beta.$$

Next if $X = \mathfrak{h}^*$ set $z_\beta = h_{\sigma(\beta)}$, and if $X = \mathbb{T}$ set $z_\beta = \mathsf{e}^{\sigma(\beta)} - 1$. Then let L_β be the ideal of $\mathcal{O}(X)$ generated by K_β and z_β.

LEMMA 6.8. *We have $\overline{U}_\beta = \mathcal{V}(K_\beta)$ and $Y_\beta = \mathcal{V}(L_\beta)$.*

PROOF. The first statement follows by [**CLO97**], Theorem 3.2.3, page 122 and the second is an immediate consequence. □

The ideal K_β is called an *elimination ideal.* Next set $V_1 = \bigcup_{\beta\in\Omega} Y_\beta$. Then $V_1 = \mathcal{V}(I_1)$ where $I_1 = \bigcap_{\beta\in\Omega} L_\beta$. Because $I = \mathcal{I}(V)$ is a radical ideal, so too are K_β, L_β and I_1. By Lemma 6.7 we have

PROPOSITION 6.9. *The set V^S contains* heart $V \cup V_1$. *So, $V^S =$* heart $V \cup V_1^S =$ $V \cup V_1^S$.

The next result shows that V_1 and I_1 are W-invariant.

LEMMA 6.10. *If V is W-invariant, then for $w \in W$, we have $w(V_\beta) = V_{w\beta}$*

(a) $w(V_\beta) = V_{w\beta}$, $w(U_\beta) = U_{w\beta}$ *and* $w(Y_\beta) = Y_{w\beta}$
(b) $w(K_\beta\mathcal{O}(X)) = K_{w\beta}\mathcal{O}(X)$ *and* $w(L_\beta) = L_{w\beta}$.

PROOF. The first statement in (a) follows from W-invariance of the bilinear form (,), the second from (6.3) and the third is an immediate consequence. By Lemma 6.3, the closed sets defined by $w(K_\beta\mathcal{O}(X))$ and $K_{w\beta}\mathcal{O}(X)$ are $w(\overline{U}_\beta)$ and $\overline{U}_{w\beta}$. Since these are equal, the defining ideals are equal. This gives the first part in (b) and the second also follows. □

Now we repeat this process. Set $A = A(q) = \{\beta_1, \ldots, \beta_q\}$ and $V_A = V \cap \bigcap_{\beta\in A} \Pi_\beta$ in case (i) or $V_A = V \cap \bigcap_{\beta\in A} \mathrm{T}_\beta$ in case (ii). Define $p_A : X \longrightarrow X$ by $p_A = p_{\beta_q} \circ \ldots \circ p_{\beta_1}$.

Next, if $X_A = \bigcap_{\beta\in A} X(\beta)$, then $\mathcal{O}(X_A) = \bigcap_{\beta\in A} \mathcal{O}(X(\beta))$. Consider the elimination ideal

$$K_A = \mathcal{I}(V_A) \cap \mathcal{O}(X_A).$$

Let L_A be the ideal of $\mathcal{O}(X)$ generated by K_A and $z_{\beta_1}, \ldots, z_{\beta_q}$. If $U_A = p_\alpha(V_A)$, then $\overline{U}_A = \mathcal{V}(K_A)$.

LEMMA 6.11. *For all $q \leq r$, V^S contains the sets*

$$\{p_{A(q)}(\lambda) + \sum_{\beta \in A(q)} \Bbbk\beta | \lambda \in V_{A(q)}\}.$$

$$\text{or } \{p_{A(q)}(\lambda) \prod_{\beta \in A(q)} \operatorname{Im} c_\alpha | \lambda \in V_{A(q)}\} \text{ in case (ii).}$$

Hence V^S also contains the closures $Y_{A(q)} = \mathcal{V}(L_{A(q)})$ of these sets.

PROOF. This follows as in the proof of Lemmas 6.7 and 6.8, by $\mathfrak{G}$-invariance. □

We define $V_q = \bigcup_{w \in W} wY_{A(q)}$ and $I_q = \bigcap_{w \in W} wL_{A(q)}$. Then $V_q = \mathcal{V}(I_q)$. By construction if $\lambda \in V_q$ then atyp $\lambda \geq q$. Suppose $I_q = \bigcap_{i=1}^{\ell} P_i$, an irredundant intersection of prime ideals, and suppose $i \in [k]$ iff $\mathcal{V}(P_i)$ contain a point λ with atyp $\lambda = q$. For $i \in [k]$, set $Q_i = P_i \cap A$. Then set $\text{heart}_q V = \mathcal{V}(\bigcap_{i=1}^{k} Q_i)$. If $k = 0$, set $\text{heart}_q V = \emptyset$.

THEOREM 6.12. *The S-closure of V is $V^S = \bigcup_{i=0}^{r} \text{heart}_i V = \bigcup_{i=0}^{r} V_i$. The set V^S is superalgebraic.*

PROOF. As in Proposition 6.9 if $q < r$, we have $V_q^S = \text{heart}_q V \cup V_{q+1}^S = V_q \cup V_{q+1}^S$ and $V_r^S = \text{heart}_r V$. This gives the equalities. Each $\text{heart}_i V$ is superalgebraic, so the result follows. □

COROLLARY 6.13. *The superalgebraic sets are exactly the closed sets that are invariant under the Weyl groupoid $\mathfrak{G}$.*

PROOF. If V is closed and $\mathfrak{G}$-invariant, then $V = V^S$ is superalgebraic. Conversely a superalgebraic set has the form $V = \mathcal{V}(I)$, with I an ideal of $\mathcal{A} = \mathcal{O}(X)^{\mathfrak{G}}$. Since elements of $\mathcal{A}$ are constant on orbits, V is a union of $\mathfrak{G}$-orbits. □

References

[CC11] C. Carmeli and G. Cassinelli, *Representations of super Lie groups: some remarks*, Supersymmetry in mathematics and physics, Lecture Notes in Math., vol. 2027, Springer, Heidelberg, 2011, pp. 45–67, DOI 10.1007/978-3-642-21744-9_3. MR2906337

[CCF11] C. Carmeli, L. Caston, and R. Fioresi, *Mathematical foundations of supersymmetry*, EMS Series of Lectures in Mathematics, European Mathematical Society (EMS), Zürich, 2011, DOI 10.4171/097. MR2840967

[CW12] S.-J. Cheng and W. Wang, *Dualities and representations of Lie superalgebras*, Graduate Studies in Mathematics, vol. 144, American Mathematical Society, Providence, RI, 2012, DOI 10.1090/gsm/144. MR3012224

[CLO97] D. Cox, J. Little, and D. O'Shea, *Ideals, varieties, and algorithms*, 2nd ed., Undergraduate Texts in Mathematics, Springer-Verlag, New York, 1997. An introduction to computational algebraic geometry and commutative algebra. MR1417938

[DK02] H. Derksen and G. Kemper, *Computational invariant theory*, Invariant Theory and Algebraic Transformation Groups, I, Springer-Verlag, Berlin, 2002. Encyclopaedia of Mathematical Sciences, 130, DOI 10.1007/978-3-662-04958-7. MR1918599

[DS05] M. Duflo and V. V. Serganova, *On associated variety for Lie superalgebras*, Preprint, `arXiv:math/0507198.`, 2005.

[Eis95] D. Eisenbud, *Commutative algebra*, Graduate Texts in Mathematics, vol. 150, Springer-Verlag, New York, 1995. With a view toward algebraic geometry, DOI 10.1007/978-1-4612-5350-1. MR1322960

[Ful89] W. Fulton, *Algebraic curves*, Advanced Book Classics, Addison-Wesley Publishing Company, Advanced Book Program, Redwood City, CA, 1989. An introduction to algebraic geometry; Notes written with the collaboration of Richard Weiss; Reprint of 1969 original. MR1042981

[Gor01] M. Gorelik, *The center of a simple P-type Lie superalgebra*, J. Algebra **246** (2001), no. 1, 414–428, DOI 10.1006/jabr.2001.8977. MR1872629

[Gor04] M. Gorelik, *The Kac construction of the centre of $U(\mathfrak{g})$ for Lie superalgebras*, J. Nonlinear Math. Phys. **11** (2004), no. 3, 325–349, DOI 10.2991/jnmp.2004.11.3.5. MR2084313

[Gor06] M. Gorelik, *Shapovalov determinants of Q-type Lie superalgebras*, IMRP Int. Math. Res. Pap. (2006), Art. ID 96895, 71. MR2282179

[Gor20] M. Gorelik, *Depths and cores in the light of DS-functors.*, Preprint, arXiv:2010.05721., 2020.

[GHSS22] M. Gorelik, C. Hoyt, V. Serganova, and A. Sherman, *The Duflo-Serganova functor, vingt ans après.*, Preprint, arXiv:2203.00529., 2022.

[HR18] C. Hoyt and S. Reif, *Grothendieck rings for Lie superalgebras and the Duflo-Serganova functor*, Algebra Number Theory **12** (2018), no. 9, 2167–2184, DOI 10.2140/ant.2018.12.2167. MR3894432

[IRS21] M. S. Im, S. Reif, and V. Serganova, *Grothendieck rings of periplectic Lie superalgebras*, Math. Res. Lett. **28** (2021), no. 4, 1175–1195, DOI 10.4310/MRL.2021.v28.n4.a8. MR4344700

[Kac77] V. G. Kac, *Lie superalgebras*, Advances in Math. **26** (1977), no. 1, 8–96, DOI 10.1016/0001-8708(77)90017-2. MR486011

[Kac84] V. G. Kac, *Laplace operators of infinite-dimensional Lie algebras and theta functions*, Proc. Nat. Acad. Sci. U.S.A. **81** (1984), no. 2, , Phys. Sci., 645–647, DOI 10.1073/pnas.81.2.645. MR735060

[Kap74] I. Kaplansky, *Commutative rings*, Revised edition, University of Chicago Press, Chicago, Ill.-London, 1974. MR345945

[Kos77] B. Kostant, *Graded manifolds, graded Lie theory, and prequantization*, Differential geometrical methods in mathematical physics (Proc. Sympos., Univ. Bonn, Bonn, 1975), Lecture Notes in Math., Vol. 570, Springer, Berlin-New York, 1977, pp. 177–306. MR580292

[Kos83] J.-L. Koszul, *Graded manifolds and graded Lie algebras*, Proceedings of the international meeting on geometry and physics (Florence, 1982), Pitagora, Bologna, 1983, pp. 71–84. MR760837

[Mus12] I. M. Musson, *Lie superalgebras and enveloping algebras*, Graduate Studies in Mathematics, vol. 131, American Mathematical Society, Providence, RI, 2012, DOI 10.1090/gsm/131. MR2906817

[Mus19] I. M. Musson, *The Nullstellensatz for supersymmetric polynomials.*, Preprint, arXiv:1905.04163., 2019.

[Mus22] I. M. Musson, *Weyl groupoids and superalgebraic sets.*, Preprint, arXiv:2211.10190., 2022.

[PT92] P. Pragacz and A. Thorup, *On a Jacobi-Trudi identity for supersymmetric polynomials*, Adv. Math. **95** (1992), no. 1, 8–17, DOI 10.1016/0001-8708(92)90042-J. MR1176151

[Rei23] S. Reif, *Grothendieck rings of queer Lie superalgebras*, Proc. Amer. Math. Soc. **151** (2023), no. 8, 3201–3211, DOI 10.1090/proc/16071. MR4591760

[Ser19] A. N. Sergeev, *On rings of supersymmetric polynomials*, J. Algebra **517** (2019), 336–364, DOI 10.1016/j.jalgebra.2018.10.003. MR3869278

[SV11] A. N. Sergeev and A. P. Veselov, *Grothendieck rings of basic classical Lie superalgebras*, Ann. of Math. (2) **173** (2011), no. 2, 663–703, DOI 10.4007/annals.2011.173.2.2. MR2776360

Department of Mathematical Sciences, University of Wisconsin-Milwaukee
Email address: musson@uwm.edu

Contemporary Mathematics
Volume **801**, 2024
https://doi.org/10.1090/conm/801/16011

A primer on twists in the noncommutative realm focusing on algebra, representation theory, and geometry

Pablo S. Ocal, Kenta Ueyama, and Padmini Veerapen

Abstract. We review several techniques that twist an algebra's multiplicative structure. We first consider twists by an automorphism, also known as Zhang twists, and we relate them to 2-cocycle twists of certain bialgebras. We then outline the classification and properties of twisted tensor products, and we examine twisted Segre products. Our exposition emphasizes clarity over generality, providing a wealth of interconnecting examples.

1. Introduction

Twisting the multiplicative and comultiplicative structure of algebras and coalgebras, respectively has long been an area of interest in mathematics; some classic examples include skew polynomial rings, skew group rings, and Ore extensions. In this paper, we present recent constructions that generalize those.

In the early 1990s, the notion of a twist of a graded algebra by an automorphism was introduced by Artin, Tate, and Van den Bergh in [**4**], and was later generalized by Zhang in [**49**]. This twist of a graded algebra became known as a Zhang twist, and it has played a pivotal role in various areas of noncommutative algebra and noncommutative projective geometry. Many fundamental properties of graded algebras are preserved under a Zhang twist, such as their graded module categories, noncommutative projective schemes, Gelfand-Kirillov dimension, global dimension, graded Ext groups, and Artin-Schelter regularity [**41**, **49**].

At the same time, similar notions were being developed for bialgebras and Hopf algebras. In the late 1980s, motivated by the tensor equivalence of two module categories, Drinfel'd [**20**] introduced a twist for the coalgebra structure of a Hopf algebra. These became known as Drinfel'd twists, and similarly to Zhang twists, they preserve important properties of Hopf algebras [**1**]. For instance, certain classes of cosemisimple unimodular, cosemisimple involutive, and cosemisimple quasitriangular finite dimensional Hopf algebras are invariant under Drinfel'd twists. The dual version of the Drinfel'd twist, called a 2-cocycle twist, was studied by Doi and Takeuchi [**18**, **19**] in the early 1990s. These modify the algebra structure of a Hopf

---

2020 *Mathematics Subject Classification.* Primary 16S80, 16S38, 16T99, 14A22.
Key words and phrases. Zhang twist, twisted tensor product, twisted Segre product.
The first author was supported by an AMS-Simons travel grant, and by the Hausdorff Research Institute for Mathematics funded by the Deutsche Forschungsgemeinschaft (DFG, German Research Foundation) under Germany's Excellence Strategy - EXC-2047/1 - 390685813. The second author was supported by JSPS KAKENHI Grant Numbers JP18K13381 and JP22K03222.

algebra and yield a tensor equivalence of the corresponding comodule categories. We focus on Zhang twists and 2-cocycle twists of $\mathbb{Z}$-graded algebras in Section 2, but there are other notions of twists of noncommutative algebras (see e.g., [**17**,**35**]).

Extending further the idea of modifying existing multiplicative structures, we consider twisted tensor products of algebras in Section 3. This notion was introduced by Čap, Schichl, and Vanžura [**13**], it is contemporary to Zhang twists and Drinfel'd twists, and when graded algebras are twisted the resulting twisted tensor product inherits a bigrading. Twisted tensor products were originally designed to provide an appropriate algebraic interpretation of the noncommutative product of topological spaces, but since they encompass vast families of algebras, they have also proven useful to study representation theory. For example, Ore extensions, noncommutative 2-tori, crossed products of C^*-algebras with groups, and universal enveloping algebras of Lie algebras, are all particular cases of twisted tensor products. Because of their generality, it is difficult to discern which properties or structures are preserved by twisted tensor products, and little is known besides Artin-Schelter regularity and Frobenius structures [**38**,**48**].

As an application of twisted tensor products, in Section 4 we consider twisted Segre products, introduced by the second author and collaborator in [**27**]. These are a noncommutative generalization of the Segre product in commutative algebra and algebraic geometry. In particular, we present a categorical relationship between twisted tensor products and twisted Segre products.

Convention. Throughout the paper, let $\Bbbk$ be a field. An *algebra* means a unital associative $\Bbbk$-algebra and unadorned $\otimes$ means $\otimes_{\Bbbk}$. For an algebra A we denote its unit map by $\eta_A : \Bbbk \to A$ and its multiplication map by $\nabla_A : A \otimes A \to A$. For a coalgebra A we denote its counit map by $\epsilon_A : A \to \Bbbk$ and its comultiplication map by $\Delta_A : A \to A \otimes A$. When A is graded, we denote by $|a|$ the degree of a homogeneous element $a \in A$. We denote the composition of maps by their concatenation.

2. Zhang twists of algebras and 2-cocycle twists of bialgebras

The notion of a twist of a $\mathbb{Z}$-graded algebra was first introduced in [**4**, Section 8]. Using the more general notion of twisting systems, Zhang proved in [**49**] that given two $\mathbb{N}$-graded algebras A and B generated in degree one, A is isomorphic to a twisted algebra of B if and only if the graded module categories of A and B are equivalent.

Definition 2.1. Let A be a $\mathbb{Z}$-graded algebra and ϕ be a graded degree-zero automorphism of A. The *right Zhang twist* A^ϕ of A by ϕ is the algebra that coincides with A as a graded vector space over $\Bbbk$, with the multiplication

$$r *_\phi s := r\phi^{|r|}(s), \quad \text{for any homogeneous elements } r, s \in A.$$

The *left Zhang twist* ${}^\phi A$ of A by ϕ is defined similarly, with the multiplication

$$r *_\phi s := \phi^{|s|}(r)s, \quad \text{for any homogeneous elements } r, s \in A.$$

For A and ϕ given in Definition 2.1, the left and right Zhang twist of A by ϕ are again graded associative algebras [**49**, Proposition 4.2]. Furthermore, there is a natural isomorphism $A^\phi \cong {}^{\phi^{-1}}A$ mapping a to $\phi^{-|a|}(a)$ [**34**, Lemma 4.1].

EXAMPLE 2.2. Consider the polynomial ring $\Bbbk[x_1, x_2, x_3, x_4]$ and define a graded degree-zero automorphism $\theta : \Bbbk[x_1, x_2, x_3, x_4] \to \Bbbk[x_1, x_2, x_3, x_4]$ by

$$\theta(x_1) = ax_1, \quad \theta(x_2) = ax_2, \quad \theta(x_3) = x_3, \quad \theta(x_4) = x_4$$

for some $a \in \Bbbk^\times$. The Zhang twist $\Bbbk[x_1, x_2, x_3, x_4]^\theta$ can be seen as the quotient of the free algebra on generators x_1, x_2, x_3, and x_4 subject to the relations

$$\begin{array}{lll} x_1x_2 - x_2x_1, & ax_1x_3 - x_3x_1, & ax_1x_4 - x_4x_1, \\ ax_2x_3 - x_3x_2, & ax_2x_4 - x_4x_2, & x_3x_4 - x_4x_3. \end{array}$$

This will be a running example throughout the paper.

Zhang twists may also be defined in the more general setting of an algebra graded by a semigroup by considering twisting systems [**49**].

DEFINITION 2.3 ([**49**, Definition 2.1]). Let $A = \bigoplus_{n\in\mathbb{Z}} A_n$ be a $\mathbb{Z}$-graded algebra and let $\tau = \{\tau_n | n \in \mathbb{Z}\}$ be a set of $\Bbbk$-linear automorphisms of A. We call τ a *twisting system* when

$$\tau_n(r\tau_m(s)) = \tau_n(r)\tau_{n+m}(s)$$

for all $n, m, \ell \in \mathbb{Z}$ and $r \in A_m, s \in A_\ell$. Given a twisting system, we may define a new graded multiplication $*$ on the underlying graded $\Bbbk$-module $\bigoplus_{n\in\mathbb{Z}} A_n$ by

$$r * s = r\tau_m(s)$$

for all $r \in A_m$, $s \in A_\ell$.

In the case of a $\mathbb{Z}$-graded algebra, such a twisting system arises from a single algebra automorphism of the associated $\mathbb{Z}$-algebra [**43**, Proposition 4.2].

EXAMPLE 2.4. [**46**] Suppose $\operatorname{char}(\Bbbk) \neq 2$. Consider the following family of quadratic algebras

$$A(\rho) = \Bbbk\langle x_1, \dots, x_4\rangle/(w_1, \dots, w_6),$$

where $\rho \in \Bbbk^\times$ and

$$\begin{array}{lll} w_1 = \rho^2 x_1x_2 + x_2x_1, & w_2 = \rho x_1x_3 - x_3x_1, & w_3 = \rho x_1x_4 + x_4x_1, \\ w_4 = x_2x_3 - \rho x_3x_2, & w_5 = x_2x_4 + \rho x_4x_2, & w_6 = x_3x_4 + x_4x_3 + x_1x_2. \end{array}$$

This is a family of quadratic Artin-Schelter regular algebras. It is noteworthy that the algebras in this family are twists of each other via a twisting system. In particular, the twisting system cannot be reduced to a twist by a single algebra automorphism.

We now consider Zhang twists on algebras and bialgebras (respectively Hopf algebras), where one should note that Zhang twists only deform the algebra structure. With this in mind, we let B be a $\mathbb{Z}$-graded bialgebra (respectively Hopf algebra). We say that an endomorphism $\phi \in \operatorname{End}_\Bbbk(B)$ is a *graded bialgebra automorphism* (respectively *graded Hopf algebra automorphism*) of B if it is a bialgebra (respectively Hopf algebra) automorphism of B preserving the grading of B, that is $\phi(B_n) \subseteq B_n$ for all $n \in \mathbb{Z}$.

The rest of the section is devoted to relating Zhang twists to 2-cocycle twists using recent work in [**28**] by the third author and collaborators. First, we review 2-cocycle twists of bialgebras and of Hopf algebras. As mentioned in the introduction, 2-cocycle twists originated from the work of Doi and Takeuchi [**18, 19**] as a dual version to the twist defined by Drinfel'd [**20**].

DEFINITION 2.5 ([**35**, Section 1]). Suppose H is a Hopf algebra. Let $\sigma : H \otimes H \to \Bbbk$ be a 2-cocycle for H. The bialgebra H^σ is called a *cocycle twist* of H with respect to σ when

(1) $H^\sigma = H$ as a coalgebra,
(2) H^σ has multiplication given by

$$h \cdot_\sigma l := \sum \sigma^{-1}(h_1, l_1) h_2 l_2 \sigma(h_3, l_3).$$

The bialgebra H^σ admits a Hopf algebra structure with the antipode

$$S^\sigma(h) := \sum \sigma^{-1}(h_1, Sh_2) Sh_3 \sigma(Sh_4, h_5).$$

We note that the notation H^σ is a modification of the one used in [**35**]. Now, to relate a Zhang twist to a 2-cocycle twist, we introduce the notion of *twisting pairs.*

DEFINITION 2.6 ([**28**, Definition D]). Let B be a bialgebra. A pair (ϕ_1, ϕ_2) of algebra automorphisms of B is said to be a *twisting pair* when the following conditions hold.

(1) $\Delta\phi_1 = (1 \otimes \phi_1)\Delta$ and $\Delta\phi_2 = (\phi_2 \otimes 1)\Delta$, and
(2) $\varepsilon\phi_1\phi_2 = \varepsilon$.

One may lift a bialgebra B to a Hopf algebra $\mathcal{H}(B)$ using Takeuchi's Hopf envelope construction [**44**]. Moreover, if B is $\mathbb{Z}$-graded as an algebra and $\Delta(B_n) \subset B_n \otimes B_n$ for all $n \in \mathbb{Z}$, then $\mathcal{H}(B)$ also satisfies these conditions. We use Takeuchi's explicit construction of $\mathcal{H}(B)$ in the result below.

THEOREM 2.7 ([**28**, Theorem 2.3.3]). *Let B be a bialgebra that is $\mathbb{Z}$-graded as an algebra and satisfying $\Delta(B_n) \subset B_n \otimes B_n$ for all $n \in \mathbb{Z}$. Given a twisting pair (ϕ_1, ϕ_2) of B there is a unique pair $(\mathcal{H}(\phi_1), \mathcal{H}(\phi_2))$ of the Hopf envelope of $\mathcal{H}(B)$ extending (ϕ_1, ϕ_2). Moreover, the 2-cocycle twist $\mathcal{H}(B)^\sigma$, with 2-cocycle $\sigma : \mathcal{H}(B) \otimes \mathcal{H}(B) \to \Bbbk$ given by*

$$\sigma(x, y) = \varepsilon(x)\varepsilon(\mathcal{H}(\phi_2)^{|x|}(y)) \quad \text{for any homogeneous elements } x, y \in \mathcal{H}(B),$$

is the right Zhang twist $\mathcal{H}(B)^{\mathcal{H}(\phi_1)\mathcal{H}(\phi_2)}$.

We now provide an application of the above construction by computing twisting pairs explicitly for $\mathcal{O}_q(M_n(\Bbbk))$, the one parameter quantization of the coordinate ring of $n \times n$ matrices.

EXAMPLE 2.8. Let $n \geq 2$ be an integer and let $q \in \Bbbk^\times$ be any scalar. As an algebra, $\mathcal{O}_q(M_n(\Bbbk))$ is generated by n^2-generators $\{x_{ij}\}_{1 \leq i,j \leq n}$ subject to the relations

$$\begin{cases} qx_{ks}x_{us} = x_{us}x_{ks} & \text{if } k < u \\ qx_{ks}x_{kv} = x_{kv}x_{ks} & \text{if } s < v \\ x_{us}x_{kv} = x_{kv}x_{us} & \text{if } s < v, k < u \\ x_{us}x_{kv} = x_{kv}x_{us} + (q - q^{-1})x_{ks}x_{uv} & \text{if } s < v, u < k. \end{cases}$$

The coalgebra structure on $\mathcal{O}_q(M_n(\Bbbk))$ given by

$$\Delta(x_{ij}) = \sum_{1 \leq k \leq n} x_{ik} \otimes x_{kj} \quad \text{and} \quad \varepsilon(x_{ij}) = \delta_{ij} \quad \text{for all } 1 \leq i, j \leq n,$$

makes $\mathcal{O}_q(M_n(\Bbbk))$ a bialgebra. Our goal is to lift $\mathcal{O}_q(M_n(\Bbbk))$ to its Hopf envelope $\mathcal{O}_q(\mathrm{GL}_n(\Bbbk))$, the one-parameter quantization of the coordinate ring of the general linear group. Following [**45**], this is achieved by first defining a coquasitriangular structure on $\mathcal{O}_q(M_n(\Bbbk))$ and then localizing at a central grouplike element. As it were, $\mathcal{O}_q(M_n(\Bbbk))$ is coquasitriangular as it is isomorphic to the coquasitriangular bialgebra $A(R_q)$ obtained via the Faddeev-Reshetikhin-Takhtajan construction (see e.g., [**12, 22**]). For any scalar $q \in \Bbbk^\times$, the classical Yang-Baxter operator R_q on V (cf. [**45**]) is given by

$$R_q(v_i \otimes v_j) = \begin{cases} qv_i \otimes v_i & i = j \\ v_i \otimes v_j & i < j \\ v_i \otimes v_j + (q - q^{-1})v_j \otimes v_i & i > j, \end{cases}$$

where V is an n-dimensional vector space with basis $\{v_1, \ldots, v_n\}$. The central group-like element g of $\mathcal{O}_q(M_n(\Bbbk))$ given by

$$g := \sum_{\sigma \in S_n} (-q)^{-l(\sigma)} x_{\sigma(1)1} \cdots x_{\sigma(n)n}$$

where $l(\sigma)$ denotes the length of the permutation σ, is called the q-determinant.

PROPOSITION 2.9 ([**28**, Proposition 3.1.7]). *All twisting pairs of $\mathcal{O}_q(\mathrm{GL}_n(\Bbbk))$ are of the form (ϕ_1, ϕ_2), where ϕ_1 and ϕ_2 are given by*

$$\begin{pmatrix} \phi_1(x_{11}) & \cdots & \phi_1(x_{1n}) \\ \vdots & \ddots & \vdots \\ \phi_1(x_{n1}) & \cdots & \phi_1(x_{nn}) \end{pmatrix} = \begin{pmatrix} x_{11} & \cdots & x_{1n} \\ \vdots & \ddots & \vdots \\ x_{n1} & \cdots & x_{nn} \end{pmatrix} \begin{pmatrix} \alpha_{11} & \cdots & \alpha_{1n} \\ \vdots & \ddots & \vdots \\ \alpha_{n1} & \cdots & \alpha_{nn} \end{pmatrix}$$

and

$$\begin{pmatrix} \phi_2(x_{11}) & \cdots & \phi_2(x_{1n}) \\ \vdots & \ddots & \vdots \\ \phi_2(x_{n1}) & \cdots & \phi_2(x_{nn}) \end{pmatrix} = \begin{pmatrix} \alpha_{11} & \cdots & \alpha_{1n} \\ \vdots & \ddots & \vdots \\ \alpha_{n1} & \cdots & \alpha_{nn} \end{pmatrix}^{-1} \begin{pmatrix} x_{11} & \cdots & x_{1n} \\ \vdots & \ddots & \vdots \\ x_{n1} & \cdots & x_{nn} \end{pmatrix},$$

with $(\alpha_{ij}) \in \mathrm{GL}_n(\Bbbk)$. Moreover, we have the following.

(1) *If $q = 1$, then the maps (ϕ_1, ϕ_2) defined above form a twisting pair for all $(\alpha_{ij}) \in \mathrm{GL}_n(\Bbbk)$.*
(2) *If* $\mathrm{char}(\Bbbk) \neq 2$ *and $q = -1$, then (α_{ij}) defines a twisting pair as above if and only if it is a generalized permutation matrix.*
(3) *If $q \neq \pm 1$, then (α_{ij}) defines a twisting pair as above if and only if it is is diagonal.*

In particular, $\phi_1(g^{-1}) = (-1)^{l(\tau)}|(\alpha_{ij})|^{-1}g^{-1}$, $\phi_2(g^{-1}) = (-1)^{l(\tau)}|(\alpha_{ij})|g^{-1}$, and $\phi_1\phi_2(g^{-1}) = g^{-1}$, where τ is trivial if $q \neq -1$, and $\tau \in S_n$ satisfied $|(\alpha_{ij})| = (-1)^{l(\tau)}\alpha_{\tau(1)1} \cdots \alpha_{\tau(n)n}$ if $q = -1$.

The above twisting pairs of $\mathcal{O}_q(\mathrm{GL}_n(\Bbbk))$ allow us to apply Theorem 2.7.

COROLLARY 2.10 ([**28**]). *The right Zhang twist of $\mathcal{O}_q(\mathrm{GL}_n(\Bbbk))$ by $\phi_1\phi_2$ is isomorphic to the 2-cocycle twist of $\mathcal{O}_q(\mathrm{GL}_n(\Bbbk))$ by the 2-cocycle σ satisfying*

$$\begin{aligned} &\sigma(x_{i_1j_1}^{p_1} \cdots x_{i_sj_s}^{p_s} g^{-r}, x_{u_1v_1}^{q_1} \cdots x_{u_kv_k}^{q_k} g^{-t}) \\ &= (-1)^{mtl(\tau)} \delta_{i_1,j_1} \cdots \delta_{i_s,j_s} \left(((\alpha_{ij})^{-m})_{u_1v_1}\right)^{q_1} \cdots \left(((\alpha_{ij})^{-m})_{u_kv_k}\right)^{q_k} |(\alpha_{ij})|^{mt}, \end{aligned}$$

where $m = p_1 + \cdots + p_s - nr$ and $1 \leq i_\bullet, j_\bullet, u_\bullet, v_\bullet \leq n$ and $p_\bullet, q_\bullet, s, k, r, t$ are non-negative integers, τ is trivial if $q \neq -1$, and $\tau \in S_n$ satisfies $|(\alpha_{ij})| = (-1)^{l(\tau)}\alpha_{\tau(1)1} \cdots \alpha_{\tau(n)n}$ if $q = -1$.

3. Twisted tensor products

When twisting the multiplication of a given algebra, a natural option is to consider noncommutative analogues of tensor products. Such an analogue was introduced by Čap, Schichl, and Vanžura [**13**], and is known as a *twisted tensor product*. These are ubiquitous in noncommutative algebra and noncommutative geometry.

Definition 3.1 ([**13**, Definition 2.3]). Let A and B be algebras. A *twisting map* $\tau : B \otimes A \to A \otimes B$ is a bijective $\Bbbk$-linear map satisfying

$$\begin{aligned} 1 \otimes \eta_B &= \tau(\eta_B \otimes 1), \\ \eta_A \otimes 1 &= \tau(1 \otimes \eta_A), \\ \tau(\nabla_B \otimes \nabla_A) &= (\nabla_A \otimes \nabla_B)(1 \otimes \tau \otimes 1)(\tau \otimes \tau)(1 \otimes \tau \otimes 1). \end{aligned}$$

The first two conditions express the compatibility of τ with units η_A and η_B, while the third ensures the compatibility of τ with the multiplication maps ∇_A and ∇_B. The commutative diagrams given by the above equalities are a convenient computational tool. Some authors allow twisting maps to be non-invertible, but that will not be our case. When working with graded algebras $A = \bigoplus_{i\in\mathbb{Z}} A_i$ and $B = \bigoplus_{i\in\mathbb{Z}} B_i$, it is often desirable to require that the twisting map be *strongly graded*, namely $\tau(B_j \otimes A_i) \subseteq A_i \otimes B_j$.

Definition 3.2 ([**13**, Definition 2.1]). Let A and B be algebras and let $\tau : B \otimes A \to A \otimes B$ be a twisting map. The *twisted tensor product* $A \otimes_\tau B$ is the $\Bbbk$-vector space $A \otimes B$ with the following unit and multiplication maps.

$$\begin{aligned} \eta_{A\otimes_\tau B} &:= (\eta_A \otimes \eta_B), \\ \nabla_{A\otimes_\tau B} &:= (\nabla_A \otimes \nabla_B)(1 \otimes \tau \otimes 1). \end{aligned}$$

Proposition 3.3 ([**13**, Proposition 2.3]). *Let A and B be algebras and let $\tau : B \otimes A \to A \otimes B$ be a twisting map. Then $(A \otimes_\tau B, \nabla_{A\otimes_\tau B}, \eta_{A\otimes_\tau B})$ is an algebra.*

In other words, the compatibility of τ with the unital associative algebra structures of A and B indeed guarantees that $A \otimes_\tau B$ inherits this structure. The compatibility of τ with ∇_A and ∇_B can be interpreted as the equalities

$$\begin{aligned} \tau(1 \otimes \nabla_A) &= (\nabla_A \otimes 1)(1 \otimes \tau)(\tau \otimes 1), \\ \tau(\nabla_B \otimes 1) &= (1 \otimes \nabla_B)(\tau \otimes 1)(1 \otimes \tau). \end{aligned}$$

Intuitively, the above equalities indicate that τ commutes with the multiplication maps. It turns out that an algebra Λ can be expressed as a twisted tensor product of two subalgebras A and B essentially when Λ is isomorphic to $A \otimes B$ as vector spaces, and surprisingly the twisting map is unique.

Theorem 3.4 ([**13**, Proposition 2.7]). *Let Λ, A, B be algebras. Then $\Lambda \cong A \otimes_\tau B$ for some twisting map $\tau : B \otimes A \to A \otimes B$ if and only if there are injective algebra morphisms $\iota_A : A \to \Lambda$ and $\iota_B : B \to \Lambda$ such that $\nabla_\Lambda(\iota_A \otimes \iota_B) : A \otimes B \to \Lambda$ is an isomorphism of $\Bbbk$-vector spaces. Moreover*

$$\tau = (\nabla_\Lambda(\iota_A \otimes \iota_B))^{-1}\nabla_\Lambda(\iota_B \otimes \iota_A)$$

whence τ is unique.

Note that $\iota_A : A \to \Lambda$ and $\iota_B : B \to \Lambda$ are then given by $\iota_A(a) = a \otimes 1_B$ and $\iota_B(b) = 1_A \otimes b$. Moreover, twisted tensor products inherit a universal property that recovers the usual universal property of the usual tensor product when the twisting map is trivial.

THEOREM 3.5. *Let A, B be algebras, let $\tau : B \otimes A \to A \otimes B$ be a twisting map, let $\iota_A : A \to A \otimes_\tau B$ and $\iota_B : B \to A \otimes_\tau B$ be such that $\tau = \nabla_{A\otimes_\tau B}(\iota_A \otimes \iota_B)$. If Λ is an algebra and $i_A : A \to \Lambda$ and $i_B : B \to \Lambda$ are algebra morphisms such that $\nabla_\Lambda(i_B \otimes i_A) = \nabla_\Lambda(i_A \otimes i_B)\tau$ then there is a unique algebra morphism $h : A \otimes_\tau B \to \Lambda$ satisfying $h\iota_A = i_A$ and $h\iota_B = i_B$.*

In other words, the triple $(A, A \otimes_\tau B, B)$ can be seen as an initial object in the class of cospans $A \to \Lambda \leftarrow B$ of algebras having varying apex Λ, fixed feet A and B, and commute with τ in the appropriate sense. As with the usual tensor product, twisted tensor products can be iterated, and the iterations will inherit the aforementioned universal property.

THEOREM 3.6 ([**30**, Theorem 2.1]). *Let A, B, C be algebras and let $\tau_{AB} : B \otimes A \to A \otimes B$, $\tau_{BC} : C \otimes B \to B \otimes C$, $\tau_{AC} : C \otimes A \to A \otimes C$ be twisting maps. Then $(1 \otimes \tau_{BC})(\tau_{AC} \otimes 1) : C \otimes (A \otimes_{\tau_{AB}} B) \to (A \otimes_{\tau_{AB}} B) \otimes C$ is a twisting map if and only if $(\tau_{AB} \otimes 1)(1 \otimes \tau_{AC}) : (B \otimes_{\tau_{BC}} C) \otimes A \to A \otimes (B \otimes_{\tau_{BC}} C)$ is a twisting map. In that case we have the isomorphism of algebras*

$$(A \otimes_{\tau_{AB}} B) \otimes_{(1\otimes\tau_{BC})(\tau_{AC}\otimes 1)} C \cong A \otimes_{(\tau_{AB}\otimes 1)(1\otimes\tau_{AC})} (B \otimes_{\tau_{BC}} C).$$

This gives one additional compatibility condition on the twisting maps, but as long as it is satisfied, the twisted tensor product of a family of algebras can be found by consecutively forming twisted tensor products.

THEOREM 3.7 ([**30**, Theorem 2.9]). *Fix $n \in \mathbb{N}$. Let A_i be algebras for all $i = 1, \ldots, n$, and let $\tau_{ij} : A_j \otimes A_i \to A_j \otimes A_i$ be twisting maps for all $1 \leq i < j \leq n$. If τ_{ij}, τ_{jk}, and τ_{ik} are compatible for all $1 \leq i < j < k \leq n$ then the iterated twisted tensor product $A_1 \otimes_{\tau_{12}} \cdots \otimes_{\tau_{i-1i}} A_i \otimes_{\tau_{ii+1}} \cdots \otimes_{\tau_{n-1n}} A_n$ is well defined.*

Twisted tensor products present a higher level of complexity when compared to their commutative counterparts. Of course, when the twisting map transposes the elements, namely $\tau : B \otimes A \to A \otimes B$ is given by $\tau(b \otimes a) = a \otimes b$ for all $a \in A$ and $b \in B$, we recover the usual tensor product, but classifying the other outcomes for most twisting maps is a gargantuan task. As a result of its universal property, a twisted tensor product is uniquely determined (whence in bijection) by a twisting map, whence it is enough to classify twisting maps [**14**, Theorem 2.4]. The first approaches to a classification consisted essentially of treating twisted tensor products on a case-by-case basis. For example, one can see that the available options for $A \otimes_\tau \Bbbk[x]$ coincide precisely with the Ore extensions of A.

EXAMPLE 3.8. Consider an algebra A, an algebra automorphism $\sigma : A \to A$, and a $\Bbbk$-linear map $\delta : A \to A$ such that $\delta(\nabla(r \otimes s)) = \nabla(\delta(r) \otimes s) + \nabla(\sigma(r) \otimes \delta(s))$ for all $r, s \in A$. We define a twisting map $\tau : \Bbbk[x] \otimes R \to R \otimes \Bbbk[x]$ by setting $\tau(x \otimes r) = \sigma(r) \otimes x + \delta(r) \otimes 1$, and extend by linearity, together with imposing the compatibility conditions to compute $\tau(x^i \otimes r)$ for all $i \in \mathbb{N}$. The resulting twisted tensor product is isomorphic to the Ore extension $A[x; \sigma, \delta]$, namely the vector space $A[x]$ with multiplication $xr = \sigma(r)x + \delta(r)$ for all $r \in A$ (see [**32**]).

The case $A \otimes_\tau \Bbbk^2$ has a similar flavor to those of Ore extensions.

THEOREM 3.9 ([**14**, Proposition 2.10]). *Let A be an algebra, let $\sigma : A \to A$ be an algebra automorphism, let ${}_\sigma A$ be the A bimodule with underlying vector space A and action $\rho : A \otimes {}_\sigma A \otimes A \to {}_\sigma A$ given by $\rho := \nabla(\sigma \otimes \nabla)$, let $\delta : A \to {}_\sigma A$ be an idempotent derivation satisfying $\delta = \delta^2 + \sigma\delta + \delta\sigma$. The twisted tensor products $A \otimes_\tau \Bbbk^2$ are in bijection with the pairs (σ, δ).*

There are only partial results for the case of algebras with truncated polynomials [**25**]. For $\Bbbk[y] \otimes_\tau \Bbbk[x]/(x^2)$ the classification is already considerably harder.

THEOREM 3.10 ([**24**, Section 4]). *Let $\Bbbk$ be a field of characteristic zero. A choice of $m \in 2\mathbb{N}$, $p_0, q_m \in \Bbbk^\times$, and $q_0, \dots, q_{m-1} \in \Bbbk$ uniquely determines the polynomials $p(y) = -y + p_0$ and $q(y) = q_0 + q_1 y + \cdots + q_m y^m$, corresponding to a unique twisting map $\tau : \Bbbk[x]/(x^2) \otimes_\tau \Bbbk[y] \to \Bbbk[y] \otimes_\tau \Bbbk[x]/(x^2)$ given as $\tau(x \otimes a(y)) = \iota_0(a(y)) \otimes 1 + \iota_1(a(y)) \otimes x$ where $\iota_0(y) = q(y)$, $\iota_1(y) = p(y)$, $\iota_1(a(y)b(y)) = \iota_1(a(y))\iota_1(b(y))$, $\iota_0(a(y)b(y)) = \iota_0(a(y))b(y) + \iota_1(a(y))\iota_0(b(y))$, $\iota_0(a(y))^2 = 0$, and $\iota_0\iota_1(a(y)) = -\iota_1\iota_0(a(y))$ for all $a(y), b(y) \in \Bbbk[y]$.*

Some partial results have been achieved for $\Bbbk^n$ with $\Bbbk^m$ where $n, m \in \mathbb{N} \setminus \{0\}$ [**2**, **29**]. These rely on rewriting the twisting map $\tau : \Bbbk^m \otimes \Bbbk^n \to \Bbbk^n \otimes \Bbbk^m$ as a family of matrices $\mathcal{M}_\tau$, translating the classification to a linear algebra problem. Given $e_1, \dots, e_m$ and $f_1, \dots, f_n$ the canonical basis of $\Bbbk^m$ and $\Bbbk^n$ respectively, we can write $\tau(e_i \otimes f_j) = \sum_{r,s} \lambda_{ij}^{rs} f_r \otimes e_s$. Set $M_\tau(i,s)_{rs} = \lambda_{ij}^{rs}$ and $\mathcal{M}_\tau = \{M_\tau(i,s)\}_{r,j}$ for $i, s = 1, \dots, m$, $j, r = 1, \dots, n$. Then τ is a twisting map if and only if $M_\tau(1,s), \dots, M_\tau(m,s)$ are orthogonal idempotent for all $s = 1, \dots, m$, and the dual condition exchanging r and s is also satisfied, and the identity matrix 1_n is in the image of $A(i,i)$ for all $i = 1, \dots, m$, and the dual condition exchanging i and j is also satisfied.

There are three main families of twisting maps arising in this way. The *standard* twisting maps τ have only entries 0 or 1 in all the columns of $M_\tau(s,s)$, and the non-zero entries of $M_\tau(i,s)$ are at most the non-zero entries of $M_\tau(s,s)$, for all $i, s = 1, \dots, m$. These give algebras isomorphic to certain square zero radical truncated quiver algebras. The *quasi-standard* twisting maps τ arise as generalizations of the standard ones where some of the conditions are relaxed, and give algebras that are formal deformations of the standard ones. The third family is given only when $n = m$, whenever there are $v_1, \dots v_n \in \Bbbk^n$ invertible such that $\det(v_1^T, \dots, v_n^T) = 1$, in which case for all $i, s = 1, \dots, n$

$$M_\tau(i,s) = (-1)^{i-1}(v_s^{-1} \cdot v_i)^T (v_l \cdot (v_1 \times \cdots \times \widehat{v_i} \times \cdots \times v_n))$$

and these give algebras isomorphic to the square matrices over the field $\Bbbk$. In general these three families of twisting maps do not give a complete description of all the possibilities, but for $\Bbbk^3$ with $\Bbbk^3$ the classification of the remaining ones was completed on a case-by-case basis [**3**].

The study of twisted planes exploits Koszulness and Artin-Schelter regularity, and techniques of linear algebra involving infinite matrices give an almost complete classification [**6**].

THEOREM 3.11 ([**16**, Theorem 1.3]). *Let $\Bbbk$ be an algebraically closed field. A quadratic twisted tensor product $\Bbbk[x] \otimes_\tau \Bbbk[y]$ is isomorphic as a graded twisted tensor product to*

$$\Bbbk\langle x, y\rangle/(yx - ax^2 - bxy - y^2) \quad \text{or} \quad \Bbbk\langle x, y\rangle/(yx - ax^2 - bxy)$$

for some $a, b \in \Bbbk$.

Note that the notion of isomorphism as graded twisted tensor products is stronger than that of isomorphism as graded algebras.

In general, little more is classified besides the above. A second typical approach would be to inquire whether twisted tensor products inherit any properties from the given algebras. For example, since connected Ore extensions preserve Artin-Schelter regularity [**50**, Theorem 0.2], one may ask what other properties are preserved. In general, this is a hard question. There is an appropriate sense in which Artin-Schelter regularity is preserved [**48**].

EXAMPLE 3.12. Let $A = \Bbbk[u,v]$ and $B = \Bbbk[x,y]$ with $|u| = |v| = |x| = |y| = 1$. Consider the strongly graded twisting map $\tau : B \otimes A \to A \otimes B$ given by

$$\tau(x\otimes u) = au\otimes x, \quad \tau(x\otimes v) = bv\otimes x, \quad \tau(y\otimes u) = cu\otimes y, \quad \tau(y\otimes v) = dv\otimes y$$

for some $a,b,c,d \in \Bbbk^\times$. The twisted tensor product $A\otimes_\tau B$ is the free algebra on generators u,v,x, and y subject to the following relations

$$\begin{array}{ccc} uv - vu, & aux - xu, & cuy - yu, \\ bvx - xv, & dvy - yv, & xy - yx, \end{array}$$

where we slightly abuse notation and write $u = u\otimes 1, v = v\otimes 1, x = 1\otimes x, y = 1\otimes y$. Observe that $A\otimes_\tau B$ has a PBW-type decomposition, namely it can be identified with the $\Bbbk$-vector space having basis $\{u^i v^j x^r y^s\}_{i,j,r,s\in\mathbb{N}}$ and multiplication

$$u^{i_1}v^{j_1}x^{r_1}y^{s_1}\cdot u^{i_2}v^{j_2}x^{r_2}y^{s_2} = a^{i_1 i_2} b^{j_1 j_2} c^{r_1 r_2} d^{s_1 s_2} u^{i_1+i_2} v^{j_1+j_2} x^{r_1+r_2} y^{s_1+s_2}$$

for all $i_1,i_2,j_1,j_2,r_1,r_2,s_1,s_2 \in \mathbb{N}$. In particular, $A\otimes_\tau B$ inherits a $\mathbb{Z}$-grading by setting $|u^i v^j x^r y^s| = i+j+r+s$ and $A\otimes_\tau B$ is a quadratic Artin-Schelter regular algebra of global dimension four. If $a = b = c = d$, then $A\otimes_\tau B$ is given by a Zhang twist of the polynomial ring $\Bbbk[u,v,x,y]$. Indeed, taking the graded algebra automorphism θ of $\Bbbk[u,v,x,y]$ defined by

$$\theta(u) = au, \quad \theta(v) = av, \quad \theta(x) = x, \quad \theta(y) = y,$$

yields $A\otimes_\tau B \cong \Bbbk[u,v,x,y]^\theta$ (see Example 2.2). Conversely, if $A\otimes_\tau B$ is a Zhang twist of $\Bbbk[u,v,x,y]$, then $bd^{-1} = ac^{-1} = dc^{-1} = ba^{-1} = 1$ by [**36**, Lemma 2.2], that is $a = b = c = d$. Therefore, Zhang twists and twisted tensor products are different but related kinds of twists, and have been shown to be useful in the study of Artin-Schelter regular algebras.

It is also known that twisted tensor products of Hopf algebras never inherit a Hopf algebra structure (as long as the twisting map is not trivial), but that twisted tensor products of Frobenius algebras always inherit a Frobenius algebra structure [**38**]. However, the following example shows that the twisted tensor product of Koszul algebras need not be Koszul.

EXAMPLE 3.13 ([**15**, Example 5.4]). Consider two polynomial rings $\Bbbk[x]$ and $\Bbbk[y]$. We can define a $\Bbbk$-linear map $\tau : \Bbbk[y]\otimes\Bbbk[x] \to \Bbbk[x]\otimes\Bbbk[y]$ by setting $\tau(y\otimes x) = qx\otimes y$ for some non-zero $q \in \Bbbk$, and extending as before. This gives a strongly graded twisting map by construction, and the twisted tensor product is isomorphic to the quantum plane $\Bbbk_q[x,y] = \Bbbk\langle x,y\rangle/(qxy + yx)$, which is Koszul.

If instead we consider $\tau : \Bbbk[y]\otimes\Bbbk[x] \to \Bbbk[x]\otimes\Bbbk[y]$ by setting $\tau(y^i\otimes x^j) = x^j\otimes y^i$ when i or j are even, $\tau(y^i\otimes x^j) = x^{j+1}\otimes y^{i-1} - x^j\otimes y^i + x^{j-1}\otimes y^{i+1}$ when i and j are both odd, then the twisted tensor product is not quadratic (whence not Koszul) and it has infinite global dimension (whence it is not Artin-Schelter regular).

Generalizations of the former construction where the twist only involves multiplying the graded generators by a scalar have been extensively considered. They are known as *twists by a bicharacter*, the resulting twisted tensor product is denoted by $A \otimes^t B$. These are constructed from algebras $A = \bigoplus_{f\in F} A_f$ and $B = \bigoplus_{g\in G} B_g$ graded by abelian groups F and G. Let $t : F \otimes_{\mathbb{Z}} G \to \Bbbk^\times$ be a $\Bbbk$-bicharacter, namely

$$t(1_F, 1_G) = 1, \quad t(f_1f_2, g) = t(f_1, g)t(f_2, g), \quad t(f, g_1g_2) = t(f, g_1)t(f, g_2),$$

then $\tau : B \otimes A \to A \otimes B$ given by linearly extending $\tau(b \otimes a) = t(|a|, |b|)a \otimes b$ for all homogeneous $a \in A$ and $b \in B$ is a twisting map. The main example are quantum complete intersections

$$\Bbbk\langle x_1, \dots, x_n\rangle / \left(x_i^{m_i}, x_ix_j - q_{ij}x_jx_i\right)_{i,j\in\{1,\dots,n\}}$$

where $n, m_1, \dots, m_n \in \mathbb{N}^n$, $n, m_1, \dots, m_n \geq 2$, $(q_{ij}) \in M_n(\Bbbk^\times)$, $q_{ii} = 1$, and $q_{ij}q_{ji} = 1$, who have been extensively studied for their relation to Happel's question [**11**], interesting support theories [**7, 40**], and (co)homologies [**8, 21, 39**]. In fact, twists by a bicharacter were introduced in [**9**] with a view to generalize the fact that the Hochschild cohomology of a tensor product is the tensor product of the Hochschild cohomologies, a result formalized in [**10**].

THEOREM 3.14 ([**33**, Theorem 3.3]). *Let A and B be algebras, one of them finite dimensional. Then there is an isomorphism of Gerstenhaber algebras*

$$\mathrm{HH}^\bullet(A \otimes B) \cong \mathrm{HH}^\bullet(A) \otimes \mathrm{HH}^\bullet(B).$$

THEOREM 3.15 ([**10**, Theorem 5.1]). *Let A and B be algebras graded by abelian groups F and G. Then there is an isomorphism of Gerstenhaber algebras*

$$\mathrm{HH}^\bullet(A \otimes^t B) \cong \bigoplus_{f\in F,\, g\in G} \mathrm{HH}^\bullet(A, A_{\hat{g}})_f \otimes \mathrm{HH}^\bullet(B, {}_{\hat{f}}B)_g$$

where $\hat{f}$ and $\hat{g}$ are automorphisms of B and A given by the bicharacter, $A_{\hat{g}}$ is the graded A bimodule with twisted right multiplication, and ${}_{\hat{f}}B$ is the graded B bimodule with twisted left multiplication.

There is a method to compute the Hochschild cohomology of a general twisted tensor product in terms of its components. Fix algebras A, B, and a twisting map τ. Given (M, ∇_M) an A bimodule and (N, ∇_N) a B bimodule together with bijective $\Bbbk$-linear maps $\tau_M : B \otimes M \to M \otimes B$ and $\tau_N : N \otimes A \to A \otimes N$ such that

$$\begin{aligned}
\tau_M(\nabla_B \otimes 1) &= (1 \otimes \nabla_B)(\tau_M \otimes 1)(1 \otimes \tau_M),\\
\tau_N(1 \otimes \nabla_A) &= (\nabla_A \otimes 1)(1 \otimes \tau_N)(\tau_N \otimes 1),\\
\tau_M(1 \otimes \nabla_M) &= (\nabla_M \otimes 1)(1 \otimes 1 \otimes \tau)(1 \otimes \tau_M \otimes 1)(\tau \otimes 1 \otimes 1),\\
\tau_N(\nabla_N \otimes 1) &= (1 \otimes \nabla_N)(\tau \otimes 1 \otimes 1)(1 \otimes \tau_N \otimes 1)(1 \otimes 1 \otimes \tau),
\end{aligned}$$

then $M \otimes N$ is an $A \otimes_\tau B$ bimodule, denoted $(M \otimes_\tau N, \nabla_{M\otimes_\tau N})$, where

$$\nabla_{M\otimes_\tau N} := (\nabla_M \otimes \nabla_N)(1 \otimes 1 \otimes \tau \otimes 1 \otimes 1)(1 \otimes \tau_M \otimes \tau_N \otimes 1).$$

In this case, M and N are said to be *compatible* with τ. Given resolutions of M and N, they are said to be *compatible* with τ when each of their modules is compatible, and the collection of resulting $\Bbbk$-linear maps lift τ_M and τ_N.

THEOREM 3.16 ([**42**, Theorem 3.10]). *Let A and B be algebras, let $\tau : B\otimes A \to A\otimes B$ be a twisting map, let M and N be A and B bimodules respectively, let $P_\bullet$ and $Q_\bullet$ be free resolutions of M and N as A and B bimodules respectively. Assume that M, N, $P_\bullet$, and $Q_\bullet$ are all compatible with τ. Then the tensor product complex $(P_\bullet \otimes Q_\bullet)_\bullet$ is a free resolution of $M \otimes N$ as a $A \otimes_\tau B$ bimodule.*

The above, together with the fact that the bar resolution is compatible with all twisting maps, allows the computation of $\mathrm{HH}^\bullet(A \otimes_\tau B)$ in general. The Gerstenhaber algebra structure can be found using the tools introduced in [**31**].

4. Twisted Segre products

From now on we assume for simplicity that $\Bbbk$ is algebraically closed. The Segre product of graded algebras plays an important role in algebraic geometry and commutative algebra (see e.g., [**23, 26**]). In this section, we discuss the notion of twisted Segre product of graded algebras.

DEFINITION 4.1. Let $A = \bigoplus_{i\in\mathbb{Z}} A_i$ and $B = \bigoplus_{i\in\mathbb{Z}} B_i$ be (not necessarily commutative) $\mathbb{Z}$-graded algebras. The *Segre product* $A \circ B$ is the $\Bbbk$-vector space

$$A \circ B := \bigoplus_{i\in\mathbb{Z}} (A_i \otimes B_i)$$

with the multiplication map

$$\nabla_{A\circ B} := (\nabla_A \otimes \nabla_B)(1 \otimes \sigma \otimes 1),$$

where $\sigma : B \otimes A \to A \otimes B$ transposes the entries. This makes $A \circ B$ a $\mathbb{Z}$-graded algebra.

Let $X \subseteq \mathbb{P}^{n-1}$ and $Y \subseteq \mathbb{P}^{m-1}$ be projective varieties with the homogeneous coordinate rings A and B, respectively. Consider the projective variety $X \times Y$ in $\mathbb{P}^{nm-1}$ via the Segre embedding. Then $A \circ B$ is the homogeneous coordinate ring of $X \times Y$.

DEFINITION 4.2 ([**27**, Section 3]). Let A, B be $\mathbb{Z}$-graded algebras and let $\tau : B \otimes A \to A \otimes B$ be a strongly graded twisting map. The *twisted Segre product* $A \circ_\tau B$ is the $\Bbbk$-vector space $A \circ B$ with the multiplication map

$$\nabla_{A\circ_\tau B} := (\nabla_A \otimes \nabla_B)(1 \otimes \tau \otimes 1).$$

This makes $A \circ_\tau B$ a $\mathbb{Z}$-graded algebra.

We give some further definitions. Let S be a Noetherian $\mathbb{Z}$-graded algebra, denote by $\operatorname{GrMod} S$ the category of graded right S-modules. We say that an element $m \in M \in \operatorname{GrMod} S$ is *torsion* if $\dim_\Bbbk mS < \infty$. A graded module $M \in \operatorname{GrMod} S$ is called *torsion* if every homogeneous element of M is torsion. We denote by $\operatorname{Tors} S$ the full subcategory of $\operatorname{GrMod} S$ consisting of torsion modules and by $\operatorname{QGr} S$ the Serre quotient category $\operatorname{GrMod} S/\operatorname{Tors} S$. The category $\operatorname{QGr} S$ is frequently thought of as the category of quasi-coherent sheaves on the imaginary projective scheme associated to S (see [**5**]).

To study $\operatorname{QGr}(A \circ_\tau B)$, where A and B are connected graded algebras finitely generated in degree 1, we use bigraded algebras. Let $S = \bigoplus_{i,j\in\mathbb{Z}} S_{(i,j)}$ be a $\mathbb{Z} \times \mathbb{Z}$-bigraded algebra, define $S_i := \bigoplus_{j\in\mathbb{Z}} S_{(i,j)}$ for each $i \in \mathbb{Z}$. Then $S = \bigoplus_{i\in\mathbb{Z}} S_i$ can be viewed as a $\mathbb{Z}$-graded algebra and $S_0 = \bigoplus_{j\in\mathbb{Z}} S_{(0,j)}$ is a $\mathbb{Z}$-graded algebra.

DEFINITION 4.3 ([**27**, Section 4.1]).

(1) A $\mathbb{Z}$-graded algebra $S = \bigoplus_{i\in\mathbb{Z}} S_i$ is called a *densely graded algebra* when $\dim_{\Bbbk}(S_{i+j}/S_iS_j) < \infty$ for all $i, j \in \mathbb{Z}$.

(2) A $\mathbb{Z}\times\mathbb{Z}$-bigraded algebra $S = \bigoplus_{i,j\in\mathbb{Z}} S_{(i,j)}$ is called a *densely bigraded algebra* when the $\mathbb{Z}$-graded algebra $S = \bigoplus_{i\in\mathbb{Z}} S_i$ is a densely graded algebra, where $S_i = \bigoplus_{j\in\mathbb{Z}} S_{(i,j)}$.

A $\mathbb{Z}$-graded algebra S is said to be *strongly graded* when $S_iS_j = S_{i+j}$ for all $i, j \in \mathbb{Z}$. Strongly graded algebras are clearly densely graded algebras. Moreover if S is a strongly graded algebra then the functor $\operatorname{GrMod} S \to \operatorname{Mod} S_0$ defined by $M \mapsto M_0$ is an equivalence by Dade's Theorem [**37**, Theorem 3.1.1].

Let S be a densely bigraded algebra satisfying the following conditions.

(D1) $S_0 := \bigoplus_{j\in\mathbb{Z}} S_{(0,\,j)}$ is a Noetherian $\mathbb{Z}$-graded algebra, and

(D2) $S_i := \bigoplus_{j\in\mathbb{Z}} S_{(i,\,j)}$ is a finitely generated graded right and left S_0-module for every $i \in \mathbb{Z}$.

Denote by $\operatorname{BiGrMod} S$ the category of bigraded right S-modules. We say that an element $m \in M \in \operatorname{BiGrMod} S$ is *locally torsion* if $\dim_{\Bbbk} mS_0 < \infty$. A bigraded module $M \in \operatorname{BiGrMod} S$ is called *locally torsion* if every homogeneous element of M is locally torsion. We denote by $\operatorname{BiLTors} S$ the full subcategory of $\operatorname{BiGrMod} S$ consisting of locally torsion modules and by $\operatorname{QBiGr_L} S$ the Serre quotient category $\operatorname{BiGrMod} S/\operatorname{BiLTors} S$. We then have the following version of Dade's Theorem for densely bigraded algebras.

THEOREM 4.4 ([**27**, Theorem 4.10]). *Let S be a densely bigraded algebra satisfying conditions* (D1) *and* (D2). *Then the functor* $(-)_0 : \operatorname{BiGrMod} S \to \operatorname{GrMod} S_0$ *given by* $M \mapsto M_0 := \bigoplus_{j\in\mathbb{Z}} M_{(0,j)}$ *induces an equivalence*

$$\operatorname{QBiGr_L} S \cong \operatorname{QGr} S_0.$$

Let us go back to consider twisted Segre products. Let A and B be $\mathbb{Z}$-graded algebras and let $\tau : B\otimes A \to A\otimes B$ be a strongly graded twisting map. We endow the twisted tensor product $A\otimes_\tau B$ with a bigraded structure by setting

$$(A\otimes_\tau B)_{(i,j)} := A_{i+j}\otimes B_j$$

for each $i, j \in \mathbb{Z}$. Then $(A\otimes_\tau B)_0$ is equal to the twisted Segre product $A\circ_\tau B$ as graded algebras.

PROPOSITION 4.5 ([**27**, Proposition 4.11]). *Let A and B be connected graded algebras finitely generated in degree* 1 *and let $\tau : B\otimes A \to A\otimes B$ be a strongly graded twisting map. Then $A\otimes_\tau B$ is a densely bigraded algebra satisfying condition* (D2).

Theorem 4.4 and Proposition 4.5 yield the following result which gives a categorical relationship between twisted tensor products and twisted Segre products.

THEOREM 4.6 ([**27**, Theorem 4.13]). *Let A and B be connected graded algebras finitely generated in degree* 1 *and let $\tau : B\otimes A \to A\otimes B$ be a strongly graded twisting map. If $A\circ_\tau B$ is Noetherian then there is an equivalence*

$$\operatorname{QBiGr_L}(A\otimes_\tau B) \cong \operatorname{QGr}(A\circ_\tau B).$$

REMARK 4.7. (1) In general, it is not known whether the twisted Segre product $A\circ_\tau B$ is Noetherian when A and B are Noetherian. As a positive example, the usual Segre product $A\circ B$ is Noetherian if A and B are commutative finitely generated connected graded algebras.

(2) A category equivalence similar to that of Theorem 4.6 was proved in [**47**, Theorem 2.4] for the usual Segre product. However, the bigradings of the tensor products $A \otimes B$ given here differ from the ones in [**47**], and hence Theorem 4.6 is not a generalization of [**47**, Theorem 2.4].

If A and B are Noetherian Koszul Artin-Schelter regular algebras and $\tau : B \otimes A \to A \otimes B$ is a strongly graded twisting map, then it follows from [**48**, Theorem 2] that $A \otimes_\tau B$ has finite global dimension. From this we see that $\mathrm{QBiGr}_{\mathrm{L}}(A \otimes_\tau B)$ has finite global dimension. Therefore, as a corollary of Theorem 4.6, we have the following.

COROLLARY 4.8 ([**27**, Theorem 4.16]). *Let A and B be Noetherian Koszul Artin-Schelter regular algebras and $\tau : B \otimes A \to A \otimes B$ be a strongly graded twisting map. If $A \circ_\tau B$ is Noetherian then $\mathrm{QGr}(A \circ_\tau B)$ has finite global dimension.*

The above corollary can be considered as a noncommutative analogue of the fact that a Segre variety $\mathbb{P}^{n-1} \times \mathbb{P}^{m-1} \cong \mathrm{Proj}(\Bbbk[x_1, \ldots, x_n] \circ \Bbbk[y_1, \ldots, y_m])$ is smooth.

EXAMPLE 4.9. We now continue Example 3.12. Consider the (ungraded) subalgebra of $A \otimes_\tau B$ generated by the basis elements ux, vx, uy, and vy, where we write $ux = u \otimes x$, $vx = v \otimes x$, $uy = u \otimes y$, and $vy = v \otimes y$, slightly abusing notation as before. Their multiplication is given by

$$
\begin{aligned}
&aux \cdot vx = bvx \cdot ux, && cux \cdot uy = auy \cdot ux, && cux \cdot vy = bvy \cdot ux,\\
&dvx \cdot uy = auy \cdot vx, && dvx \cdot vy = bvy \cdot vx, && cuy \cdot vy = dvy \cdot uy,\\
&aux \cdot vy = bvx \cdot uy.
\end{aligned}
$$

Denote $X = uv$, $Y = vx$, $Z = uy$, $W = vy$ and regrade $|X| = |Y| = |Z| = |W| = 1$. The twisted Segre product $A \circ_\tau B$ is the $\mathbb{Z}$-graded algebra freely generated by X, Y, Z, and W subject to the relations

$$
\begin{aligned}
&aXY - bYX, && cXZ - aZX, && cXW - bWX,\\
&dYZ - aZY, && dYW - bWY, && cZW - dWZ, && aXW - bYZ.
\end{aligned}
$$

It is straightforward to check that $A \circ_\tau B$ is a quotient ring of a skew polynomial ring, so it is Noetherian. Hence $\mathrm{QGr}(A \circ_\tau B)$ has finite global dimension by Corollary 4.8.

An interesting future research topic would be to further explore properties of twisted Segre products.

Acknowledgments

We thank the organizers of the conferences "Recent Advances and New Directions in the Interplay of Noncommutative Algebra and Geometry" (Seattle 2022) and "Noncommutative Geometry and Noncommutative Invariant Theory" (BIRS 2022) for enabling this collaboration. In particular, we effusively thank Michaela Vancliff and James J. Zhang for their continued support.

References

[1] Eli Aljadeff, Pavel Etingof, Shlomo Gelaki, and Dmitri Nikshych, *On twisting of finite-dimensional Hopf algebras*, J. Algebra **256** (2002), no. 2, 484–501, DOI 10.1016/S0021-8693(02)00092-3. MR1939116

[2] Jack Arce, Jorge A. Guccione, Juan J. Guccione, and Christian Valqui, *Twisted tensor products of K^n with K^m*, Algebr. Represent. Theory **22** (2019), no. 6, 1599–1651, DOI 10.1007/s10468-018-9833-1. MR4034796

[3] Jack Arce, Jorge A. Guccione, Juan J. Guccione, and Christian Valqui, *Twisted tensor products of K^3 with K^3*, Comm. Algebra **49** (2021), no. 8, 3614–3634, DOI 10.1080/00927872.2021.1901911. MR4283170

[4] M. Artin, J. Tate, and M. Van den Bergh, *Modules over regular algebras of dimension* 3, Invent. Math. **106** (1991), no. 2, 335–388, DOI 10.1007/BF01243916. MR1128218

[5] M. Artin and J. J. Zhang, *Noncommutative projective schemes*, Adv. Math. **109** (1994), no. 2, 228–287, DOI 10.1006/aima.1994.1087. MR1304753

[6] Ricardo Bances and Christian Valqui, *On the Classification of Graded Twisted Planes*, Algebr. Represent. Theory **26** (2023), no. 4, 1231–1270, DOI 10.1007/s10468-022-10131-8. MR4627265

[7] David J. Benson, Karin Erdmann, and Miles Holloway, *Rank varieties for a class of finite-dimensional local algebras*, J. Pure Appl. Algebra **211** (2007), no. 2, 497–510, DOI 10.1016/j.jpaa.2007.02.006. MR2341266

[8] Petter Andreas Bergh and Karin Erdmann, *Homology and cohomology of quantum complete intersections*, Algebra Number Theory **2** (2008), no. 5, 501–522, DOI 10.2140/ant.2008.2.501. MR2429451

[9] Petter Andreas Bergh and Steffen Oppermann, *Cohomology of twisted tensor products*, J. Algebra **320** (2008), no. 8, 3327–3338, DOI 10.1016/j.jalgebra.2008.08.005. MR2450729

[10] Benjamin Briggs and Sarah Witherspoon, *Hochschild cohomology of twisted tensor products*, Math. Z. **301** (2022), no. 2, 1237–1257, DOI 10.1007/s00209-021-02949-7. MR4418319

[11] Ragnar-Olaf Buchweitz, Edward L. Green, Dag Madsen, and Øyvind Solberg, *Finite Hochschild cohomology without finite global dimension*, Math. Res. Lett. **12** (2005), no. 5-6, 805–816, DOI 10.4310/MRL.2005.v12.n6.a2. MR2189240

[12] Stefaan Caenepeel, Gigel Militaru, and Shenglin Zhu, *Frobenius and separable functors for generalized module categories and nonlinear equations*, Lecture Notes in Mathematics, vol. 1787, Springer-Verlag, Berlin, 2002, DOI 10.1007/b83849. MR1926102

[13] Andreas Čap, Hermann Schichl, and Jiří Vanžura, *On twisted tensor products of algebras*, Comm. Algebra **23** (1995), no. 12, 4701–4735, DOI 10.1080/00927879508825496. MR1352565

[14] Claude Cibils, *Non-commutative duplicates of finite sets*, J. Algebra Appl. **5** (2006), no. 3, 361–377, DOI 10.1142/S0219498806001776. MR2235816

[15] Andrew Conner and Peter Goetz, *The Koszul property for graded twisted tensor products*, J. Algebra **513** (2018), 50–90, DOI 10.1016/j.jalgebra.2018.07.030. MR3849878

[16] Andrew Conner and Peter Goetz, *Classification, Koszulity and Artin-Schelter regularity of certain graded twisted tensor products*, J. Noncommut. Geom. **15** (2021), no. 1, 41–78, DOI 10.4171/jncg/395. MR4248207

[17] Andrew Davies, *Cocycle twists of algebras*, Comm. Algebra **45** (2017), no. 3, 1347–1363, DOI 10.1080/00927872.2016.1178271. MR3573384

[18] Yukio Doi, *Braided bialgebras and quadratic bialgebras*, Comm. Algebra **21** (1993), no. 5, 1731–1749, DOI 10.1080/00927879308824649. MR1213985

[19] Yukio Doi and Mitsuhiro Takeuchi, *Multiplication alteration by two-cocycles—the quantum version*, Comm. Algebra **22** (1994), no. 14, 5715–5732, DOI 10.1080/00927879408825158. MR1298746

[20] V. G. Drinfel′d, *Quantum groups*, Proceedings of the International Congress of Mathematicians, Vol. 1, 2 (Berkeley, Calif., 1986), Amer. Math. Soc., Providence, RI, 1987, pp. 798–820. MR934283

[21] Karin Erdmann and Magnus Hellstrøm-Finnsen, *Hochschild cohomology of some quantum complete intersections*, J. Algebra Appl. **17** (2018), no. 11, 1850215, 22, DOI 10.1142/S0219498818502158. MR3879091

[22] L. D. Faddeev, N. Yu. Reshetikhin, and L. A. Takhtajan, *Quantization of Lie groups and Lie algebras*, Algebraic analysis, Vol. I, Academic Press, Boston, MA, 1988, pp. 129–139. MR992450

[23] Shiro Goto and Keiichi Watanabe, *On graded rings. I*, J. Math. Soc. Japan **30** (1978), no. 2, 179–213, DOI 10.2969/jmsj/03020179. MR494707

[24] Jorge A. Guccione, Juan J. Guccione, and Christian Valqui, *Twisted planes*, Comm. Algebra **38** (2010), no. 5, 1930–1956, DOI 10.1080/00927870903023105. MR2642035

[25] Jorge A. Guccione, Juan J. Guccione, and Christian Valqui, *Non commutative truncated polynomial extensions*, J. Pure Appl. Algebra **216** (2012), no. 11, 2315–2337, DOI 10.1016/j.jpaa.2012.01.021. MR2927170

[26] Joe Harris, *Algebraic geometry: A first course*, Graduate Texts in Mathematics, vol. 133, Springer-Verlag, New York, 1992, DOI 10.1007/978-1-4757-2189-8. MR1182558

[27] Ji-Wei He and Kenta Ueyama, *Twisted Segre products*, J. Algebra **611** (2022), 528–560, DOI 10.1016/j.jalgebra.2022.08.017. MR4478354

[28] H. Huang, V. C. Nguyen, C. Ure, P. Veerapen, K. B. Vashaw, and X. Wang. Twisting of graded quantum groups and solutions to the quantum yang-baxter equation. *Transform. Groups*, 2022.

[29] P. Jara, J. López Peña, G. Navarro, and D. Ştefan, *On the classification of twisting maps between K^n and K^m*, Algebr. Represent. Theory **14** (2011), no. 5, 869–895, DOI 10.1007/s10468-010-9222-x. MR2832263

[30] Pascual Jara Martínez, Javier López Peña, Florin Panaite, and Freddy van Oystaeyen, *On iterated twisted tensor products of algebras*, Internat. J. Math. **19** (2008), no. 9, 1053–1101, DOI 10.1142/S0129167X08004996. MR2458561

[31] Tekin Karadağ, Dustin McPhate, Pablo S. Ocal, Tolulope Oke, and Sarah Witherspoon, *Gerstenhaber brackets on Hochschild cohomology of general twisted tensor products*, J. Pure Appl. Algebra **225** (2021), no. 6, Paper No. 106597, 14, DOI 10.1016/j.jpaa.2020.106597. MR4168969

[32] Christian Kassel, *Quantum groups*, Graduate Texts in Mathematics, vol. 155, Springer-Verlag, New York, 1995, DOI 10.1007/978-1-4612-0783-2. MR1321145

[33] Jue Le and Guodong Zhou, *On the Hochschild cohomology ring of tensor products of algebras*, J. Pure Appl. Algebra **218** (2014), no. 8, 1463–1477, DOI 10.1016/j.jpaa.2013.11.029. MR3175033

[34] Cesar Lecoutre and Susan J. Sierra, *A new family of Poisson algebras and their deformations*, Nagoya Math. J. **233** (2019), 32–86, DOI 10.1017/nmj.2017.29. MR3901632

[35] Susan Montgomery, *Algebra properties invariant under twisting*, Hopf algebras in noncommutative geometry and physics, Lecture Notes in Pure and Appl. Math., vol. 239, Dekker, New York, 2005, pp. 229–243. MR2106932

[36] Manizheh Nafari and Michaela Vancliff, *Graded skew Clifford algebras that are twists of graded Clifford algebras*, Comm. Algebra **43** (2015), no. 2, 719–725, DOI 10.1080/00927872.2013.847949. MR3274033

[37] Constantin Năstăsescu and Freddy Van Oystaeyen, *Methods of graded rings*, Lecture Notes in Mathematics, vol. 1836, Springer-Verlag, Berlin, 2004, DOI 10.1007/b94904. MR2046303

[38] Pablo S. Ocal and Amrei Oswald, *A dichotomy between twisted tensor products of bialgebras and Frobenius algebras*, J. Algebra **644** (2024), 351–380, DOI 10.1016/j.jalgebra.2023.12.039. MR4695623

[39] Steffen Oppermann, *Hochschild cohomology and homology of quantum complete intersections*, Algebra Number Theory **4** (2010), no. 7, 821–838, DOI 10.2140/ant.2010.4.821. MR2776874

[40] Julia Pevtsova and Sarah Witherspoon, *Varieties for modules of quantum elementary abelian groups*, Algebr. Represent. Theory **12** (2009), no. 6, 567–595, DOI 10.1007/s10468-008-9100-y. MR2563183

[41] Daniel Rogalski, *Noncommutative projective geometry*, Noncommutative algebraic geometry, Math. Sci. Res. Inst. Publ., vol. 64, Cambridge Univ. Press, New York, 2016, pp. 13–70. MR3618472

[42] Anne Shepler and Sarah Witherspoon, *Resolutions for twisted tensor products*, Pacific J. Math. **298** (2019), no. 2, 445–469, DOI 10.2140/pjm.2019.298.445. MR3936025

[43] Susan J. Sierra, *G-algebras, twistings, and equivalences of graded categories*, Algebr. Represent. Theory **14** (2011), no. 2, 377–390, DOI 10.1007/s10468-009-9193-y. MR2776790

[44] Mitsuhiro Takeuchi, *Free Hopf algebras generated by coalgebras*, J. Math. Soc. Japan **23** (1971), 561–582, DOI 10.2969/jmsj/02340561. MR292876

[45] Mitsuhiro Takeuchi, *A short course on quantum matrices*, New directions in Hopf algebras, Math. Sci. Res. Inst. Publ., vol. 43, Cambridge Univ. Press, Cambridge, 2002, pp. 383–435. Notes taken by Bernd Strüber. MR1913443

[46] H. V. Tran, *Twisting Systems and some Quantum $\mathbb{P}^3$s with Point Scheme a Rank-2 Quadric*, PhD thesis, University of Texas Arlington, 2022.

[47] Kristel Van Rompay, *Segre product of Artin-Schelter regular algebras of dimension 2 and embeddings in quantum* $\mathbf{P}^3$*'s*, J. Algebra **180** (1996), no. 2, 483–512, DOI 10.1006/jabr.1996.0078. MR1378541

[48] Xin Wang and Yuan Shen, *Artin-Schelter regularity of twisted tensor products*, Front. Math. China **13** (2018), no. 5, 1141–1167, DOI 10.1007/s11464-018-0726-8. MR3866681

[49] J. J. Zhang, *Twisted graded algebras and equivalences of graded categories*, Proc. London Math. Soc. (3) **72** (1996), no. 2, 281–311, DOI 10.1112/plms/s3-72.2.281. MR1367080

[50] James J. Zhang and Jun Zhang, *Double Ore extensions*, J. Pure Appl. Algebra **212** (2008), no. 12, 2668–2690, DOI 10.1016/j.jpaa.2008.05.008. MR2452318

Department of Mathematics, UCLA, Los Angeles, California 90095-1555
Email address: socal@math.ucla.edu

Department of Mathematics, Faculty of Science, Shinsu University, Matsumoto, Nagano 390-8621, Japan
Email address: ueyama@shinshu-u.ac.jp

Department of Mathematics, Tennessee Tech University, Cookeville, Tennessee 38505
Email address: pveerapen@tntech.edu

Contemporary Mathematics
Volume **801**, 2024
https://doi.org/10.1090/conm/801/16084

Artin-Schelter regular algebras

Daniel Rogalski

To S. Paul Smith on the occasion of his 65th Birthday

Contents

1. Introduction

There are two aspects to this survey. First, in Sections 1-3 we give an introduction to the main ideas and fundamental results in the theory of Artin-Schelter regular algebras, suitable for readers new to the subject who have some basic background in noncommutative algebra, homological algebra and algebraic geometry. For those looking for more details, in particular for beginning graduate students, we recommend starting with the course notes in [**88**] instead. Second, in the remainder of the paper we give an overview of the most important themes and results of the past 30 years of research in the subject. This literature review, mostly designed to point readers to the original papers where they can find more information, will not be able to linger long on any one topic or idea. It could not possibly be exhaustive given the vast literature in this area. We apologize in advance to the authors of papers we do not discuss.

Throughout these notes, let $\Bbbk$ be a base field, which for convenience we assume is algebraically closed, though some results do not require this assumption. A polynomial algebra in finitely many variables over $\Bbbk$, say $A = \Bbbk[x_1, \dots, x_m]$, is one of the most basic examples in commutative ring theory. Noncommutative ring theory has long been influenced by the beautiful and strong results about commutative rings, motivated by number theory and algebraic geometry, that formed part of the core of the development of modern algebra. Because polynomial rings are so important, it is natural to ask what noncommutative algebras over a field play a similar role in the noncommutative theory as commutative polynomial algebras do in the commutative theory.

As phrased, this question is vague, since any answer depends on what aspect of the polynomial ring one wants to generalize. For example, a polynomial ring $\Bbbk[x_1, \dots, x_m]$ is a free object on a set with m elements in the category of commutative $\Bbbk$-algebras. The noncommutative analog in this sense is the free object on a set with m elements in the category of all associative $\Bbbk$-algebras, namely, the free associative algebra $\Bbbk\langle x_1, \dots, x_m\rangle$. However, free algebras in more than one variable are unwieldy and have very different ring-theoretic properties from a commutative polynomial ring. For example, such algebras are not noetherian. What we seek is a class of rings which are the same size as polynomial rings in some sense, and which have similarly good properties.

One reasonable answer to the question, which goes back to the earliest study of noncommutative rings, is the notion of an iterated Ore extension. Note that we can think of (or define) a polynomial ring via iterating a one variable extension:

$$\Bbbk[x_1, \dots, x_m] = (\dots((\Bbbk[x_1])[x_2])\dots)[x_m].$$

Ore defined a noncommutative analog of a single variable extension.

DEFINITION 1.1 (Øystein Ore; 1933). Let R be a ring, $\sigma : R \to R$ an automorphism, and $\delta : R \to R$ a σ-derivation, that is, a linear map satisfying $\delta(ab) = \delta(a)b + \sigma(a)\delta(b)$ for all $a, b \in R$. The *Ore extension* $R[x; \sigma, \delta]$ is the left free R-module $R \oplus Rx \oplus Rx^2 \oplus \dots$ with a new multiplication induced by the rule $xr = \sigma(r)x + \delta(r)$ for all $r \in R$.

The reader can find more details on this construction in [**44**, Chapter 2]. Since an Ore extension is a noncommutative polynomial extension, *iterated Ore extensions* of the form

$$\Bbbk[x_1; \sigma_1, \delta_1][x_2; \sigma_2, \delta_2]\dots[x_m; \sigma_m, \delta_m], \tag{1.2}$$

where all of the σ_i and δ_i are $\Bbbk$-linear, give a class of noncommutative $\Bbbk$-algebras which can be thought of as noncommutative analogs of commutative polynomial rings over $\Bbbk$. These rings give one possible answer to our original question. Indeed, these rings have many good properties, for example they are noetherian by a noncommutative version of the Hilbert basis theorem. In addition, (1.2) has the same size as a polynomial algebra in m variables, in the sense that the set $\{x_1^{i_1} \dots x_m^{i_m} | i_1, \dots, i_m \geq 0\}$ is a $\Bbbk$-basis for the algebra. Many important examples, such as certain quantum groups, can be realized as iterated Ore extensions. There are many interesting and deep questions about them, such as how to understand the structure of their prime spectra.

In the late 1980's, Artin and Schelter introduced a different class of noncommutative analogs of polynomial rings, motivated instead by the *homological* properties polynomial rings have as graded rings. To study some of these examples, completely new techniques coming from algebraic geometry were required. This was the seed of an entire subject of noncommutative projective algebraic geometry that has flowered ever since.

To describe the class of algebras Artin and Schelter studied, let us first recall the basic definitions of graded rings.

DEFINITION 1.3. A $\Bbbk$-algebra is $\mathbb{N}$-*graded* if there is a $\Bbbk$-space decomposition $A = A_0 \oplus A_1 \oplus \ldots \oplus A_n \oplus \ldots$ such that $A_i A_j \subseteq A_{i+j}$ for all i, j. If in addition $A_0 = \Bbbk$, then A is *connected.*

In this paper our primary interest is *finitely generated* connected $\mathbb{N}$-graded $\Bbbk$-algebras. Note that a commutative polynomial ring $A = \Bbbk[x_1, \ldots, x_m]$ has various $\mathbb{N}$-gradings, by assigning arbitrary degrees $\deg x_i = d_i$ and letting A_n be the $\Bbbk$-span of $\{x_1^{i_1} x_2^{i_2} \ldots x_m^{i_m} | d_1 i_1 + d_2 i_2 + \cdots + d_m i_m = n\}$. In fact, such graded polynomial rings can be characterized as the commutative graded algebras of finite global dimension. The following result was proved by Smith and Zhang (unpublished), see [**118**, Remark after Corollary 3.2].

THEOREM 1.4. *A connected graded finitely generated commutative $\Bbbk$-algebra A with* $\operatorname{gl.dim}(A) = m$ *is isomorphic to $\Bbbk[x_1, \ldots, x_m]$ with some grading.*

For a connected graded algebra, let $A_{\geq 1} = \bigoplus_{n \geq 1} A_n$ and write $\Bbbk$ for the *trivial module* $\Bbbk = A/A_{\geq 1}$. This is both a left and right A-module, and we write ${}_A\Bbbk$ for its left structure and $\Bbbk_A$ for the right structure.

DEFINITION 1.5 (Artin-Schelter, 1987). Let A be a connected finitely generated graded $\Bbbk$-algebra. The algebra A is *regular* if

(1) A has finite global dimension $d < \infty$;
(2) A has polynomial growth, i.e. $f(n) = \dim_{\Bbbk} A_n$ is bounded above by a polynomial function of n; and
(3) A is *Gorenstein*, i.e. $\operatorname{Ext}^i_A({}_A\Bbbk, A) \cong \begin{cases} 0 & i \neq d \\ \Bbbk_A & i = d \end{cases}$.

Algebras satisfying Definition 1.5 are now called Artin-Schelter regular or AS regular. There are other noncommutative notions of regular ring, but in this paper if we just say "regular" we will mean "AS regular". Also, while it is not required in the definition that A be *generated in degree* 1, that is, generated as a $\Bbbk$-algebra by the elements of A_1, all of the initial results in the subject have focused on that case, and we will tacitly assume it unless we say otherwise.

There is a related more general notion: A is called *AS Gorenstein* if it satisfies (2) and (3) of the definition above but (1) is replaced by the weaker condition that A has finite injective dimension as an A-module. AS Gorenstein rings will only be mentioned in passing in this paper.

While commutative graded rings of finite global dimension are polynomial rings, by Theorem 1.4, noncommutative graded rings of finite global dimension do not behave like analogs of polynomial rings without further restrictions. For example, a free associative algebra $\Bbbk\langle x_1, \ldots, x_m \rangle$ has global dimension 1, regardless of m, and is not noetherian when $m \geq 2$. But this example satisfies neither (2) nor (3) in the definition of AS regular when $m \geq 2$.

Here are some other examples that show the importance of conditions (2) and (3) in the definition of regularity. We don't justify the claims made here, but they all follow using methods similar to those reviewed in the next section (but easier).

EXAMPLE 1.6. Let $A = \Bbbk\langle x, y \rangle / (yx)$. Then A has a $\Bbbk$-basis consisting of words $\{x^i y^j | i, j \geq 0\}$, the same as for a polynomial ring in two variables. In particular,

A has polynomial growth. Also, one may show that $\text{gl.dim}(A) = 2$. However A is patently not a domain, and it is easy to show it is not noetherian. One may also see that A is not Gorenstein, that is, condition (3) fails.

EXAMPLE 1.7. Let $A = \Bbbk\langle x, y, z\rangle/(x^2 + y^2 + z^2)$. Then A is a graded domain with global dimension 2 and the Gorenstein property (3), but with $f(n) = \dim_{\Bbbk} A_n$ having exponential growth. Again, A is non-noetherian.

On the other hand, regular algebras of global dimension at most 2 are easy to classify, and do have good properties similar to those of commutative polynomial rings.

EXAMPLE 1.8. If A is a $\Bbbk$-algebra which is regular of global dimension 1, then $A \cong \Bbbk[x]$.

EXAMPLE 1.9. Let A be a regular algebra of global dimension 2. Then either $A \cong A_q = \Bbbk\langle x, y\rangle/(yx - qxy)$ for some $0 \neq q \in \Bbbk$ (the *quantum plane*), or $A \cong A_J = \Bbbk\langle x, y\rangle/(yx - xy - x^2)$ (the *Jordan plane*).

If R is an $\mathbb{N}$-graded ring, a general Ore extension $S = R[x; \sigma, \delta]$ need not have a natural grading. However, if σ is a graded automorphism ($\sigma(R_n) \subseteq R_n$ for all n) and δ is a σ-derivation of degree 1 ($\delta(R_n) \subseteq R_{n+1}$ for all n), then S is again graded, with $S_n = \sum_{i+j=n} R_i x^j$. One may check that in fact $A_q \cong \Bbbk[x][y; \sigma, \delta]$ as a graded ring, where $\sigma(x) = qx$ and $\delta(x) = 0$, and similarly $A_J \cong \Bbbk[x][y; \sigma, \delta]$, where $\sigma(x) = x$ and $\delta(x) = x^2$. It is known that a graded iterated Ore extension is indeed regular, but the notion of regular algebra turns out to be far more general. Already in dimension 3 there are interesting examples of regular algebras which cannot be constructed as iterated Ore extensions.

2. Regular algebras of dimension 3

The classification of regular algebras of dimension 3 was the goal of the earliest work in the subject, starting with the original paper by Artin and Schelter, and continuing in two papers of Artin, Tate, and Van den Bergh. In this section we review the main ideas of these papers, which have influenced all subsequent work.

2.1. The work of Artin and Schelter. We would like to give the reader a sense of why the axioms of an AS regular algebra are restrictive enough to give any hope of classifying those of low dimension. In fact, we will see that all of the properties in the definition of regular can be encoded in terms of the properties of the free resolution of the trivial module $\Bbbk$. The reference for this entire subsection is the paper [**1**].

Let us first make a comment on morphisms between free modules. An A-module morphism between free left modules of finite rank, say $f : A^{\oplus r} \to A^{\oplus s}$, can be represented by an $r \times s$ matrix M with entries in A, where we represent the elements of the free modules by row vectors and f by right multiplication by the matrix M. Similarly, a morphism between free right modules $g : A^{\oplus s} \to A^{\oplus r}$ can also be represented by an $r \times s$-matrix M, where we now represent the free modules as columns and g by left multiplication by the matrix. The contravariant functor $\text{Hom}_A(-, A)$ interchanges the categories of left and right free modules of finite rank, and leaves the matrices representing maps the same if we use these conventions.

Let A be a connected graded finitely generated $\Bbbk$-algebra. We work in the category of $\mathbb{Z}$-graded left A-modules $M = \bigoplus_{n \in \mathbb{Z}} M_n$. For any such module M and

$i \in \mathbb{Z}$, let $M(i)$ be the graded module with the same ungraded module structure M, but with new grading $M(i)_j = M_{i+j}$. Every finitely generated graded module M has a graded free resolution

$$\to F_n \to \dots \xrightarrow{\partial_1} F_1 \xrightarrow{\partial_0} F_0 \to M \to 0, \tag{2.1}$$

where each F_i is a direct sum of various graded free modules $A(j)$ and the maps preserve degree. Let us assume that each F_i has finite rank, which we can always achieve if A is noetherian. A resolution such as (2.1) is *minimal* if the matrices representing the maps ∂_i have all of their entries in $A_{\geq 1}$. The minimal resolution is unique up to isomorphism of complexes, and as the word minimal suggests, it is the resolution for which the ranks of the free modules F_i are smallest.

For any connected graded finitely generated $\Bbbk$-algebra A, it is well known that the global dimension of the algebra A is the same as the projective dimension of the trivial module $\Bbbk$ (as a left or right module). Thus $\operatorname{gl.dim}(A) = d$ if and only if $\operatorname{proj.dim}({}_A\Bbbk) = d$, which is if and only if the minimal resolution of $\Bbbk$ as in (2.1) has $F_i = 0$ for all $i \geq d+1$.

Now let A be regular of dimension 3, and generated in degree 1. Let A have a minimal presentation $A \cong \Bbbk\langle x_1, \dots, x_n\rangle/(r_1, \dots, r_m)$, where $\deg(x_i) = 1$ and each r_i is a homogeneous relation with $\deg(r_i) = d_i$, say. The degrees of the generators and relations determine the ranks and shifts of the first few graded free modules in the minimal resolution of ${}_A\Bbbk$. Specifically, since $\operatorname{proj.dim}(k) = 3$, the minimal graded free resolution of the trivial left module $\Bbbk = A/A_{\geq 1}$ has the form

$$0 \to F \to \bigoplus_{i=1}^{m} A(-d_i) \to \bigoplus_{i=1}^{n} A(-1) \to A \to \Bbbk \to 0,$$

for some graded free module F.

Next, consider the Gorenstein condition. In the definition of regular we ignored shifts, but $\operatorname{Ext}^i_A(\Bbbk, A)$ is naturally a $\mathbb{Z}$-graded right A-module, and so we can actually assume that $\operatorname{Ext}^3_A(\Bbbk, A) \cong \Bbbk_A(\ell)$ as graded right A-modules, for some shift of grading ℓ. Calculating Ext by applying $\operatorname{Hom}_A(-, A)$ to the (deleted) free resolution above, $\operatorname{Ext}^i(\Bbbk, A)$ is the ith cohomology group of a complex

$$0 \to A \to \bigoplus_{i=1}^{n} A(1) \to \bigoplus_{i=1}^{m} A(d_i) \to \operatorname{Hom}(F, A) \to 0.$$

Since $\operatorname{Ext}^i(\Bbbk, A) = 0$ for $i \neq 3$, the complex is exact in every spot except at $\operatorname{Hom}(F, A)$. We also know $\operatorname{Ext}^3(\Bbbk, A) = \Bbbk(\ell)$, so there is an exact sequence of right modules

$$0 \to A \to \bigoplus_{i=1}^{n} A(1) \to \bigoplus_{i=1}^{m} A(d_i) \to \operatorname{Hom}(F, A) \to \Bbbk(\ell) \to 0.$$

Shifting all terms by $-\ell$ we get

$$0 \to A(-\ell) \to \bigoplus_{i=1}^{n} A(1-\ell) \to \bigoplus_{i=1}^{m} A(d_i - \ell) \to \operatorname{Hom}(F, A)(-\ell) \to \Bbbk \to 0.$$

This must be a minimal graded free resolution of the right trivial module $\Bbbk_A$, since the matrices representing the maps have not changed, so they still have entries in $A_{\geq 1}$. Also, the ranks and shifts of its first few terms are again determined by the number of generators and the degrees of the relations, so we conclude that

$F \cong A(-\ell)$; $d_i - \ell = -1$ for all i; and the minimal number m of relations is the same as the minimal number n of generators of A. The original minimal resolution of ${}_A\Bbbk$ now must look like

$$0 \to A(-\ell) \to \bigoplus_{i=1}^{m} A(-\ell+1) \to \bigoplus_{i=1}^{m} A(-1) \to A \to \Bbbk \to 0.$$

To summarize, the Gorenstein condition implies that the minimal resolutions of the left and right trivial modules are interchanged (up to shift of grading) by the operation $\operatorname{Hom}_A(-, A)$. This forces the minimal resolution to have a rather symmetric form.

Finally, we can relate the form of the minimal free resolution of $\Bbbk$ to the growth of the graded pieces of the algebra. The *Hilbert series* of a connected finitely generated $\mathbb{N}$-graded algebra is the power series $h_A(t) = \sum_{n=0}^{\infty} (\dim_{\Bbbk} A_n) t^n$, which can be manipulated in the power series ring $\mathbb{Z}[[t]]$. Using the form of the minimal free resolution of $\Bbbk$, it is straightforward to see that

$$h_A(t) = \sum_{n=0}^{\infty} (\dim_{\Bbbk} A_n) t^n = \frac{1}{(1 - mt + mt^{\ell} - t^{\ell+1})}.$$

Similarly, the Hilbert series of any connected finitely graded algebra of finite global dimension is of the form $\frac{1}{p(t)}$, where the coefficients of the polynomial $p(t)$ can be determined in a simple way from the ranks and shifts of the graded free modules in the minimal free resolution of $\Bbbk$. Some basic complex analysis can be used to show that if $p(t)$ is a monic polynomial in $\mathbb{Z}[t]$ and $\frac{1}{p(t)} = 1 + a_1 t + a_2 t^2 + \dots$ in $\mathbb{Z}[[t]]$, then $f(n) = a_n$ is a polynomially bounded function of n if and only if all zeroes of $p(t)$ in $\mathbb{C}$ are roots of unity [**3**, Proposition 2.14]. Thus, the polynomial growth condition (2) in the definition of regularity forces $1 - mt + mt^{\ell} - t^{\ell+1}$ to have only roots of unity for zeroes. A little calculus shows this happens if and only if either $m = 3, \ell = 2$ or $m = 2, \ell = 3$. In conclusion, either A has 3 generators and 3 quadratic relations, with $h_A(t) = \frac{1}{(1-t)^3}$, or A has 2 generators and 2 cubic relations, with $h_A(t) = \frac{1}{(1-t)^2(1-t^2)}$. These are the same as the Hilbert series for a polynomial ring $\Bbbk[x, y, z]$ with variables of degrees $1, 1, 1$ or $1, 1, 2$, respectively. (Note, however, that the polynomial ring $\Bbbk[x, y, z]$ with generators of degrees $1, 1, 2$ is not generated in degree 1, and so is not an example of the type considered in this section.) Based on the degrees of their relations, the two kinds of regular algebras of dimension 3 are referred to as quadratic or cubic.

Both kinds of Hilbert series were treated uniformly by Artin and Schelter, but here let us focus on the quadratic case, where $A = \Bbbk\langle x_1, x_2, x_3\rangle/(r_1, r_2, r_3)$ has 3 minimal generators and 3 minimal relations of degree 2. Let us consider the maps in the free resolution

$$0 \to A(-\ell) \xrightarrow{\partial_3} \bigoplus_{i=1}^{m} A(-\ell+1) \xrightarrow{\partial_2} \bigoplus_{i=1}^{m} A(-1) \xrightarrow{\partial_1} A \to \Bbbk \to 0.$$

The map ∂_i is given by right multiplication by some matrix M_i, and we identify ∂_i and M_i. It is standard that we can take ∂_1 to be the 3×1 matrix $(x_1, x_2, x_3)^t$. Because of the symmetry coming from the Gorenstein condition, the three entries of ∂_3 also must be a basis of A_1. By making a change of basis in the $\bigoplus_{i=1}^{m} A(-\ell+1)$-spot, we can assume that $\partial_3 = (x_1, x_2, x_3)$. This also changes the matrix representing ∂_2, but no assumptions about it have been made. The free resolution now looks like

$$0 \to A(-3) \overset{(x_1,x_2,x_3)}{\to} \bigoplus_{i=1}^{3} A(-2) \overset{M}{\to} \bigoplus_{i=1}^{3} A(-1) \overset{(x_1,x_2,x_3)^t}{\to} A \to \Bbbk \to 0.$$

Note that the entries of the 3×3 matrix M are in $\Bbbk x_1 + \Bbbk x_2 + \Bbbk x_3$. By the standard way the resolution of $\Bbbk$ begins, when calculated in the free algebra $\Bbbk\langle x_1, x_2, x_3\rangle$ the product $M(x_1, x_2, x_3)^t$ is equal to $(r_1, r_2, r_3)^t$ for the three minimal relations r_1, r_2, r_3. Again by the symmetry coming from the Gorenstein condition, the product of the final two matrices (taken in $\Bbbk\langle x_1, x_2, x_3\rangle$) is $(x_1, x_2, x_3)M = (s_1, s_2, s_3)$, where s_1, s_2, s_3 is also a basis for the span of the relations $\Bbbk r_1 + \Bbbk r_2 + \Bbbk r_3$. So there is a matrix $Q \in \mathrm{GL}_3(\Bbbk)$ with $(x_1, x_2, x_3)M = M(x_1, x_2, x_3)^t Q$.

Consider the product $w = (x_1, x_2, x_3)M(x_1, x_2, x_3)^t$, again taken in the free algebra $\Bbbk\langle x_1, x_2, x_3\rangle$. Note that w determines the relations r_1, r_2, r_3, since w has a unique expression $w = x_1 r_1 + x_2 r_2 + x_3 r_3$ in the free algebra. Using the equation $(x_1, x_2, x_3)M = M(x_1, x_2, x_3)^t Q$, one checks that there is a graded automorphism σ of $\Bbbk\langle x_1, x_2, x_3\rangle$ (given by the matrix Q) such that under the linear map τ which operates on degree 3 monomials by $x_{i_1} x_{i_2} x_{i_3} \to \sigma(x_{i_3}) x_{i_1} x_{i_2}$, one has $\tau(w) = w$. This says that w is a *twisted superpotential*, in more modern terminology (not used by Artin and Schelter) we explain later in Section 10.

Using some simple representation theory, Artin and Schelter wrote down a list of possible twisted superpotentials corresponding to diagonalizable Q's. They were able to show certain superpotentials led to regular algebras by direct calculation, namely the theory of noncommutative Gröbner bases, which played a large role in this early work. Schelter's Affine computer algebra package was a key tool. For the reader unfamiliar with noncommutative Gröbner bases, we refer to Bergman's survey [**12**] for a general introduction, or see the author's course notes in [**88**, Section 1]. In particular, the Gröbner basis method allows one to compute the Hilbert series of a finitely presented algebra in some cases, and can also aid in other calculations, such as proving that elements are non-zero-divisors. Such information is helpful in proving that a proposed free resolution of $\Bbbk$ is actually exact.

Some algebro-geometric analysis showed that, up to isomorphism, all regular algebras of dimension 3 whose twisted superpotentials have a diagonalizable Q belong to one of a finite number of "types", where each type is a parametrized family of superpotentials. Conversely, each type gives a regular algebra for generic choices of the parameters. The types of quadratic algebras were named A, B, H, S_1, S_1', S_2. Let us give an example of one of the simpler types.

EXAMPLE 2.2. Type S_2 is given by the twisted superpotential

$$w = x_1 x_3 x_1 + \alpha x_3 x_1^2 + \alpha^{-1} x_1^2 x_3 + x_2 x_3 x_2 - \alpha x_3 x_2^2 - \alpha^{-1} x_2^2 x_3$$

for any $0 \neq \alpha \in \Bbbk$. This is a twisted superpotential where $\sigma(x_1) = \alpha^{-1} x_1$, $\sigma(x_2) = -\alpha^{-1} x_2$, $\sigma(x_3) = \alpha^2 x_3$. Since $w = x_1 r_1 + x_2 r_2 + x_3 r_3$ for the corresponding relations r_1, r_2, r_3, we have a family of algebras $A(\alpha)$ with parametrized relations:

$$A(\alpha) = \Bbbk\langle x_1, x_2, x_3\rangle/(\alpha x_3 x_1 + x_1 x_3,\ \alpha x_3 x_2 - x_2 x_3,\ x_1^2 - x_2^2).$$

The family S_2 is quite simple, because it depends on only one parameter and each relation involves only two of the variables. In fact, an algebra in this family is easily represented as an iterated Ore extension $\Bbbk[x_1 + x_2][x_1 - x_2; \sigma][x_3; \tau]$, and one can use this to see that $A(\alpha)$ is regular for all nonzero α. On the other hand, for some of the other families in the classification, the question of which values of the parameters give a regular algebra turned out to be quite complicated, and was not answered until the later work of Artin, Tate, and Van den Bergh in [**2**].

2.2. Linear modules. The following type of module plays a crucial role in the theory of AS regular algebras.

DEFINITION 2.3. Let A be a connected $\mathbb{N}$-graded $\Bbbk$-algebra, generated in degree 1. A cyclic graded A-module M, generated in degree 0, is called *linear* if it has Hilbert series $h_M(t) = \bigoplus_{n=0}^{\infty} \dim_{\Bbbk}(M_n)t^n = \frac{1}{(1-t)^d}$ for some $d \geq 0$. If $d = 1$, M is called a *point module*, and if $d = 2$, M is called a *line module*.

Point modules were essential to Artin, Tate, and Van den Bergh's approach, which we review in the next section, to the classification of AS regular algebras of dimension 3. To help the reader picture such modules, we note that since $\frac{1}{(1-t)} = 1+t+t^2+t^3+\dots$, a point module P has $\dim_{\Bbbk} P_n = 1$ for all $n \geq 0$. The polynomial ring $\Bbbk[x]$ is a point module over itself, and over any connected graded algebra that maps onto $\Bbbk[x]$. Similarly, a line module L has $\dim_{\Bbbk} L_n = n+1$ for all $n \geq 0$, and $\Bbbk[x, y]$ is a line module over any connected graded algebra mapping onto it.

If $A = \Bbbk[x_1, \dots, x_n]$ is a commutative polynomial ring, then one may prove that a linear module M with $h_M(t) = \frac{1}{(1-t)^d}$ is isomorphic to A/I for $I = \operatorname{ann}_A(M)$, where I is generated as an ideal by any $\Bbbk$-subspace V of A_1 with $\dim_{\Bbbk} V = n - d$. Thus such linear modules over A are in bijective correspondence with the projective linear subspaces of $\operatorname{Proj} A \cong \mathbb{P}^{n-1}$ that are isomorphic to $\mathbb{P}^{d-1}$. In particular, the isomorphism classes of point modules over A are simply in bijection with the $\Bbbk$-points of $\mathbb{P}^{n-1}$. More generally, when $A = \Bbbk[x_1, \dots, x_n]$ the isomorphism classes of point modules over any factor ring A/I by a homogeneous ideal I are in bijection with the $\Bbbk$-points of $\operatorname{Proj} A/I$.

Now let $A = \Bbbk\langle x_1, \dots, x_n\rangle/(r_1, \dots, r_m)$ be a finitely presented connected graded $\Bbbk$-algebra. For any d, define a *truncated point module of length* $d+1$ to be a graded A-module, generated in degree 0, with Hilbert series $1 + t + \dots + t^d$ (length here refers to the composition length). Standard arguments show that the isomorphism classes of truncated point modules of length $d+1$ for A are parametrized by a closed subscheme X_d of $(\mathbb{P}^{n-1})^{\times d}$. In fact, the scheme X_d can be easily calculated as the common zero locus in $(\mathbb{P}^{n-1})^{\times d}$ of multilinearizations of the relations r_i. We will speak of parametrization informally in these notes. It means that these truncated point modules, up to isomorphism, are in bijection with the $\Bbbk$-points of the scheme X_d, in a natural way which is compatible with base ring extension; see [**2**, Section 3].

For each $d \geq 1$ there is a morphism of schemes $\phi_d : X_d \to X_{d-1}$, which corresponds to taking a truncated point module M of length $d+1$ and truncating it further to get a truncated point module $M_{\leq(d-1)}$ of length d. Then the inverse limit of schemes $X = \varprojlim X_d$ parametrizes the isomorphism classes of full point modules for A. In specific examples, one often sees through direct calculation that this inverse system stabilizes, i.e. that there is $\ell \geq 1$ such that ϕ_d is an isomorphism for all $d > \ell$. In this case $X = X_\ell$ is a projective scheme which parametrizes the point modules for A, and is called the *point scheme* for A. If the inverse limit should happen not to stabilize, then we can still treat X as an inverse limit of schemes, but it may not be an object in the category of schemes itself. If M is point module for A, then we can lop off the zero-degree piece of M and shift everything over, obtaining another point module $M_{\geq 1}(1)$. This operation translates into a map $\sigma : X \to X$ of the parametrizing space X. The map σ is a morphism of schemes in the case that X is a scheme, and is typically an automorphism of X.

Here are a few simple examples of the theory above.

EXAMPLE 2.4. If $A = A_q = \Bbbk\langle x, y\rangle/(yx - qxy)$ for $q \neq 0$, then the (left) point modules up to isomorphism are the modules $A/A(\alpha x+\beta y)$, parameterized by points $(\alpha : \beta) \in \mathbb{P}^1$. The induced map $\sigma : \mathbb{P}^1 \to \mathbb{P}^1$ is given by $\sigma(\alpha : \beta) = (q\alpha : \beta)$.

EXAMPLE 2.5. As in Example 2.2, consider the family S_2 of regular algebras of dimension 3, given by the presentation

$$A(\alpha) = \Bbbk\langle x_1, x_2, x_3\rangle/(\alpha x_3x_1 + x_1x_3,\ \alpha x_3x_2 - x_2x_3,\ x_1^2 - x_2^2)$$

for nonzero α. The scheme X_2, which parametrizes truncated point modules of length 3, is the subscheme of $\mathbb{P}^2 \times \mathbb{P}^2$ defined by the vanishing of the multilinearized relations, that is

$$\{((x_1 : x_2 : x_3), (y_1 : y_2 : y_3)) \,|\, \alpha x_3y_1 + x_1y_3 = \alpha x_3y_2 - x_2y_3 = x_1y_1 - x_2y_2 = 0\} \\ = X_2 \subseteq \mathbb{P}^2 \times \mathbb{P}^2.$$

One can check that the projection of X_2 onto either coordinate is the subscheme E of $\mathbb{P}^2$ given by the union of the three coordinate lines, and $X_2 = \{(\sigma(p), p) | p \in E\}$ for an automorphism $\sigma : E \to E$ which interchanges two of the lines. It is known that the inverse system already stabilizes at this point, so that $X = X_2 \cong E$ is the point scheme, and the truncation shift on point modules corresponds to the automorphism σ.

For regular algebras of dimension 3, the only interesting linear modules are the point modules. For example, if A is regular of dimension 3 with Hilbert series $\frac{1}{(1-t)^3}$, then A is known to be a domain, and it easily follows that the line modules of A are simply given by the modules A/Ay for any nonzero $y \in A_1$. In particular, the line modules are parametrized by $\mathbb{P}^2$. Line modules, and other higher-dimensional linear modules, become more relevant in the investigation of regular algebras of larger dimension.

2.3. The work of Artin-Tate-Van den Bergh. Artin, Tate and Van den Bergh gave a geometric description of regular algebras of dimension 3, which refined the original classification results of Artin and Schelter. We state their results only in the quadratic case, for simplicity. However, their work also fully handles the cubic case, in a completely analogous manner.

THEOREM 2.6 ([**2**, Theorems 1,2]). *Let A be an AS regular algebra of dimension 3 with three degree one generators and three quadratic relations.*

(1) *The point modules for A are parametrized by either $E = \mathbb{P}^2$ or a degree three divisor $E \subseteq \mathbb{P}^2$, and translation-shift gives an automorphism $\sigma : E \to E$;*
(2) *Defining the sheaf $\mathcal{L} = \mathcal{O}_{\mathbb{P}^2}(1)|_E$ with corresponding element $\lambda \in \operatorname{Pic} E$, then $(\sigma - 1)^2\lambda = 0$ always, while $(\sigma - 1)\lambda = 0$ if and only if $E = \mathbb{P}^2$;*
(3) *$B = B(E, \mathcal{L}, \sigma) = \bigoplus_{n\geq 0} H^0(E, \mathcal{L}\otimes\sigma^*(\mathcal{L})\otimes\cdots\otimes(\sigma^{n-1})^*(\mathcal{L}))$ has a natural ring structure, and there is a normal element $g \in A_3$ such that $A/gA \cong B$. When $E = \mathbb{P}^2$ then $g = 0$; otherwise, g is a non-zero-divisor in A.*

The ring B appearing in part (3) is called a *twisted homogeneous coordinate ring*. The multiplication in the ring is defined for $x \in B_m$, $y \in B_n$, by $x \otimes (\sigma^m)^*(y)$; here, for $y \in H^0(\mathcal{L} \otimes \sigma^*(\mathcal{L}) \otimes \cdots \otimes (\sigma^{n-1})^*(\mathcal{L}))$ we note that pulling back by σ^m yields an element in $H^0((\sigma^m)^*\mathcal{L} \otimes \cdots \otimes (\sigma^{m+n-1})^*(\mathcal{L}))$, so that $x \otimes (\sigma^m)^*(y)$ gives an element in $H^0(\mathcal{L} \otimes \sigma^*(\mathcal{L}) \otimes \cdots \otimes (\sigma^{n+m-1})^*(\mathcal{L}))$, that is, an element of B_{m+n}. Twisted homogeneous coordinate rings $B(X, \mathcal{L}, \sigma)$ can be defined for any projective scheme X, invertible sheaf $\mathcal{L}$, and automorphism σ, and are interesting rings in their own right, apart from their applications to regular algebras [**4**]. Their ring-theoretic properties can often be understood in terms of the associated geometry. There is a robust understanding of when they are noetherian, for example [**54**]. The particular rings $B(E, \mathcal{L}, \sigma)$ occurring in part (3) of the theorem are always noetherian. A general lemma for connected graded rings [**2**, Section 8] implies that because $A/gA \cong B$, the ring A is also noetherian.

The converse of Theorem 2.6 also holds:

THEOREM 2.7 ([**2**, Theorem 3]). *Consider the twisted homogeneous coordinate rings $B(X, \mathcal{L}, \sigma)$ as above.*

(1) *The algebra $B(\mathbb{P}^2, \mathcal{O}_{\mathbb{P}^2}(1), \sigma)$ is regular of dimension 3 for any automorphism σ of $\mathbb{P}^2$.*
(2) *For any degree 3 divisor $E \subseteq \mathbb{P}^2$ and automorphism $\sigma : E \to E$ such that the element $\lambda \in \operatorname{Pic} E$ corresponding to $\mathcal{L} = \mathcal{O}_{\mathbb{P}^2}(1)|_E$ satisfies the properties $(\sigma - 1)^2\lambda = 0$ and $(\sigma - 1)\lambda \neq 0$, the ring $B = B(E, \mathcal{L}, \sigma)$ has a minimal presentation with 3 generators of degree 1 and minimal relations of degrees 2, 2, 2, 3. Removing the degree 3 relation produces an regular algebra A of dimension 3 with a normal element $g \in A_3$ such that $A/gA \cong B$.*

The theorem shows that a regular algebra of dimension 3 with Hilbert series $\frac{1}{(1-t)^3}$ can be recovered from the geometric data $(E, \mathcal{L}, \sigma)$ coming from the point scheme, and thus the choices of such triples satisfying Theorem 2.6(1)(2), up to a natural notion of equivalence of triples, are in bijective correspondence with the isomorphism classes of quadratic regular algebras of dimension 3.

EXAMPLE 2.8. Consider the following family of algebras for $a, b, c \in \Bbbk$:

$$S(a, b, c) = \Bbbk\langle x_1, x_2, x_3\rangle/(ax_i^2 + bx_{i+1}x_{i+2} + cx_{i+2}x_{i+1} \mid i \in \mathbb{Z}_3),$$

where the indices are taken modulo 3. This family was known as "type A" in Artin and Schelter's original classification. Any member of the family is now known as a *Sklyanin algebra of dimension* 3, for reasons we describe in Section 4.2 below. The computational methods used in [**1**] involving Gröbner bases did not succeed to understand this family. Using the geometric approach of [**2**], on the other hand, it is relatively simple to calculate the point modules for S, and to use the theorems above to show that an algebra in this family is regular precisely when at least two of the parameters a, b, c are nonzero and the equation $a^3 = b^3 = c^3$ does not hold. When S is regular, for generic values of a, b, c the point modules E are parametrized by a smooth elliptic curve E, and the associated automorphism $\sigma : E \to E$ is translation by a point in the group structure of E.

We remark that Iyudu and Shkarin [**50**], [**52**] showed many years later the surprising result that a (necessarily infinite) Gröbner basis for the ideal of relations in the Sklyanin algebra can in fact be analyzed effectively, and used to find the

Hilbert series of this algebra computationally. In particular, they recovered in a different way the classification of exactly which parameters a, b, c give a regular algebra.

Point modules are not the only way to approach the geometry associated to regular algebras of dimension 3. Bondal and Polishchuk [**15**] gave a different point of view on the classification of quadratic regular algebras of dimension 3, by considering the notion of the $\mathbb{Z}$-algebra associated to a helix. Ultimately, this still involves considering line bundles on cubic divisors in $\mathbb{P}^2$. This other approach turns out to be especially interesting when applied in the cubic case, where work of Van den Bergh showed that "cubic regular $\mathbb{Z}$-algebras" are a bit more general than cubic regular algebras of dimension 3 [**123**].

Given a parametrized family whose generic member is AS regular, it is also interesting to ask what happens for the special values of the parameters where the algebra is not regular. In the case of the Sklyanin family in Example 2.8, the non-regular members are known as *degenerate* Sklyanin algebras. Their properties were first studied by Walton [**135**], who showed that these algebras S all have the same Hilbert series, where $\dim_{\Bbbk} S_n$ grows exponentially, and that they are neither noetherian nor domains. She also calculated the truncated point schemes X_d for these algebras, and showed that the inverse limit $\varprojlim X_d$ does not stabilize. However, there is still a surjection $S \to B$, for a kind of generalized twisted homogeneous coordinate ring B formed from the point scheme data. Smith [**102**] noted that all of these degenerate Sklyanin algebras actually have monomial relations after a change of basis of the generators, and that they are all Zhang twists of each other, in the sense we define below in Section 3.5. This helps to explain why the degenerate Sklyanin algebras all have essentially the same properties. See also [**36**].

3. Basic properties of regular algebras

As soon as the classification of regular algebras of dimension 3 was completed, the obvious question was whether similar ideas could be used to classify regular algebras of dimension 4. The dimension 4 case has proven to be much more complicated, and so far has resisted any sort of general classification. However, the search for a better understanding of dimension 4 (and higher-dimensional) regular algebras has led to the discovery of many interesting examples and techniques. In this section we review some additional background and describe some general questions about regular algebras. Then in the succeeding sections we describe some of the known results, organized by general theme.

One consequence of the classification of regular algebras of dimension 3 is that such algebras have good ring-theoretic properties. For example, we have already remarked that they are noetherian, and they also must be domains [**3**, Section 3]. There are a number of other properties that also hold for all known examples of regular algebras, and thus it is natural to conjecture that they hold in general. It is true that the vast majority of specific examples that have been examined in detail have global dimension at most 5, so it remains possible that the general higher-dimensional picture is different somehow.

3.1. GK-dimension. First let us recall the GK-dimension of a module. The reader can find more details in [**58**].

Definition 3.1. Let A be a finitely generated $\Bbbk$-algebra, generated by a finite-dimensional $\Bbbk$-subspace V. Let M be a finitely generated right A-module, generated by a finite-dimensional subspace W. The *Gelfand-Kirillov dimension* (GK-dimension) of M is $\operatorname{GKdim}(M) = \limsup_{n\to\infty} \log_n \dim_{\Bbbk}(WV^n)$ (which one may prove is independent of V and W).

Essentially, $\operatorname{GKdim}(M) = \alpha$ means that the sequence $f(n) = \dim_{\Bbbk} WV^n$ grows roughly at the same rate as n^α. When A is a connected, finitely generated $\mathbb{N}$-graded algebra, and M is a finitely generated $\mathbb{Z}$-graded A-module, then

$$\operatorname{GKdim}(M) = \limsup_{n\to\infty} \log_n d(n), \text{ where } d(n) = \dim_{\Bbbk} \bigoplus_{i=-n}^{n} M_i = \sum_{i=-n}^{n} \dim_{\Bbbk} M_i.$$

In particular, applying this to the ring A itself, it is easy to see that $\operatorname{GKdim}(A) < \infty$ if and only if the sequence $\dim_{\Bbbk} A_n$ has polynomially bounded growth. So condition (2) in Definition 1.5 is equivalent to $\operatorname{GKdim}(A) < \infty$.

Now suppose that A is a connected $\mathbb{N}$-graded algebra with finite global dimension. As mentioned in the discussion of the global dimension 3 case, by examining the graded free resolution of the trivial module $\Bbbk$, one sees that the Hilbert series of A has the form $h_A(t) = \frac{1}{p(t)}$ for some monic polynomial $p(t) \in \mathbb{Z}[t]$. Moreover, $\operatorname{GKdim}(A) < \infty$ if and only if $p(t)$ has only roots of unity for zeroes. If instead $\operatorname{GKdim}(A) = \infty$, then one can prove in fact that A has *exponential growth*, i.e. that $\dim_{\Bbbk} A_n$ grows asymptotically like some function β^n with $\beta > 1$. Stephenson and Zhang showed in [**118**, Theorem 1.2] that a connected graded algebra with exponential growth cannot be noetherian (this paper is also a good place to find proofs of the other facts mentioned above). Thus, finite GK-dimension is a necessary condition for a connected graded algebra of finite global dimension to be noetherian. This certainly justifies why the polynomial growth condition is needed in the definition of AS regular, if one is interested in noetherian rings. On the other hand, finite GK-dimension is not a *sufficient* condition for an algebra of finite global dimension to be noetherian, as Example 1.6 shows. All known examples of regular algebras are noetherian, so the Gorenstein condition seems to be absolutely crucial in order to guarantee good ring-theoretic properties, in a way that is still not well-understood.

We mention that in contrast to the results above for algebras of finite global dimension, it is unknown if there exists a connected graded noetherian algebra A (necessarily of infinite global dimension) with subexponential growth, yet with $\operatorname{GKdim}(A) = \infty$. Such examples do exist if one removes the graded or noetherian assumptions [**118**, Remark after Example 1.4].

3.2. Auslander conditions. The study of the interaction between homological conditions on an algebra and its ring-theoretic properties long predates the concept of an AS regular algebra. In commutative algebra, one has such results as far back as the 1950's, such as the Auslander-Buchsbaum theorem that a regular local ring is a UFD. The following is an important earlier notion of regularity for noncommutative rings.

Definition 3.2. Let A be a (not necessarily graded) noetherian $\Bbbk$-algebra A with $\operatorname{gl.dim}(A) = d < \infty$. For an A-module M, set $j(M) = \inf\{i \mid \operatorname{Ext}^i_A(M, A) \neq 0\}$, the *j-number* or *grade* of the module M. A is called *Auslander regular* if for every finitely generated module M and $0 \leq i \leq d$, all A-submodules N of $\mathrm{Ext}^i_A(M, A)$ have $j(N) \geq i$.

If $\mathrm{GKdim}(A) < \infty$, the algebra A is *GK-Cohen-Macaulay* if for all finitely generated modules M, $\mathrm{GKdim}(M) + j(M) = \mathrm{GKdim}(A)$.

At the time that AS regular algebras were introduced, the notion of Auslander regularity had already been important in ring theory for investigating examples related to rings of differential operators and Lie theory, for example. Levasseur showed that if A is a connected graded algebra with $\mathrm{GKdim}(A) < \infty$, then if A is Auslander regular, it must be Artin-Schelter regular [**64**, Theorem 6.3]. Conversely, there is no known example of an AS regular algebra that is not also Auslander regular.

3.3. Ring-theoretic properties of AS regular algebras. Here are some basic questions about properties that AS regular algebras might satisfy.

QUESTION 3.3. Let A be an AS regular algebra of global dimension d.

(1) Is A noetherian?
(2) Is A a domain?
(3) Does $\mathrm{GKdim}(A) = d$?
(4) Is the minimal number of generators of A as an algebra at most d?
(5) Does A have Hilbert series of the form $h_A(t) = \frac{1}{(1-t^{d_1})\dots(1-t^{d_m})}$ for some $d_1, \dots d_m$ (the same as some graded commutative polynomial ring)?
(6) Is A Auslander regular? If so, is A GK-Cohen Macaulay?
(7) Is A a maximal order inside its Goldie quotient ring?

The answer to each one of these questions is yes for every known example of regular algebra for which the property has been checked, and so the the answer is conjectured to be yes in every case. Note that a maximal order is the noncommutative analog of an integrally closed ring in commutative algebra.

The following notion will appear occasionally below. A $\Bbbk$ algebra R is a *polynomial identity (PI) algebra* if there is some nonzero $f \in \Bbbk\langle x_1, \dots, x_n\rangle$ such that $f(r_1, \dots, r_n) = 0$ for all $r_1, \dots, r_n \in R$. If R is commutative, then R is a PI algebra by taking $f = x_1x_2 - x_2x_1$, and in general PI algebras have special properties that make them "close to commutative." For example, an algebra R that is finitely generated as a module over its center $Z(R)$ is PI, but the converse is not true in general.

There are known connections between some of the questions above. Levasseur proved that if A is connected graded and Auslander regular, then A is a domain [**64**, Theorem 4.8]. In [**3**, Section 3], Artin, Tate and Van den Bergh proved that a regular algebra of dimension at most 4 which is noetherian is a domain. Stafford proved that if an AS-regular algebra A is noetherian, Auslander regular and GK-Cohen-Macaulay, then A is indeed a maximal order. Moreover, such an A is a PI algebra if and only if it is a finite module over its noetherian center [**105**, Theorem 2.10]. Stafford and Zhang examined the homological properties of graded PI algebras more generally in [**109**]. They proved that if A is a PI noetherian AS regular algebra, then A is Auslander regular and GK-Cohen-Macaulay (so in particular, a domain and maximal order, as noted above).

One may also ask the opposite sort of question: are there conditions weaker than those in the definition of AS regular that imply AS regularity, perhaps for special kinds of algebras? Stephenson and Zhang noted that connected graded noetherian algebras of global dimension 2 are AS regular [**118**, Theorem 0.4]. In other words, the Gorenstein condition comes for free in the presence of the noetherian property (we already noted earlier that finite GK-dimension is forced). They extended this result to some extent to dimension 3, proving that any connected graded noetherian algebra A of global dimension 3 which is generated by 2 elements, has a graded free resolution of $\Bbbk$ of the expected form, or has relations all in the same degree, must be AS regular [**119**, Theorem 0.1].

Based on this evidence, one might conjecture that connected graded noetherian algebras of finite global dimension must be AS regular. In [**89**], however, the author and Sierra found examples of algebras A with $\operatorname{gl.dim}(A) = 4$ and $h_A(t) = \frac{1}{(1-t)^4}$ which are noetherian domains, but for which the Gorenstein condition fails. These examples are even Koszul, as defined in Section 7 below.

3.4. Noncommutative Proj. When $A = \Bbbk[x_1, \dots, x_n]$ is a polynomial ring, with some choice of grading given by assigning positive degrees (or *weights*) to the variables, then the associated scheme $\operatorname{Proj} A$ is a called a *weighted projective space.* For this reason, intuitively a regular algebra A can be thought of as a coordinate ring of a *noncommutative* weighted projective space. But what is the space?

Artin and Zhang showed that instead of a space, one can attach a category to A which serves many of the same functions [**5**]. We loosely describe the construction here. Let A be a noetherian connected graded finitely generated $\Bbbk$-algebra. Let $\operatorname{gr} A$ be the abelian category of finitely generated $\mathbb{Z}$-graded right A-modules, where the morphisms are graded (degree-preserving) module homomorphisms. Let $\operatorname{tors} A$ be the full subcategory of $\operatorname{gr} A$ consisting of modules which are finite dimensional over $\Bbbk$. Using a construction of quotient category due to Gabriel, one defines $\operatorname{proj} A = \operatorname{gr} A/\operatorname{tors} A$ and calls this category the "noncommutative proj" associated to A. If A is commutative, generated in degree 1, then the category $\operatorname{proj} A$ is equivalent to the category of coherent sheaves on the usual scheme $X = \operatorname{Proj} A$. So really we are generalizing the idea of a category of sheaves rather than a space to the noncommutative setting, but this turns out to be enough in order to develop a rich theory. For example, the image of a point module M for A is a simple object in the category $\operatorname{proj} A$, and so does behave geometrically as a "point".

Noncommutative projective geometry is its own large subject. See the survey article [**107**]; the author's course notes [**88**] also give an introduction to the noncommutative proj construction. The categories $\operatorname{proj} A$ for regular algebras A are very important in noncommutative geometry, because they represent noncommutative weighted projective spaces. In particular, if A is a quadratic regular algebra of dimension 3, then $\operatorname{proj} A$ is a noncommutative $\mathbb{P}^2$, and if A is a cubic regular algebra of dimension 3, then $\operatorname{proj} A$ is a noncommutative $\mathbb{P}^1 \times \mathbb{P}^1$. In this article, we will focus on the algebraic aspects of regular algebras, and will not attempt to survey the work that is focused primarily on their associated noncommutative projective spaces. The interaction between the algebra A and its category $\operatorname{proj} A$ is an essential part of the story, however, and we will want to refer to the categories $\operatorname{proj} A$ occasionally below.

3.5. Twists. Graded twists are a fundamental construction for graded rings that have played a role in much of the work on regular algebras, including the original work of Artin, Tate, and Van den Bergh.

Let G be a semigroup. Suppose that A is a G-graded $\Bbbk$-algebra, so $A = \bigoplus_{g \in G} A_g$ with $A_g A_h \subseteq A_{gh}$ for all $g, h \in G$. Assume that for each $g \in G$ we have a $\Bbbk$-linear bijection $\tau_g : A \to A$. The collection $\tau = \{\tau_g | g \in G\}$ is called a *twisting system* if $\tau_g(y\tau_h(z)) = \tau_g(y)\tau_{gh}(z)$ for all $g, h, l \in G$ and $y \in A_h$, $z \in A_l$. Given a twisting system τ on A, one can define a new G-graded algebra A^τ which is called a *twist* of A. The algebra A^τ has the same underlying graded vector space as A, but a new multiplication $*$, defined on homogeneous elements $x \in A_g, y \in A_h$ by $x * y = x\,\tau_g(y)$, and extended linearly. The condition defining a twisting system is exactly what is needed to ensure that A^τ is associative. Zhang defined and studied this notion in [**138**], and these twists are often called Zhang twists. Zhang proved that for any twisting system $\{\tau_i\}$ on A, one has equivalent categories of graded modules $\operatorname{Gr} A \sim \operatorname{Gr} A^\tau$.

Consider in particular the case that $G = \mathbb{N}$, so A is $\mathbb{N}$-graded. In this case there is an easy way to construct twisting systems: for any algebra automorphism $\sigma : A \to A$, if $\tau_n = \sigma^n$ for all n then $\tau = \{\tau_n\}$ is a twisting system. The new multiplication in this case is given by $x * y = x\sigma^m(y)$ for all $x \in A_m, y \in A_n$. As a converse to the result above in the $\mathbb{N}$-graded case, if A and B are connected $\mathbb{N}$-graded and $A_1 \neq 0$, then $\operatorname{Gr} A \sim \operatorname{Gr} B$ implies that there exists a twisting system $\{\tau_i\}$ on A such that $B \cong A^\tau$ as graded algebras [**138**, Theorems 3.1, 3.5].

Because the noncommutative projective scheme $\operatorname{proj} A$ is a quotient category of the category of finitely generated graded A-modules, it is immediate that twisting does not change the underlying noncommutative scheme $\operatorname{proj} A$. However, in general twisting certainly does change the algebra up to isomorphism, so one can get a lot of different algebras as twists A^τ, which can be thought of as different noncommutative coordinate rings of the same noncommutative scheme. Zhang also proved that many properties are preserved under twist: in particular, if A is AS regular, then so is A^τ for any twisting system $\tau = \{\tau_i\}$ [**138**, Theorem 5.11]. Thus twisting gives an equivalence relation on the the set of AS regular algebras. In classification problems for regular algebras, it is often useful to work up to twist equivalence.

A related kind of twist is a cocycle twist. Suppose that A is any algebra with a grading by a group G. One may attempt to deform the multiplication of A to a new one $*$ by choosing for each pair of elements $g, h \in G$ a nonzero constant $\mu(g, h) \in \Bbbk$ such that for $a \in A_g, b \in A_h$, we have $a * b = \mu(g, h)ab$. This gives a new associative multiplication on A if and only if $\mu : G \times G \to \Bbbk^\times$ is a 2-cocycle. The new algebra $(A, *)$ is written as A_μ. One may see that this is in fact a Zhang twist where each τ_g scales A_h by $\mu(g, h)$, and in this case the 2-cocycle condition is equivalent to the twisting system axiom.

Now suppose that we have a connected graded algebra A that has an additional grading by a group G which refines the given $\mathbb{N}$-grading. Then given a 2-cocycle μ for G, the new algebra A_μ will still be $\mathbb{N}$-graded. However, the preservation of nice properties under this kind of twist no longer follows from Zhang's results. Davies proved that AS regularity and other good properties are in fact invariant under twisting when G is a finite abelian group [**34**, Theorem 1.2]. He also showed that some families of examples of 4-dimensional regular algebras classified by the author and Zhang in [**90**] are related by non-obvious twists of this kind.

4. Algebras related to an elliptic curve

The Sklyanin algebras of dimension 3, defined above in Example 2.8, are the most interesting family of 3-dimensional regular algebras, due to the nontrivial and beautiful connections between these algebras and the geometry of elliptic curves. In fact, they belong to a large family of examples related to elliptic curves, as we describe in this section.

4.1. The Sklyanin algebras of dimension 4. The first regular algebras of dimension 4 to be studied in detail were the 4-dimensional Sklyanin algebras.

EXAMPLE 4.1. Let $\alpha, \beta, \gamma \in \Bbbk$ with $\alpha + \beta + \gamma + \alpha\beta\gamma = 0$.
The 4-dimensional Sklyanin algebra is defined by

$$S(\alpha, \beta, \gamma) = \Bbbk\langle x_0, x_1, x_2, x_3\rangle/(r_1, \dots, r_6)$$

where the six relations are

$$\begin{aligned} r_1 &= x_0x_1 - x_1x_0 - \alpha(x_2x_3 + x_3x_2), & r_2 &= x_0x_1 + x_1x_0 - (x_2x_3 - x_3x_2), \\ r_3 &= x_0x_2 - x_2x_0 - \beta(x_3x_1 + x_1x_3), & r_4 &= x_0x_2 + x_2x_0 - (x_3x_1 - x_1x_3), \\ r_5 &= x_0x_3 - x_3x_0 - \gamma(x_1x_2 + x_2x_1), & r_6 &= x_0x_3 + x_3x_0 - (x_1x_2 - x_2x_1). \end{aligned}$$

The algebra $S(\alpha, \beta, \gamma)$ was proved to be AS regular precisely when $\{\alpha, \beta, \gamma\}$ does not contain the subset $\{1, -1\}$, by Smith and Stafford [**104**]. They proved that such a regular S is a noetherian domain with Hilbert series $H_S(t) = \frac{1}{(1-t)^4}$. The proof uses a similar idea to the proof that the Sklyanin algebras of dimension 3 are regular in [**2**]. In more detail, S has two linearly independent central elements of degree two, g_1 and g_2, say, such that $A/(g_1, g_2) \cong B = B(E, \mathcal{L}, \sigma)$ is a twisted homogeneous coordinate ring on an elliptic curve E. The properties of B can be determined geometrically, and good properties can be lifted from B to S. In this case E is not the entire point scheme X, which also has 4 isolated points in addition to E. Further properties of S were proved by Staniszkis [**110**] and Levasseur and Smith [**65**].

The 4-dimensional Sklyanin algebras S have also been a good starting place for constructing other interesting examples. Since S is presented by 4 generators and 6 quadratic relations, B is presented by 4 generators and 8 quadratic relations. Stafford examined another class of rings obtained by removing a different choice of 2 relations from the presentation of B, and showed these are also AS regular of dimension 4 [**106**]. This showed that the ring B does not uniquely determine the 4-dimensional regular algebra mapping onto it, unlike the situation for regular algebras of dimension 3.

In [**35**], Davies constructed regular cocycle twists of 4-dimensional Sklyanin algebras S that have quite different properties from S. For example, these twists have only finitely many point modules up to isomorphism, unlike S which has an entire elliptic curve of point modules, demonstrating how a cocycle twist can change the geometry associated to an algebra dramatically, unlike a usual Zhang twist. Chirvasitu and Smith investigated many additional properties of these and more general kinds of twists in a series of papers [**31**] [**30**].

4.2. Feigin and Odesskii's elliptic algebras. Roughly at the same time that Artin, Schelter, Tate, and Van den Bergh were laying down the foundations for the classification of regular algebras of dimension 3, Feigin and Odesskii defined a class of algebras built out of the geometry of an elliptic curve [**79**]. Serendipitously, these algebras turned out to be important examples of AS regular algebras.

There are several minor variations on the basic definition that have appeared in the literature; we follow here the presentation given by Chirvasitu, Kanda, and Smith in [**26**].

Let E be a complex elliptic curve, defined as $\mathbb{C}/\Lambda$ for a lattice $\Lambda = \mathbb{Z}+\mathbb{Z}\eta$, where η is in the upper half of the complex plane. Fix $1 \leq k < n$ with $\gcd(k, n) = 1$ and a point $\tau \in \mathbb{C} - (1/n)\Lambda$. A *theta function* is a holomorphic function which is quasi-periodic with respect to the lattice, in the sense that translating by one of the lattice generators 1 or τ scales the function by a specified complex exponential function. More specifically, here we are interested in holomorphic functions $f : \mathbb{C} \to \mathbb{C}$ which satisfy

$$f(z+1) = f(z) \qquad \text{and} \qquad f(z+\eta) = -e^{-2\pi i n z} f(z).$$

These are called *theta functions of order* n as they have n zeros inside every period parallelogram of the lattice. The set of all such theta functions of order n is a vector space over $\mathbb{C}$ of dimension n denoted $\Theta_n(\Lambda)$, which is an irreducible representation of the finite Heisenberg group H_n of order n^3 in a natural way. One chooses a basis $\theta_0, \dots, \theta_n$ of $\Theta_n(\Lambda)$ which is well-suited to the action of H_n; see [**26**, Proposition 2.6]. In the next definition we take the indices of the generators x_i and the theta functions θ_i modulo n.

Definition 4.2. Define

$$Q_{n,k}(E, \tau) = \Bbbk\langle x_0, x_1, \dots, x_{n-1}\rangle/(r_{ij} | 1 \leq i, j \leq n)$$

where

$$r_{ij} = \sum_{r \in \mathbb{Z}_n} \frac{\theta_{j-i+(k-1)r}(0)}{\theta_{j-i-r}(-\tau)\theta_{kr}(\tau)} x_{j-r}\, x_{i+r},$$

for $\theta_0, \dots \theta_{n-1}$ the special basis of $\Theta_n(\Lambda)$ chosen above.

Clearly there is a lot to unpack in this definition, and in a survey article like this we cannot give much of a feel for these algebras. The already mentioned paper of Chirvasitu, Kanda, and Smith [**26**] is the recommended starting point, as it is the first paper in a large project by these authors to provide more explicit details of the basic properties of these algebras, many originally stated by Feigin and Odesskii with brief or no proofs. While we assumed $\tau \notin (1/n)\Lambda$ above, there is a natural way to extend the definition so that an algebra $Q_{n,k}(E, \tau)$ is defined for every $\tau \in \mathbb{C}$.

Feigin and Odesskii were motivated by algebras defined by Sklyanin in papers predating any of the work on regular algebras [**97**], [**98**], which studied certain algebras that were defined in terms of Baxter's "elliptic solution" to the quantum Yang-Baxter equation. In the more general setup of Feigin and Odesskii, the algebras defined by Sklyanin are the algebras $Q_{4,1}(E, \tau)$, and of course this is why the general construction is named after Sklyanin. For generic τ, these are Sklyanin algebras of dimension 4 whose presentation was given in Example 4.1, in the case $\Bbbk = \mathbb{C}$. Similarly, for generic τ the algebras $Q_{3,1}(E, \tau)$ are the 3-dimensional Sklyanin algebras in Example 2.8. Thus the algebras $Q_{n,1}(E, \tau)$ are direct generalizations of the algebras in Sklyanin's papers, and are commonly referred to as Sklyanin algebras of dimension n.

In related work, Tate and Van den Bergh gave an essentially different approach to the algebras $Q_{n,1}(E, \tau)$ from Feigin and Odesskii's [**120**]. In particular, they showed these algebras are noetherian domains. See also [**121**]. Smith studied the point modules of these algebras in [**99**]. More generally, all linear modules of $Q_{n,1}(E,\tau)$ were classified by Staniszkis [**111**]. As in the case of $n = 3$ that appeared in [**2**], considering point modules for $Q_{n,1}(E,\tau)$ gives a natural embedding of E inside $\mathbb{P}^{n-1}$. Staniszkis showed that the linear modules of $Q_{n,1}(E,\tau)$ correspond to those linear subspaces of $\mathbb{P}^{n-1}$ which relate to the embedded elliptic curve E in a certain way. She proved that these linear modules are GK-Cohen-Macaulay and critical with respect to GK-dimension. Zhang [**139**] showed that the d-linear modules of $Q_{n,1}(E,\tau)$ have projective dimension $n-d$ and have linear free resolutions.

As we see above, the Sklyanin algebras—that is, the $Q_{n,k}(E,\tau)$ with $k = 1$—have been the most studied among Feigin and Odesskii's algebras. As already mentioned, Chirvasitu, Kanda, and Smith have a recent series of papers [**24**] [**26**] [**27**] [**28**] [**29**] revisiting the work of Feigin and Odesskii on the algebras $Q_{n,k}(E,\tau)$, with more attention on the case $k \neq 1$. In more detail, after an introductory paper [**26**] they define and study the characteristic variety $X_{n/k}$, which is a quotient of E^g by a finite group—here g is a positive integer depending on a continued fraction expansion of n/k [**24**]. When $k = 1$, $X_{n/k} = E$. There is a map from $Q_{n,k}(E,\tau)$ to a certain canonical twisted homogeneous coordinate ring $B(X_{n/k}, \mathcal{L}, \sigma)$ which is surjective in many cases [**27**]. In particular, they verify carefully that $Q_{n,k}(E,\tau)$ is a regular algebra with Hilbert series $\frac{1}{(1-t)^n}$ for all but countably many τ [**29**]. There are still many basic open questions in this area, and we close this section with just a few. Are the regular algebras of the form $Q_{n,k}(E,\tau)$ noetherian domains when $k \neq 1$? What are the finite-dimensional representations over $Q_{n,k}(E,\tau)$? Can one define analogs of the algebras $Q_{n,k}(E,\tau)$ for any abelian variety E?

5. Linear modules over regular algebras

Point modules, and the scheme parametrizing them, were crucial to the geometric classification of AS regular algebras of dimension 3 given in [**2**]. Naturally, many investigations of regular algebras of dimension 4 have also considered the point scheme, with a hope that one might be able to recover the algebra from geometric data on the point scheme in some cases, for example. Starting in dimension 4, larger dimensional linear modules such as line modules are also natural objects of study.

5.1. Hilbert schemes. We first review the technical theory underpinning the generalization of the point scheme to parameter spaces of higher-dimensional modules. As we saw earlier, for any connected graded algebra A generated in degree 1, the point modules are in bijection with an inverse limit of schemes $\varprojlim X_d$, where X_d is a projective scheme parametrizing truncated point modules of length $d+1$. This construction is most useful when this inverse limit stabilizes, in which case there is a projective scheme parametrizing the point modules.

More generally, one would like to find a nice moduli space for the isomorphism classes of cyclic graded modules with a given fixed Hilbert series—a *Hilbert scheme*. Schemes which parametrize point modules, line modules, and other linear modules are then just a special case. Artin and Zhang developed in [**6**] such a theory of noncommutative Hilbert schemes. They showed that if a connected graded algebra A, finitely generated in degree 1, is *strongly noetherian*, that is, $A \otimes_{\Bbbk} R$ is noetherian for all commutative noetherian $\Bbbk$-algebras R, then the Hilbert schemes are projective schemes. In particular, in this case there is a projective point scheme, line scheme, and so on. Commutative noetherian finitely generated graded algebras are automatically strongly noetherian, and Artin and Zhang's result simply recovers the commutative theory of Hilbert schemes in that case. However, noncommutative noetherian connected graded algebras may fail to be strongly noetherian, and the inverse limit parametrizing their point modules may fail to stabilize [**53**] [**87**]. A large class of such examples called *naïve blowups* can be constructed as certain subalgebras of twisted homogeneous coordinate rings [**55**].

All known regular algebras are strongly noetherian. In particular examples, though, one tends to show that the point modules or line modules are parametrized by a projective scheme directly, without first attempting to prove the algebra is strongly noetherian. In fact, in regular algebras the inverse limits which parametrize point and line modules tend to stabilize in the lowest possible degree, similar to the behavior seen in regular algebras of dimension 3. For simplicity, let us focus on the case of a regular algebra A of dimension d with quadratic relations and Hilbert series $\frac{1}{(1-t)^d}$. In this case, in all known examples the inverse limit $\varprojlim X_d$ which parametrizes point modules stabilizes already at $X_2 \subseteq \mathbb{P}^{m-1} \times \mathbb{P}^{m-1}$, and moreover, X_2 is the graph of an automorphism $\sigma : X \to X$ for some $X \subseteq \mathbb{P}^{m-1}$. When this happens, $X \cong X_2$ parametrizes the point modules, and truncation shift is an automorphism on the set of point modules (given by σ).

Shelton and Vancliff gave an extensive study of line modules for AS regular algebras A of dimension 4 with quadratic relations. They showed that if A is Auslander regular and GK-Cohen Macaulay, then the line modules of A are parameterized by a projective scheme that can easily calculated as a subscheme of a Grassmannian [**92**, Lemma 2.4, Remark 2.10]. Under the same condition, they showed that this line scheme has dimension at least 1, and that every point module is a quotient of some line module [**92**, Proposition 3.1]. They computed the line scheme explicitly in a number of interesting examples [**93**]. In more recent work, Chirvasitu, Smith, and Vancliff studied the geometric properties of the scheme parametrizing line modules when it has dimension 1, proving that it always is a subscheme of degree 20 inside an ambient $\mathbb{P}^5$ [**32**].

5.2. Questions about the point scheme. Here are some questions related to the point modules of regular algebras. Again, for convenience we focus on the case of algebras with quadratic relations, but one can ask similar questions in general.

QUESTION 5.1. Let A be an AS regular algebra, generated in degree 1 with $\dim_{\Bbbk} A_1 = m$.

(1) Assuming A is noetherian, is A strongly noetherian?
(2) Suppose that A has quadratic relations, and let $X_2 \subseteq \mathbb{P}^{m-1} \times \mathbb{P}^{m-1}$ be the subscheme defined by the multilinearizations of the relations. Does the inverse limit $\varprojlim X_d$ parametrizing the point modules already stabilize at X_2? If so, is X_2 the graph of an automorphism of some subscheme of $\mathbb{P}^{m-1}$?
(3) When is the algebra A determined by its point scheme data? If it is not, what if we include the data of some other Hilbert schemes?

(4) Is whether or not A is a polynomial identity algebra (PI) related in a nice way to the point scheme data? If it is not, what if we include the data of some other Hilbert schemes?

These questions are inspired by what happens in dimension 3. For any AS regular algebra A of dimension 3, the answer to (1) is yes by a similar argument as the proof that A is noetherian. For a quadratic AS regular algebra A of dimension 3, the scheme X_2 is indeed the graph of an automorphism $\sigma : E \to E$ as in (2), and for (3) the data $(E, \mathcal{O}_{\mathbb{P}^2}|_E, \sigma)$ determines the algebra, as we saw earlier in Theorem 2.7. For (4), [**3**, Theorem 7.1] proved that A is PI if and only if the automorphism σ of the point scheme has finite order.

The answer is also yes to questions (1) and (2) in all known higher-dimensional examples for which it has been checked. Assume in the following discussion that A is regular of dimension 4, generated in degree 1 with quadratic relations and Hilbert series $\frac{1}{(1-t)^4}$. It is known that if A is Auslander regular and GK-Cohen Macaulay, then (2) holds; this was first proved by Vancliff and Van Rompay in [**130**, Theorem 1.10], under an additional minor hypothesis which was later removed by Shelton and Vancliff [**91**, Theorem 1.4]. If (2) holds, one may consider the subspace of $A_1 \otimes_{\Bbbk} A_1$ consisting of elements whose multilinearizations vanish along $X_2 \subseteq \mathbb{P}^{m-1} \times \mathbb{P}^{m-1}$. In nice cases, this may be equal to the span of the multilinearlizations of the degree 2 relations of A, and hence A is determined by X_2, giving an answer to (3). Similarly, it is possible for A to be completely determined by the line scheme data of A; in particular, this happens if the line scheme is of dimension 1, which is known to be the minimal possible dimension for the line scheme [**92**, Theorem 4.1].

5.3. The dimension of the point scheme of regular algebras of dimension 4. Suppose that A is a regular algebra of dimension 4 with quadratic relations and Hilbert series $\frac{1}{(1-t)^4}$. In general, the larger the point scheme X of the regular algebra A, the more information it is expected to provide. At one extreme, it could be that $X \cong \mathbb{P}^3$. This case is rather trivial, as A will then be a twisted homogeneous coordinate ring $B(\mathbb{P}^3, O(1), \sigma)$, which is just a Zhang twist of a polynomial ring in 4 variables.

The next is case where $\dim X = 2$. One might expect this case to bear some similarity to the most typical situation for quadratic regular algebras of dimension 3, in which the point scheme is a degree three divisor inside $\mathbb{P}^2$. A number of cases with $\dim X = 2$ have been studied in detail. In [**126**], Vancliff defined a natural class of such algebras A with a normal element $x \in A_2$ such that $A/xA \cong B(Q, \mathcal{L}, \sigma)$, for some nonsingular quadric hypersurface Q in $\mathbb{P}^3$ with automorphism σ and invertible sheaf $\mathcal{L} = \mathcal{O}(1)|_Q$. The quadric Q is not quite the point scheme X of A, but rather $X = Q \cup L$ for some line L intersecting Q in two points. Vancliff wrote down the relations of all such algebras; they include interesting examples such as the coordinate ring of quantum 2×2 matrices $\mathcal{O}_q(M_2(\Bbbk))$ for $q^2 \neq 1$. She also examined the line modules and their incidence relations with the point modules.

Van Rompay [**125**] constructed similar examples using Segre products. Then in [**130**], Vancliff and Van Rompay classified all regular algebras A of dimension 4 which surject onto a twisted homogeneous coordinate ring $B(Q, \mathcal{L}, \sigma)$ for a nonsingular quadric Q; in many cases the point scheme of A is of the form $X = Q \cup L$, where in general the line L can be tangent to Q or embedded in Q as a line of multiplicity 2. They showed that it is also possible to have $X = X_2 = Q$, and in this case there is a whole family of regular algebras that surject onto $B(Q, \mathcal{L}, \sigma)$; that is, the point scheme data does not determine the relations of the algebra [**130**]. However, in [**131**], they give a further analysis of this class of algebras using line modules, and prove that the schemes of point and line modules together determine the relations of the algebra, giving a kind of positive answer to Question 5.1(3) in this case.

Shelton and Vancliff [**91**] extended this program to singular quadrics Q of rank 3 in $\mathbb{P}^3$, classifying all regular algebras A of dimension 4 which surject onto $B(Q, \mathcal{L}, \sigma)$ for such Q.

We are unaware of any systematic study of 4-dimensional regular algebras with point scheme X such that $\dim X = 1$, though of course there are many examples, notably the 4-dimensional Sklyanin algebras.

The final case is when the point scheme X has $\dim X = 0$, so that X is a collection of points, possibly with non-reduced scheme structure. In contrast to the examples where the point scheme is a quadric surface, Shelton and Vancliff proved the surprising result that in fact the point scheme X data determines the relations of a regular algebra A of dimension 4 as above, whenever $\dim X = 0$ [**92**, Theorem 4.1, Remark 4.2]. So Question 5.1(3) has a nice answer in this case. There is a useful construction, which we review next, that produces many examples of this type with X of various different configurations.

5.4. (Skew) graded Clifford algebras. Fix $n \geq 2$. Fix nonzero scalars $\mu_{ij} \in \mathbb{k}$ such that $\mu_{ii} = 1$ for all i and $\mu_{ij}\mu_{ji} = 1$ for all $i \neq j$. A matrix $M \in M_n(\mathbb{k})$ is *μ-symmetric* if $M_{ij} = \mu_{ij} M_{ji}$ for all i, j.

Suppose we are given a collection of $n \times n$ μ-symmetric matrices $M_1, \dots, M_n$ in $M_n(\mathbb{k})$. We define an algebra $A(\mu, M_1, \dots, M_n)$ by the following presentation:

$$\mathbb{k}\langle x_1, \dots, x_n, y_1, \dots, y_n\rangle/(x_i x_j + \mu_{ij} x_j x_i - \sum_{k=1}^{n} (M_k)_{ij}\, y_k \,|\, 1 \leq i, j \leq n).$$

Here the generators x_i are of degree 1 and the y_i are of degree 2; so the algebra is naturally connected $\mathbb{N}$-graded. Recall that an element x in a ring R is *normal* if $xR = Rx$, so that the right ideal xR is an ideal. Following Cassidy and Vancliff, $A = A(\mu, M_1, \dots, M_n)$ is called a *graded skew Clifford algebra* if in addition there is a $\mathbb{k}$-basis of $\mathbb{k}y_1 + \cdots + \mathbb{k}y_n$, say $r_1, \dots, r_n$, which is a normalizing sequence in A in the sense that each r_i is normal in the factor ring $A/(r_1, \dots, r_{i-1})$. While by definition these algebras have generators in degrees 1 and 2, the point of this construction is to find interesting regular algebras that are generated in degree 1. As long as the matrices $M_1, \dots, M_n$ are linearly independent over $\mathbb{k}$, one may easily check that the generators y_i may be eliminated from the presentation, and so A is generated by $x_1, \dots x_n$.

If $\mu_{ij} = 1$ for all i, j, then the matrices M_i are symmetric. In this case we omit μ from the notation, writing $A = A(M_1, \dots, M_n)$, and call this a *graded Clifford algebra* if in addition $y_1, \dots, y_n$ are central in A for all i. This was the first case to be studied. In this case, each matrix M_k determines a quadric hypersurface Q_k in $\mathbb{P}^{n-1} = \operatorname{Proj} \mathbb{k}[z_1, \dots, z_n]$ given by the vanishing of $\sum_{i,j} (M_k)_{ij} z_i z_j = 0$. The system of quadrics $Q_1, \dots, Q_n$ is called *basepoint free* if $Q_1 \cap \cdots \cap Q_n = \emptyset$. Le Bruyn [**59**] proved that $A(M_1, \dots, M_n)$ is a regular algebra with Hilbert series $\frac{1}{(1-t)^n}$ if and only if the associated system of quadrics is basepoint free.

EXAMPLE 5.2 (Section 2, [**132**]). Let $M_1, M_2, M_3, M_4 \in M_n(\Bbbk)$ be the symmetric matrices such that

$$M(\alpha_1,\alpha_2,\alpha_3,\alpha_4) = \sum_{i=1}^{4} \alpha_i M_i = \begin{bmatrix} 2\alpha_1 & \alpha_3 & \alpha_4 & \alpha_2 \\ \alpha_3 & 2\alpha_2 & \alpha_1 & \alpha_3+\alpha_4 \\ \alpha_4 & \alpha_1 & 2\alpha_3 & \alpha_1 \\ \alpha_2 & \alpha_3+\alpha_4 & \alpha_1 & 2\alpha_4 \end{bmatrix}$$

for indeterminates α_i. In this case the relations of the Clifford algebra $A = A(M_1, M_2, M_3, M_4)$ include $y_i = x_i^2$ for all i, so eliminating the generators y_i we see that A has the following presentation:

$$A = \Bbbk\langle x_1, x_2, x_3, x_4\rangle/(r_1, r_2, r_3, r_4, r_5, r_6) \text{ where}$$
$$r_1 = x_1x_4 + x_4x_1 - x_2^2, \qquad r_2 = x_2x_4 + x_4x_2 - x_3^2 - x_4^2,$$
$$r_3 = x_3x_4 + x_4x_3 - x_1^2, \qquad r_4 = x_1x_3 + x_3x_1 - x_4^2,$$
$$r_5 = x_2x_3 + x_3x_2 - x_1^2, \qquad r_6 = x_1x_2 + x_2x_1 - x_3^2.$$

The basepoint-free condition on the associated quadratic forms is easily checked, so the Clifford algebra A is regular.

Cassidy and Vancliff defined and studied the more general notion of graded skew Clifford algebra in [**21**]. Replacing $\Bbbk[z_1, \ldots, z_n]$ by the skew polynomial ring $S = \Bbbk\langle z_1, \ldots, z_n\rangle/(z_iz_j + \mu_{ij}z_jz_i | 1 \leq i < j \leq n)$, they defined an analog of the system of quadrics associated to $M_1, \ldots, M_n$, where the quadric Q_k associated to M_k is now a hypersurface inside the point scheme of S. They proved that the graded skew Clifford algebra $A(\mu, M_1, \ldots, M_n)$ is AS regular if and only if the associated system of quadrics is base point free in a suitable sense, generalizing Le Bruyn's result.

As evidence that the skew graded Clifford algebra construction is quite general, Nafari, Vancliff, and Jun Zhang revisited the classification of regular algebras of dimension 3, and showed that the majority of these algebras are, up to twist equivalence, either skew graded Clifford algebras of dimension 3, or Ore extensions of skew graded Clifford algebras of dimension 2 [**78**]. The reader can find a survey with more details on skew graded Clifford algebras in [**134**].

5.5. 0-dimensional point schemes. Consider a regular algebra A of dimension 4 with quadratic relations and Hilbert series $\frac{1}{(1-t)^4}$. Van den Bergh observed that when the point scheme X of A is 0-dimensional, then X consists of 20 points, counted with multiplicity [**122**]. An outline of the argument can be found in [**128**, p. 377]. See also [**23**], where Chirvasitu and Kanda generalize the argument to apply to regular algebras of dimension 4 with other Hilbert series.

Van den Bergh's result raised the question of what kinds of finite schemes of length 20 actually occur for regular algebras with a 0-dimensional point scheme. For a regular graded Clifford algebra $A(M_1, \ldots, M_4)$ as described above, the number m of distinct closed points in the point scheme X of A can be efficiently calculated from the properties of the matrices M_i [**132**, Theorem 1.7]. Namely, $m = 2r_2 + r_1$, where r_i is the number of matrices of rank i in the projectivized linear system $\mathbb{P}(\Bbbk M_1 + \cdots + \Bbbk M_4)$ spanned by the defining matrices inside $M_4(\Bbbk)$. For example, in Example 5.2 this amounts to finding the points $(\alpha_1 : \alpha_2 : \alpha_3 : \alpha_4) \in \mathbb{P}^3$ where the matrix $M(\alpha_1, \alpha_2, \alpha_3, \alpha_4)$ has rank ≤ 2. in [**132**, Section 2] the authors show there are 20 distinct points in this example. They also give an example which has one single point of multiplicity 20.

To calculate the values where a symmetric matrix with indeterminate entries has rank ≤ 2 generally involves intersecting 10 cubic equations given by the vanishing of all of the 3×3 minors. Stephenson and Vancliff showed that in the setting of regular graded Clifford algebras, one can replace this calculation with the intersection of just 2 cubic divisors [**117**]. Using this idea, they fully analyzed the possible numbers m of distinct points occurring for graded Clifford algebras with finite point scheme, showing that all $1 \leq m \leq 20$ are possible except for $m = 2, 15, 17, 19$.

Much of this story generalizes nicely to skew graded Clifford algebras. Cassidy and Vancliff found a family of skew Clifford algebras $A(\mu, M_1, M_2, M_3, M_4)$, which as μ only is varied, produces examples of skew graded Clifford algebras with various different numbers of distinct points [**21**, Table 1], demonstrating the considerable additional generality of the skew case. In particular, they give examples in this way with 15 or 17 distinct point modules. Vancliff and Veerapen gave a general formula for the number of distinct points in terms of the defining data for graded skew Clifford algebras, analogous to the formula for graded Clifford algebras [**133**].

5.6. Other results on points. In addition to point modules, so-called "fat" point modules of regular algebras A can also give interesting information. A graded module M over a connected graded algebra A is called a *fat point module* of multiplicity d if M is generated in degree 0, has $\dim_{\Bbbk} M_n = d$ for all $n \geq 0$, and every proper factor module of M is finite-dimensional. Just like point modules, fat point modules correspond to simple objects in the category $\operatorname{proj} A$. The moduli theory of fat point modules was studied by Daniel Chan [**22**].

Consider Question 5.1(4). It was certainly not expected that the PI property for a regular algebra A of dimension 4 would always be determined just by its point scheme data. Indeed, Stephenson and Vancliff gave examples with 0-dimensional point scheme whose associated truncation shift automorphism has finite order, yet A is not finite over its center, in contrast to what happens in dimension 3 [**116**]. Goetz and Shelton [**43**] found fat point modules of multiplicity 2 for some of these regular algebras, and showed that truncation shift has infinite order on the set of fat points of multiplicity 2. This suggests that question (4) might still have some reasonable answer for regular algebras of dimension 4 if fat point schemes are also included. Goetz studied fat points for some other regular algebras of dimension 4 as well [**42**].

For regular algebras of dimension higher than 4, there are few general results about point or line modules. Already in dimension 5, something rather different happens with the point scheme. Recall that for regular algebras of dimension 4 with quadratic relations, the minimal dimension of the point scheme X is $\dim X = 0$, and the point scheme data contains enough structure in this case to recover the relations of the algebra [**92**, Theorem 4.1, Remark 4.2]. It has long been known, on the other hand, that the same arguments used by Van den Bergh in [**122**] show that a generic quadratic regular algebra of dimension 5 has a point scheme which is *empty*. The referee of this article noticed that an explicit example of a dimension 5 regular algebra with empty point scheme was missing from the literature. Vancliff has provided the construction of a regular Clifford algebra with such a property in a companion article for this volume [**129**].

Of course, point modules may still be interesting for special kinds of higher-dimensional regular algebras. Belmans, De Laet, and Le Bruyn showed [**8**] that the point scheme for a quantum polynomial algebra $A = \Bbbk\langle x_1, \dots, x_n\rangle/(x_j x_i - q_{ij} x_i x_j)$ has an interesting combinatorial structure, for arbitrary dimension n. They prove it is a union of linear subspaces of $\mathbb{P}^{n-1}$ and determine its irreducible components. Not all unions of linear subspaces can occur as such a point scheme; they give an additional necessary condition which is sufficient for small n.

6. Extensions

As we mentioned already in the introduction, an easy way to produce examples of regular algebras is by taking iterated graded Ore extensions. This construction does not capture many interesting examples of regular algebras, however. In this section we describe some more general constructions which have wider scope.

6.1. Normal extensions. We say that a ring B is a *normal extension* of A if there is a normal non-zero-divisor $x \in B$ such that $B/xB \cong A$. If in fact x is central in B, we call B a *central extension* of A.

Suppose that B is any connected $\mathbb{N}$-graded algebra with a normal non-zero-divisor $x \in B_n$ such that $A \cong B/xB$. For example, we could have $B = A[x;\sigma]$ where σ is a graded automorphism of a connected graded algebra A and $\deg(x) = n$. The normal extension B of A is again AS regular, with Hilbert series $h_B(t) = h_A(t)/(1-t^n)$ and $\operatorname{gl.dim}(B) = \operatorname{gl.dim}(A) + 1$. Conversely, if B is an AS regular algebra with normal element $x \in B_1$, then $A = B/xB$ will again be AS regular. In general, however, if $x \in B_n$ is a normal element in an AS regular algebra, then $A = B/xB$ will be AS Gorenstein but usually not regular–this is already clear by taking x^n for $n \geq 2$ in $\Bbbk[x]$. For proofs see, for example, [**90**, Section 1]. There are similar statements for the Auslander regular condition [**64**].

Le Bruyn, Smith, and Van den Bergh studied central extensions in [**61**]. Starting with a regular algebra A of dimension 3 with Hilbert series $\frac{1}{(1-t)^3}$, their goal was to classify the possible central extensions B with a central element $x \in B_1$ such that $B/xB \cong A$. They focused on central extensions rather than the more general normal extensions, because if $x \in B_1$ is normal, there is a Zhang twist B^τ of B such that x becomes central in B^τ. So up to twist equivalence, considering central extensions is sufficient. For each family of algebras A in the Artin-Schelter classification, the potential relations of central extensions B of A can be written down, and after some reductions, the possible isomorphism types of central extensions were obtained. There are interesting nontrivial families of central extensions of the 3-dimensional Sklyanin algebras, for example.

Cassidy took up this theme in [**17**], considering normal extensions of regular algebras of dimension 3 (including the cubic ones) by normal elements of any degree. He gave an example of a normal extension by a degree 2 element which is not twist equivalent to a central extension. In [**18**], Cassidy pushed the ideas even further to consider normal extensions of those regular algebras A which are not necessarily generated in degree 1, as had been classified by Stephenson (see Section 9.1 below).

As mentioned above, factoring out a normal element of degree greater than 1 in a regular algebra does not necessarily yield another regular algebra; it may be only AS Gorenstein. However, this also means that certain special classes of AS Gorenstein but non-regular algebras A could have normal extensions which are regular. In [**7**], Bauwen and Van den Bergh classify regular algebras A of dimension 4 with $h_A(t) = \frac{1}{(1-t)^4}$ with a normal element $x \in A_2$ such that $B = A/xA$ is isomorphic to the 2-Veronese ring $\bigoplus_{n\geq 0} C_{2n}$ of a cubic regular algebra C of dimension 3; here B is AS Gorenstein only. Geometrically, these extensions can be seen as noncommutative analogs of the Segre embedding of $\mathbb{P}^1 \times \mathbb{P}^1$ into $\mathbb{P}^3$.

6.2. Double Ore extensions. James Zhang and Jun Zhang defined the notion of a double Ore extension in [**141**]. The general idea is to form a new algebra from a given one by adding two variables both at the same time, with various axioms to ensure that the resulting algebra behaves like a noncommutative version of a polynomial extension in two variables. Because the relation between the two new variables also needs to be specified, it is a quite complicated object in general. Here we only define a special case which is the most relevant to the theory of AS regular algebras, following the presentation in [**142**].

Let $A \subseteq B$ be an extension of connected graded $\Bbbk$-algebras, both generated in degree 1. Suppose further that (i) $A_{\geq 1}B = BA_{\geq 1}$; (ii) there is a set $S \subseteq B$ such that B is a free left and right A-module with basis S; and (iii) the image of S in the quotient ring $C = B/BA_{\geq 1}$ is a $\Bbbk$-basis of C. In this setting, we say that B is an *n-extension* of A if C is AS regular of dimension n, and we say that B is an *n-coextension* of A if A is AS regular of dimension n. These notions unify some common constructions already described. In particular, an extension of connected graded $\Bbbk$-algebras $A \subseteq B$ is a 1-extension if and only if $B \cong A[x; \sigma, \delta]$ is a graded Ore extension of A, and $A \subseteq B$ is a 1-coextension if and only if B is a normal extension of A [**142**, Lemma 1.2].

In this language, we are interested in 2-extensions $A \subseteq B$. This is equivalent by [**142**, Lemma 1.4] to B being a graded double Ore extension of A as defined in [**141**]. So we assume that $A_{\geq 1}B = BA_{\geq 1}$ and that $C = B/A_{\geq 1}B$ is regular of dimension 2. Without loss of generality, the set S can be taken to have the form $\{y_1^i y_2^j | i, j \geq 0\}$ for some elements $y_1, y_2 \in B$, satisfying a relation of the form $y_2 y_1 = p_{12} y_1 y_2 + p_{11} y_1^2 + \tau_1 y_1 + \tau_2 y_2 + y_0$ for some $p_{11}, p_{12} \in \Bbbk$, $\tau_1, \tau_2 \in A_1$, $y_0 \in A_2$, with $p_{12} \neq 0$. Thus the image $\{\overline{y_1}^i \overline{y_2}^j | i, j \geq 0\}$ of this basis in C satisfies the relation $\overline{y_2 y_1} = p_{12} \overline{y_1 y_2} + p_{11} \overline{y_1}^2$, which one can check does give C the structure of a regular algebra of dimension 2.

One of the advantages of an Ore extension is that given an algebra A, we can describe all possible Ore extensions $B = A[x; \sigma, \delta]$ by finding the automorphisms σ and σ-derivations δ of A. Describing the 2-extensions B of A in terms of certain functions of A is part of what is done in [**141**]. Given the algebra A, it requires choosing an algebra homomorphism $\sigma : A \to M_2(A)$ and a σ-derivation $\delta : A \to A \oplus A$. Here, we write the elements in the image of δ as columns, so the usual definition of σ-derivation requiring $\delta(ab) = \sigma(a)\delta(b) + \delta(a)b$ for all $a, b \in A$ makes sense. The multiplication in B is then determined by the matrix equation

$$\begin{pmatrix} y_1 \\ y_2 \end{pmatrix} a = \sigma(a) \begin{pmatrix} y_1 \\ y_2 \end{pmatrix} + \delta(a)$$

for $a \in A$, together with the relation $y_2 y_1 = p_{12} y_1 y_2 + p_{11} y_1^2 + \tau_1 y_1 + \tau_2 y_2 + \tau_0$. The resulting algebra is written as $A_P[y_1, y_2; \sigma, \delta, \tau]$. Unfortunately, in this external description of a double Ore extension there are six additional rather complicated consistency equations involving the data that need to be satisfied, in order to ensure that $A_P[y_1, y_2; \sigma, \delta, \tau]$ is actually a 2-extension of A [**142**, Proposition 1.6].

Double Ore extensions extend the reach of the Ore extension concept enormously, as they turn out to be considerably more general than iterated Ore extensions where one adds 2 variables, one at a time. Also, Zhang and Zhang prove that if B is a 2-extension of A, and A is also regular of dimension n (that is, B is also an n-coextension of A), then B is regular of dimension $n+2$ [**141**, Theorem 0.2]. So this really is an effective way to create new examples of regular algebras. The disadvantage of double Ore extensions is the intricacy of the data defining them, as well as the compatibility conditions required. Also, it is not known if other basic ring theoretic properties, such as the noetherian condition, are preserved by the construction, unlike the case of usual Ore extensions.

The most extensive use of the concept so far has been in [**142**]. In this paper, Zhang and Zhang consider double Ore extensions $B = A_P[y_1, y_2; \sigma, \delta, \tau]$ where A is regular of dimension 2. Those that are isomorphic to a Ore extension $C[y; \sigma, \delta]$ of a regular algebra C of dimension 3 are excluded; since regular algebras of dimension 3 are classified, in theory those Ore extensions can already be understood. They then completely classify the associated *trimmed* double Ore extensions $A_P[y_1, y_2; \sigma]$ of a regular algebra A of dimension 2, obtained by setting $\delta = \tau = 0$, obtaining 26 parametrized families. The trimmed double Ore extension $A_P[y_1, y_2; \sigma]$ is an associated graded ring of $A_P[y_1, y_2; \sigma, \delta, \tau]$. In this particular case, the authors show that all of the trimmed extensions have good ring-theoretic properties (strongly noetherian, Auslander regular, and GK-Cohen Macaulay), and so the same is true by standard arguments for the arbitrary double Ore extensions.

To give the reader a feel for the concept, here is one family of double Ore extensions from [**142**] (this is family $\mathbb{C}$ among families labeled $\mathbb{A}$ through $\mathbb{Z}$). Examples like this seem unlikely to have been discovered without the double Ore extension concept, or something similarly involved.

EXAMPLE 6.1. The algebra

$$\Bbbk\langle x_1, x_2, y_1, y_2\rangle/(r_1, r_2, r_3, r_4, r_5, r_6)$$

for p a primitive third root of unity, where the six relations are

$$\begin{aligned} r_1 &= y_2y_1 - py_1y_2, \qquad r_2 = x_2x_1 - px_1x_2, \\ r_3 &= y_1x_1 + x_1y_1 - p^2x_2y_1 - x_1y_2 + px_2y_2, \\ r_4 &= y_1x_2 + px_1y_1 - x_2y_1 - x_1y_2 + px_2y_2, \\ r_5 &= y_2x_1 + px_1y_1 + 2p^2x_2y_1 - px_1y_2 + px_2y_2, \\ r_6 &= y_2x_2 + px_1y_1 - p^2x_2y_1 - x_1y_2 + x_2y_2, \end{aligned}$$

is a trimmed double Ore extension $A_P[y_1, y_2; \sigma]$ of $A = \Bbbk\langle x_1, x_2\rangle/(x_2x_1 - px_1x_2)$. One may observe how the last four relations each mix all of the generators together, unlike what one would expect to see in an iterated Ore extension.

6.3. Additional grading. While an AS regular algebra is by definition $\mathbb{N}$-graded, some examples have an additional grading that gives them extra structure. For example, recall the S_2 family of regular algebras of dimension 3 from Example 2.2, where

$$A = A(\alpha) = \Bbbk\langle x_1, x_2, x_3\rangle/(\alpha x_3x_1 + x_1x_3,\ \alpha x_3x_2 - x_2x_3,\ x_1^2 - x_2^2).$$

We see that A is actually $\mathbb{N}\times\mathbb{N}$-graded, if we assign $\deg(x_1) = \deg(x_2) = (1, 0)$ and $\deg(x_3) = (0, 1)$. The relations are then homogeneous of degrees $(1, 1), (1, 1)$, and

$(2,0)$. Conversely, of course, any connected $\mathbb{N}\times\mathbb{N}$-graded algebra is naturally $\mathbb{N}$-graded with $A_n = \bigoplus_{i+j=n} A_{(i,j)}$. Given a connected $\mathbb{N}$-graded algebra, generated in degree 1, we will call it *bigraded* if it has a grading by $\mathbb{N}\times\mathbb{N}$ which refines the given one in the sense that $A_n = \bigoplus_{i+j=n} A_{(i,j)}$, and which is nontrivial in the sense that both $A_{(1,0)} \neq 0$ and $A_{(0,1)} \neq 0$.

While the classification of general regular algebras of dimension 4 remains out of reach, the special class of bigraded regular algebras of dimension 4 has proven more tractable. It is easy to see that Example 6.1 is bigraded by taking $\deg(x_1) = \deg(x_2) = (1,0)$ and $\deg(y_1) = \deg(y_2) = (0,1)$. In fact, this is a standard feature of a trimmed double extension $A_P[y_1, y_2; \sigma]$. Conversely, in [**142**], Zhang and Zhang proved that if B is a regular algebra of dimension 4 with Hilbert series $\frac{1}{(1-t)^4}$ which is bigraded, then B is either a trimmed double Ore extension $A_P[y_1, y_2; \sigma]$ of a regular algebra A of dimension 2, or else an Ore extension $A[y;\sigma]$ for a regular algebra A of dimension 3. This shows that bigraded regular algebras with this Hilbert series are essentially understood.

In the next sections we will see that bigraded regular algebras of dimension 4 with other Hilbert series have also been classified.

7. The Ext-algebra of a regular algebra

7.1. The Ext-algebra. Let A be any connected $\mathbb{N}$-graded finitely presented $\Bbbk$-algebra which is generated in degree 1. Considering the trivial left module ${}_A\Bbbk = A/A_{\geq 1}$, we can construct the associated *Ext-algebra* $E = E(A) = \bigoplus_{i=0}^{\infty} \operatorname{Ext}^i_A(\Bbbk, \Bbbk)$. This has a well-known multiplication given by the Yoneda product, and as such it is again a connected $\mathbb{N}$-graded $\Bbbk$-algebra. The study of Ext-algebras and their properties is an important general topic in homological algebra. In this section we review results which link the AS regularity of a connected graded algebra to properties of its Ext-algebra.

7.2. Koszul algebras. There is a special case where the connection between a connected graded algebra A and its Ext-algebra $E = E(A)$ is especially tight. Let

$$\cdots \to P_n \to \cdots \to P_2 \to P_1 \to P_0 \to \Bbbk \to 0$$

be the minimal graded free resolution of the trivial module $\Bbbk$. We say that A is *Koszul* if P_i is generated in degree i, and so is a direct sum of copies of $A(-i)$, for each $i \geq 0$. We know that $P_0 = A$, $P_1 = A(-1)^n$ where $n = \dim_\Bbbk A_1$ is the minimal number of generators of A, and $P_2 = \bigoplus_{i=1}^s A(-d_i)$, where the minimal homogeneous relations are $r_1, \dots, r_s$ with $\deg r_i = d_i$. In particular, if A is Koszul then it is quadratic, that is, its minimal relations all have degree 2.

For any connected $\mathbb{N}$-graded $\Bbbk$-algebra A generated in degree 1 with quadratic relations, say $A = \Bbbk\langle x_1, \dots, x_n\rangle/(r_1, \dots, r_m)$, with $\deg r_i = 2$ for all i, the *Koszul dual* of A is an algebra $A^!$ defined as follows. Let $V = A_1$ with its $\Bbbk$-basis $x_1, \dots, x_n$, and let $x_1^*, \dots, x_n^*$ be a dual basis of V^*. Thinking of $r_1, \dots, r_m$ as elements of $V \otimes_\Bbbk V$, say with $\Bbbk$-span R, we let $R^\perp \in V^* \otimes V^*$ be the orthogonal complement of R under the canonical pairing between $V^* \otimes V^*$ and $V \otimes V$. Then we define $A^! = \Bbbk\langle x_1^*, \dots, x_n^*\rangle/(R^\perp)$. For example, it is easy to calculate that if $A = \Bbbk[x_1, \dots, x_n]$ is a commutative polynomial ring, then $A^!$ is isomorphic to an exterior algebra in n generators. If A is Koszul then so is $A^!$. A basic result is that a connected graded $\Bbbk$-algebra A, generated in degree 1 with quadratic relations, is Koszul if and only if $A^! \cong E(A) = \bigoplus_{i=0}^{\infty} \operatorname{Ext}^i_A(k,k)$ as graded $\Bbbk$-algebras. Note that by definition $(A^!)^! \cong A$ always, so one consequence of this is that if A is Koszul, then $E(E(A)) \cong A$.

Given an AS regular algebra, whether or not it is Koszul can generally be easily recognized by its Hilbert series. If A is AS regular of global dimension d with Hilbert series $h_A(t) = \frac{1}{(1-t)^d}$ (the same as a polynomial ring in d variables of degree 1), then in fact A is Koszul, as noted by Zhang [**100**, Theorem 5.11]. Thus among regular algebras of dimension 3, for example, the quadratic ones are Koszul, and the cubic ones are not, because Koszul algebras must have quadratic relations. It seems very likely that, conversely, any Koszul AS regular algebra of dimension d must have Hilbert series $\frac{1}{(1-t)^d}$, but this seems to be unknown in general. This is true when $d \leq 5$ under mild additional hypotheses, by the classification of possible Hilbert series we will review in Section 5.1 below.

Suppose now that A is an AS regular algebra, not necessarily Koszul. Since $\operatorname{gl.dim}(A) = d < \infty$, we have $E = E(A) = \bigoplus_{i=0}^{d} \operatorname{Ext}^i_A(\Bbbk,\Bbbk)$. Even if we do not necessarily assume that A is noetherian, for a regular algebra each P_i in the minimal graded free resolution of $\Bbbk$ is finitely generated [**118**, Propostion 3.1], and this implies that $E_i = \operatorname{Ext}^i_A(\Bbbk,\Bbbk)$ is finite-dimensional over $\Bbbk$ for all i. So E is a finite-dimensional connected $\mathbb{N}$-graded algebra. A finite dimensional algebra E is called *Frobenius* if $E^* = \operatorname{Hom}_{\Bbbk}(E,\Bbbk)$ is isomorphic to E as a left E-module.

We can now see how the regularity of an algebra is reflected in its Ext-algebra.

THEOREM 7.1 ([**69**, Corollary D]). *Let A be a connected $\mathbb{N}$-graded algebra, generated in degree 1, with $\operatorname{gl.dim}(A) < \infty$. Then A satisfies the AS Gorenstein condition if and only if $E(A)$ is Frobenius. In particular, if $\operatorname{GKdim}(A) < \infty$ then A is regular if and only if $E(A)$ is Frobenius.*

We have stated the final, most general version of this theorem, but important special cases of the theorem were proved earlier. In particular, Smith considered the case where A is Koszul and proved Theorem 7.1 in this case [**100**]. Inspired partly by a desire to put cubic regular algebras of dimension 3 on a similar footing as the quadratic ones, Berger defined a notion of *m-Koszul* algebra [**9**]. This is an algebra A presented by homogeneous relations, all of some degree m, for which the term P_i in the minimal graded free resolution of $\Bbbk$ is generated in degree d_i for all i, where for all even i we have $d_i = mi/2$ and $d_{i+1} = (mi/2) + 1$. Thus Koszul algebras are 2-Koszul, and cubic regular algebras of dimension 3 are 3-Koszul. Berger and Marconnet showed that Theorem 7.1 holds for m-Koszul algebras [**10**, Theorem 1.2]. In addition to cubic regular algebras this applies to some examples with infinite GK-dimension (see also Section 9.2 below).

We are unaware of any explicit examples of AS regular algebras which are m-Koszul for $m > 2$, other than the cubic regular algebras of dimension 3; in fact, non-quadratic regular algebras of higher dimension seldom have relations all in one degree, as we will see in Section 8. Another generalization of Koszul, which subsumes m-Koszulity, is the notion of $\mathcal{K}_2$-algebra defined by Cassidy and Shelton, which does hold for certain regular algebras of dimension 4 with relations in multiple degrees [**20**, Theorem 4.7].

In order to obtain the full generality of Theorem 7.1, Lu, Palmieri, Wu, and Zhang introduced new methods, which we describe next.

7.3. A_∞-algebras. A_∞-algebras were defined by Stasheff. The original motivation for their introduction came from topology, but as often happens, they have found many applications in an algebraic setting far removed from topology.

DEFINITION 7.2. An A_∞-algebra is a $\mathbb{Z}$-graded vector space $\bigoplus_{i\in\mathbb{Z}} A^i$ together with a family of $\mathbb{k}$-linear "multiplication maps" $m_n : A^{\otimes n} \to A$ for $n \geq 1$, where m_n is graded of degree $2-n$, such that Stasheff's identities are satisfied:

$$\sum(-1)^{r+st} m_{r+1+t}(\mathrm{id}^r \otimes m_s \otimes \mathrm{id}^t) = 0$$

for all $n \geq 1$, where we sum over all $n = r+s+t$ with $r, t \geq 0$ and $s \geq 1$.

We give the definition here explicitly in order to demonstrate that the relations that must hold among the m_n are quite subtle. The map $m_2 : A \otimes A \to A$ is a degree 0 multiplication operation, while $m_1 : A \to A$ is of degree -1 and is a differential of $\bigoplus_{i\in\mathbb{Z}} A^i$ considered as a complex. In fact, when $m_n = 0$ for $n \geq 3$, an A_∞-algebra is precisely a differential graded algebra. The maps m_n for $n \geq 3$ are thought of as "higher multiplications", and in general the multiplication m_2 is associative only up to homotopy (although it is actually associative in case $m_1 = 0$ or $m_3 = 0$). For a general introduction to A_∞-algebras see [**56**], and for an introduction geared towards their application to noncommutative algebra see [**67**].

Suppose now that A is a connected $\mathbb{N}$-graded $\mathbb{k}$-algebra, finitely generated in degree 1. Then the Ext-algebra $E = E(A) = \bigoplus_{i\geq 0} \mathrm{Ext}^i_A(\mathbb{k}, \mathbb{k})$ can be given the structure of an A_∞-algebra for which $m_1 = 0$ (such A_∞-algebras with 0 differential m_1 are called *minimal*). Moreover, the relations of the algebra A can be recovered from knowledge of E together with its entire A_∞-algebra structure, in a straightforward way [**70**, Corollary B]. If A is Koszul, then in fact $m_n = 0$ for $n \neq 2$, so E is just an associative algebra and this is just the statement that A is recoverable from the algebra E (via $E(E(A)) \cong A$). If A is not Koszul, on the other hand, one needs to remember the higher multiplications in order to recover A from E.

Using these ideas, Lu, Palmieri, Wu, and Zhang proved Theorem 7.1 in its full stated generality [**69**, Corollary D]. Thus, for any connected graded algebra of finite global dimension, the Gorenstein condition is equivalent to the condition that the Ext-algebra $E(A)$ is Frobenius. Because one cannot recover A from $E = E(A)$ without invoking the A_∞-algebra structure on E in general, it is not surprising that one needs A_∞-techniques to characterize when A is Gorenstein in terms of the properties of E, even though in the end the Frobenius property of E does not depend on the higher multiplications.

EXAMPLE 7.3 (Example 13.4, [**67**]). Consider the algebras

$$A = A(\alpha, \beta) = \mathbb{k}\langle x, y\rangle/(xy^2 + \alpha y^2 x + \beta yxy,\ x^2 y + \alpha yx^2 + \beta xyx).$$

with $\alpha, \beta \in \mathbb{k}$. These are the graded algebras among the well-studied class of *down-up algebras* which were first defined by Benkart and Roby.

The Ext-algebra $E = E(A)$ has Hilbert series $h_E(t) = 1+2t+2t^2+t^3$; choosing a basis as follows:

$$E = E^0 \oplus E^1 \oplus E^2 \oplus E^3 = \mathbb{k}1 \oplus (\mathbb{k}a \oplus \mathbb{k}b) \oplus (\mathbb{k}c \oplus \mathbb{k}d) \oplus \mathbb{k}e,$$

where 1 is the identity, then the multiplication is given by the rules $ac = e$, $bd = \alpha e$, $ca = \alpha e$, $db = e$, with all other products of pairs of basis elements equal to 0. It is easy to see directly that E is Frobenius exactly when $\alpha \neq 0$, so A is regular of dimension 3 under the same condition.

The higher multiplications on E, when restricted to elements in E^1, are determined from the relations in a straightforward way [**70**, Corollary B]. In this case $m_n = 0$ for $n \geq 4$, and $m_3(a,b,b) = c$, $m_3(b,b,a) = \alpha c$, $m_3(b,a,b) = \beta c$, $m_3(a,a,b) = d$, $m_3(b,a,a) = \alpha d$, $m_3(a,b,a) = \beta d$, and $m_3(a,a,a) = m_3(b,b,b) = 0$.

Note that the multiplication m_2 of the Ext-algebra E only involves α, while β does not appear. As β varies, one gets a whole family of regular algebras with isomorphic Ext-algebras. On the other hand, β appears in the formulas for the higher multiplication m_3, which is why A can be recovered from E with its entire A_∞-structure.

7.4. Using the Ext-algebra. One of the main motivations in relating properties of a graded ring and its Ext-algebra, as in Theorem 7.1, is the hope that one might be able to understand or classify regular algebras by instead considering finite-dimensional Frobenius algebras. There has been some success in this regard.

In [**68**], the authors apply this idea to regular algebras A of dimension 4, in particular those with Hilbert series $h_A(t) = \frac{1}{(1-t)^2(1-t^2)(1-t^3)}$. These algebras have 2 degree 1 generators and 2 minimal relations, one each of degrees 3 and 4. We discuss the possible Hilbert series of regular algebras of dimension 4 in the next section; the case at hand is the one in which the Ext-algebra $E = E(A)$ is smallest, specifically $h_E(t) = 1 + 2t + 2t^2 + 2t^3 + t^4$. The authors show that the only nonzero higher multiplications of E are m_2, m_3, m_4. Starting with a Frobenius algebra E of this shape with unknown parameters and arbitrary m_3, m_4 satisfying the Shasheff relations, they classify those that are "generic" in the sense that the maps m_3, m_4 satisfy certain Zariski open conditions. Then, relations for the corresponding algebras A such that $E = E(A)$ can be determined. Finally, which of these A are actually noetherian and regular is decided. The analysis is quite sensitive, but leads in the end to just 4 parametrized families of regular algebras [**68**, Theorem A]. This classifies all noetherian regular algebras of dimension 4 with this Hilbert series whose Ext-algebra has generic higher multiplications.

As a sample of what these algebras can look like, here is one of the families:

$$A(p) = \Bbbk\langle x, y\rangle/(xy^2 - p^2y^2x,\ x^3y + px^2yx + p^2xyx^2 + p^3yx^3), \qquad \text{where } 0 \neq p \in \Bbbk.$$

For this family, $A(p)$ is indeed regular for all nonzero p (in fact, all algebras in the family are Zhang twists of $A(1)$).

Note that all of the members of this family $A(p)$ are bigraded with $\deg(x) = (1,0)$ and $\deg(y) = (0,1)$. This is not an accident, as it turns out that all of the algebras classified in [**68**, Theorem A] are bigraded. The authors prove that the same four families also classify regular algebras of dimension 4 with Hilbert series $\frac{1}{(1-t)^2(1-t^2)(1-t^3)}$ that are bigraded [**68**, Theorem B].

More recent work has pushed this A_∞-algebra method a bit farther. Many authors have studied PBW deformations of regular algebras—these are algebras which have a nice filtration such that the associated graded algebra is AS regular. Shen, Zhou, and Lu exploit this idea to construct new regular algebras closely related to given ones, produced by "adding terms" to the relations [**94**] [**95**]. This allows them to extend the technique of A_∞-algebras from [**68**] to classify certain regular algebras with the same Hilbert series where the A_∞-structure on the Ext-algebra is non-generic.

In another direct generalization of the work of **[68]**, Wang and Wu looked at regular algebras of dimension 5 generated by two degree 1 generators, with relations of degrees $(4,4,4)$ **[136]**, which have Hilbert series $\frac{1}{(1-t)^2(1-t^2)(1-t^3)^2}$. They performed an analysis similar to **[68]**, and classified all such algebras which are domains, under a genericity condition on the structure of the associated A_∞-algebra. Similarly as in **[68]**, the algebras that arise under this generic condition are also bigraded.

8. Hilbert series

Let A be a regular algebra, generated in degree 1, with global dimension d. Much of the work on regular algebras we described above has concentrated on those which are Koszul with $h_A(t) = \frac{1}{(1-t)^d}$, and many of the most prominent examples, such as the Sklyanin algebras, are in this category. But even for $d = 3$, in addition to the Koszul quadratic regular algebras there are also the cubic regular algebras with Hilbert series $h_A(t) = \frac{1}{(1-t)^2(1-t^2)}$, and the work of Artin, Tate, and Van den Bergh managed to treat both cases in roughly the same way.

For higher d, an obvious general question is what Hilbert series are possible for regular algebras. We will see momentarily that there are not many possibilities for $d = 4$, but starting with $d = 5$ there are a considerable number of cases. It seems unlikely that any eventual understanding or classification will be able to treat these all uniformly.

Already in **[1]**, Artin and Schelter gave examples of regular algebras of dimension 4 with Hilbert series $\frac{1}{(1-t)^4}$, $\frac{1}{(1-t)^3(1-t^2)}$, and $\frac{1}{(1-t)^2(1-t^2)(1-t^3)}$. Lu, Palmieri, Wu, and Zhang proved in **[68]** that if A is regular, generated in degree 1, with $\operatorname{gl.dim}(A) = 4$, and either $\operatorname{GKdim}(A) \geq 3$ or A is a domain, then A has one of these three Hilbert series (and in fact $\operatorname{GKdim}(A) = 4$). Of course, conjecturally a regular algebra A is a domain and satisfies $\operatorname{GKdim}(A) = \operatorname{gl.dim}(A)$ (as in Question 3.3). So the hypothesis that $\operatorname{GKdim}(A) \geq 3$ or A is a domain is almost surely unnecessary, and one expects that these three Hilbert series are the only possible ones in dimension 4.

8.1. Enveloping algebras and graded Ore extensions.

The easiest way to construct examples of higher-dimensional regular algebras with various Hilbert series is to use graded Ore extensions. If A is AS regular, then an Ore extension $B = A[x; \sigma, \delta]$ is again connected graded and regular by taking $\deg(x) = n$, as long as σ preserves degree and δ is a σ-derivation of degree n. We are still primarily interested in regular algebras which are generated in degree 1; the key observation is that B may be generated in degree 1 even if A is not.

For instance, let L be a finite-dimensional Lie algebra over $\Bbbk$. Assume that L is positively graded, so $L = \bigoplus_{i=1}^m L_i$ as vector spaces, with $[L_i, L_j] \subseteq L_{i+j}$ for all i, j. Let $A = U(L)$ be the universal enveloping algebra of L. Then because L is graded, A is naturally connected $\mathbb{N}$-graded. Under these conditions $U(L)$ is Artin-Schelter regular, with Hilbert series $h_{U(L)}(t) = \frac{1}{\prod_{i=1}^m (1-t^i)^{\dim_\Bbbk L_i}}$. If L is generated as a Lie algebra by L_1, then $U(L)$ will be generated as an algebra in degree 1. The algebra $U(L)$ can always be expressed as an iterated Ore extension of the form $U(L) = \Bbbk[x_1][x_2; \delta_2] \dots [x_m; \delta_m]$, by choosing x_i to be a homogeneous basis of L in decreasing order of degree.

EXAMPLE 8.1. Let L have $\Bbbk$-basis x, y, w, z with $\deg(x) = \deg(y) = 1$, $\deg(w) = 2$, and $\deg(z) = 3$, where $[x, y] = w = -[y, x]$, $[x, w] = z = -[w, x]$, and all other brackets between pairs of variables 0. This is easily seen to be a graded Lie algebra generated in degree 1. Then $U(L)$ is regular with Hilbert series $\frac{1}{(1-t)^2(1-t^2)(1-t^3)}$, generated by only x and y because of the relations $w = xy - yx$ and $z = xw - wx$. We can also write $U(L) = \Bbbk[z][w][y][x; \delta]$ where $\delta(y) = w$, $\delta(w) = z$, and $\delta(z) = 0$.

As an aside, we mention that there is also a different way to use Lie algebras to construct interesting examples of regular algebras. When L is not a graded Lie algebra, $U(L)$ will not be graded either. However, in this case one can form a *homogenized enveloping algebra* by adding an additional variable z and defining

$$H(L) = \Bbbk\langle x_1, \dots, x_n, z\rangle/(x_j x_i - x_i x_j - [x_i, x_j]z \,|\, 1 \le i < j \le n)$$

for a basis $x_1, \dots, x_n$ of L. These algebras need not be iterated Ore extensions. Le Bruyn and Van den Bergh proved that homogenized enveloping algebras $H(L)$ are AS regular, and they gave some information about their linear modules [**62**]. Le Bruyn and Smith further investigated $H(L)$ for the special case of $L = \mathfrak{sl}_2$ in [**60**].

8.2. Hilbert series in dimension 5. For each of the three possible Hilbert series for regular algebras of dimension 4, there is a graded Lie algebra L of dimension 4 such that $U(L)$ is regular (generated in degree 1) of that Hilbert series. In fact the examples initially given by Artin and Schelter are of that form [**1**, Example 1.21].

This raises the question of whether universal enveloping algebras already achieve all of the possible Hilbert series of regular algebras in higher dimension. Actually, already in global dimension 5 there is an additional consideration; one may ask for the finer information of the possible *Betti numbers* of the minimal graded free resolution of $\Bbbk$. That is, one wants to know the non-negative integers β_{ij} such that in the minimal graded resolution $0 \to P_5 \to \cdots \to P_2 \to P_1 \to P_0 \to \Bbbk$, one has $P_i = \bigoplus_{j=1}^{d_j} A(-\beta_{ij})$. The Hilbert series can be calculated from the Betti numbers, but starting in global dimension 5 it is possible for regular algebras to have different Betti numbers but the same Hilbert series. Assuming we have m degree 1 generators and minimal relations $r_1, \dots, r_t$ of degrees $\deg(r_i) = d_i$, then $P_1 = A(-1)^m$ and $P_2 = \bigoplus_{i=1}^{r} A(-d_i)$. By the symmetry of the free resolution coming from the Gorenstein condition, there is an ℓ such that $P_5 = A(-\ell)$ (the number ℓ is sometimes called the *AS-index* or the *Gorenstein parameter*). Once ℓ is fixed the free resolution must look like

$$0 \to A(-\ell) \to A(-\ell+1)^m \to \bigoplus_{i=1}^{r} A(-\ell + d_i)$$

$$\to \bigoplus_{i=1}^{r} A(-d_i) \to A(-1)^m \to A \to \Bbbk \to 0,$$

so the Betti numbers are determined. Then the Hilbert series is equal to $h_A(t) = \frac{1}{p(t)}$ where $p(t) = 1 - mt + \sum_{i=1}^{r} t^{d_i} - \sum_{i=1}^{r} t^{\ell - d_i} + mt^{\ell-1} - t^{\ell}$.

Floystad and Vatne considered the question of the possible Hilbert series of dimension 5 regular algebras A (generated in degree 1). It is relatively straightforward to calculate all of the Hilbert series of universal enveloping algebras $U(L)$ where L is a graded Lie algebra of dimension 5 generated in degree 1, which is done in [**39**, Proposition 3.1]. The authors then proved that if A is regular of global dimension 5, generated by $m = 2$ degree 1 elements, has $\mathrm{GKdim}(A) \geq 4$, and A is a domain, then the only possibilities for the degrees $(d_1, \ldots, d_t)$ of the minimal relations are $(4,4,4)$, $(4,4,4,5)$, $(4,4,4,5,5)$, $(3,5,5)$, or $(3,4,7)$. In each case there is only one possible ℓ (respectively $\ell = 10, 10, 10, 11$ or 12) [**39**, Theorem 5.6]. The respective Hilbert series are $\frac{1}{(1-t)^2(1-t^2)(1-t^3)(1-t^3)}$ (for the first three possibilities), $\frac{1}{(1-t)^2(1-t^2)(1-t^3)(1-t^4)}$, and $\frac{1}{(1-t)^2(1-t^2)(1-t^3)(1-t^5)}$. Note that certainly all of the Hilbert series which occur here are the same as a commutative graded polynomial ring with variables of some degrees, consistent with a positive answer to Question 3.3(5).

The authors gave examples of regular algebras with each of the types with three relations (leaving open the question of whether any regular algebras with two generators and minimal relations of degrees $(4,4,4,5)$ or $(4,4,4,5,5)$ exist). They refer to the case with relations of degree $(3,4,7)$ as "extremal" since it has the largest AS-index of 12. From their earlier calculation of the possible Hilbert series of universal enveloping algebras $U(L)$, they conclude that there is no enveloping algebra of this extremal type. Thus enveloping algebras of graded Lie algebras are not in fact general enough to achieve all possible Hilbert series of regular algebras of dimension 5.

Since enveloping algebras of positively graded Lie algebras are just a special case of iterated Ore extensions, one may also ask whether graded iterated Ore extensions exhibit all possible types of regular algebras of dimension 5. Elle studied iterated graded Ore extensions of dimension 5 which are generated in degree 1 [**38**]. She showed that for all types in dimension 5 for which there is some known example there is an example of that type given by an iterated Ore extension. In particular, this is true for the extremal type which Floystad and Vatne showed cannot be obtained via a universal enveloping algebra of a graded Lie algebra. She also exhibited a similar type with three generators and relations of degrees $(2,2,3)$ which can be obtained via an iterated Ore extension but not via an enveloping algebra. There is still no known example to our knowledge of a Hilbert series achieved by some regular algebra but not by a graded iterated Ore extension.

This type of analysis becomes harder as the dimension d of the regular algebra increases. Relatively little is known about the full range of possible Hilbert series and Betti numbers for regular algebras of dimension d with $d \geq 6$. It would be good to have a more theoretical understanding of the possibilities, apart from brute force computation.

8.3. Gröbner basis methods. The original work of Artin and Schelter made heavy use of the methods of noncommutative Gröbner bases to construct examples of regular algebras of dimension 3 [**1**, Sections 5-10]. For certain special kinds of regular algebras of dimension 4 and 5, Gröbner basis methods are in fact sufficient to achieve a classification.

Recall that bigraded regular algebras of dimension 4 with Hilbert series $\frac{1}{(1-t)^4}$ were classified in [**142**] by Zhang and Zhang using double Ore extensions, while bigraded regular algebras with Hilbert series $\frac{1}{(1-t)^2(1-t^2)(1-t^3)}$ were classified by Lu, Palmieri, Wu, and Zhang by using A_∞-algebras [**68**]. That leaves algebras with Hilbert series $\frac{1}{(1-t)^3(1-t^2)}$, which have 3 degree 1 generators, and 4 relations, 2 each of degrees 2 and 3. The author and Zhang considered bigraded regular algebras of this type in [**90**]. It turns out that after some reductions, including eliminating Ore extensions of regular algebras of dimension 3, all such algebras can be presented with relations with the same leading terms, and one can show that for the Hilbert series to be correct, the Gröbner basis must have a specific form (namely, the overlaps must lead to just one additional relation with a particular leading term, after which all overlaps must resolve). One can then use a computer calculation to find families of algebras for which the Gröbner basis has this form. In [**90**] the authors show that such examples fall into one of 8 parameterized families. Computational methods are also used to help check that the algebras in these families really are regular.

Thus bigraded algebras of dimension 4 of all three possible Hilbert series are understood. One goal of classifying algebras with additional special properties is to produce a large portfolio of examples on which to test the potential validity of conjectures about regular algebras. Using the various classification results, all bigraded regular algebras of dimension 4 are known to be strongly noetherian, Auslander regular, and GK-Cohen Macaulay, for example, giving further evidence that these properties might hold for all regular algebras.

Zhou and Lu [**143**] classified all nontrivially bigraded regular algebras of dimension 5 with two generators, using a Gröbner basis method similar to that of [**90**]. There are such examples of most of the possible relation types $(4,4,4)$, $(3,5,5)$, $(3,4,7)$ and $(4,4,4,5,5)$ in Floystad and Vatne's work. In particular, these were the first examples of the final relation type, but it is still open if the relation type $(4,4,4,5)$ can occur. The authors also note that in all of their examples, there is a choice of Gröbner basis for the ideal of relations whose leading terms consist of *Lyndon words*: words w that are maximal in the lexicographical ordering among all of the cyclic permutations of their letters. A Gröbner basis consisting of Lyndon words has good combinatorial properties, and so the authors ask if for bigraded regular algebras of arbitrary dimension with 2 generators, it is always true that there is a Gröbner basis of Lyndon words. This would imply several of the conjectured properties of regular algebras in this special case, for example it would follow that $\operatorname{GKdim}(A) = \operatorname{gl.dim}(A)$, and that the Hilbert series of A is the same as that of some graded polynomial ring [**143**, Theorem 1.9].

In higher dimension, Gateva-Ivanova wrote down all regular algebras occurring as universal enveloping algebras of certain kinds of Lie algebras of dimension 6 and 7—those with a Gröbner basis of relations with leading terms consisting of Lyndon words [**40**].

9. Generalizations of the AS regular condition

In this section we discuss what happens when some of the hypotheses in the definition of AS regular are relaxed. Then in the final section we will return to some of the recent work on AS regular algebras.

9.1. Generators in arbitrary degrees. Artin and Schelter did not require generation in degree 1 in their original definition of regular algebra, but almost all of the first papers in the subject only considered regular algebras under this additional assumption, as we have in this survey until now. It was clear from the outset that allowing generators in arbitrary degrees would introduce many significant complications. Recall that point modules, their parametrizing scheme, and the associated automorphism coming from the truncation-shift operation played a large role in the classification of regular algebras of dimension 3. If A is a connected $\mathbb{N}$-graded $\Bbbk$-algebra with generators in arbitrary degrees, and M is a point module for A as defined above, then there is no reason the truncation shift $M_{\geq 1}(1)$ must be a point module any longer—it still has the correct Hilbert series, but it may not be cyclic. In [**108**], Stafford and Zhang presented an interesting example of a (non-regular) idealizer ring B, which has generators of degree 1 and 2 inside of a Jordan plane. This ring has many exotic properties in the context of noncommutative projective geometry. In particular, it has point modules M for which $M_{\geq 1}(1)$ is not a point module.

The basic analysis of how the axioms of a regular algebra restrict the form of the free resolution of $\Bbbk$, as we outlined earlier in Section 2.1, does go through with minor adjustments for algebras with generators in arbitrary degrees. For instance, the classification of regular algebras of dimension 2 generated in arbitrary degrees is very similar:

PROPOSITION 9.1 (Proposition 3.3, Remark 3.4, [**118**]). *Let A be an AS regular algebra of global dimension* 2. *Then either*

(1) $A \cong \Bbbk\langle x, y\rangle/(yx - qxy)$ *for some* $0 \neq q \in \Bbbk$, *where* $\deg x = a$, $\deg y = b$, *for any* $a, b \geq 1$; *or*
(2) $A \cong \Bbbk\langle x, y\rangle/(yx - xy - x^m)$ *where* $\deg x = a$, $\deg y = (m-1)a$, *for some* $a \geq 1$ *and* $m \geq 2$.

In fact, many aspects of the classification of regular algebras of dimension 3 also go through for generators in arbitrary degrees. The theory was worked out by Stephenson in a series of papers [**112**] [**113**]. In particular, it is still possible to use point module techniques, but this requires care because of the issues mentioned above. For these regular algebras, it turns out that if one considers only point modules which are generated in degrees 0 and 1 and have no socle, then there is a nice scheme X parametrizing them for which the methods of [**2**] can then be used. For instance, there are Sklyanin-like examples for which X is a smooth elliptic curve.

Stephenson also investigated the interesting geometry associated with these regular algebras. The particular regular algebras A with 3 generators in degrees $(1, 1, n)$ for some $n \geq 2$ are examined more closely in [**115**] [**114**]. The associated category $\operatorname{proj} A$ can be thought of as a noncommutative analog of a weighted projective space $Z = \operatorname{Proj} k[x, y, z]$, where $\deg(x) = \deg(y) = 1$, and $\deg(z) = n$. The commutative geometry of Z is well-known: Z can be obtained by blowing down the unique exceptional curve on a rational ruled surface with invariant n. In terms of the point module techniques used in [**113**], Stephenson shows that the point modules for A are parametrized by X where either $X \cong Z$ or else X is isomorphic to a subscheme of Z defined by a single equation of degree $n+2$. In the latter case, A has a normal element g of degree $n+2$ such that A/gA is isomorphic to a twisted homogeneous coordinate ring on X. Thus, for these algebras the analogy with the geometric approach used in [**2**] is very tight. Ingalls and Patrick [**47**] made the geometric analogy even more explicit in terms of the noncommutative projective schemes, by showing there is a way to define a blowup of $\operatorname{Proj} A$ at a point using a Rees ring, such that the resulting category is a quantum ruled surface in a sense studied earlier by Patrick [**80**] [**81**].

Smith studied another regular algebra A of dimension 3 with two generators of degrees 1 and 2, and showed that $\operatorname{proj} A$ is equivalent to the category of coherent sheaves on a degree 6 Del Pezzo surface [**103**].

One important reason why one might be interested in regular algebras generated in arbitrary degrees is their occurrence as invariant rings in noncommutative invariant theory (see [**57**]).

9.2. Omitting the condition on growth. Consider the definition of an AS regular algebra A, but remove axiom (2) that the graded pieces of A have polynomially bounded growth (or equivalently, that $\operatorname{GKdim}(A)$ is finite). We call such an algebra *weakly* AS regular. There is no standard term; some papers in the literature simply call such examples AS regular again, making it clear at the outset that they are considering a version of the definition in which finite GK dimension is not required, unlike Artin and Schelter's original definition.

Suppose that A is weakly regular (and not regular), so that $\operatorname{GKdim}(A) = \infty$. As we remarked earlier in Section 3.1, in this case A has exponential growth, and therefore cannot be noetherian by a theorem of Stephenson and Zhang [**118**]. Thus a weakly regular algebra certainly does not closely resemble a commutative polynomial ring, and is a different kind of animal. Techniques coming from point modules are also likely to be difficult to apply: one does not expect the inverse system of truncated point modules defining the point scheme to stabilize in a non-noetherian algebra. Nevertheless, weakly regular algebras are interesting and seem to be well-behaved in certain other ways. For example, it is still true that all known examples are domains.

Once again, the examples of global dimension 2 are special enough to be classified, as was done by Zhang. For simplicity we state the result for those algebras generated in degree 1, though the general classification is not really any harder.

THEOREM 9.2 ([**140**]). *Let A be a weakly AS regular algebra of global dimension 2 which is generated in degree 1. Then*

$$A \cong \Bbbk\langle x_1, \dots, x_n\rangle/(\sum_{i=1}^{n} x_i\tau(x_i))$$

For some $\Bbbk$-linear bijection τ from $\Bbbk x_1 + \cdots + \Bbbk x_n$ to itself and some $n \geq 2$. Conversely, all such algebras are weakly regular.

Zhang also examined the basic properties of such algebras. For $n > 2$, these are all non-noetherian domains of exponential growth which contain a free associative algebra, and so do not have a classical quotient division ring of fractions. They have no normal elements.

While for $n > 2$ these algebras are non-noetherian, Piontkovski proved that these algebras (and in fact all algebras defined by a single degree 2 relation in a free algebra) are graded coherent—that is, every finitely generated graded left or right ideal is finitely presented [**82**]. Coherence is a weak substitute for the noetherian property, which implies at least that the subcategory of finitely presented modules is abelian. Just as it is conjectured that all regular algebras are noetherian, Bondal has conjectured that all weakly AS-regular algebras are graded coherent [**82**, Conjecture 1.3]. For A as in Theorem 9.2, Piontkovski also studied the associated category $\operatorname{proj} A$, which is defined here as the category of finitely presented graded A-modules modulo finite-dimensional modules. These categories can be interpreted as noncommutative projective lines. Sisodia and Smith studied some further properties of these categories, including their Grothendieck groups [**96**]. Another point of view on the graded coherent property for these algebras (and more generally, preprojective algebras) was given by Minamoto [**73**].

As far as weakly regular algebras of dimension greater than 2, Smith gave an interesting such example of dimension 3 with 7 generators and 7 relations, which is constructed from the geometry of the octonions [**101**]. We mentioned earlier in Section 7 the m-Koszul algebras defined by Berger. Berger gives a family of examples defined by relations given by anti-symmetrizers [**10**, Theorem 1.1], which include weakly AS-regular algebras which are m-Koszul, for unbounded choices of m. Further examples of m-Koszul weakly regular algebras of dimension 3 are the Yang-Mills algebras studied by Connes and Dubois-Violette [**33**].

Other examples have been studied in the more general context of twisted Calabi-Yau algebras, which we review in the next section.

9.3. Twisted Calabi-Yau algebras. Suppose that A, B are any $\Bbbk$-algebras (not necessarily graded). We write B^{op} for the opposite algebra of B. Recall that an (A, B)-bimodule M is the same as an $A \otimes_{\Bbbk} B^{op}$-module, where $(a \otimes b) \cdot m = amb$ for $m \in M$, $a, b \in A$. The algebra $A^e = A \otimes_{\Bbbk} A$ is called the *enveloping algebra* of A, and a left A^e-module is the same as an (A, A)-bimodule. Because $(A^e)^{op} \cong A^{op} \otimes_k A \cong A \otimes_k A^{op}$ as algebras, we can also identify (A, A)-bimodules with right A^e-modules.

DEFINITION 9.3. Let A be a $\Bbbk$-algebra. Suppose that

(1) A is *homologically smooth*: that is, the A^e-module A has a length d projective resolution over A^e by finitely generated projective left A^e-modules; and

(2) $\operatorname{Ext}^i_{A^e}(A, A^e) \cong \begin{cases} 0 & i \neq d \\ A^{\mu} & i = d \end{cases}$ as right A^e-modules, for some automorphism $\mu : A \to A$.

then A is called *twisted Calabi-Yau.*

Here, the module A^{μ} has its right action by A twisted by the automorphism μ, i.e. b acts on the right of $a \in A^{\mu}$ by $a * b = a\mu(b)$. The automorphism μ is called the *Nakayama automorphism* of A. It is determined only up to composition with an inner automorphism of A. If one may take $\mu = 1$ in the definition, then one says that A is *Calabi-Yau.*

Ginzburg defined the notion above and studied some properties of Calabi-Yau algebras in [**41**]. He was inspired by Kontsevich's notion of a Calabi-Yau category, which itself was related to investigations of mirror symmetry involving Calabi-Yau varieties in algebraic geometry. Thus Calabi and Yau ended up with their names on these algebraic objects via a rather roundabout route. While there was reason to be particularly interested in Calabi-Yau algebras from this original motivation, from the algebraic perspective the generalization to twisted Calabi-Yau algebras is very natural, as we will see.

We now abbreviate Calabi-Yau by CY. Note that the two parts of the definition of twisted CY algebra have clear formal similarities to the first and third axioms of the definition of AS regular algebra, where the special (A, A)-bimodule A acts as a kind of generalized trivial module. Indeed, algebras such as Sklyanin algebras were among Ginzburg's initial examples of CY algebras. In [**85**, Lemma 1.2], Reyes, the author, and Zhang proved that a connected $\mathbb{N}$-graded $\Bbbk$-algebra is weakly AS regular if and only if A is twisted CY. Technically, this result is about a version of the twisted CY property stated in the graded category, but this doesn't make any difference [**84**, Theorem 4.2]. One direction of [**85**, Lemma 1.2] is quite straightforward: if A is connected graded twisted CY, by tensoring the minimal A^e-projective resolution of A with the trivial module $\Bbbk$ on each side, one gets good properties for the minimal projective A-resolution of $\Bbbk$, which implies that A is weakly AS regular. This direction was also shown in the special case of Calabi-Yau algebras in [**11**, Proposition 4.3]. The converse uses the theory of dualizing complexes and relies on an earlier paper of Yekutieli and Zhang [**137**].

From this point of view, we see that twisted CY algebras provide a vast generalization of AS regular algebras to arbitrary, possibly ungraded algebras. The main difference is the lack of a condition on growth in the twisted CY concept. This simply reflects the fact that there was less attention initially from those interested in CY algebras on ring-theoretic properties such as the noetherian condition, which typically fails without a restriction on growth.

The notion of twisted CY unifies other possible generalizations of the (weakly) AS regular condition. For example, suppose that Q is a finite quiver with n vertices, and let $\Bbbk Q$ be its path algebra, graded by path length. Let $A = \Bbbk Q/I$ for some graded ideal I generated by homogeneous elements of degree at least 2. Martinez-Villa and Solberg [**72**] defined a notion of generalized AS regular for such an algebra A, which essentially treats the n simple left and right modules associated to the vertices of Q as a set of trivial modules. Namely, the AS Gorenstein condition is replaced by a requirement that $\operatorname{Ext}^j_A(-, A)$ acts as 0 on all simple graded left modules for $0 \leq j < d = \operatorname{gl.dim} A$, while $\operatorname{Ext}^d(-, A)$ gives a bijection from the simple graded left modules to the simple graded right modules (as a set). It turns out that generalized Artin Schelter regular in this sense is the same as twisted CY for those $\mathbb{N}$-graded algebras of the form $\Bbbk Q/I$ [**74**, Theorem 3.19]. More generally, the author and Reyes characterized the finitely generated $\mathbb{N}$-graded twisted CY algebras A with $\dim_\Bbbk A_i < \infty$ over any base field $\Bbbk$, in terms of another variation on the definition of AS regular [**84**, Theorem 1.5].

Twisted CY algebras are their own subject, with a large and growing literature we cannot hope to address further in this survey. However, in the next section we describe some of the ways in which the language and ideas of CY algebras have played back into the theory of regular algebras in recent years.

10. From twisted CY algebras back to AS regular algebras

10.1. Superpotentials. Let $F = \Bbbk\langle x_1, \dots x_m\rangle$ be a free algebra, graded with $\deg(x_i) = 1$ for all i, and suppose that τ is some graded automorphism of F. Fixing a degree n, there is a unique $\Bbbk$-linear map $\rho : F_n \to F_n$ which acts on the basis of monomials via $x_{i_1} \dots x_{i_n} \mapsto \tau(x_{i_n})x_{i_1} \dots x_{i_{n-1}}$. A *$\tau$-twisted superpotential* is a homogeneous element $w \in F_n$ such that $\rho(w) = w$. For each $1 \leq i \leq m$ there is an associated linear operator ∂_i which acts on monomials in F by $\partial_i(v) = u$ if $v = x_i u$ for some monomial u and $\partial_i(v) = 0$ otherwise. Given some $k \leq n-2$, one may define an associated *derivation quotient algebra* $D(w, k) = F/I$, where I is generated by the set of elements

$$\{\partial_{i_1} \circ \partial_{i_2} \circ \cdots \circ \partial_{i_k}(w) | 1 \leq i_1, \dots, i_k \leq m\}.$$

This setup also generalizes easily to the quiver setting, by replacing F with a path algebra $\Bbbk Q$ and the variables with arrows. Two overarching questions about such algebras are (i) whether every twisted CY algebra is a derivation quotient algebra, in graded or local settings; and (ii) for which w, k the algebra $D(w, k)$ is twisted CY of dimension $k + 2$, in which case the corresponding potential w is called *good* for the given k.

Let us first address progress on question (i). Note that Artin and Schelter's original work already showed that regular algebras of dimension 3 come from twisted superpotentials (just not in that language). More exactly, as we explained in Section 2, any regular algebra of dimension 3 either is equal to $D(w, 1)$ for some twisted potential $w \in \Bbbk\langle x_1, x_2, x_3\rangle_3$; or else equal to $D(w, 1)$ for some twisted potential $w \in \Bbbk\langle x_1, x_2\rangle_4$. The restrictions on the number of generators and the degree of the potential are due to the polynomial growth condition.

Bocklandt showed that Calabi-Yau algebras of dimension 3 of the form $\Bbbk Q/I$ are derivation quotient algebras $D(w, 1)$ [**13**]. Then Bocklandt, Schedler and Wemyss [**14**] showed that any m-Koszul graded twisted CY algebra of the form $\Bbbk Q/I$ with global dimension $d \geq 2$ is a derivation quotient algebra of the form $D(w, d-2)$. Finally, Van den Bergh proved a very general result on the existence of superpotentials for complete CY algebras, which includes the graded case [**124**].

Question (ii) is much more sensitive. It was already hard for regular algebras of dimension 3, where it amounts to determining which values of the parameters in each family of the classification yield a regular algebra. The deep geometric methods of [**2**] were introduced to answer this. So the question about which potentials are good in an even more general setting is expected to be very difficult.

In any case, the language and point of view of superpotentials and the associated concepts from CY algebras, in particular the Nakayama automorphism, have come to be increasingly important in the theory of regular algebras. An early example is the paper [**37**], in which Dubois-Violette reinterprets some of the theory of weakly regular algebras of dimension 2 and 3 using the language of potentials and the techniques of m-Koszul algebras.

10.2. Superpotentials of regular algebras. The details of the classification of regular algebras of dimension 3 have been reexamined in recent years, with more of a focus on the twisted superpotentials. First, the parametrized superpotentials listed explicitly in [**1**] do not produce all regular algebras of dimension 3 up to isomorphism; they are "generic forms", in the sense that they are restricted to those whose associated twist τ is diagonalizable. Because diagonalizable matrices form an open subset of the general linear group, the superpotentials of other regular algebras lie in the closures of the orbits of the given superpotentials under the general linear group. Second, although which twisted superpotentials are good can be checked, in theory, from the geometric results in [**2**], a more direct understanding in terms of the properties of the superpotential is desirable.

Mori and Smith considered the quadratic Calabi-Yau algebras of dimension 3 in [**76**]. Given a superpotential w of degree 3 with associated derivation quotient algebra $A(w) = D(w, 1)$, the image of w in the symmetric algebra has some zero locus in $\mathbb{P}^2$. A complete classification of which w are good is given, primarily in terms of the geometry of that zero locus (which is different from the point scheme of $A(w)$).

Mori and Ueyama extended similar methods to cubic Calabi-Yau algebras of dimension 3 [**77**]. Extensions to twisted superpotentials, therefore recovering the classification of general AS regular algebras from the point of view of superpotentials, followed in several papers of Itaba and Matsuno [**48**] [**49**]. In related work, Iyudu and Shkarin [**51**] gave a classification of the isomorphism classes of the derivation quotient algebras $D(w, 1)$ for the w's occurring in this setting (good or not), using more purely computational methods.

In a different direction, Chirvasitu, Kanda, and Smith gave a new take on constructing normal extensions of regular algebras using twisted superpotentials, which works equally well in the setting of weakly regular algebras [**25**].

10.3. The Nakayama automorphism and the CY case. The Nakayama automorphism μ of a regular algebra has received new attention as an important invariant. In general, given a regular algebra defined by generators and relations, it may be hard to calculate the Nakayama automorphism. There are number of papers which show how to calculate μ_A for certain classes of algebras, see for instance [**66**], [**144**], [**71**].

In the m-Koszul case, the Nakayama automorphism is closely related to the superpotential. Suppose that A is an AS regular derivation quotient algebra $A = D(w, k)$, for a μ-twisted superpotential w. Here μ was defined as an automorphism of the free algebra, but one may see that by construction it preserves the relations of A, and so descends to an automorphism of A. Importantly, this is essentially the same as the Nakayama automorphism μ of A [**75**, Theorem 4.4]. (Technically, this is true only up to sign, and possibly up to replacing μ by μ^{-1} depending on one's conventions. There are sign conventions which motivated the "super" part of the term superpotential, which we have ignored for simplicity).

Notably, some of the regular algebras that have long attracted the most interest, such as Sklyanin algebras, are CY (where $\mu = 1$), and so there is reason to focus on this case in particular. An important example is the work of Pym, who studied the flat deformations of the polynomial ring $\Bbbk[x_1, \dots, x_4]$—which automatically stay CY—and examined the resulting AS-regular algebras, which fall into 6 families [**83**]. Any such algebra gives a Poisson structure to the polynomial ring as a semi-classical limit, and the proof relies heavily on Poisson geometry. Vancliff had studied earlier a more limited family of flat deformations coming from twists [**127**]. In related work, Lecoutre and Sierra generalized one of Pym's 6 families to give examples of CY regular algebras in all dimensions [**63**].

One would like to relate an arbitrary twisted CY algebra to a CY one in some way. Reyes, the author, and Zhang studied the behavior of the Nakayama automorphism under Zhang twist [**85**]. Suppose that $\Bbbk$ is algebraically closed of characteristic 0. In this case, for a noetherian AS regular A with Nakayama automorphism μ there is a always a graded automorphism σ such that the associated Zhang twist A^σ has Nakayama automorphism of finite order [**86**, Theorem 5.3], [**85**, Corollary 0.7]. Moreover, if μ has finite order, then the skew group algebra $A \rtimes \langle \mu \rangle$ is CY of the same global dimension as A (but is no longer connected graded) [**86**, Theorem 5.3], [**85**, Corollary 0.6].

Alternatively, if one is willing to bump the global dimension up by one, the Ore extension $A[x; \mu]$ is always CY [**86**, Theorem 5.3],[**85**, Corollary 0.6]. This was proved independently in the Koszul case in [**46**], and Goodman and Krähmer proved a more general result for arbitrary (ungraded) algebras A [**45**]. These results provide a potential framework for reducing the amount of work needed to prove that regular algebras have good properties. For example, if all CY regular algebras of dimension $d+1$ are domains, then all regular algebras of dimension d are domains.

We close the paper with a reminder that the origin of the subject—the paper [**1**] by Artin and Schelter—proceeded by considering twisted superpotentials and classifying geometrically the generic forms of these that could lead to a regular algebra of dimension 3 (though not in that language). A direct analog of this for 4-dimensional regular algebras is still lacking. There is not a good understanding of the parameter space of twisted superpotentials and how the known classes of 4-dimensional regular algebras fit into it and relate to each other. Some initial work on generalizing [**1**] to dimension 4 was done by Caines in his PhD thesis [**16**], but it was just a start. This idea seems ripe for further development.

Acknowledgments

We thank Paul Smith and Michaela Vancliff for helpful comments and suggestions. We especially thank the referee for their thoughtful advice which led to many significant improvements to this survey. We also thank our many mentors and friends with more expertise than ourselves, including Paul Smith to whom this article is dedicated, for conversations over the years from which we learned about many of the beautiful results of this subject. Finally, we thank Michaela Vancliff for writing the companion article [**129**] in response to a comment from the referee.

References

[1] Michael Artin and William F. Schelter, *Graded algebras of global dimension* 3, Adv. in Math. **66** (1987), no. 2, 171–216, DOI 10.1016/0001-8708(87)90034-X. MR917738

[2] M. Artin, J. Tate, and M. Van den Bergh, *Some algebras associated to automorphisms of elliptic curves*, The Grothendieck Festschrift, Vol. I, Progr. Math., vol. 86, Birkhäuser Boston, Boston, MA, 1990, pp. 33–85. MR1086882

[3] M. Artin, J. Tate, and M. Van den Bergh, *Modules over regular algebras of dimension* 3, Invent. Math. **106** (1991), no. 2, 335–388, DOI 10.1007/BF01243916. MR1128218

[4] M. Artin and M. Van den Bergh, *Twisted homogeneous coordinate rings*, J. Algebra **133** (1990), no. 2, 249–271, DOI 10.1016/0021-8693(90)90269-T. MR1067406

[5] M. Artin and J. J. Zhang, *Noncommutative projective schemes*, Adv. Math. **109** (1994), no. 2, 228–287, DOI 10.1006/aima.1994.1087. MR1304753

[6] M. Artin and J. J. Zhang, *Abstract Hilbert schemes*, Algebr. Represent. Theory **4** (2001), no. 4, 305–394, DOI 10.1023/A:1012006112261. MR1863391

[7] Kristien Bauwens and Michel Van den Bergh, *Normalizing extensions of the two-Veronese of a three-dimensional Artin-Schelter regular algebra on two generators*, J. Algebra **205** (1998), no. 2, 368–390, DOI 10.1006/jabr.1997.7210. MR1632812

[8] Pieter Belmans, Kevin De Laet, and Lieven Le Bruyn, *The point variety of quantum polynomial rings*, J. Algebra **463** (2016), 10–22, DOI 10.1016/j.jalgebra.2016.06.013. MR3527537

[9] Roland Berger, *Koszulity for nonquadratic algebras*, J. Algebra **239** (2001), no. 2, 705–734, DOI 10.1006/jabr.2000.8703. MR1832913

[10] Roland Berger and Nicolas Marconnet, *Koszul and Gorenstein properties for homogeneous algebras*, Algebr. Represent. Theory **9** (2006), no. 1, 67–97, DOI 10.1007/s10468-005-9002-1. MR2233117

[11] Roland Berger and Rachel Taillefer, *Poincaré-Birkhoff-Witt deformations of Calabi-Yau algebras*, J. Noncommut. Geom. **1** (2007), no. 2, 241–270, DOI 10.4171/JNCG/6. MR2308306

[12] George M. Bergman, *The diamond lemma for ring theory*, Adv. in Math. **29** (1978), no. 2, 178–218, DOI 10.1016/0001-8708(78)90010-5. MR506890

[13] Raf Bocklandt, *Graded Calabi Yau algebras of dimension 3*, J. Pure Appl. Algebra **212** (2008), no. 1, 14–32, DOI 10.1016/j.jpaa.2007.03.009. MR2355031

[14] Raf Bocklandt, Travis Schedler, and Michael Wemyss, *Superpotentials and higher order derivations*, J. Pure Appl. Algebra **214** (2010), no. 9, 1501–1522, DOI 10.1016/j.jpaa.2009.07.013. MR2593679

[15] A. I. Bondal and A. E. Polishchuk, *Homological properties of associative algebras: the method of helices* (Russian, with Russian summary), Izv. Ross. Akad. Nauk Ser. Mat. **57** (1993), no. 2, 3–50, DOI 10.1070/IM1994v042n02ABEH001536; English transl., Russian Acad. Sci. Izv. Math. **42** (1994), no. 2, 219–260. MR1230966

[16] Ian Caines, *New examples of four dimensional AS-regular algebras*, ProQuest LLC, Ann Arbor, MI, 2005. Thesis (Ph.D.)–Massachusetts Institute of Technology. MR2717252

[17] Thomas Cassidy, *Global dimension 4 extensions of Artin-Schelter regular algebras*, J. Algebra **220** (1999), no. 1, 225–254, DOI 10.1006/jabr.1999.7902. MR1713429

[18] Thomas Cassidy, *Central extensions of Stephenson's algebras*, Comm. Algebra **31** (2003), no. 4, 1615–1632, DOI 10.1081/AGB-120018499. MR1972883

[19] Thomas Cassidy, *Homogenized down-up algebras*, Comm. Algebra **31** (2003), no. 4, 1765–1775, DOI 10.1081/AGB-120018507. MR1972891

[20] Thomas Cassidy and Brad Shelton, *Generalizing the notion of Koszul algebra*, Math. Z. **260** (2008), no. 1, 93–114, DOI 10.1007/s00209-007-0263-8. MR2413345

[21] Thomas Cassidy and Michaela Vancliff, *Generalizations of graded Clifford algebras and of complete intersections*, J. Lond. Math. Soc. (2) **81** (2010), no. 1, 91–112, DOI 10.1112/jlms/jdp057. MR2580455

[22] Daniel Chan, *Twisted rings and moduli stacks of "fat" point modules in noncommutative projective geometry*, Adv. Math. **229** (2012), no. 4, 2184–2209, DOI 10.1016/j.aim.2011.12.006. MR2880219

[23] Alex Chirvasitu and Ryo Kanda, *Flat families of point schemes for connected graded algebras*, Michigan Math. J. **73** (2023), no. 1, 195–208, DOI 10.1307/mmj/20205939. MR4555226

[24] A. Chirvasitu, R. Kanda, and S. P. Smith, *The characteristic variety for Feigin and Odesskii's elliptic algebras*, 2019, Preprint, `arXiv:1903.11798`.

[25] Alex Chirvasitu, Ryo Kanda, and S. Paul Smith, *New Artin-Schelter regular and Calabi-Yau algebras via normal extensions*, Trans. Amer. Math. Soc. **372** (2019), no. 6, 3947–3983, DOI 10.1090/tran/7672. MR4009424

[26] Alex Chirvasitu, Ryo Kanda, and S. Paul Smith, *Feigin and Odesskii's elliptic algebras*, J. Algebra **581** (2021), 173–225, DOI 10.1016/j.jalgebra.2021.04.009. MR4249759

[27] Alex Chirvasitu, Ryo Kanda, and S. Paul Smith, *Maps from Feigin and Odesskii's elliptic algebras to twisted homogeneous coordinate rings*, Forum Math. Sigma **9** (2021), Paper No. e4, 50, DOI 10.1017/fms.2020.60. MR4202489

[28] A. Chirvasitu, R. Kanda, and S. P. Smith, *Modular properties of elliptic algebras*, 2021. Preprint, `arXiv:2108.09143`.

[29] Alex Chirvasitu, Ryo Kanda, and S. Paul Smith, *Elliptic R-matrices and Feigin and Odesskii's elliptic algebras*, Selecta Math. (N.S.) **29** (2023), no. 2, Paper No. 31, 81, DOI 10.1007/s00029-023-00827-0. MR4563694

[30] Alex Chirvasitu and S. Paul Smith, *Exotic elliptic algebras of dimension 4*, Adv. Math. **309** (2017), 558–623, DOI 10.1016/j.aim.2017.01.010. With an appendix by Derek Tomlin. MR3607286

[31] Alex Chirvasitu and S. Paul Smith, *Exotic elliptic algebras*, Trans. Amer. Math. Soc. **371** (2019), no. 1, 279–333, DOI 10.1090/tran/7341. MR3885145

[32] Alex Chirvasitu, S. Paul Smith, and Michaela Vancliff, *A geometric invariant of 6-dimensional subspaces of* 4×4 *matrices*, Proc. Amer. Math. Soc. **148** (2020), no. 3, 915–928, DOI 10.1090/proc/14294. MR4055923

[33] Alain Connes and Michel Dubois-Violette, *Yang-Mills algebra*, Lett. Math. Phys. **61** (2002), no. 2, 149–158, DOI 10.1023/A:1020733628744. MR1936574

[34] Andrew Davies, *Cocycle twists of 4-dimensional Sklyanin algebras*, J. Algebra **457** (2016), 323–360, DOI 10.1016/j.jalgebra.2016.01.046. MR3490085

[35] Andrew Davies, *Cocycle twists of algebras*, Comm. Algebra **45** (2017), no. 3, 1347–1363, DOI 10.1080/00927872.2016.1178271. MR3573384

[36] Kevin De Laet, *Quotients of degenerate Sklyanin algebras*, J. Algebra Appl. **16** (2017), no. 5, 1750085, 26, DOI 10.1142/S0219498817500852. MR3634090

[37] Michel Dubois-Violette, *Multilinear forms and graded algebras*, J. Algebra **317** (2007), no. 1, 198–225, DOI 10.1016/j.jalgebra.2007.02.007. MR2360146

[38] Susan Elle, *Classification of relation types of Ore extensions of dimension 5*, Comm. Algebra **45** (2017), no. 3, 1323–1346, DOI 10.1080/00927872.2016.1177830. MR3573383

[39] Gunnar Fløystad and Jon Eivind Vatne, *Artin-Schelter regular algebras of dimension five*, Algebra, geometry and mathematical physics, Banach Center Publ., vol. 93, Polish Acad. Sci. Inst. Math., Warsaw, 2011, pp. 19–39, DOI 10.4064/bc93-0-2. MR2884420

[40] Tatiana Gateva-Ivanova, *Algebras defined by Lyndon words and Artin-Schelter regularity*, Trans. Amer. Math. Soc. Ser. B **9** (2022), 648–699, DOI 10.1090/btran/89. MR4446800

[41] V. Ginzburg, *Calabi-Yau algebras*, 2006. Preprint, arXiv:math/0612139.

[42] Pete Goetz, *Fat point modules over generalized Laurent polynomial rings*, J. Algebra **313** (2007), no. 2, 657–671, DOI 10.1016/j.jalgebra.2007.03.042. MR2329562

[43] Pete Goetz and Brad Shelton, *Representation theory of two families of quantum projective 3-spaces*, J. Algebra **295** (2006), no. 1, 141–156, DOI 10.1016/j.jalgebra.2005.05.010. MR2188854

[44] K. R. Goodearl and R. B. Warfield Jr., *An introduction to noncommutative Noetherian rings*, 2nd ed., London Mathematical Society Student Texts, vol. 61, Cambridge University Press, Cambridge, 2004, DOI 10.1017/CBO9780511841699. MR2080008

[45] Jake Goodman and Ulrich Krähmer, *Untwisting a twisted Calabi-Yau algebra*, J. Algebra **406** (2014), 272–289, DOI 10.1016/j.jalgebra.2014.02.018. MR3188338

[46] Ji-Wei He, Fred Van Oystaeyen, and Yinhuo Zhang, *Skew polynomial algebras with coefficients in Koszul Artin-Schelter regular algebras*, J. Algebra **390** (2013), 231–249, DOI 10.1016/j.jalgebra.2013.05.023. MR3072121

[47] Colin Ingalls and David Patrick, *Blowing up quantum weighted projective planes*, J. Algebra **254** (2002), no. 1, 92–114, DOI 10.1016/S0021-8693(02)00036-4. MR1927433

[48] Ayako Itaba and Masaki Matsuno, *Defining relations of 3-dimensional quadratic AS-regular algebras*, Math. J. Okayama Univ. **63** (2021), 61–86. MR4174859

[49] Ayako Itaba and Masaki Matsuno, *AS-regularity of geometric algebras of plane cubic curves*, J. Aust. Math. Soc. **112** (2022), no. 2, 193–217, DOI 10.1017/S1446788721000070. MR4389808

[50] Natalia Iyudu and Stanislav Shkarin, *Three dimensional Sklyanin algebras and Gröbner bases*, J. Algebra **470** (2017), 379–419, DOI 10.1016/j.jalgebra.2016.08.023. MR3565439

[51] Natalia Iyudu and Stanislav Shkarin, *Potential algebras with few generators*, Glasg. Math. J. **62** (2020), no. S1, S28–S76, DOI 10.1017/s0017089520000233. MR4173125

[52] N. Iyudu and S. Shkarin, *Sklyanin algebras and a cubic root of 1*, preprint, arXiv:2108.06290.

[53] David A. Jordan, *The graded algebra generated by two Eulerian derivatives*, Algebr. Represent. Theory **4** (2001), no. 3, 249–275, DOI 10.1023/A:1011481028760. MR1851999

[54] Dennis S. Keeler, *Criteria for σ-ampleness*, J. Amer. Math. Soc. **13** (2000), no. 3, 517–532, DOI 10.1090/S0894-0347-00-00334-9. MR1758752

[55] D. S. Keeler, D. Rogalski, and J. T. Stafford, *Naïve noncommutative blowing up*, Duke Math. J. **126** (2005), no. 3, 491–546, DOI 10.1215/S0012-7094-04-12633-8. MR2120116

[56] Bernhard Keller, *Introduction to A-infinity algebras and modules*, Homology Homotopy Appl. **3** (2001), no. 1, 1–35, DOI 10.4310/hha.2001.v3.n1.a1. MR1854636

[57] Ellen E. Kirkman, *Invariant theory of Artin-Schelter regular algebras: a survey*, Recent developments in representation theory, Contemp. Math., vol. 673, Amer. Math. Soc., Providence, RI, 2016, pp. 25–50, DOI 10.1090/conm/673/13489. MR3546709

[58] Günter R. Krause and Thomas H. Lenagan, *Growth of algebras and Gelfand-Kirillov dimension*, Revised edition, Graduate Studies in Mathematics, vol. 22, American Mathematical Society, Providence, RI, 2000, DOI 10.1090/gsm/022. MR1721834

[59] Lieven Le Bruyn, *Central singularities of quantum spaces*, J. Algebra **177** (1995), no. 1, 142–153, DOI 10.1006/jabr.1995.1290. MR1356364

[60] Lieven Le Bruyn and S. P. Smith, *Homogenized* $\mathfrak{sl}(2)$, Proc. Amer. Math. Soc. **118** (1993), no. 3, 725–730, DOI 10.2307/2160112. MR1136235

[61] Lieven Le Bruyn, S. P. Smith, and Michel Van den Bergh, *Central extensions of three-dimensional Artin-Schelter regular algebras*, Math. Z. **222** (1996), no. 2, 171–212, DOI 10.1007/PL00004532. MR1429334

[62] Lieven Le Bruyn and Michel Van den Bergh, *On quantum spaces of Lie algebras*, Proc. Amer. Math. Soc. **119** (1993), no. 2, 407–414, DOI 10.2307/2159921. MR1149975

[63] Cesar Lecoutre and Susan J. Sierra, *A new family of Poisson algebras and their deformations*, Nagoya Math. J. **233** (2019), 32–86, DOI 10.1017/nmj.2017.29. MR3901632

[64] Thierry Levasseur, *Some properties of noncommutative regular graded rings*, Glasgow Math. J. **34** (1992), no. 3, 277–300, DOI 10.1017/S0017089500008843. MR1181768

[65] Thierry Levasseur and S. Paul Smith, *Modules over the 4-dimensional Sklyanin algebra* (English, with English and French summaries), Bull. Soc. Math. France **121** (1993), no. 1, 35–90. MR1207244

[66] Liyu Liu, Shengqiang Wang, and Quanshui Wu, *Twisted Calabi-Yau property of Ore extensions*, J. Noncommut. Geom. **8** (2014), no. 2, 587–609, DOI 10.4171/JNCG/165. MR3275042

[67] D. M. Lu, J. H. Palmieri, Q. S. Wu, and J. J. Zhang, *A_∞-algebras for ring theorists*, Proceedings of the International Conference on Algebra, Algebra Colloq. **11** (2004), no. 1, 91–128. MR2058967

[68] D.-M. Lu, J. H. Palmieri, Q.-S. Wu, and J. J. Zhang, *Regular algebras of dimension 4 and their A_∞-Ext-algebras*, Duke Math. J. **137** (2007), no. 3, 537–584, DOI 10.1215/S0012-7094-07-13734-7. MR2309153

[69] Di Ming Lu, John H. Palmieri, Quan Shui Wu, and James J. Zhang, *Koszul equivalences in A_∞-algebras*, New York J. Math. **14** (2008), 325–378. MR2430869

[70] D.-M. Lu, J. H. Palmieri, Q.-S. Wu, and J. J. Zhang, *A-infinity structure on Ext-algebras*, J. Pure Appl. Algebra **213** (2009), no. 11, 2017–2037, DOI 10.1016/j.jpaa.2009.02.006. MR2533303

[71] J.-F. Lü, X.-F. Mao, and J. J. Zhang, *The Nakayama automorphism of a class of graded algebras*, Israel J. Math. **219** (2017), no. 2, 707–725, DOI 10.1007/s11856-017-1495-x. MR3649604

[72] Roberto Martinéz-Villa and Øyvind Solberg, *Artin-Schelter regular algebras and categories*, J. Pure Appl. Algebra **215** (2011), no. 4, 546–565, DOI 10.1016/j.jpaa.2010.06.006. MR2738371

[73] Hiroyuki Minamoto, *A criterion for graded coherence of tensor algebras and applications to higher dimensional AR-theory*, Eur. J. Math. **4** (2018), no. 2, 612–621, DOI 10.1007/s40879-017-0202-0. MR3799159

[74] Hiroyuki Minamoto and Izuru Mori, *The structure of AS-Gorenstein algebras*, Adv. Math. **226** (2011), no. 5, 4061–4095, DOI 10.1016/j.aim.2010.11.004. MR2770441

[75] Izuru Mori and S. Paul Smith, *m-Koszul Artin-Schelter regular algebras*, J. Algebra **446** (2016), 373–399, DOI 10.1016/j.jalgebra.2015.09.016. MR3421098

[76] Izuru Mori and S. Paul Smith, *The classification of 3-Calabi-Yau algebras with 3 generators and 3 quadratic relations*, Math. Z. **287** (2017), no. 1-2, 215–241, DOI 10.1007/s00209-016-1824-5. MR3694675

[77] Izuru Mori and Kenta Ueyama, *The classification of 3-dimensional noetherian cubic Calabi-Yau algebras*, J. Pure Appl. Algebra **223** (2019), no. 5, 1946–1965, DOI 10.1016/j.jpaa.2018.08.009. MR3906535

[78] Manizheh Nafari, Michaela Vancliff, and Jun Zhang, *Classifying quadratic quantum $\mathbb{P}^2$s by using graded skew Clifford algebras*, J. Algebra **346** (2011), 152–164, DOI 10.1016/j.jalgebra.2011.07.034. MR2842075

[79] A. V. Odesskiĭ and B. L. Feĭgin, *Sklyanin's elliptic algebras* (Russian), Funktsional. Anal. i Prilozhen. **23** (1989), no. 3, 45–54, 96, DOI 10.1007/BF01079526; English transl., Funct. Anal. Appl. **23** (1989), no. 3, 207–214 (1990). MR1026987

[80] David M. Patrick, *Noncommutative ruled surfaces*, ProQuest LLC, Ann Arbor, MI, 1997. Thesis (Ph.D.)–Massachusetts Institute of Technology. MR2716732

[81] David Patrick, *Noncommutative symmetric algebras of two-sided vector spaces*, J. Algebra **233** (2000), no. 1, 16–36, DOI 10.1006/jabr.2000.8445. MR1793588

[82] Dmitri Piontkovski, *Coherent algebras and noncommutative projective lines*, J. Algebra **319** (2008), no. 8, 3280–3290, DOI 10.1016/j.jalgebra.2007.07.010. MR2408318

[83] Brent Pym, *Quantum deformations of projective three-space*, Adv. Math. **281** (2015), 1216–1241, DOI 10.1016/j.aim.2015.06.005. MR3366864

[84] Manuel L. Reyes and Daniel Rogalski, *Graded twisted Calabi-Yau algebras are generalized Artin-Schelter regular*, Nagoya Math. J. **245** (2022), 100–153, DOI 10.1017/nmj.2020.32. MR4413364

[85] Manuel Reyes, Daniel Rogalski, and James J. Zhang, *Skew Calabi-Yau algebras and homological identities*, Adv. Math. **264** (2014), 308–354, DOI 10.1016/j.aim.2014.07.010. MR3250287

[86] Manuel Reyes, Daniel Rogalski, and James J. Zhang, *Skew Calabi-Yau triangulated categories and Frobenius Ext-algebras*, Trans. Amer. Math. Soc. **369** (2017), no. 1, 309–340, DOI 10.1090/tran/6640. MR3557775

[87] Daniel Rogalski, *Generic noncommutative surfaces*, Adv. Math. **184** (2004), no. 2, 289–341, DOI 10.1016/S0001-8708(03)00147-6. MR2054018

[88] Daniel Rogalski, *Noncommutative projective geometry*, Noncommutative algebraic geometry, Math. Sci. Res. Inst. Publ., vol. 64, Cambridge Univ. Press, New York, 2016, pp. 13–70. MR3618472

[89] D. Rogalski and Susan J. Sierra, *Some projective surfaces of GK-dimension 4*, Compos. Math. **148** (2012), no. 4, 1195–1237, DOI 10.1112/S0010437X12000188. MR2956041

[90] D. Rogalski and J. J. Zhang, *Regular algebras of dimension 4 with 3 generators*, New trends in noncommutative algebra, Contemp. Math., vol. 562, Amer. Math. Soc., Providence, RI, 2012, pp. 221–241, DOI 10.1090/conm/562/11139. MR2905562

[91] Brad Shelton and Michaela Vancliff, *Embedding a quantum rank three quadric in a quantum* $\mathbf{P}^3$, Comm. Algebra **27** (1999), no. 6, 2877–2904, DOI 10.1080/00927879908826599. MR1687269

[92] Brad Shelton and Michaela Vancliff, *Schemes of line modules. I*, J. London Math. Soc. (2) **65** (2002), no. 3, 575–590, DOI 10.1112/S0024610702003186. MR1895734

[93] Brad Shelton and Michaela Vancliff, *Schemes of line modules. II*, Comm. Algebra **30** (2002), no. 5, 2535–2552, DOI 10.1081/AGB-120003569. MR1904652

[94] Y. Shen, G.-S. Zhou, and D.-M. Lu, *Homogeneous PBW deformation for Artin-Schelter regular algebras*, Bull. Aust. Math. Soc. **91** (2015), no. 1, 53–68, DOI 10.1017/S0004972714000628. MR3294259

[95] Y. Shen, G.-S. Zhou, and D.-M. Lu, *Regularity criterion and classification for algebras of Jordan type*, Glasg. Math. J. **58** (2016), no. 1, 69–95, DOI 10.1017/S0017089515000075. MR3426429

[96] Gautam Sisodia and S. Paul Smith, *The Grothendieck group of non-commutative non-noetherian analogues of* $\mathbb{P}^1$ *and regular algebras of global dimension two*, J. Algebra **420** (2015), 188–210, DOI 10.1016/j.jalgebra.2014.11.024. MR3301905

[97] E. K. Sklyanin, *Some algebraic structures connected with the Yang-Baxter equation* (Russian), Funktsional. Anal. i Prilozhen. **16** (1982), no. 4, 27–34, 96. MR684124

[98] E. K. Sklyanin, *Some algebraic structures connected with the Yang-Baxter equation. Representations of a quantum algebra* (Russian), Funktsional. Anal. i Prilozhen. **17** (1983), no. 4, 34–48. MR725414

[99] S. P. Smith, *Point modules over Sklyanin algebras*, Math. Z. **215** (1994), no. 2, 169–177, DOI 10.1007/BF02571706. MR1259453

[100] S. Paul Smith, *Some finite-dimensional algebras related to elliptic curves*, Representation theory of algebras and related topics (Mexico City, 1994), CMS Conf. Proc., vol. 19, Amer. Math. Soc., Providence, RI, 1996, pp. 315–348. MR1388568

[101] S. P. Smith, *A 3-Calabi-Yau algebra with* G_2 *symmetry constructed from the octonions*, 2011. Preprint, `arXiv:1104.3824`.

[102] S. Paul Smith, *"Degenerate" 3-dimensional Sklyanin algebras are monomial algebras*, J. Algebra **358** (2012), 74–86, DOI 10.1016/j.jalgebra.2012.01.039. MR2905019

[103] S. Paul Smith, *A non-commutative homogeneous coordinate ring for the degree six del Pezzo surface*, J. Algebra **354** (2012), 95–109, DOI 10.1016/j.jalgebra.2011.11.030. MR2879225

[104] S. P. Smith and J. T. Stafford, *Regularity of the four-dimensional Sklyanin algebra*, Compositio Math. **83** (1992), no. 3, 259–289. MR1175941

[105] J. T. Stafford, *Auslander-regular algebras and maximal orders*, J. London Math. Soc. (2) **50** (1994), no. 2, 276–292, DOI 10.1112/jlms/50.2.276. MR1291737

[106] J. T. Stafford, *Regularity of algebras related to the Sklyanin algebra*, Trans. Amer. Math. Soc. **341** (1994), no. 2, 895–916, DOI 10.2307/2154589. MR1148046

[107] J. T. Stafford and M. van den Bergh, *Noncommutative curves and noncommutative surfaces*, Bull. Amer. Math. Soc. (N.S.) **38** (2001), no. 2, 171–216, DOI 10.1090/S0273-0979-01-00894-1. MR1816070

[108] J. T. Stafford and J. J. Zhang, *Examples in non-commutative projective geometry*, Math. Proc. Cambridge Philos. Soc. **116** (1994), no. 3, 415–433, DOI 10.1017/S0305004100072716. MR1291750

[109] J. T. Stafford and J. J. Zhang, *Homological properties of (graded) Noetherian* PI *rings*, J. Algebra **168** (1994), no. 3, 988–1026, DOI 10.1006/jabr.1994.1267. MR1293638

[110] Joanna M. Staniszkis, *The 4-dimensional Sklyanin algebra*, J. Algebra **167** (1994), no. 1, 104–115, DOI 10.1006/jabr.1994.1178. MR1282819

[111] Joanna M. Staniszkis, *Linear modules over Sklyanin algebras*, J. London Math. Soc. (2) **53** (1996), no. 3, 464–478, DOI 10.1112/jlms/53.3.464. MR1396711

[112] D. R. Stephenson, *Artin-Schelter regular algebras of global dimension three*, J. Algebra **183** (1996), no. 1, 55–73, DOI 10.1006/jabr.1996.0207. MR1397387

[113] Darin R. Stephenson, *Algebras associated to elliptic curves*, Trans. Amer. Math. Soc. **349** (1997), no. 6, 2317–2340, DOI 10.1090/S0002-9947-97-01769-8. MR1390046

[114] Darin R. Stephenson, *Corrigendum: "Quantum planes of weight* $(1,1,n)$*" [J. Algebra* **225** *(2000), no. 1, 70–92; MR1743651 (2001a:16044)]*, J. Algebra **234** (2000), no. 1, 277–278, DOI 10.1006/jabr.2000.8626. MR1799486

[115] Darin R. Stephenson, *Quantum planes of weight* $(1,1,n)$, J. Algebra **225** (2000), no. 1, 70–92, DOI 10.1006/jabr.1999.8093. MR1743651

[116] Darin R. Stephenson and Michaela Vancliff, *Some finite quantum* $\mathbb{P}^3$*s that are infinite modules over their centers*, J. Algebra **297** (2006), no. 1, 208–215, DOI 10.1016/j.jalgebra.2005.04.005. MR2206855

[117] Darin R. Stephenson and Michaela Vancliff, *Constructing Clifford quantum* $\mathbb{P}^3$*s with finitely many points*, J. Algebra **312** (2007), no. 1, 86–110, DOI 10.1016/j.jalgebra.2007.02.015. MR2320448

[118] Darin R. Stephenson and James J. Zhang, *Growth of graded Noetherian rings*, Proc. Amer. Math. Soc. **125** (1997), no. 6, 1593–1605, DOI 10.1090/S0002-9939-97-03752-0. MR1371143

[119] Darin R. Stephenson and James J. Zhang, *Noetherian connected graded algebras of global dimension 3*, J. Algebra **230** (2000), no. 2, 474–495, DOI 10.1006/jabr.2000.8323. MR1775800

[120] John Tate and Michel van den Bergh, *Homological properties of Sklyanin algebras*, Invent. Math. **124** (1996), no. 1-3, 619–647, DOI 10.1007/s002220050065. MR1369430

[121] Kok-Ming Teo, *Homological properties of Sklyanin algebras*, Comm. Algebra **24** (1996), no. 9, 3027–3035, DOI 10.1080/00927879608825728. MR1396871

[122] M. Van den Bergh, *An example with 20 points*, unpublished note, 1988.

[123] Michel Van den Bergh, *Noncommutative quadrics*, Int. Math. Res. Not. IMRN **17** (2011), 3983–4026, DOI 10.1093/imrn/rnq234. MR2836401

[124] Michel Van den Bergh, *Calabi-Yau algebras and superpotentials*, Selecta Math. (N.S.) **21** (2015), no. 2, 555–603, DOI 10.1007/s00029-014-0166-6. MR3338683

[125] Kristel Van Rompay, *Segre product of Artin-Schelter regular algebras of dimension 2 and embeddings in quantum* $\mathbf{P}^3$*'s*, J. Algebra **180** (1996), no. 2, 483–512, DOI 10.1006/jabr.1996.0078. MR1378541

[126] M. Vancliff, *Quadratic algebras associated with the union of a quadric and a line in* $\mathbf{P}^3$, J. Algebra **165** (1994), no. 1, 63–90, DOI 10.1006/jabr.1994.1098. MR1272579

[127] Michaela Vancliff, *Primitive and Poisson spectra of twists of polynomial rings*, Algebr. Represent. Theory **2** (1999), no. 3, 269–285, DOI 10.1023/A:1009914728281. MR1715749

[128] Michaela Vancliff, *The interplay of algebra and geometry in the setting of regular algebras*, Commutative algebra and noncommutative algebraic geometry. Vol. I, Math. Sci. Res. Inst. Publ., vol. 67, Cambridge Univ. Press, New York, 2015, pp. 371–390. MR3525477

[129] M. Vancliff, *An example of a quadratic AS-Regular algebra without any point modules*, Contemporary Mathematics, to appear.

[130] Michaela Vancliff and Kristel Van Rompay, *Embedding a quantum nonsingular quadric in a quantum* $\mathbf{P}^3$, J. Algebra **195** (1997), no. 1, 93–129, DOI 10.1006/jabr.1997.7077. MR1468885

[131] Michaela Vancliff and Kristel Van Rompay, *Four-dimensional regular algebras with point scheme, a nonsingular quadric in* $\mathbf{P}^3$, Comm. Algebra **28** (2000), no. 5, 2211–2242, DOI 10.1080/00927870008826955. MR1757458

[132] M. Vancliff, K. Van Rompay, and L. Willaert, *Some quantum* $\mathbf{P}^3$*s with finitely many points*, Comm. Algebra **26** (1998), no. 4, 1193–1208, DOI 10.1080/00927879808826193. MR1612220

[133] Michaela Vancliff and Padmini P. Veerapen, *Point modules over regular graded skew Clifford algebras*, J. Algebra **420** (2014), 54–64, DOI 10.1016/j.jalgebra.2014.08.003. MR3261452

[134] Padmini Veerapen, *Graded Clifford algebras and graded skew Clifford algebras and their role in the classification of Artin-Schelter regular algebras*, Adv. Appl. Clifford Algebr. **27** (2017), no. 3, 2855–2871, DOI 10.1007/s00006-017-0795-x. MR3688849

[135] Chelsea Walton, *Degenerate Sklyanin algebras and generalized twisted homogeneous coordinate rings*, J. Algebra **322** (2009), no. 7, 2508–2527, DOI 10.1016/j.jalgebra.2009.02.024. MR2553692

[136] S.-Q. Wang and Q.-S. Wu, *A class of AS-regular algebras of dimension five*, J. Algebra **362** (2012), 117–144, DOI 10.1016/j.jalgebra.2012.04.012. MR2921634

[137] Amnon Yekutieli and James J. Zhang, *Homological transcendence degree*, Proc. London Math. Soc. (3) **93** (2006), no. 1, 105–137, DOI 10.1017/S0024611505015698. MR2235944

[138] J. J. Zhang, *Twisted graded algebras and equivalences of graded categories*, Proc. London Math. Soc. (3) **72** (1996), no. 2, 281–311, DOI 10.1112/plms/s3-72.2.281. MR1367080

[139] James J. Zhang, *Linear modules over Sklyanin algebras are Cohen-Macaulay*, Comm. Algebra **26** (1998), no. 6, 1979–1983, DOI 10.1080/00927879808826253. MR1621724

[140] James J. Zhang, *Non-Noetherian regular rings of dimension* 2, Proc. Amer. Math. Soc. **126** (1998), no. 6, 1645–1653, DOI 10.1090/S0002-9939-98-04480-3. MR1459158

[141] James J. Zhang and Jun Zhang, *Double Ore extensions*, J. Pure Appl. Algebra **212** (2008), no. 12, 2668–2690, DOI 10.1016/j.jpaa.2008.05.008. MR2452318

[142] James J. Zhang and Jun Zhang, *Double extension regular algebras of type (14641)*, J. Algebra **322** (2009), no. 2, 373–409, DOI 10.1016/j.jalgebra.2009.03.041. MR2529094

[143] G.-S. Zhou and D.-M. Lu, *Artin-Schelter regular algebras of dimension five with two generators*, J. Pure Appl. Algebra **218** (2014), no. 5, 937–961, DOI 10.1016/j.jpaa.2013.10.011. MR3149644

[144] Can Zhu, Fred Van Oystaeyen, and Yinhuo Zhang, *Nakayama automorphisms of double Ore extensions of Koszul regular algebras*, Manuscripta Math. **152** (2017), no. 3-4, 555–584, DOI 10.1007/s00229-016-0865-8. MR3608305

Department of Mathematics, University of California, San Diego, La Jolla, California 92093

Contemporary Mathematics
Volume **801**, 2024
https://doi.org/10.1090/conm/801/16040

Twisting systems and some quantum $\mathbb{P}^3$s with point scheme a rank-2 quadric

Hung V. Tran and Michaela Vancliff

ABSTRACT. In 1996, J. J. Zhang introduced the concept of twisting a graded algebra by a twisting system, which generalizes the concept of twisting a graded algebra by an automorphism (the latter concept having been introduced in an article by M. Artin, J. Tate and M. Van den Bergh in 1991). Zhang proved that twisting using a twisting system is an equivalence relation and that certain important algebraic properties are invariant under twisting. We call a twisting system *nontrivial* if it is not given by an automorphism. However, there are very few examples of nontrivial twisting systems in the literature.

In 1997, the second author and K. Van Rompay and, in 1999, B. Shelton and the second author were successful in finding one example each of a nontrivial twisting system. Their twisting systems were constructed on certain quadratic algebras A (on four generators) using two invertible linear maps t and τ of A_1 that satisfy $t^2 =$ identity and $\tau^2 \in \operatorname{Aut}(A)$. We extend their work on twisting systems to any finitely generated quadratic algebra B using analogous maps that satisfy $t^n =$ identity and $\tau^n \in \operatorname{Aut}(B)$, for some $n \in \mathbb{N}$.

We illustrate our new method for producing a nontrivial twisting system on a certain quadratic algebra that is a quantum $\mathbb{P}^3$ and whose point scheme is isomorphic to a rank-2 quadric in $\mathbb{P}^3$. We prove that our algebra is not determined by the zero locus of its defining relations.

Introduction

In 1996, J. J. Zhang introduced the concept of twisting a graded algebra by a twisting system, which generalizes the concept of twisting a graded algebra by an automorphism ([**2**, Section 8], [**13**]). Zhang proved that twisting using a twisting system is an equivalence relation and that certain algebraic properties (such as noetherian, regularity, Hilbert series, etc) are invariant under twisting. This notion of twisting has been used extensively in the attempted classification of AS-regular algebras, such as in [**2**–**4**, **9**, **12**]. In [**10**], S. J. Sierra uses $\mathbb{Z}$-algebras to generalize some of Zhang's results from [**13**] in order to address the problem of finding necessary and sufficient conditions on two graded algebras that determine when their categories of graded modules are equivalent.

We call a twisting system *nontrivial* if it is not given by an automorphism. However, there are very few examples of nontrivial twisting systems in the literature.

2020 *Mathematics Subject Classification.* Primary 16W50, 16S80, 16S38, 16S37.
Key words and phrases. twist, twisting system, regular algebra, point scheme, quadric, quadratic.

In [**12**], the second author and K. Van Rompay and, in [**9**], B. Shelton and the second author were successful in finding one example each of a nontrivial twisting system. Their twisting systems were constructed on certain quadratic algebras A (on four generators) using two invertible linear maps t and τ of A_1 that satisfy $t^2 =$ identity and $\tau^2 \in \text{Aut}(A)$. That work was dependent on certain geometric data associated to the algebras.

In this article, we extend the ideas in [**9**] and [**12**] for producing a twisting system to any finitely generated quadratic algebra B using analogous maps that satisfy $t^n =$ identity and $\tau^n \in \text{Aut}(B)$, for some $n \in \mathbb{N}$. We illustrate our new method for producing a nontrivial twisting system on a quadratic algebra that is a quantum $\mathbb{P}^3$ whose point scheme is isomorphic to a rank-2 quadric in $\mathbb{P}^3$. We prove that our algebra is not determined by the zero locus of its defining relations.

The paper is outlined as follows. Section 1 recalls the definition of twisting system from [**13**] and some related concepts. Also, in that section, we describe pertinent details of the nontrivial twisting system that was constructed in [**12**]. In Section 2, in Definition 2.1, we introduce our generalization of the twisting systems introduced in [**9, 12**], and provide, in Theorem 2.7, conditions that guarantee that our generalization is indeed a twisting system. We also prove, in Corollary 2.10, some equivalent conditions regarding such a twisting system. In the next (final) section, we define a family of quadratic AS-regular algebras of global dimension four on which we illustrate our twisting system can be applied, and we prove, in Theorem 3.5, that such an algebra has point scheme isomorphic to a rank-2 quadric in $\mathbb{P}^3$ and, in Corollary 3.6, that the algebra is not determined by the zero locus of its defining relations. To our knowledge, this algebra is the first quantum $\mathbb{P}^3$ known to have its point scheme be isomorphic to a rank-2 quadric in $\mathbb{P}^3$.

This article is based on the Ph.D. dissertation, [**11**], of the first author, but some results herein use methods of proof that differ from those in [**11**]. Moreover, some details are omitted from this article that can be found in [**11**].

1. Preliminaries

In this section, we recall the definition of a twisting system given in [**13**] and discuss relevant details of the example of a twisting system that was presented in [**12**].

Throughout this article, the notation $\mathbb{N}$ denotes the set of all positive integers, $\mathbb{N}_0$ denotes the set $\mathbb{N} \cup \{0\}$, $\mathbb{Z}$ denotes the set of all integers, $\Bbbk$ denotes a field, and $\Bbbk^\times$ denotes the set of all nonzero elements in $\Bbbk$. A $\Bbbk$-algebra C is said to be $\mathbb{Z}$-graded if $C = \bigoplus_{j \in \mathbb{Z}} C_j$ where C_j is a $\Bbbk$-subspace of C for all $j \in \mathbb{Z}$ and $C_i C_j \subset C_{i+j}$ for all $i, j \in \mathbb{Z}$.

Definition 1.1 ([**2**, Section 8]). Let $A = \bigoplus_{j \in \mathbb{Z}} A_j$ be a $\mathbb{Z}$-graded $\Bbbk$-algebra, and suppose $\phi \in \text{Aut}(A)$ is a graded automorphism of degree zero. We form a new graded algebra B from these data as follows:

(a) as vector spaces over $\Bbbk$, $B = A$, and
(b) we define a multiplication $*$ on B by defining $*$ on homogeneous elements $a_B, b_B \in B$ by
$$a_B * b_B = (a\phi^d(b))_B,$$
where $c_B \in B$ denotes the counterpart in B of the element $c \in A$ and $d = \deg(a_B)$, and extending $*$ to B by using the distributive property.

In particular, if A is generated by A_1 and if $a_1, \dots, a_m \in A_1$, then

$$(a_1)_B * (a_2)_B * \cdots * (a_m)_B = a_1\phi(a_2)\phi^2(a_3)\cdots\phi^{m-1}(a_m).$$

The algebra B is called the *twist of A by the automorphism ϕ*.

Twisting a $\mathbb{Z}$-graded $\Bbbk$-algebra by an automorphism is reflexive and symmetric, but usually it is not transitive. To remedy this, twisting a graded algebra by a *twisting system* (see Definitions 1.2 and 1.5) was introduced by J. J. Zhang in [**13**]. Although Zhang's definition was for algebras graded by a semigroup, we state his definitions in this section for $\mathbb{Z}$-graded $\Bbbk$-algebras A that are generated by A_1, since our focus herein will be quadratic algebras.

DEFINITION 1.2 ([**13**, Definition 2.1]). Let $A = \bigoplus_{j\in\mathbb{Z}} A_j$ be a $\mathbb{Z}$-graded $\Bbbk$-algebra. A set $\{t_n\}_{n\in\mathbb{Z}}$ of degree-0, graded, $\Bbbk$-linear bijections of A is called a *twisting system* of A if

$$t_n(at_m(b)) = t_n(a)t_{n+m}(b),$$

or, equivalently, if

$$t_n(ab) = t_n(a)t_{n+m}t_m^{-1}(b),$$

for all $a \in A_m$, $b \in A_i$, m, i, $n \in \mathbb{Z}$.

REMARK 1.3 ([**13**, Page 284]). If A is a $\mathbb{Z}$-graded $\Bbbk$-algebra and if ϕ is a degree-0, graded automorphism of A, then $\{\phi^n\}_{n\in\mathbb{Z}}$ is a twisting system of A.

DEFINITION 1.4. Considering Remark 1.3, we refer to a twisting system that is not given by an automorphism as being *nontrivial*.

DEFINITION 1.5 ([**13**, Definition 2.3]). Let A and $\{t_n\}_{n\in\mathbb{Z}}$ be defined as in Definition 1.2, and suppose A is generated by A_1. We form a new graded algebra B from these data as follows:

(a) as vector spaces over $\Bbbk$, $B = A$, and
(b) we define a multiplication $*$ on B by

$$a_1 * a_2 * \cdots * a_m = a_1t_1(a_2)t_2(a_3)\cdots t_{m-1}(a_m),$$

for all $a_1, \dots, a_m \in A_1$, $m \in \mathbb{N}$, and use the distributive property to extend $*$ to the rest of B.

In [**13**], the algebra B is called the *twist of A by the twisting system* $\{t_n\}_{n\in\mathbb{Z}}$, but some authors refer to such a twist as a *Zhang twist of A* (cf. [**10**]).

We remind the reader that Definition 1.5 is a special case of Zhang's definition in [**13**], in which the algebra is graded using a semigroup and $\Bbbk$ is not necessarily a field and the algebra is not necessarily generated by elements of the same degree. In [**13**], Zhang proved that the multiplication $*$ induced on a twisted algebra by a twisting system is well defined and associative.

By Remark 1.3, a twist of a $\mathbb{Z}$-graded $\Bbbk$-algebra A by a degree-0 graded automorphism is a special case of twisting by a twisting system.

Moreover, by [**13**, Proposition 2.5], twisting an algebra by a twisting system is an equivalence relation. Furthermore, by [**13**, Proposition 5.7], if A is a connected, finitely generated, $\mathbb{Z}$-graded $\Bbbk$-algebra, then various properties of A are invariant under twisting A by a twisting system, such as being a domain or being noetherian, and the values of GKdim(A), Kdim(A) and gldim(A) are preserved by twisting. Additionally, by [**13**, Theorem 5.11], if A is also noetherian, then the property of being AS-regular is invariant under twisting. In fact, the same result proves that if A is also noetherian, Auslander-regular and satisfies the Cohen-Macaulay property, then these properties are invariant under twisting.

By the transitivity property of twisting systems, twisting a graded $\Bbbk$-algebra A by an automorphism and twisting the resulting algebra by an automorphism yields a twist of A by a (potentially nontrivial) twisting system. Nevertheless, very few examples of nontrivial twisting systems are provided in the literature. Among these few examples are those constructed in [**9**, **12**] on certain quadratic algebras. We next remind the reader of the twisting system constructed in [**12**]; the construction in [**9**] used a similar method to that in [**12**].

EXAMPLE 1.6 ([**12**, Section 4]). In this example, let A denote an algebra from [**12**, Section 4]. Such an algebra is quadratic and satisfies $A = T(V)/\langle W\rangle = \bigoplus_{j=0}^{\infty} A_j$, where $T(V)$ is the tensor algebra of a vector space V of dimension four and W is a subspace of $V \otimes V$ of dimension six. In the setting of [**12**, Section 4], there exist invertible linear maps $\tau : A_1 \to A_1$ and $t : A_1 \to A_1$ that induce automorphisms of $T(V)$, but do not induce automorphisms of A. Nevertheless, $\tau^2 \in \operatorname{Aut}(A)$ and t^2 is the identity map on A. Let t_0 denote the identity map on A and let $t_1 : A \to A$ denote the linear map given by $t_1|_{\Bbbk} = t_0|_{\Bbbk}$ and

$$t_1(a_1 \cdots a_m) = ((\tau^{-1}t)(a_1))((\tau^{-2}t\tau)(a_2)) \cdots ((\tau^{-m}t\tau^{m-1})(a_m)),$$

where $m \in \mathbb{N}$ and $a_i \in A_1$ for all $i = 1, \ldots, m$. It is proved in [**12**], by using certain geometric data specific to the algebra A, that the set

$$\{\tau^{r-n}t_r : n \in \mathbb{Z},\ r \in \{0, 1\},\ n \equiv r \pmod 2\}$$

is a twisting system of A.

The purpose of the next section is to generalize the type of twisting system occurring in Example 1.6.

2. Twisting systems of type $\mathbb{T}_\delta$

In this section, we generalize the twisting-system construction discussed in Example 1.6. Explicitly, in Definition 2.8, we introduce the notion of a twisting system of type $\mathbb{T}_\delta$, where $\delta \in \mathbb{N}$. However, unlike the methods in [**9**, **12**], our method does not entail the use of any geometric data. The main objective of this section is to prove Theorem 2.7; namely, that, under certain conditions, the twisting system of type $\mathbb{T}_\delta$ is well defined on any finitely generated quadratic algebra. We prove, in Corollary 2.10, that any twisting system of type $\mathbb{T}_\delta$ is also of type $\mathbb{T}_{k\delta}$, for any $k \in \mathbb{N}$.

In this section, let $T(V)$ denote the tensor algebra of a finite-dimensional vector space V and give $T(V)$ the standard grading; that is, $T(V)_1 = V$. Let $A = T(V)/\langle W\rangle = \bigoplus_{j=0}^{\infty} A_j$, where W is a finite-dimensional subspace of $V \otimes V$. It follows that A is a quadratic algebra, where $A_1 = V$.

DEFINITION 2.1. With the above notation, let t and τ be $\Bbbk$-linear bijections on V, and extend them to be graded automorphisms of degree zero on $T(V)$. For $n \in \mathbb{Z}$, we define maps $t_n : T(V) \to T(V)$ by $t_n(\alpha) = \alpha$ for all $\alpha \in T(V)_0$ and

$$t_n(a_1 \cdots a_m) = \prod_{j=1}^{m} \tau^{1-j-n}t^n\tau^{j-1}(a_j),$$

where $m \in \mathbb{N}$, $a_j \in T(V)_1$ for $j = 1, \ldots, m$, and extend t_n to $T(V)$ by linearity.

Clearly, t_0 is the identity map on $T(V)$ and each t_n is a graded $\Bbbk$-linear bijection on $T(V)$.

The goal of this section is to prove that, under certain hypotheses, for each $n \in \mathbb{Z}$, t_n induces a well-defined graded $\Bbbk$-linear bijection on A, and that $\{t_n\}_{n\in\mathbb{Z}}$ induces a twisting system of A. With a slight abuse of notation, we refer to such an induced map on A as t_n.

LEMMA 2.2. *Suppose that t_n is defined as in Definition 2.1 and that, for all $n \in \mathbb{Z}$, t_n induces a well-defined graded $\Bbbk$-linear bijection of degree zero on A.*

(a) *If $t\tau = \tau t$ on V, then $t_n = t_1^n \in \mathrm{Aut}(A)$ for all $n \in \mathbb{Z}$.*
(b) *If $t_1 \in \mathrm{Aut}(A)$, then $t\tau = \tau t$ on V.*

PROOF. (a) Suppose $t\tau = \tau t$ on V. It follows that

$$t_n(a_1 \cdots a_m) = \prod_{j=1}^{m} \tau^{-n} t^n(a_j) = \prod_{j=1}^{m} (\tau^{-1}t)^n(a_j)$$

where $m \in \mathbb{N}$, $a_j \in T(V)_1$ for $j = 1, \ldots, m$. Hence, $t_1 \in \mathrm{Aut}(A)$ and $t_n = t_1^n$ for all $n \in \mathbb{Z}$.

(b) If $t_1 \in \mathrm{Aut}(A)$, then

$$t_1(a_1)t_1(a_2) = t_1(a_1a_2) = \tau^{-1}t(a_1)\tau^{-2}t\tau(a_2)$$

for all a_1, $a_2 \in A_1$, which implies that $\tau^{-2}t\tau = \tau^{-1}t$ on V. Thus, (b) follows. □

REMARK 2.3. By Remark 1.3 and Lemma 2.2, if we are to use the maps t_n from Definition 2.1 to build nontrivial twisting systems of A, then we require $t\tau \neq \tau t$ on V.

LEMMA 2.4. *Suppose t_n is defined as in Definition 2.1 for all $n \in \mathbb{Z}$. If $a \in T(V)_m$ and $b \in T(V)_i$, where m, $i \in \mathbb{N}_0$, then*

(a) *$t_n(ab) = t_n(a)t_{n+m}t_m^{-1}(b)$, for all $n \in \mathbb{Z}$, so the set $\{t_n\}_{n\in\mathbb{Z}}$ is a twisting system of $T(V)$, and*
(b) *$t_n(T(V)_m b) \subset T(V)_m t_{n+m}t_m^{-1}(b)$, for all $n \in \mathbb{Z}$.*

PROOF. Part (b) follows from part (a), so it suffices to prove part (a) only. By linearity of t_n, it suffices to consider $a = a_1a_2 \cdots a_m$ and $b = b_1b_2 \cdots b_i$, where a_j, $b_j \in T(V)_1$, for all j. In particular, for all $n \in \mathbb{Z}$, we have

$$\begin{aligned} t_n(a)t_{n+m}t_m^{-1}(b) &= t_n(a) \prod_{j=1}^{i} \tau^{1-j-(n+m)}t^{n+m}\tau^{j-1}\tau^{1-j}t^{-m}\tau^{m+j-1}(b_j) \\ &= \prod_{k=1}^{m} \tau^{1-k-n}t^n\tau^{k-1}(a_k) \prod_{j=1}^{i} \tau^{1-(m+j)-n}t^n\tau^{m+j-1}(b_j) \\ &= \prod_{k=1}^{m+i} \tau^{1-k-n}t^n\tau^{k-1}(c_k), \end{aligned}$$

where

$$c_k = \begin{cases} a_k & \text{for } 1 \leq k \leq m, \\ b_{k-m} & \text{for } m+1 \leq k \leq m+i, \end{cases}$$

It follows that $t_n(a)t_{n+m}t_m^{-1}(b) = t_n(ab)$ for all $n \in \mathbb{Z}$, which completes the proof. □

For $\delta \in \mathbb{N}$, we write $\mathbb{Z}_\delta = \{0, 1, 2, \ldots, \delta - 1\}$.

LEMMA 2.5. *Suppose t_n is defined as in Definition 2.1 for all $n \in \mathbb{Z}$. Let $\delta \in \mathbb{N}$ and suppose that t^δ is the identity map on V. If $n \equiv r \in \mathbb{Z}_\delta \pmod{\delta}$, then $t_n = \tau^{r-n} t_r$ on $T(V)$.*

PROOF. Suppose $n \in \mathbb{Z}$ and write $n \equiv r \in \mathbb{Z}_\delta \pmod{\delta}$. In particular, $t^n = t^r$, as t^δ is the identity map. Since $\tau \in \mathrm{Aut}(T(V))$, it follows that

$$\begin{aligned}
t_n(a_1 \cdots a_m) &= \prod_{j=1}^{m} \tau^{1-j-n} t^n \tau^{j-1}(a_j) \\
&= \prod_{j=1}^{m} \tau^{r-n} \tau^{1-j-r} t^r \tau^{j-1}(a_j) \\
&= \tau^{r-n} \prod_{j=1}^{m} \tau^{1-j-r} t^r \tau^{j-1}(a_j) \\
&= \tau^{r-n} t_r(a_1 \cdots a_m),
\end{aligned}$$

where $m \in \mathbb{N}$ and $a_j \in T(V)_1$ for $j = 1, \ldots, m$. The result follows by linearity. □

PROPOSITION 2.6. *Suppose t_n is defined as in Definition 2.1 for all $n \in \mathbb{Z}$. Let $\delta \in \mathbb{N}$ and suppose that t^δ is the identity map on V and that $\tau^\delta \in \mathrm{Aut}(A)$. If $t_r(W) \subset W$ for all $r \in \mathbb{Z}_\delta$, then t_n is defined on A for all $n \in \mathbb{Z}$.*

PROOF. We need to prove that $t_n(\langle W \rangle) \subset \langle W \rangle$ for all $n \in \mathbb{Z}$. Let $m, i \in \mathbb{N}_0$. By Lemma 2.4, we have

$$(*) \qquad t_n(T(V)_m W T(V)_i) \subset T(V)_m (t_{n+m} t_m^{-1}(W)) T(V)_i,$$

for all $n \in \mathbb{Z}$. By Lemma 2.5, $t_n = \tau^{r-n} t_r$ on $T(V)$, where $n \equiv r \in \mathbb{Z}_\delta \pmod{\delta}$. Writing $m \equiv \mu \in \mathbb{Z}_\delta \pmod{\delta}$ and $n + m \equiv \nu \in \mathbb{Z}_\delta \pmod{\delta}$, we have

$$(\dagger) \qquad t_{n+m} t_m^{-1} = \tau^{\nu-(n+m)} t_\nu t_\mu^{-1} \tau^{m-\mu}.$$

By assumption, $\tau(W) \subset W$ and $t_r(W) \subset W$ for all $r \in \mathbb{Z}_\delta$. Moreover, $t_\mu^{-1}(W) \subset W$ (since $\dim(W) < \infty$). Thus, $(\dagger)$ implies that $t_{n+m} t_m^{-1}(W) \subset W$, so the result follows from $(*)$. □

THEOREM 2.7. *Suppose t_n is defined as in Definition 2.1 for all $n \in \mathbb{Z}$. Let $\delta \in \mathbb{N}$ and suppose that t^δ is the identity map on V and that $\tau^\delta \in \mathrm{Aut}(A)$. If $t_r(W) \subset W$ for all $r \in \mathbb{Z}_\delta$, then $\{t_n\}_{n \in \mathbb{Z}}$ is a twisting system of A.*

PROOF. By Proposition 2.6, t_n is defined on A for all $n \in \mathbb{Z}$. Therefore, the result follows from Lemma 2.4(a). □

DEFINITION 2.8. We say that the twisting system given by Theorem 2.7 is of *type* $\mathbb{T}_\delta$.

With this terminology, the twisting systems constructed in [**9**,**12**] are of type $\mathbb{T}_2$. We present a new example in the next section of a twisting system of type $\mathbb{T}_2$.

REMARK 2.9. In Definition 2.8, if $\{t_n\}_{n\in\mathbb{Z}}$ is a twisting system of A of type $\mathbb{T}_1$, then t is the identity map on A and $\tau \in \mathrm{Aut}(A)$. In particular, in this case, Remark 1.3 and Lemma 2.2 apply, so the twisting system $\{t_n\}_{n\in\mathbb{Z}}$ is $\{t_1^n\}_{n\in\mathbb{Z}}$, where $t_1 = \tau^{-1} \in \mathrm{Aut}(A)$. Hence, a twist of A by this twisting system is, in fact, a twist of A by an automorphism.

COROLLARY 2.10. *Let t_n be defined as in Definition* 2.1 *for all $n \in \mathbb{Z}$. If $\delta \in \mathbb{N}$, then the following conditions are equivalent:*

(a) *$\{t_n\}_{n\in\mathbb{Z}}$ is a twisting system of A of type $\mathbb{T}_\delta$,*
(b) *$\{t_n\}_{n\in\mathbb{Z}}$ is a twisting system of A of type $\mathbb{T}_{k\delta}$, for all $k \in \mathbb{N}$,*
(c) *$\{t_n\}_{n\in\mathbb{Z}}$ is a twisting system of A of type $\mathbb{T}_{\mu\delta}$, for some $\mu \in \mathbb{N}$, t^δ is the identity map on V and $\tau^\delta \in \mathrm{Aut}(A)$,*
(d) *$\{t_n\}_{n\in\mathbb{Z}}$ is a twisting system of A of type $\mathbb{T}_\alpha$, for some $\alpha \in \mathbb{N}$, $\alpha \geq \delta$, t^δ is the identity map on V and $\tau^\delta \in \mathrm{Aut}(A)$,*
(e) *$\{t_n\}_{n\in\mathbb{Z}}$ is a twisting system of A simultaneously of types $\mathbb{T}_{\alpha_1}, \ldots, \mathbb{T}_{\alpha_m}$, where $\alpha_1, \ldots, \alpha_m \in \mathbb{N}$, $m \in \mathbb{N}$, and δ is the greatest common divisor of $\alpha_1, \ldots, \alpha_m$.*

PROOF. If (b) holds, then (a), (c) and (e) hold. Moreover, if (c) holds, then (d) holds.

Suppose that (a) holds. We will prove that (b) holds. Assume that $\{t_n\}_{n\in\mathbb{Z}}$ is a twisting system of A of type $\mathbb{T}_\delta$. In particular, t^δ is the identity map on V and $\tau^\delta(W) \subset W$ and $t_s(W) \subset W$ for all $s \in \mathbb{Z}_\delta$. Let $k \in \mathbb{N}$ and write $\beta = k\delta$. Hence, t^β is the identity map on V and $\tau^\beta(W) \subset W$. For $r \in \mathbb{Z}_\beta$, write $r \equiv s \in \mathbb{Z}_\delta \pmod{\delta}$. By Lemma 2.5, it follows that $t_r = \tau^{s-r} t_s$, so $t_r(W) \subset W$ for all $r \in \mathbb{Z}_\beta$. By Theorem 2.7, $\{t_n\}_{n\in\mathbb{Z}}$ is a twisting system of A of type $\mathbb{T}_\beta$, so (b) holds.

We now prove that if (d) holds then (a) holds. Assume (d) holds, so t^δ is the identity map on V and $\tau^\delta \in \mathrm{Aut}(A)$. As sets, we may view $\mathbb{Z}_\delta \subset \mathbb{Z}_\alpha$, so the fact that $t_r(W) \subset W$ for all $r \in \mathbb{Z}_\alpha$ implies that $t_s(W) \subset W$ for all $s \in \mathbb{Z}_\delta$. Hence, Theorem 2.7 implies that (a) holds.

At this stage, it suffices to prove that if (e) holds, then (a) holds. Assume that (e) holds. If $m = 1$, then the result is immediate. Suppose $m \geq 2$. Since δ is the greatest common divisor of $\alpha_1, \ldots, \alpha_m$, there exist $k_1, \ldots, k_m \in \mathbb{Z}$ such that $\delta = k_1\alpha_1 + \cdots + k_m\alpha_m$. It follows that $t^\delta = t^{k_1\alpha_1} \cdots t^{k_m\alpha_m}$ is the identity map on V, since t^{α_i} is the identity map on V for all i. Similarly, $\tau^\delta \in \mathrm{Aut}(A)$. Moreover, we may view $\mathbb{Z}_\delta \subset \mathbb{Z}_{\alpha_i}$, as sets, for all i. Thus, since $t_r(W) \subset W$ for all $r \in \mathbb{Z}_{\alpha_i}$, it follows that $t_s(W) \subset W$ for all $s \in \mathbb{Z}_\delta$. Consequently, Theorem 2.7 implies that (a) holds, which completes the proof. □

By Corollary 2.10, if $\{t_n\}_{n\in\mathbb{Z}}$ is a twisting system of A of types $\mathbb{T}_\alpha$ and $\mathbb{T}_\beta$ simultaneously, where α, $\beta \in \mathbb{N}$ are relatively prime, then $\{t_n\}_{n\in\mathbb{Z}}$ is a twisting system of A of type $\mathbb{T}_1$. Hence, in this case, by Remark 2.9, a twist of A by $\{t_n\}_{n\in\mathbb{Z}}$ is a twist by an automorphism.

3. A quantum $\mathbb{P}^3$ whose point scheme is a rank-2 quadric

In this section, we illustrate our method from the previous section for producing a nontrivial twisting system on an algebra that is a quadratic AS-regular algebra of global dimension four (sometimes called a quantum $\mathbb{P}^3$) whose point scheme is isomorphic to a rank-2 quadric in $\mathbb{P}^3$. To our knowledge, this algebra is the first quantum $\mathbb{P}^3$ known to have its point scheme be isomorphic to a rank-2 quadric in $\mathbb{P}^3$.

Our work in this section is in three parts. In Section 3.1, we introduce the algebra and the twisting system, and prove in Theorem 3.1 that the twisting system is defined on the algebra and is of type $\mathbb{T}_2$. In Section 3.2, we prove in Theorem 3.3 that the algebra is a quantum $\mathbb{P}^3$ and satisfies many properties that are satisfied by polynomial rings. In the final subsection, we prove in Theorem 3.5 that the point scheme of the algebra is isomorphic to a rank-2 quadric in $\mathbb{P}^3$.

Throughout this section, we impose the additional assumption that $\text{char}(\Bbbk) \neq 2$. Moreover, for Theorem 3.1, we also require $\text{char}(\Bbbk) \neq 3$.

3.1. The algebra $A(q)$ and the twisting system. Let $q \in \Bbbk^\times$ and let $A(q)$ denote the $\Bbbk$-algebra $T(V)/\langle W\rangle$ on four generators $x_1, \dots, x_4$ where $V = \Bbbk x_1 \oplus \cdots \oplus \Bbbk x_4$ and $W = \Bbbk\omega_1 \oplus \cdots \oplus \Bbbk\omega_6$, with

$$\omega_1 = q^2x_1x_2 + x_2x_1, \qquad \omega_2 = qx_1x_3 - x_3x_1, \qquad \omega_3 = qx_1x_4 + x_4x_1,$$

$$\omega_4 = x_2x_3 - qx_3x_2, \qquad \omega_5 = x_2x_4 + qx_4x_2, \qquad \omega_6 = x_3x_4 + x_4x_3 + x_1x_2.$$

Let t and τ be the graded $\Bbbk$-linear bijections of degree zero on $T(V)$ given by the matrices $[t]$ and $[\tau]$ respectively, where

$$(**) \qquad [t] = \begin{bmatrix} 1 & 0 & 0 & 0 \\ 0 & 1 & 0 & 0 \\ 0 & 0 & 1 & 0 \\ 0 & 0 & 1 & -1 \end{bmatrix} \qquad \text{and} \qquad [\tau] = \begin{bmatrix} 1 & 0 & 0 & 0 \\ 0 & 4 & 0 & 0 \\ 0 & 0 & 2 & 0 \\ 0 & 0 & 3 & -2 \end{bmatrix}$$

and we use row i to indicate the image of x_i for all i (e.g., $[t](x_4) = x_3 - x_4$). (The matrices act in the usual manner on the dual V^* of V.)

THEOREM 3.1. *Suppose* $\text{char}(\Bbbk) \notin \{2, 3\}$. *Using the maps t and τ defined in* $(**)$, *the maps t_n from Definition* 2.1 *yield a nontrivial twisting system* $\{t_n\}_{n\in\mathbb{Z}}$ *of $A(q)$ of type* $\mathbb{T}_2$.

PROOF. Clearly t and τ are invertible linear bijections on V since their matrices are invertible as $\text{char}(\Bbbk) \neq 2$. A simple calculation shows that t^2 is the identity map on V. Since $\text{char}(\Bbbk) \notin \{2, 3\}$, we have that $t\tau \neq \tau t$ and $t(\omega_3) \notin W$ and $\tau(\omega_3) \notin W$. On the other hand, another simple calculation using $\text{char}(\Bbbk) \neq 2$ reveals that $\tau^2(\omega_i) \in \Bbbk^\times\omega_i$ for all i, so $\tau^2 \in \text{Aut}(A(q))$ for all $q \in \Bbbk^\times$.

Since $t_1(x_ix_j) = \tau^{-1}t(x_i)\tau^{-2}t\tau(x_j)$ for all i, j, a computation shows that

$$t_1(\omega_1) = \tfrac{1}{4}\omega_1, \qquad t_1(\omega_2) = \tfrac{1}{2}\omega_2, \qquad t_1(\omega_3) = \tfrac{1}{4}(\omega_2 + 2\omega_3),$$

$$t_1(\omega_4) = \tfrac{1}{8}\omega_4, \qquad t_1(\omega_5) = \tfrac{1}{16}(\omega_4 + 2\omega_5), \qquad t_1(\omega_6) = \tfrac{1}{4}\omega_6,$$

so $t_1(W) \subset W$. The result follows from Remark 2.3 and Theorem 2.7. □

3.2. Regularity of the algebra $A(q)$. In this subsection, we prove that the algebra $A(q)$ introduced in Section 3.1 is an AS-regular algebra of global dimension four. Such an algebra is often called a quantum $\mathbb{P}^3$. In fact, we will prove that $A(q)$ shares many desirable properties with the polynomial ring on four variables, such as having the same Hilbert series, satisfying the Cohen-Macaulay property, and being a noetherian, Auslander-regular domain. Our method is to show first that, for all $q \in \Bbbk^\times$, $A(q)$ is an Ore extension of a well-behaved subring. (In [**11**], an alternative approach using the Koszul dual of $A(q)$ is used.)

PROPOSITION 3.2. *Let $S(q)$ denote the $\Bbbk$-algebra on generators x_1, x_2, x_4 with defining relations given by ω_1, ω_3, ω_5 from the start of Section 3.1. For all $q \in \Bbbk^\times$, $A(q)$ is an Ore extension of $S(q)$; more precisely, $A(q) = (S(q))[x_3; \sigma, d]$, where $x_3 s = \sigma(s)x_3 + d(s)$ for all $s \in S(q)$, and $\sigma \in \mathrm{Aut}(S(q))$ is given by*

$$\sigma(x_1) = qx_1, \quad \sigma(x_2) = q^{-1}x_2 \quad \text{and} \quad \sigma(x_4) = -x_4,$$

and d is the σ-derivation of $S(q)$ given by

$$d(x_1) = 0, \quad d(x_2) = 0 \quad \text{and} \quad d(x_4) = -x_1x_2.$$

PROOF. Clearly, the map σ in the statement is an automorphism of $S(q)$. Using the formula $d(ab) = \sigma(a)d(b) + d(a)b$, for all $a, b \in S(q)$, we have immediately that $d(\omega_1) = 0$. Similarly, $d(\omega_3) = q^2x_1(-x_1x_2) - x_1x_2x_1$, which is zero in $S(q)$. Moreover, $d(\omega_5) = q^{-1}x_2(-x_1x_2) - qx_1x_2^2$, which, in $S(q)$, yields $d(\omega_5) = -q^{-1}(-q^2x_1x_2)x_2 - qx_1x_2^2 = 0$. The result follows. □

THEOREM 3.3. *For all $q \in \Bbbk^\times$, the algebra $A(q)$ is a noetherian, Auslander-regular, AS-regular, Koszul algebra of global dimension four that is a domain and satisfies the Cohen-Macaulay property, and has Hilbert series $H(t) = (1-t)^{-4}$.*

PROOF. The result is a consequence of $A(q)$ being the Ore extension described in Proposition 3.2. In particular, the claim that $A(q)$ is a noetherian domain of global dimension four can be obtained from [**7**, Theorem 1.2.9 and Corollary 7.9.18]. The Hilbert series of $S(q)$ is the same as that of the polynomial ring on three variables, and $A(q)$ is a free module over $S(q)$ with basis $\{1, x_3, x_3^2, \ldots\}$, so the claim regarding Hilbert series follows. By the lemma in [**6**], $A(q)$ is Auslander-regular and satisfies the Cohen-Macaulay property. It follows that $A(q)$ is AS-regular by applying [**5**, Theorem 6.3]. Lastly, from [**8**, Theorem 2.2], we have that $A(q)$ is a Koszul algebra. □

3.3. The point scheme of the algebra $A(q)$. In this subsection, we show that the point scheme of $A(q)$ is isomorphic to a rank-2 quadric in $\mathbb{P}^3$. By [**2**, **13**], the point scheme of any quadratic algebra is invariant under isomorphism and twisting. Hence, in this subsection, we simplify our computations by working with a certain quadratic algebra $\mathcal{A}$ that is isomorphic to a twist of $A(q)$ by an automorphism.

LEMMA 3.4. *For all $q \in \Bbbk^\times$, the $\Bbbk$-algebra $\mathcal{A}$ on four generators, $x_1, \ldots, x_4$, with defining relations*

$$\begin{aligned} x_2x_1 &= x_1x_2, & x_3x_1 &= x_1x_3, & x_4x_1 &= x_1x_4, \\ x_3x_2 &= -x_2x_3, & x_4x_2 &= -x_2x_4, & x_4x_3 &= x_3x_4 + x_1x_2, \end{aligned}$$

is isomorphic to a twist of $A(q)$, and so has point scheme isomorphic to the point scheme of $A(q)$.

PROOF. The last statement is a consequence of [**2**, Section 8] and [**13**]. Twisting $A(q)$ by the map $\theta \in \mathrm{Aut}(A(q))$ given by

$$\theta(x_1) = x_1, \qquad \theta(x_2) = -q^2x_2, \qquad \theta(x_3) = qx_3, \qquad \theta(x_4) = -qx_4,$$

yields the $\Bbbk$-algebra $B(q)$ on four generators $x_1, \ldots, x_4$, with defining relations

$$\begin{aligned} x_2x_1 &= x_1x_2, & x_3x_1 &= x_1x_3, & x_4x_1 &= x_1x_4, \\ x_3x_2 &= -x_2x_3, & x_4x_2 &= -x_2x_4, & x_4x_3 &= x_3x_4 + q^{-1}x_1x_2. \end{aligned}$$

The map that sends $x_1 \mapsto qx_1$ and fixes x_i, for $2 \le i \le 4$, is an isomorphism from $B(q)$ to the algebra $\mathcal{A}$ in the statement, which completes the proof. □

By [**1**], the standard procedure for finding a subscheme of $\mathbb{P}(V^*)$ that is isomorphic to the point scheme of a quadratic factor algebra of $T(V)$ entails writing the defining relations of the factor algebra in matrix form, where the coefficient matrix M has entries that are elements of V, and computing the zero locus in $\mathbb{P}(V^*)$ of the maximal minors of M. Although, there are at least two ways to write the defining relations of a quadratic algebra in matrix form, where M has entries that are elements of V, the presentations yield the same subscheme of $\mathbb{P}(V^*)$ if the algebra satisfies the properties in Theorem 3.3.

For the algebra $\mathcal{A}$ in Lemma 3.4, writing the defining relations of $\mathcal{A}$ in the form Mx, where M is a 6×4 matrix and x is the transpose of the 1×4 matrix $\begin{bmatrix} x_1 & x_2 & x_3 & x_4 \end{bmatrix}$, yields

$$M = \begin{bmatrix} -x_2 & x_1 & 0 & 0 \\ -x_3 & 0 & x_1 & 0 \\ -x_4 & 0 & 0 & x_1 \\ 0 & x_3 & x_2 & 0 \\ 0 & x_4 & 0 & x_2 \\ 0 & x_1 & -x_4 & x_3 \end{bmatrix}.$$

Since $\operatorname{char}(\Bbbk) \neq 2$, the following ten polynomials determine the zero locus of the fifteen 4×4 minors of M:

$$\begin{array}{ccccc} x_1^2x_2^2, & x_1^3x_2, & x_1x_2^3, & x_1^2x_2x_3, & x_1^2x_2x_4, \\ x_1x_2^2x_3, & x_1x_2^2x_4, & x_1x_2x_3^2, & x_1x_2x_4^2, & x_1x_2x_3x_4. \end{array}$$

It follows that the 4×4 minors of M generate the ideal

$$\langle x_1, x_2, x_3, x_4\rangle^2\langle x_1x_2\rangle,$$

which has saturation $\langle x_1x_2\rangle$.

THEOREM 3.5. *Suppose* $\operatorname{char}(\Bbbk) \neq 2$. *For all* $q \in \Bbbk^\times$, *the point scheme of the algebra* $A(q)$ *is isomorphic to a rank-2 quadric in* $\mathbb{P}^3$.

PROOF. By Lemma 3.4, for all $q \in \Bbbk^\times$, the point scheme of $A(q)$ is isomorphic to the point scheme of the algebra $\mathcal{A}$. The preceding discussion proves that $\mathcal{A}$ has point scheme isomorphic to the zero locus in $\mathbb{P}(V^*)$ of the ideal generated by the polynomial x_1x_2, which is a rank-2 quadratic form. Hence, the zero locus is a rank-2 quadric, and is the union of two distinct planes in $\mathbb{P}(V^*)$. Since $\dim(V) = 4$, one can identify $\mathbb{P}(V^*)$ with $\mathbb{P}^3$. □

Moreover, if $\operatorname{char}(\Bbbk) \neq 2$, then the zero locus $\mathcal{Z}$ in $\mathbb{P}(V^*) \times \mathbb{P}(V^*)$ of the defining relations of $\mathcal{A}$ is the set

$$\mathcal{Z} = \{((\alpha_1, \alpha_2, \alpha_3, \alpha_4), (\alpha_1, -\alpha_2, \alpha_3, \alpha_4)) \in \mathbb{P}(V^*) \times \mathbb{P}(V^*) : \alpha_1\alpha_2 = 0\},$$

and so the element $x_1 \otimes x_2 \in T(V)_2$ vanishes on $\mathcal{Z}$, even though $x_1 \otimes x_2$ does not belong to the ideal of defining relations of $\mathcal{A}$. This means that the zero locus of the defining relations of $\mathcal{A}$ does not determine $\mathcal{A}$.

COROLLARY 3.6. *Suppose* $\operatorname{char}(\Bbbk) \neq 2$. *For all* $q \in \Bbbk^\times$, *the zero locus in* $\mathbb{P}(V^*) \times \mathbb{P}(V^*)$ *of the defining relations of* $A(q)$ *does not determine* $A(q)$.

PROOF. The preceding discussion shows that the statement holds for the algebra $\mathcal{A}$. Since $\mathcal{A}$ is isomorphic to a twist of $A(q)$ by an automorphism, the result follows by [**2**, Section 8]. □

The situation in Corollary 3.6 is analogous to the situation that occurs for the algebras in the last section of each of [**9**] and [**12**] on which twisting systems were constructed. In those cases, the point scheme of the respective algebra is isomorphic to a rank-3 quadric (in [**9**]), or a rank-4 quadric (in [**12**]), and the zero locus in $\mathbb{P}(V^*) \times \mathbb{P}(V^*)$ of the defining relations also does not determine the algebra.

The results in [**9,12**] classify the quadratic quantum $\mathbb{P}^3$s whose point scheme is isomorphic to a rank-3, or rank-4, quadric in $\mathbb{P}^3$. The classification of the quadratic quantum $\mathbb{P}^3$s whose point scheme is isomorphic to a rank-2 quadric in $\mathbb{P}^3$ remains an open problem. The authors currently do not know of the existence of any quantum $\mathbb{P}^3$ whose point scheme is isomorphic to a rank-1 quadric in $\mathbb{P}^3$.

Acknowledgment

The authors warmly thank the organizers of the conference *Recent Advances and New Directions in the Interplay of Noncommutative Algebra and Geometry*, held in Seattle, Washington, USA, in June 2022, for the opportunity to display a poster presenting the results of this article.

References

[1] M. Artin, J. Tate, and M. Van den Bergh, *Some algebras associated to automorphisms of elliptic curves*, The Grothendieck Festschrift, Vol. I, Progr. Math., vol. 86, Birkhäuser Boston, Boston, MA, 1990, pp. 33–85, DOI 10.1007/978-0-8176-4574-8_3. MR1086882

[2] M. Artin, J. Tate, and M. Van den Bergh, *Modules over regular algebras of dimension* 3, Invent. Math. **106** (1991), no. 2, 335–388, DOI 10.1007/BF01243916. MR1128218

[3] Thomas Cassidy and Michaela Vancliff, *Generalizations of graded Clifford algebras and of complete intersections*, J. Lond. Math. Soc. (2) **81** (2010), no. 1, 91–112, DOI 10.1112/jlms/jdp057. (Corrigendum: 90 No. 2 (2014), 631-636. MR3263968). MR2580455

[4] Lieven Le Bruyn, S. P. Smith, and Michel Van den Bergh, *Central extensions of three-dimensional Artin-Schelter regular algebras*, Math. Z. **222** (1996), no. 2, 171–212, DOI 10.1007/PL00004532. MR1429334

[5] Thierry Levasseur, *Some properties of noncommutative regular graded rings*, Glasgow Math. J. **34** (1992), no. 3, 277–300, DOI 10.1017/S0017089500008843. MR1181768

[6] T. Levasseur and J. T. Stafford, *The quantum coordinate ring of the special linear group*, J. Pure Appl. Algebra **86** (1993), no. 2, 181–186, DOI 10.1016/0022-4049(93)90102-Y. MR1215645

[7] J. C. McConnell and J. C. Robson, *Noncommutative Noetherian rings*, Revised edition, Graduate Studies in Mathematics, vol. 30, American Mathematical Society, Providence, RI, 2001. With the cooperation of L. W. Small, DOI 10.1090/gsm/030. MR1811901

[8] Brad Shelton and Craig Tingey, *On Koszul algebras and a new construction of Artin-Schelter regular algebras*, J. Algebra **241** (2001), no. 2, 789–798, DOI 10.1006/jabr.2001.8781. MR1843325

[9] Brad Shelton and Michaela Vancliff, *Embedding a quantum rank three quadric in a quantum* $\mathbf{P}^3$, Comm. Algebra **27** (1999), no. 6, 2877–2904, DOI 10.1080/00927879908826599. MR1687269

[10] Susan J. Sierra, *G-algebras, twistings, and equivalences of graded categories*, Algebr. Represent. Theory **14** (2011), no. 2, 377–390, DOI 10.1007/s10468-009-9193-y. MR2776790

[11] H. V. Tran, *Twisting systems and some quantum* $\mathbb{P}^3$*s with point scheme a rank-2 quadric*, Ph.D. Thesis, Univ. of Texas at Arlington, `https://rc.library.uta.edu/uta-ir/handle/10106/30854` 2022.

[12] Michaela Vancliff and Kristel Van Rompay, *Embedding a quantum nonsingular quadric in a quantum* $\mathbf{P}^3$, J. Algebra **195** (1997), no. 1, 93–129, DOI 10.1006/jabr.1997.7077. MR1468885

[13] J. J. Zhang, *Twisted graded algebras and equivalences of graded categories*, Proc. London Math. Soc. (3) **72** (1996), no. 2, 281–311, DOI 10.1112/plms/s3-72.2.281. MR1367080

Department of Mathematics, University of Texas at Arlington, Arlington, Texas 76019-0408

Email address: hung.tran2@mavs.uta.edu

Department of Mathematics, University of Texas at Arlington, Arlington, Texas 76019-0408

Email address: vancliff@uta.edu

Contemporary Mathematics
Volume **801**, 2024
https://doi.org/10.1090/conm/801/16041

An example of a quadratic AS-regular algebra without any point modules

Michaela Vancliff

ABSTRACT. Although it has been known for decades that a quadratic AS-regular algebra of global dimension n need not have any point modules if $n \geq 5$, an explicit example appears to be absent, prior to the present article, from the published literature. This situation was brought to the attention of the author by the anonymous referee of D. Rogalski's survey article *Artin-Schelter Regular Algebras*, who encouraged the author of the present article to remedy the situation. Consequently, an explicit example is presented herein of a quadratic AS-regular algebra of global dimension five that is a graded Clifford algebra having no point modules.

Introduction

D. Rogalski's article, [**9**], provides a thorough survey of results regarding AS-regular algebras. While reading that article, the anonymous referee of [**9**] noticed that an example of a quadratic AS-regular algebra that does not have any point modules appeared to be absent from the published literature, in spite of the existence of such algebras being well known to many researchers for decades. Consequently, the referee of [**9**] encouraged the author of the present article to provide such an example in the published literature, and that is our objective below. For definitions of many of the terms used herein, the reader should consult [**9**].

Let $\Bbbk$ denote an algebraically closed field, and let A denote a quadratic AS-regular $\Bbbk$-algebra of global dimension n. By [**1, 2**], if $n \in \{2, 3\}$, then A has infinitely many point modules, and, by [**2**], a point module over a quadratic (not necessarily AS-regular) $\Bbbk$-algebra determines a point in the zero locus of the defining relations of the algebra. If a quadratic algebra has four generators and six defining relations, then there exists a point in the zero locus of the defining relations of the algebra, and so the algebra has a truncated point module of length three (this was originally proved by M. Van den Bergh – see the dimension discussion in [**12**, Section 1D]). By [**10**, Lemma 1.3], an AS-regular $\Bbbk$-algebra of global dimension four on four generators is Koszul (so quadratic) and has Hilbert series $H(t) = (1-t)^{-4}$. Moreover, by [**7**], if the reduced variety of the zero locus of the defining relations of an AS-regular $\Bbbk$-algebra with Hilbert series $H(t) = (1-t)^{-4}$ is the graph of an isomorphism, then the zero locus of the defining relations is the graph of an

2020 *Mathematics Subject Classification.* 16S38, 16E65, 16W50, 16S37, 15A66.
Key words and phrases. Regular algebra, point module, graded Clifford algebra, quadric system.

automorphism. Additionally, by [**10**, Theorem 1.4], the zero locus of the defining relations of a quadratic *Auslander-regular* $\Bbbk$-algebra of global dimension four that satisfies the Cohen-Macaulay property and has Hilbert series $H(t) = (1-t)^{-4}$ is the graph of an automorphism. In the latter two scenarios, any point belonging to the graph of the automorphism determines a point module over the algebra. In addition, by [**8**], quadratic Auslander-regular $\Bbbk$-algebras with polynomial growth are AS-regular.

Although these results alone do not rule out the conceivable existence of a quadratic AS-regular $\Bbbk$-algebra, with $n = 4$, that has no point modules, they do suggest that the case with $n = 5$ is more likely to provide a candidate. Hence, we present below a quadratic AS-regular $\Bbbk$-algebra of global dimension five, where $\mathrm{char}(\Bbbk) \neq 2$, that has no point modules; the algebra is also Auslander-regular and satisfies the Cohen-Macaulay property and has Hilbert series $H(t) = (1-t)^{-5}$.

The construction

Our construction makes use of unpublished notes, [**15**], of M. Van den Bergh regarding a method for calculating point modules over graded Clifford algebras. Given symmetric $n \times n$ matrices, $M_1, \ldots, M_n$, with entries in a field $\Bbbk$, let C denote the graded $\Bbbk$-algebra on generators $x_1, \ldots, x_n$ (of degree 1) and on generators $y_1, \ldots, y_n$ (of degree 2) with defining relations given by

$$x_i x_j + x_j x_i = \sum_{k=1}^{n} (M_k)_{ij} y_k \qquad \text{and} \qquad y_\ell c = c y_\ell$$

for all $i, j, \ell \in \{1, \ldots, n\}$ and for all $c \in C$, and call C a *graded Clifford algebra* (cf. [**6**]). Such algebras C arise in the context of the universal enveloping algebra of an (n, n)-quadratic Lie algebra ([**3**]), and need not be quadratic $\Bbbk$-algebras. Let $\mathfrak{Q}$ denote the linear system of quadrics in $\mathbb{P}^{n-1}$ given by $M_1, \ldots, M_n$. If C is quadratic, then $\mathfrak{Q}$ is intimately related to the Koszul dual of C (cf. [**4, 13**]). A base point of $\mathfrak{Q}$ is a point that belongs to every quadric in $\mathfrak{Q}$; that is, a point that belongs to the quadrics determined by $M_1, \ldots, M_n$.

Henceforth, suppose that $\Bbbk$ is algebraically closed and that $\mathrm{char}(\Bbbk) \neq 2$. By [**6**, Proposition 7], the graded Clifford algebra C is quadratic and Auslander-regular of global dimension n and satisfies the Cohen-Macaulay property if and only if $\mathfrak{Q}$ has no base points. Moreover, by [**13**, Theorem 1.7], if $\mathfrak{Q}$ has no base points, then the number of isomorphism classes of point modules over C is $2r_2 + r_1 \in \mathbb{N} \cup \{0, \infty\}$, where r_j denotes the number of matrices in $\mathbb{P}(\sum_{k=1}^{n} \Bbbk M_k)$ that have rank j. However, if $n = 5$, then a generic candidate for $\mathfrak{Q}$ yields $r_1 = 0 = r_2$, and this fact is a consequence of the following dimension argument inspired by [**15**]. Identify the projective scheme of all symmetric $n \times n$ matrices with $\mathbb{P}^{(n/2)(n+1)-1}$ and let X denote the subscheme of all symmetric $n \times n$ matrices that have rank at most two. We have that

$$\begin{aligned}\dim(X \cap \mathbb{P}(\sum_{k=1}^{n} M_k)) &\geq \dim(X) + \dim(\mathbb{P}(\sum_{k=1}^{n} M_k)) - \dim(\mathbb{P}^{(n/2)(n+1)-1}) \\ &= (2n-2) + (n-1) - (\frac{n}{2}(n+1) - 1) \\ &= \frac{(n-1)(4-n)}{2},\end{aligned}$$

so, if $n \geq 5$ and if $\mathbb{P}(\sum_{k=1}^{n} M_k)$ lies in general position with respect to X, then $r_1 = 0 = r_2$. In contrast, if $n = 4$ and $\mathbb{P}(\sum_{k=1}^{4} M_k)$ lies in general position with respect to X, then the dimension of the intersection is zero, so $2r_2 + r_1$ is a positive integer (indeed, $r_1 = 0$ and $r_2 = 10$ by [**15**] – see the proof of [**13**, Theorem 2.1]).

In order to construct an explicit example with $n = 5$, consider the quadric system given by the zero locus in $\mathbb{P}^4$ of the polynomials

$$q_1 = z_1^2 + z_2 z_4, \quad q_2 = z_2^2 + z_1 z_3, \quad q_3 = z_3^2 + z_1 z_4, \quad q_4 = z_4^2 + z_1 z_5, \quad q_5 = z_5^2 + z_1 z_2,$$

in commuting variables $z_1, \ldots, z_5$. Solving $q_k = 0$ for all k reveals that this quadric system has no base points; in fact, $z_1 = 0$ if and only if $z_k = 0$ for all k, and setting $z_1 = -1$ implies that $z_3 z_5 = z_2^2 z_4^2 = 1$, so $1 = z_3^2 z_5^2 = z_4 z_2 = -1$, which violates the assumption that $\text{char}(\Bbbk) \neq 2$. In order to compute r_1 and r_2, we consider the 3×3 minors of the symmetric matrix

$$\sum_{k=1}^{5} a_k M_k = \begin{bmatrix} 2a_1 & a_5 & a_2 & a_3 & a_4 \\ a_5 & 2a_2 & 0 & a_1 & 0 \\ a_2 & 0 & 2a_3 & 0 & 0 \\ a_3 & a_1 & 0 & 2a_4 & 0 \\ a_4 & 0 & 0 & 0 & 2a_5 \end{bmatrix},$$

where, for all k, M_k is the symmetric 5×5 matrix associated to q_k and $a_k \in \Bbbk$. We claim that the zero locus in $\mathbb{P}^4$ of the 3×3 minors of this matrix is empty. In fact, setting the minors

$$\begin{aligned} &-2(a_2 a_4^2 - 4a_1 a_2 a_5 + a_5^3), && -2(a_1^3 + a_2 a_3^2 - 4a_1 a_2 a_4 - a_1 a_3 a_5 + a_4 a_5^2), \\ &-2(a_4^3 + a_3^2 a_5 - 4a_1 a_4 a_5), && -2(a_3^3 + a_2^2 a_4 - 4a_1 a_3 a_4) \end{aligned}$$

equal to zero implies that, if $a_2 = 0$, then $a_k = 0$ for all k. On the other hand, if $a_2 \neq 0$, then setting the minors

$$\begin{aligned} &-2a_2^2 a_4, \quad -4a_2^2 a_5, && -2(a_3^3 + a_2^2 a_4 - 4a_1 a_3 a_4), \\ &a_1 a_2^2 - 4a_1^2 a_3 + 2a_3^2 a_5, && -2(a_2^3 - 4a_1 a_2 a_3 + a_3 a_5^2) \end{aligned}$$

equal to zero contradicts the assumption that $\text{char}(\Bbbk) \neq 2$. Hence, for the given quadric system, we have that $r_1 = 0 = r_2$. By the above remarks, it follows that the graded Clifford algebra C associated to $q_1, \ldots, q_5$ is quadratic and AS-regular of global dimension five and has no point modules. Writing out C explicitly reveals that C is the $\Bbbk$-algebra on generators $x_1, \ldots, x_5$ with defining relations:

$$\begin{array}{rclcrcl} x_1x_2 + x_2x_1 &=& x_5^2, & & x_1x_3 + x_3x_1 &=& x_2^2, \\ x_1x_4 + x_4x_1 &=& x_3^2, & & x_1x_5 + x_5x_1 &=& x_4^2, \\ x_2x_3 + x_3x_2 &=& 0, & & x_2x_4 + x_4x_2 &=& x_1^2, \\ x_2x_5 + x_5x_2 &=& 0, & & x_3x_4 + x_4x_3 &=& 0, \\ x_3x_5 + x_5x_3 &=& 0, & & x_4x_5 + x_5x_4 &=& 0. \end{array}$$

In this algebra, $y_k = x_k^2$ for all k.

The reader interested in results concerning point modules over graded Clifford algebras is referred to [**11, 13**]. Some results regarding generalizations of graded Clifford algebras and point modules over such algebras can be found in [**4, 5, 14**]. The reader is also encouraged to consult D. Rogalski's recent survey article, [**9**].

Acknowledgments

This article is published in a volume of *Contemporary Mathematics* that honors S. Paul Smith on the occasion of his 65th birthday. The author would like to take this opportunity to thank S. Paul Smith for introducing her to many wonderful algebra topics, especially the application of geometric techniques to study noncommutative algebras. Additionally, the author would like to thank M. Van den Bergh for sharing and discussing his unpublished notes, [**15**], with her in the mid-1990s. Lastly, the author is grateful to her fellow editors of the volume of *Contemporary Mathematics* in which this article appears and to the anonymous referee of [**9**] for their encouragement to write this article.

References

[1] Michael Artin and William F. Schelter, *Graded algebras of global dimension* 3, Adv. in Math. **66** (1987), no. 2, 171–216, DOI 10.1016/0001-8708(87)90034-X. MR917738

[2] M. Artin, J. Tate, and M. Van den Bergh, *Some algebras associated to automorphisms of elliptic curves*, The Grothendieck Festschrift, Vol. I, Progr. Math., vol. 86, Birkhäuser Boston, Boston, MA, 1990, pp. 33–85, DOI 10.1007/978-0-8176-4574-8_3. MR1086882

[3] Marc Aubry and Jean-Michel Lemaire, *Zero divisors in enveloping algebras of graded Lie algebras*, J. Pure Appl. Algebra **38** (1985), no. 2-3, 159–166, DOI 10.1016/0022-4049(85)90006-4. MR814174

[4] Thomas Cassidy and Michaela Vancliff, *Generalizations of graded Clifford algebras and of complete intersections*, J. Lond. Math. Soc. (2) **81** (2010), no. 1, 91–112, DOI 10.1112/jlms/jdp057. (Corrigendum: 90 No. 2 (2014), 631-636. MR3263968). MR2580455

[5] Thomas Cassidy and Michaela Vancliff, *Skew Clifford algebras*, J. Pure Appl. Algebra **223** (2019), no. 12, 5091–5105, DOI 10.1016/j.jpaa.2019.03.012. MR3975057

[6] Lieven Le Bruyn, *Central singularities of quantum spaces*, J. Algebra **177** (1995), no. 1, 142–153, DOI 10.1006/jabr.1995.1290. MR1356364

[7] Lieven Le Bruyn, S. P. Smith, and Michel Van den Bergh, *Central extensions of three-dimensional Artin-Schelter regular algebras*, Math. Z. **222** (1996), no. 2, 171–212, DOI 10.1007/PL00004532. MR1429334

[8] Thierry Levasseur, *Some properties of noncommutative regular graded rings*, Glasgow Math. J. **34** (1992), no. 3, 277–300, DOI 10.1017/S0017089500008843. MR1181768

[9] D. Rogalski, *Artin-Schelter Regular Algebras*, Recent Advances and New Directions in the Interplay of Noncommutative Algebra and Geometry, Eds. K. A. Brown, T. J. Hodges, M. Vancliff and J. J. Zhang, Contemporary Math., to appear.

[10] Brad Shelton and Michaela Vancliff, *Embedding a quantum rank three quadric in a quantum* $\mathbf{P}^3$, Comm. Algebra **27** (1999), no. 6, 2877–2904, DOI 10.1080/00927879908826599. MR1687269

[11] Darin R. Stephenson and Michaela Vancliff, *Constructing Clifford quantum $\mathbb{P}^3$s with finitely many points*, J. Algebra **312** (2007), no. 1, 86–110, DOI 10.1016/j.jalgebra.2007.02.015. MR2320448

[12] Michaela Vancliff, *The interplay of algebra and geometry in the setting of regular algebras*, Commutative algebra and noncommutative algebraic geometry. Vol. I, Math. Sci. Res. Inst. Publ., vol. 67, Cambridge Univ. Press, New York, 2015, pp. 371–390. https://library.slmath.org/books/Book67/files/150122-Vancliff.pdf. MR3525477

[13] M. Vancliff, K. Van Rompay, and L. Willaert, *Some quantum $\mathbf{P}^3$s with finitely many points*, Comm. Algebra **26** (1998), no. 4, 1193–1208, DOI 10.1080/00927879808826193. MR1612220

[14] Michaela Vancliff and Padmini P. Veerapen, *Point modules over regular graded skew Clifford algebras*, J. Algebra **420** (2014), 54–64, DOI 10.1016/j.jalgebra.2014.08.003. MR3261452

[15] M. Van den Bergh, *An Example with 20 Points*, unpublished notes, 1988.

Department of Mathematics, University of Texas at Arlington, Arlington, Texas 76019-0408

Email address: vancliff@uta.edu

SELECTED PUBLISHED TITLES IN THIS SERIES

801 **K. A. Brown, T. J. Hodges, M. Vancliff, and J. J. Zhang, Editors,** Recent Advances in Noncommutative Algebra and Geometry, 2024

798 **Marat V. Markin, Igor V. Nikolaev, and Carsten Trunk, Editors,** Advances in Functional Analysis and Operator Theory, 2024

796 **John Cremona, John Jones, Jennifer Paulhus, Andrew V. Sutherland, and John Voight, Editors,** LuCaNT: LMFDB, Computation, and Number Theory, 2024

795 **Michael A. Jones, David McCune, and Jennifer M. Wilson, Editors,** Mathematical Analyses of Decisions, Voting and Games, 2024

794 **Jean-Pierre Magnot, Editor,** Recent Advances in Diffeologies and Their Applications, 2024

793 **Abba Gumel, Editor,** Mathematical and Computational Modeling of Phenomena Arising in Population Biology and Nonlinear Oscillations, 2024

792 **Shaoming Guo, Zane Kun Li, and Brian Street, Editors,** Recent Developments in Harmonic Analysis and its Applications, 2024

791 **Mee Seong Im, Bach Nguyen, and Arik Wilbert, Editors,** Algebraic and Topological Aspects of Representation Theory, 2024

790 **Peter Crooks and Alexandru I. Suciu, Editors,** Compactifications, Configurations, and Cohomology, 2023

789 **I. S. Krasil′shchik, A. B. Sossinsky, and A. M. Verbovetsky, Editors,** The Diverse World of PDEs, 2023

788 **I. S. Krasil′shchik, A. B. Sossinsky, and A. M. Verbovetsky, Editors,** The Diverse World of PDEs, 2023

787 **Hemen Dutta, Editor,** Mathematical Modelling, 2023

786 **Hemen Dutta, Editor,** Mathematical Modelling, 2023

785 **A. Leroy and S. K. Jain, Editors,** Algebra and Coding Theory, 2023

784 **Dinh-Liem Nguyen, Loc Hoang Nguyen, and Thi-Phong Nguyen, Editors,** Advances in Inverse Problems for Partial Differential Equations, 2023

783 **Alla Detinko, Michael Kapovich, Alex Kontorovich, Peter Sarnak, and Richard Schwartz, Editors,** Computational Aspects of Discrete Subgroups of Lie Groups, 2023

782 **Galina Filipuk, Alberto Lastra, and Sławomir Michalik, Editors,** Recent Trends in Formal and Analytic Solutions of Diff. Equations, 2023

781 **Emanuel Indrei, Diego Marcon, and Levon Nurbekyan, Editors,** Geometric and Functional Inequalities and Recent Topics in Nonlinear PDEs, 2023

780 **Erik Koelink, Stefan Kolb, Nicolai Reshetikhin, and Bart Vlaar, Editors,** Hypergeometry, Integrability and Lie Theory, 2022

779 **Samuele Anni, Valentijn Karemaker, and Elisa Lorenzo García, Editors,** Arithmetic, Geometry, Cryptography, and Coding Theory 2021, 2022

778 **Carlos Galindo, Alejandro Melle Hernández, Julio José Moyano-Fernández, and Wilson A. Zúñiga-Galindo, Editors,** p-Adic Analysis, Arithmetic and Singularities, 2022

777 **Bang-Yen Chen, Nicholas D. Brubaker, Takashi Sakai, Bogdan D. Suceavă, Makiko Sumi Tanaka, Hiroshi Tamaru, and Mihaela B. Vajiac, Editors,** Differential Geometry and Global Analysis, 2022

776 **Aaron Wootton, S. Allen Broughton, and Jennifer Paulhus, Editors,** Automorphisms of Riemann Surfaces, Subgroups of Mapping Class Groups and Related Topics, 2022

775 **Fernando Galaz-García, Cecilia González-Tokman, and Juan Carlos Pardo Millán, Editors,** Mexican Mathematicians in the World, 2021

For a complete list of titles in this series, visit the
AMS Bookstore at **www.ams.org/bookstore/conmseries/**.